Heating Services Design

3230011449

£22.50

Heating Services Design

Ronald K. McLaughlin, BSc, MCIBS
Senior Lecturer in Building Services,
Glasgow College of Building and Printing

R. Craig McLean, BSc, MCIBS
Lecturer in Environmental Engineering,
University of Strathclyde, Glasgow

W. John Bonthron, BSc, CEng, MIMechE, MCIBS, MASHRAE
Associate, Hulley and Kirkwood,
Consulting Engineers, Glasgow

BUTTERWORTHS
London · Boston · Sydney · Wellington · Durban · Toronto

First published 1981

British Library Cataloguing in Publication Data

McLaughlin, R K
Heating services design.
1. Heating
I. Title II. McLean, R C
III. Bonthron, W J
697 TH7222 79-42812

ISBN 0-408-00380-4

Typeset by Reproduction Drawings Ltd., Sutton, Surrey.
Printed in Scotland by Thomson Litho Ltd., East Kilbride.

Preface

The 1970 edition of the I.H.V.E. Guide heralded not only a change to the S.I. system of units but more importantly, a significant step towards the strengthening of the technological basis of design methods and calculations for building services. This reflected a trend in which standards at all levels of education and training throughout the industry were being raised. More recently, the granting of a Royal Charter to the C.I.B.S. (I.H.V.E.) has led to the recognition of the building services engineer as a Chartered Technologist, educated to degree level. These developments have led to the growth of university and polytechnic departments teaching courses in building services and environmental engineering. Research activities in these departments now supplement the work done by B.R.E., B.S.R.I.A. (H.V.R.A.) and others.

During this period of rapid development in industry, education and research, there has been, unfortunately, little parallel development in the availability of higher-level books to meet the needs of the designer. This has aggravated a situation in which, compared to other branches of engineering, there was already a lack of suitable books.

The aim of this book is to go part of the way towards rectifying this situation within the topic area of heating services design. It has been written primarily for professional engineers and undergraduate students in building services and environmental engineering and in architecture. It should also prove useful to building services technicians and students in related disciplines.

The early chapters are concerned with a rigorous treatment of fundamentals which are used subsequently in the presentation of the design methods and calculation routines set out in later chapters. These methods and routines have been presented deliberately in a form *suitable for manual solution* because the authors believe that this is a necessary pre-requisite to the application of computer-based methods. It is evident from Chapters 5 and 6 that many of the new design methods are intrinsically computer orientated and it is hoped that the discussion and explanations given will prove useful to those who wish to employ computer-based design methods.

It is important to note that this book is not intended to be a self-contained design manual but has been written for use primarily with the C.I.B.S. Guide and the other sources referenced. Indeed, it should be considered as complementing rather than replacing existing textbooks.

The authors wish to express their thanks to Bill Boyle, Dr Joe Clarke, Professor Tom Maver, Don Stewart and Roy Veitch, all of the University of Strathclyde, for their help and encouragement; to Mrs Pat Gray for her patience in typing the text; and to the members of three families for their fortitude.

Finally, thanks are due to all those who permitted the reproduction of material in the book, in particular the C.I.B.S.

R.K. McL.
R.C. McL.
W.J.B.

Contents

1 The Fundamentals of Fluid Flow 1

2 The Fundamentals of Heat and Mass Transfer and Thermodynamics 39

3 Fundamentals of Automatic Control 83

4 Thermal Comfort 138

5 Heating System Design 200

6 Estimating the Heating System Load and Energy Consumption 222

7 Steady-state Heat Losses 265

8 Heat Emission and Emitter Selection 301

9 Water Heating Systems 330

10 Water Supply and the Design of Cold and Hot Water Services 390

Index 441

1 The Fundamentals of Fluid Flow

1.1 INTRODUCTION

A highly significant aspect of the work undertaken by the heating services engineer is concerned with the design of the fluid distribution systems for both heating and water supply purposes. It is undoubtedly the case that in many straightforward situations workable flow systems may be devised with little awareness on the part of the 'designer' of the fundamental engineering principles involved. However, the problems which can be treated successfully at such a level are limited and if the engineer is to be equipped to tackle competently the whole range of situations likely to be encountered in practice (and indeed be involved in the initiation of design improvements and new design techniques), then a firm understanding of fluid flow and the nature of its associated mechanisms is required.

The aim of this chapter is to go part way towards meeting this requirement. It is by no means intended as an all-embracing introduction to fluid mechanics but seeks rather to provide a suitable treatment of the fundamentals which are relevant to the design problems considered in the later chapters. Thus, although some of the material covered in the earlier part of the chapter has applications in many areas of fluid mechanics, the subject matter of the chapter as a whole is concentrated principally on the study of incompressible fluid flow in pipes.

1.2 FLUID PROPERTIES

1.2.1 Introduction

Whilst the *ideal fluid,* which constitutes the simplest model for flow analysis, is considered to be incompressible and to offer no resistance to deformation, all real fluids display to varying degrees both compressibility and viscous effects. In addition, vaporisation effects are displayed in liquids which have a free surface. These three important characteristics will now be reviewed briefly.

1.2.2 Compressibility

All fluids may be compressed by the application of pressure forces. The degree of compressibility of a fluid is characterised by defining the *bulk modulus*

$$K_E = -\Delta p \frac{V'}{\Delta V'} \tag{1.1}$$

Here Δp represents the increase in pressure necessary to decrease a given volume V' by the amount $-\Delta V'$. All liquids have a high value of bulk modulus and are compressible only to a small extent. For example, the bulk modulus of water is quoted as 20.085×10^5 kN/m^2, and thus a decrease of only 0.2% in a given volume requires a pressure increase of

$$\Delta p = K_E \frac{\Delta V'}{V'} = \frac{20.685 \times 10^5}{500} = 41.37 \times 10^2 \text{ kN/m}^2$$

In the study of fluid mechanics it is necessary to make the distinction between flows in which compressibility effects may be ignored and flows in which they require to be taken into account. As the change in the density of a liquid with an increase in pressure is small even for very large pressure changes, the density of a liquid is consequently taken as constant in most flow situations. The analysis of problems involving liquids is thereby greatly simplified. Some exceptions to this general simplification do exist however, and in certain special flow problems the compressibility of liquids is an important factor—as in the case of water-hammer, where the fluid is subjected to a very high rate of velocity change.

Unlike liquids, gases are highly compressible. However, in flows where a gas is subjected to relatively small changes in pressure (e.g. ventilation and air conditioning systems), the correspondingly small density variations are generally ignored and the gas is treated as an incompressible fluid. On the other hand, in high-speed flows, where the fluid velocity approaches that at which sound is propagated through the medium, compressibility effects become important and must be taken into account. Theory developed on the assumption of incompressibility would lead to very serious errors if applied to such situations.

1.2.3 Viscosity

Fluids resist any force which tends to cause the motion of one layer of fluid relative to another. This frictional phenomenon is attributed to the *viscosity* of the fluid and is manifested as a shearing stress between the layers acting to oppose the relative motion.

Consider the two-dimensional flow of a fluid adjacent to a solid boundary as described in Figure 1.1. Two infinitesimally thin fluid layers are shown in relative motion. A frictional stress τ will exist at the interface of these two layers, such as to retard the faster moving layer and accelerate the slower moving one. This shear stress is defined by Newton's viscosity law

Figure 1.1

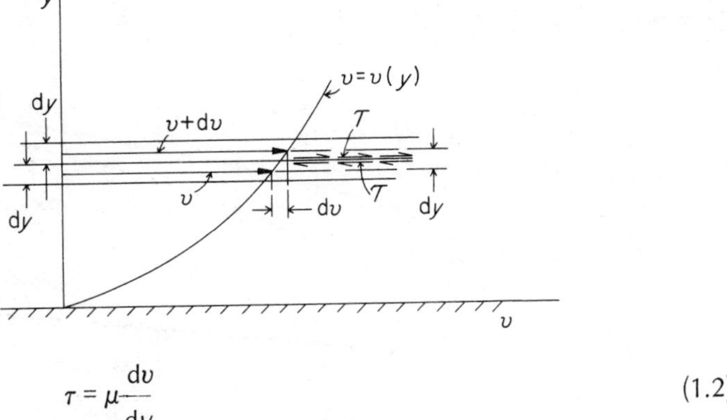

$$\tau = \mu \frac{dv}{dy} \tag{1.2}$$

where dv/dy is the *velocity gradient* and μ is known as the *coefficient of viscosity* (also termed the *absolute viscosity, dynamic viscosity,* or simply, the *viscosity* of the fluid). This coefficient, being the ratio of a shear stress to a velocity gradient, is dimensionally given by

$$\left[\frac{FT}{L^2}\right]$$

Since $[F] \equiv [ML/T^2]$, it is evident that this dimensional combination is equivalent to

$$\left[\frac{M}{LT}\right]$$

The basic S.I. unit of viscosity may therefore be quoted as Ns/m^2 or kg/ms. As we shall see in 1.5.1. and 2.3.4, the second dimensional form of μ given above is the one which is employed usually in dimensional analysis procedures. The c.g.s. units of viscosity, the *poise* (dyne s/cm^2) and the *centipoise*, are still widely used

$$1\ Ns/m^2 = 10 \text{ poise} = 10^3 \text{ centipoise}$$

The viscosity of a fluid has its origins in two fundamental molecular mechanisms, both of which are highly dependent on temperature–intermolecular attraction and molecular momentum exchange between fluid layers. In *gases,* the dominant mechanism is the momentum exchange which results from the thermal agitation of molecules normal to the direction of motion. As this activity increases with temperature, it therefore follows that the viscosity of gases will also increase with temperature (Figure 1.2a). In *liquids,* the effects of molecular momentum exchange between layers are insignificant compared with the forces of intermolecular cohesion and the viscosity depends primarily on the magnitude of these forces. Since they decrease rapidly with temperature, the viscosity of liquids decreases as the temperature rises (Figure 1.2b).

In order to maintain relative motion in a fluid, it is obvious

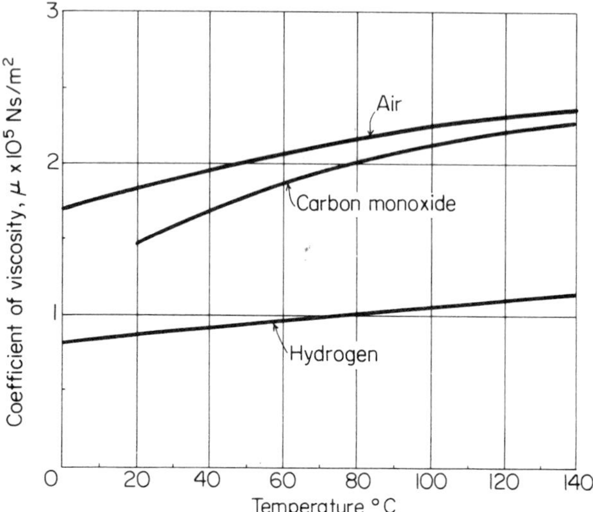

Figure 1.2a
Viscosities of some common gases

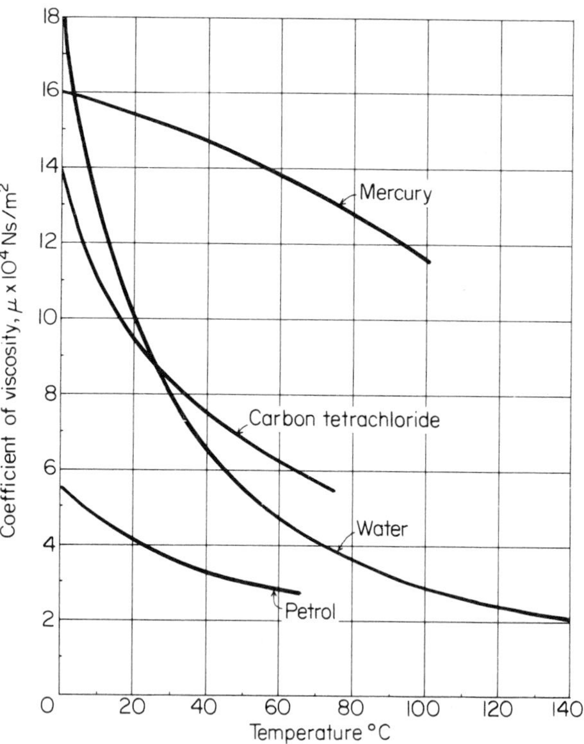

Figure 1.2b
Viscosities of some common liquids

that work must continuously be done in overcoming the viscous forces opposing the motion. This work is manifested wholly as an irreversible addition of heat to the fluid. The maintenance of the relative motion is therefore accompanied by a unidirectional flow of energy from the mechanical forms generating the motion to the internal energy of the fluid. Most fluids, including those encountered in heating services applications, behave in the manner described by equation 1.2, with shear stress being proportional to velocity gradient. These are known as *Newtonian* fluids. Some other fluids, such as plastics, are not characterised by this proportionality. Such fluids are described as *non-Newtonian* (Figure 1.3).

Figure 1.3

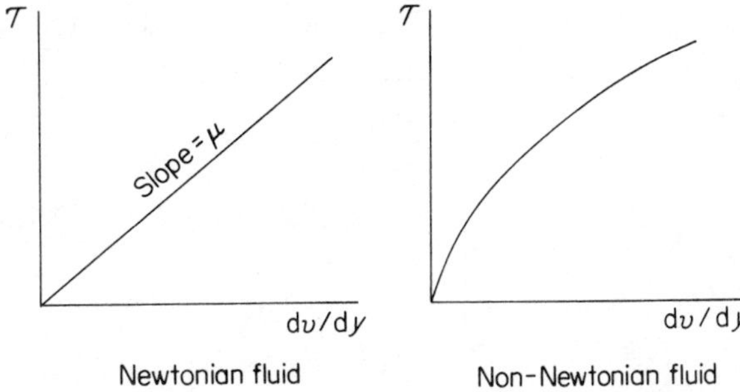

In many fluid flow problems it is frequently convenient to use the ratio of the viscosity to the density. This ratio,

$$\nu = \frac{\mu}{\rho} \tag{1.3}$$

is known as the *kinematic viscosity*. Dimensionally, ν is given by

$$[M/LT]\,/[M/L^3] \equiv [L^2/T],$$

and the basic S.I. unit of ν is therefore m^2/s. More generally, however, the mm^2/s (= 10^{-6} m^2/s) is employed. The c.g.s. units of kinematic viscosity are the *stoke* (cm^2/s) and the *centistoke*, thus

1 centistoke = 10^{-2} stokes = 1 mm^2/s

1.2.4 Vaporisation

All liquids having a free surface tend to vaporise because of the continuous projection of molecules through the free surface and out of the body of the liquid. Some of these molecules re-enter the liquid so that there is, in fact, a constant interchange of molecules between the liquid and the space above it. The molecules which are free of the liquid, being gaseous, exert their own partial pressure, known as the *vapour pressure.* This pressure, together with the pressure of any other gases present, make up the total pressure above the liquid.

For a liquid with an enclosed space above it, the vapour pressure will increase until a maximum value is attained. At this equilibrium condition the rate of molecules leaving the liquid is equal to the rate at which molecules are returning to it. This maximum value of vapour pressure is termed the *saturation vapour pressure* and any gas above the liquid is said to be *saturated* with the vapour. Saturation vapour pressure depends only on temperature (Table 1.1) and is independent of the presence of any other gas(es). If the total pressure above the liquid becomes less than the saturation vapour pressure then bubbles of vapour develop inside the liquid and rise to the surface. This is the phenomenon of boiling and it is, of course, associated with an extremely rapid escape of molecules from the liquid.

Table 1.1 Saturation Vapour Pressure of Water

Temperature (° C)	*Saturation Vapour Pressure* (N/m^2)
0	611
10	1227
20	2337
40	7375
60	19920
80	47360
100	101325

Even in the absence of a free surface, the vaporisation characteristics of liquids can still be of great importance. If a liquid enters a region of a flow system where the pressure falls locally to a value below the prevailing saturation vapour pressure, bubbles of vapour form within the fluid and are carried along by the fluid stream. It is interesting to note that a similar process may occur at pressures greater than the saturation vapour pressure if dissolved gases are present in the liquid, because as the pressure is reduced these gases come out of solution as bubbles. The production of bubbles (whether of vapour or gas) in a flow system under low pressure conditions, is known as *cavitation.* When the fluid stream enters a region of higher pressure, cavitation bubbles undergo a sudden collapse. As the liquid rushes in to fill the cavities, extremely high pressures and temperatures are generated and if the collapse occurs adjacent to a solid surface, serious mechanical damage can result. Anti-cavitation precautions will be considered in Chapter 9.

1.3 VISCOUS FLUIDS IN MOTION

1.3.1 The two regimes of flow

In regions of flow adjacent to solid surfaces (pipes, boundary layers), velocity gradients are large and fluid viscosity is an important parameter. The effects of viscosity allow the possibility of two physically different types of flow, known as *flow regimes.* The essential characteristics of these regimes were first clearly demonstrated by Reynolds(1) in 1883. It is instructive to consider briefly the nature of the experiments which Reynolds performed.

Reynolds' apparatus is schematically illustrated in Figure 1.4.

Figure 1.4

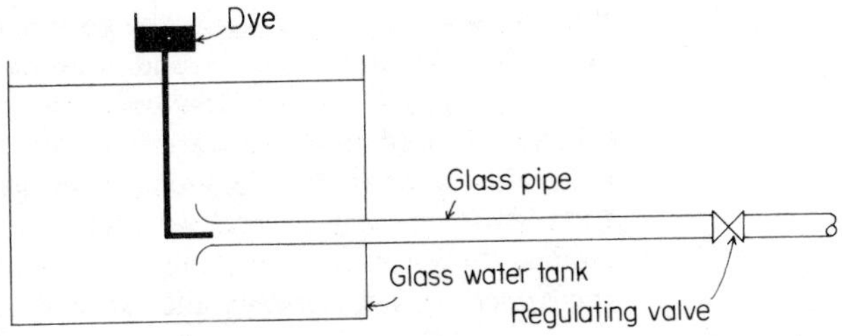

Water from the tank was allowed to flow through the bell-mouthed glass pipe at a rate controlled by the regulating valve. Dye was introduced into the flow through a small diameter tube. By observing the behaviour of the dye, Reynolds was able to study the characteristics of the flow inside the glass pipe. The experiments showed that, for small average velocities of flow, the dye remained as a straight and stable filament parallel to the axis of the pipe. As the flow rate was increased this type of flow persisted until a velocity was attained (the *critical velocity*) at which the dye filament wavered and broke up, mixing completely with the flowing water in the pipe.

In the first type of flow, fluid particles evidently move in straight lines parallel to the pipe axis. Since the flow may thus be envisaged as a series of concentric cylindrical layers flowing through one another, this regime of flow is known as *laminar flow*. The second regime of flow, in which the dye was diffused over the entire pipe, is known as *turbulent flow*. In this case, fluid particles are no longer constrained to move in layers and instead move in a haphazard manner throughout the flow, causing a general mixing of the fluid to take place. Turbulent flow is consequently characterised by a rapid and irregular pulsation of velocity at all points in the flow (Figure 1.5). When such a flow occurs in a pipeline only the mean motion of the fluid is parallel to the pipe axis. Superimposed upon this mean motion is a three-dimensional random eddying motion. As the worked examples in the next section will show (Examples 1.1. and 1.2.), turbulent flow is the flow regime normally encountered by the engineer in heating services design problems.

An interesting fact noted by Reynolds was that the velocity at which the dye filament broke up was dependent on the extent of the initial disturbances in the flow. The longer the water in

Figure 1.5

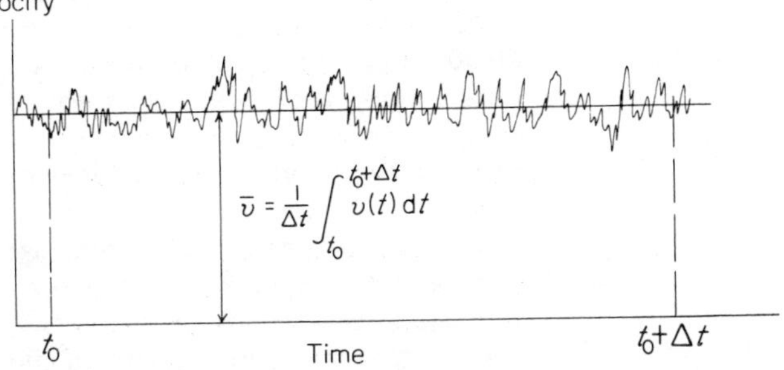

the tank was allowed to settle, the greater was the critical velocity attainable. Reynolds also found that once turbulent flow had been established, laminar flow could be restored by reducing the velocity. Indeed, this restoration always took place at approximately the same velocity. Laminar flow does in fact break up into turbulence at some critical velocity greater than that at which laminar flow can be re-established; the former velocity is called the *upper critical velocity*, the latter the *lower critical velocity.*

1.3.2 The criterion of flow

As laminar flow and turbulent flow differ widely in their characteristics and in their effects on engineering processes (e.g. convection), it is important for the engineer to understand which type of flow may be expected to exist under any given set of conditions. The fundamental criterion which determines whether a flow will be of a laminar or turbulent nature was established by Reynolds as a result of his dye stream experiments. These experiments showed that the character of the flow in a pipe was dependent on the pipe diameter D, the average velocity of flow V (i.e. the volume flow rate divided by the pipe cross-sectional area), the density ρ, and the viscosity μ. Reynolds summarised his results by combining these four variables into a dimensionless parameter (the *Reynolds number, Re*) defined by

$$Re = \frac{\rho VD}{\mu} = \frac{VD}{\nu} \tag{1.4}$$

(A detailed examination of this parameter reveals that it does in fact represent the ratio of the inertia forces to the viscous forces in a flow). It was ascertained by Reynolds that certain critical Reynolds numbers were sufficient to specify the upper and lower critical velocities for the flow of all fluids in pipes of any size. The consequence of this is that the limits of laminar and turbulent pipe flow may be expressed in terms of single values of Re.

As the upper limit of laminar pipe flow depends to a great extent on the initial degree of disturbance within the flow, the upper critical Reynolds number is consequently a highly variable quantity. Reynolds himself was able to reach values of Re of about 12 000 before turbulence set in. In subsequent experiments where exceptional care was taken to eliminate all sources of disturbance and vibration, upper critical Reynolds numbers of over 40 000 were obtained. However, the experimental conditions under which such high values have been achieved are of little practical importance and for engineering purposes it is usual to take the upper critical Reynolds number as lying between 2500 and 4000.

Of perhaps greater engineering significance is the lower limit of turbulent flow, defined by the lower critical Reynolds number. This specifies a condition below which the flow in pipes will always be laminar, irrespective of the initial nature of the flow.

The generally accepted value of the lower critical Reynolds number is 2000. Combining these results, we may predict that for *Re* <2000, the flow will be laminar; and that for *Re* >4000, the flow will be turbulent. Between these limits is a region known as the *critical zone* within which the flow may be of either regime, the actual nature of the flow depending on the flow situation itself.

Some worked examples will help to illustrate the conditions under which the two types of flow are likely to exist in practice.

Example 1.1
For water at temperatures of 10 °C and 75 °C (representative of cold water supply and low temperature heating systems respectively), investigate the lower critical velocities in pipes ranging from 15 mm to 150 mm in diameter.

At 10 °C, $\rho = 1000$ kg/m^3; $\mu = 1.306 \times 10^{-3}$ Ns/m^2
At 75 °C, $\rho = 975$ kg/m^3; $\mu = 3.78 \times 10^{-4}$ Ns/m^2

Noting that

$$\text{lower crit. } Re = 2000 = \frac{V_{\text{crit}}\,\rho D}{\mu}$$

we see that the critical velocity is given by

$$V_{\text{crit}} = \frac{2000\,\mu}{\rho D}$$

Consider as an illustration a 40 mm diameter pipe.

$$10\,°\text{C}: V_{\text{crit}} = \frac{2000 \times 1.306 \times 10^{-3}}{1000 \times 0.04} = 0.065 \text{ m/s}$$

$$75\,°\text{C}: V_{\text{crit}} = \frac{2000 \times 3.78 \times 10^{-4}}{975 \times 0.04} = 0.019 \text{ m/s}$$

If this calculation is repeated for different pipe sizes, the critical velocities obtained are as tabulated below.

Temperature (°C)	*Pipe Diameter* (mm)						
	15	25	40	50	80	100	150
10	0.174	0.104	0.065	0.052	0.033	0.026	0.017
75	0.052	0.031	0.019	0.016	0.01	0.008	0.005

The results of the foregoing example are instructive in two ways. Firstly, for both temperatures the computed critical velocities are generally well below the water flow velocities typical of normal engineering practice (this will be made evident in Chapters 9 and 10). It may therefore be predicted that the flow of water in pipes will invariably be turbulent. Secondly, the example serves as a demonstration of the significant effect of temperature on the behaviour of a fluid, by indicating that the critical velocities for water at 10 °C are more than three times those at 75 °C.

If the procedure followed in Example 1.1 is repeated for other low-viscosity fluids such as steam and air, the conclusion is again reached that the laminar pipe flow of these fluids is unlikely in practice. If, however, a high-viscosity fluid such as heavy fuel oil is considered, then the situation may well be very different. Both these points can be illustrated by the following simple example.

Example 1.2

Determine the lower critical velocity in a 50 mm diameter pipe for (i) steam having a kinematic viscosity of 7 mm^2/s (dry saturated at 150 °C) and (ii) fuel oil having a kinematic viscosity of 300 mm^2/s (class F at 35 °C).

In terms of ν, the critical velocity is evaluated from

$$V_{crit} = \frac{2000\,\nu}{D}$$

(i) steam: $$V_{crit} = \frac{2000 \times 7}{0.05 \times 10^6} = 0.28 \text{ m/s}$$

(ii) oil: $$V_{crit} = \frac{2000 \times 300}{0.05 \times 10^6} = 12 \text{ m/s}$$

The critical velocity for the oil obtained in this example is far in excess of velocities normally encountered in practice. We note also that even in a 150 mm diameter pipe a velocity exceeding 4.0 m/s is required before turbulent flow could be sustained. It is therefore likely that in most practical situations the flow of this particular fluid will be laminar.

1.3.3 Laminar and turbulent flow

The differences in the characteristics of laminar and turbulent flow may be illustrated conveniently by considering briefly the following three aspects of fluid motion:

(i) Velocity profiles.
(ii) Shear stresses.
(iii) Energy dissipation.

(i) *Velocity profiles*

For the laminar flow of a fluid in a pipe the velocity distribution is defined by the parabolic relationship

$$v = v_m \left(1 - \frac{r^2}{R^2}\right) \qquad (1.5)$$

where v_m is the velocity at the pipe axis (Figure 1.6). It will be demonstrated later (in 1.5.2, where the above equation is derived) that this maximum velocity is twice the average velocity of flow, V. For turbulent pipe

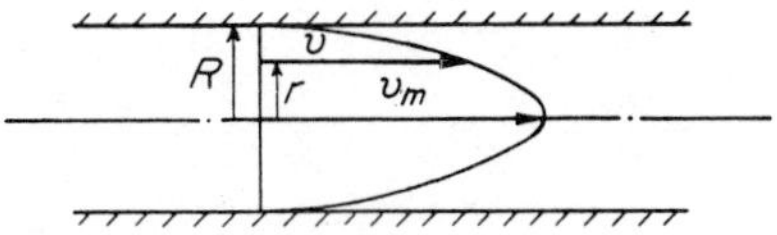

Figure 1.6
Laminar velocity distribution

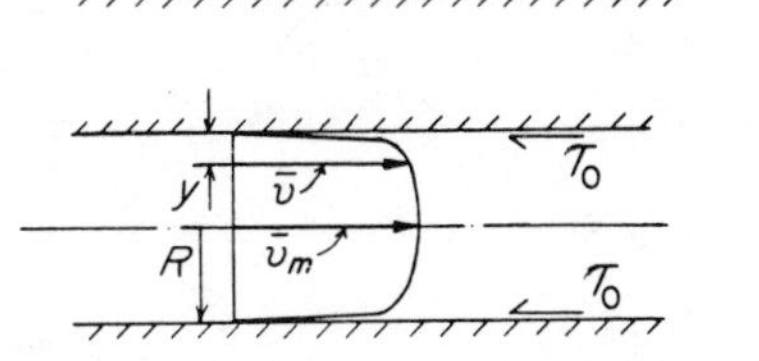

Figure 1.7
Turbulent velocity distribution

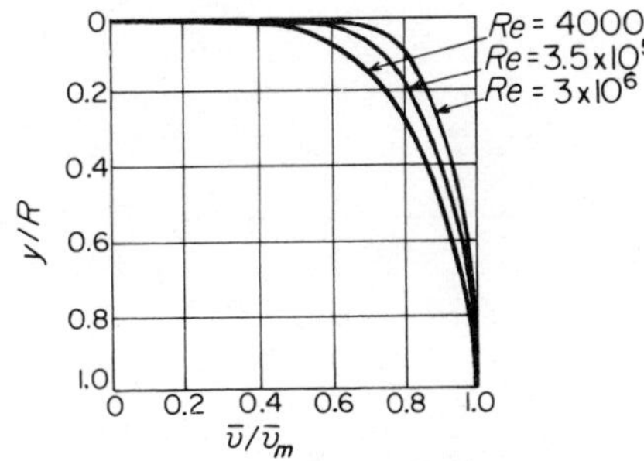

Figure 1.8

flow, the velocity distribution is defined in terms of the temporal mean velocity $\bar{v}$ (Figure 1.5). The turbulent profile is much flatter in appearance than the laminar flow parabola and is generally characterised by the following logarithmic equation (Figure 1.7)

$$\frac{\bar{v}_m - \bar{v}}{\sqrt{\tau_o/\rho}} = -2.5 \ \ln \frac{y}{R} \tag{1.6}$$

where τ_o is the boundary shear stress. As the dimensionless curves of Figure 1.8 clearly illustrate, the turbulent velocity profile becomes flatter as the Reynolds number is increased.

(ii) *Shear stresses*

For the two-dimensional laminar flow described in Figure 1.1, the shear stress (the origins of which are purely molecular in nature) is defined by equation 1.2. In turbulent flow the velocity fluctuations which are superimposed on the mean motion (v'_x, v'_y, v'_z) give rise to a system of stresses which far exceed those due to viscosity alone. (The instantaneous velocity v at a point in a turbulent flow is the vector sum of the temporal mean velocity $\bar{v}$ and these fluctuating velocity components. Typical examples of this vector addition are represented in Figure 1.9).

Consider the points A and B in the turbulent flow cross-section shown in Figure 1.10. At point A the temporal mean velocity is $\bar{v}$, at B it is $\bar{v} + \Delta\bar{v}$. If these velocities are now considered to be mean velocities of

Figure 1.9

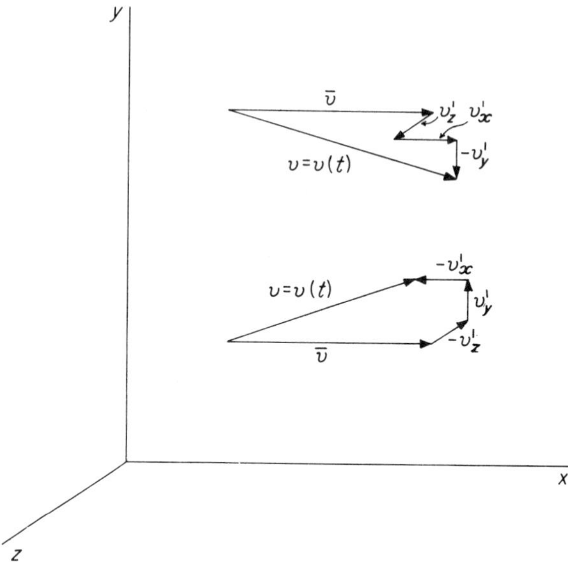

Figure 1.10

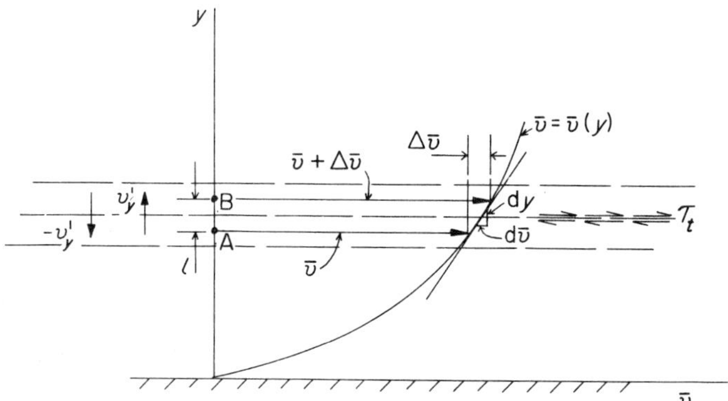

two *imaginary* fluid layers, the fluctuating component v'_y is associated with a movement of fluid volumes transverse to the layers. Due to the gradient of mean velocity, this will result in a momentum exchange between the layers. As this exchange must act to accelerate the slower layer and decelerate the faster one, it will be manifested as the turbulent shear stress τ_t at the layer interface.

The satisfactory specification of such a turbulent stress in terms of the mean velocity field forms one of the central problems of turbulence. Introducing the concept of the *eddy viscosity*, η, τ_t may conveniently be expressed as

$$\tau_t = \eta \frac{d\bar{v}}{dy} \qquad (1.7)$$

The eddy viscosity differs from μ in that it is not a property of the fluid but is a parameter which is dependent on the structure of the turbulence. In addition to τ_t, there will also be a molecular interaction between the layers giving rise to a viscous shear stress. The total stress acting may be written as

$$\tau = (\mu + \eta)\frac{d\bar{v}}{dy} \tag{1.8}$$

(iii) *Energy dissipation*

Consider the experimental arrangement shown in Figure 1.11, in which a fluid may be passed through a pipe with a varying velocity. For each average velocity of flow, the pressure drop over the length L is measured by a manometer. By plotting log $\Delta p/L$ against log V, two distinct forms of relationship become readily apparent (Figure 1.12).

For velocities giving laminar flow in the pipe, a straight line having a slope of unity is obtained, demonstrating that for this flow regime the pressure drop per unit length is directly proportional to the velocity. However, after the transition to turbulent flow this law is no longer valid and instead the line will be found to

Figure 1.11

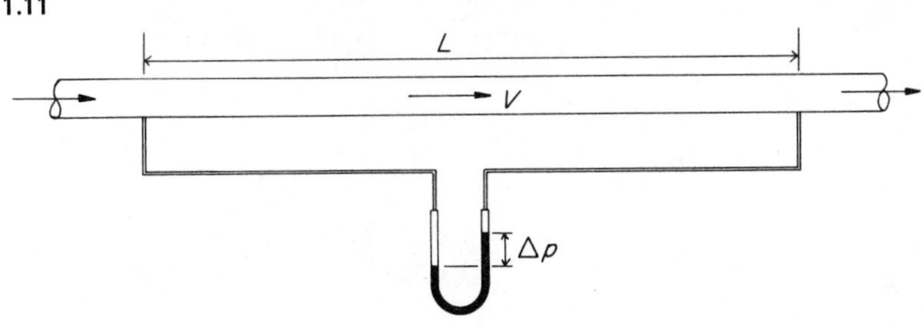

Figure 1.12

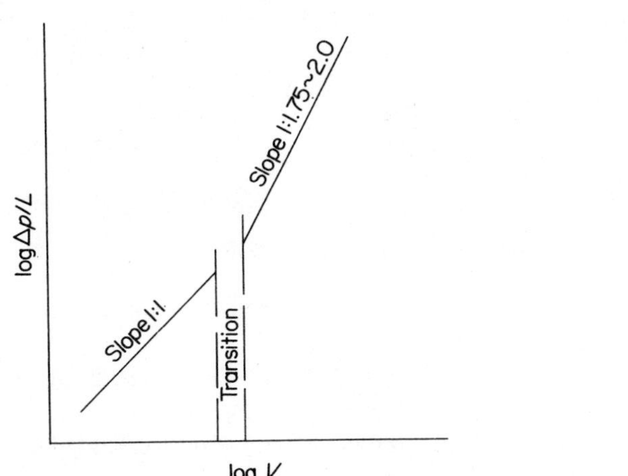

have a slope which lies between 1.75 and 2.0. This upper value does of course correspond to a parabolic relationship between $\Delta p/L$ and V. As will be evident later, the actual slope within the turbulent regime depends upon the surface roughness of the pipe.

1.4 THE EULER AND BERNOULLI EQUATIONS

1.4.1 Introduction

Two equations of fundamental importance in the study of fluid mechanics are the *Euler* and *Bernoulli* equations, which follow from the application of Newton's Second Law to the motion of an ideal fluid. As we are concerned here with a hypothetical fluid of zero viscosity, there will be no frictional effects manifested and consequently no dissipation of mechanical energy to heat.

1.4.2 Euler's Equation

Consider an element of an ideal fluid moving along a streamline as shown in Figure 1.13. Let us examine the forces acting on this element in the direction of motion. We have

(i) the net pressure force

$$p\mathrm{d}A - (p + \mathrm{d}p)\mathrm{d}A = -\mathrm{d}p\ \mathrm{d}A$$

Figure 1.13

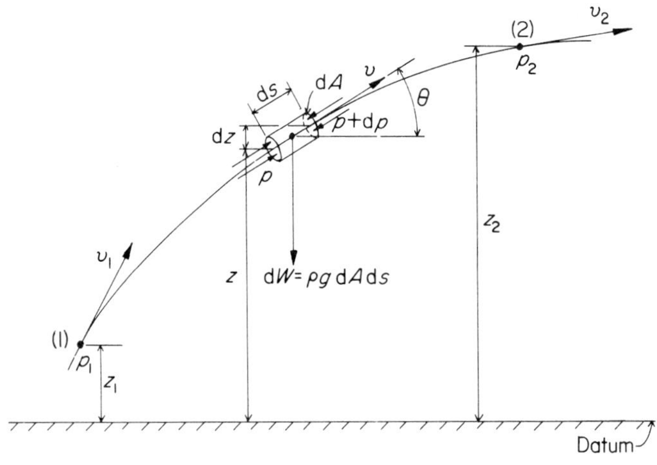

(ii) the component of the gravity force

$$-\mathrm{d}W \sin\theta = -\rho g \mathrm{d}s\ \mathrm{d}A \frac{\mathrm{d}z}{\mathrm{d}s} = -\rho g \mathrm{d}A\ \mathrm{d}z$$

The differential mass being accelerated by the action of these forces is

$$dM = \rho ds\, dA$$

Newton's second law of motion states that the resultant force is equal to the product of the mass and the acceleration. Thus

$$-dp\, dA - \rho g dA\, dz = \rho ds\, dA \frac{dv}{dt} \tag{1.9}$$

Expressing the acceleration as

$$\frac{dv}{dt} = \frac{dv}{ds}\frac{ds}{dt} = v\frac{dv}{ds}$$

and dividing equation 1.9 throughout by ρdA yields the Euler equation

$$\frac{dp}{\rho} + v\, dv + g\, dz = 0 \tag{1.10}$$

In the derivation of this differential relationship no restriction was placed on the density ρ and it therefore applies to both incompressible and compressible flows of an inviscid fluid.

1.4.3 Bernoulli's Equation

Equation 1.10 can be integrated along the length of the streamline between sections (1) and (2) (Figure 1.13) to give

$$\int_{p_1}^{p_2} \frac{dp}{\rho} + \frac{v_2^2 - v_1^2}{2} + g(z_2 - z_1) = 0$$

The complete integration of Euler's equation is possible only if ρ is constant or some known function of p. For an *incompressible* flow, the result of this integration is the Bernoulli equation

$$\frac{p_1}{\rho} + \frac{v_1^2}{2} + gz_1 = \frac{P_2}{\rho} + \frac{v_2^2}{2} + gz_2 \tag{1.11}$$

The dimensions of the terms in this equation are energy per unit mass and the units are $J/kg \equiv m^2/s^2$.

Equation 1.11 may also be written as

$$\frac{p}{\rho} + \frac{v^2}{2} + gz = \text{CONSTANT} \tag{1.12}$$

When expressed in this indefinite form Bernoulli's equation states that, in the absence of viscous effects, the sum of the pressure energy, the kinetic energy and the potential energy is constant along a given streamline. Depending upon the conditions under which the motion is taking place, this constant may assume different values for the other streamlines of the flow. However, for general pipe and duct flow situations, all the streamlines can be taken as having the same total energy. Hence the application of Bernoulli's equation may be extended to cover such stream-tubes of finite cross-sectional area. If we consider, for example, the typical conduit flow problem depicted in Figure 1.14, then equation 1.11 may be applied directly to the flow as a whole between sections (1) and (2). Note that in the absence of friction the distribution of velocity across the sections will be uniform.

Figure 1.14

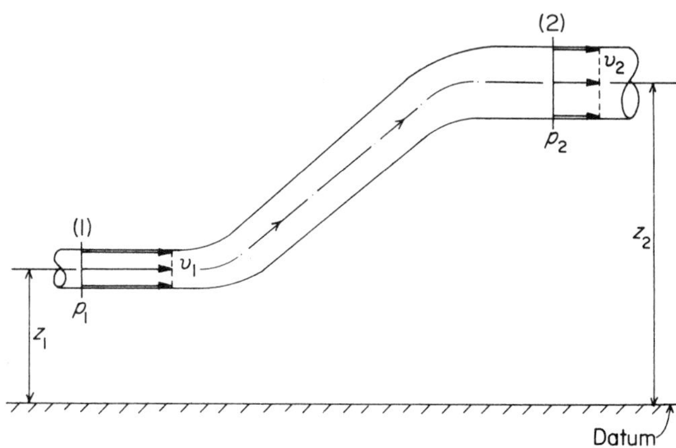

Dividing equation 1.12 throughout by g (noting that $\rho g = \omega$, the specific weight), we obtain another common form of Bernoulli's equation

$$\frac{p_1}{\omega} + \frac{v_1^2}{2g} + z_1 = \frac{p_2}{\omega} + \frac{v_2^2}{2g} + z_2 \tag{1.13}$$

Here each term has the dimensions of energy per unit weight and the units of length (J/N ≡ m). It is therefore customary, as an extension of the terminology of hydrostatics, to refer to the term p/ω as the pressure head, the term $v^2/2g$ as the velocity head and the term z as the potential head. Their sum is called the *total head.*

1.4.4 Modification of Bernoulli's Equation

The Bernoulli equation as it has been derived applies only to frictionless flows. Although all real fluids are indeed viscous, it should not be concluded that the assumption of zero friction leads to a theoretical abstraction which is of no practical use. On the contrary, in many real flow situations viscous effects are small and in such problems the application of Bernoulli's equation yields solutions which are sufficiently accurate for normal engineering purposes. The equation has therefore a surprisingly wide application, an important example being its use in the theory of flow measurement. However, before it can be applied to flow situations in which non-ideal effects are of significance, as is generally the

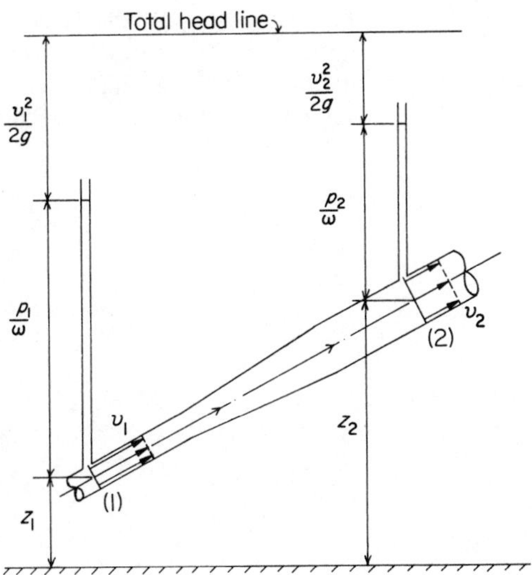

Figure 1.15a
Ideal fluid

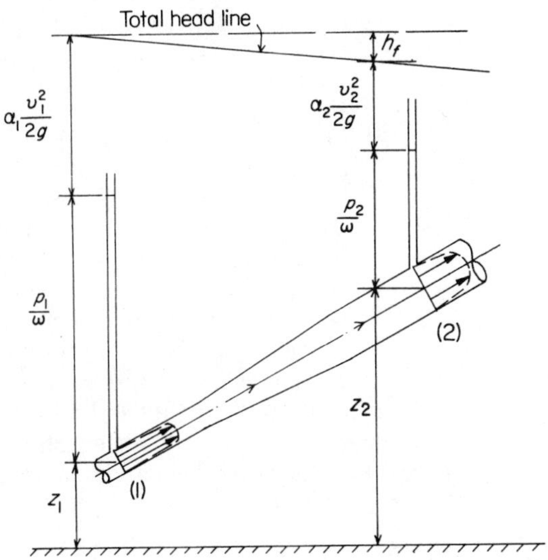

Figure 1.15b
Real fluid

case in pipe and duct flow problems, Bernoulli's equation must clearly be modified in some appropriate manner.

The modifications which require to be made to the equation may be described conveniently with reference to Figure 1.15, which compares the flow of an ideal fluid to that of a real fluid in the same conduit system. We will examine in particular the modification of equation 1.13, as dealing in terms of 'head' allows the energy variations in the system to be graphically represented as shown in the figure. From the figure the differences in the flows are readily apparent.

(i) For the real fluid the velocity profiles are no longer uniform at the sections considered. The kinetic energy in this case is calculated as a function of the average velocity of flow V, and correspondingly we write the velocity head as

$$\alpha \frac{V^2}{2g}$$

Here α is a kinetic energy correction factor

$$\alpha = \frac{1}{V^2} \frac{\int_0^R v^3 \, rdr}{\int_0^R v \, rdr} ; \alpha > 1.0,$$

the magnitude of which depends on the extent to which the actual velocity distribution deviates from the uniform ideal distribution. Comparing the parabolic profile of laminar flow with the much flatter profile of turbulent flow, it is to be expected that the α value for the former will exceed that for the latter.

(ii) In the real flow situation the shear stresses generated act to oppose the motion of the fluid. The work done in overcoming these stresses is transformed into heat and hence represents a loss of useful mechanical energy from the system. Thus in this case the total energy line between sections (1) and (2) does not remain horizontal, but drops by some amount h_f, which is referred to as the 'head loss due to friction'. To maintain the concept of energy conservation, this loss term must be added to the right-hand side of equation 1.13.

By modifying the Bernoulli equation to take account of the above two effects, the equation of motion for incompressible flow in pipes becomes

$$\frac{p_1}{\omega} + \alpha_1 \frac{V_1^2}{2g} + z_1 = \frac{p_2}{\omega} + \alpha_2 \frac{V_2^2}{2g} + z_2 + h_f \tag{1.14}$$

Some comments on the importance of the kinetic energy correction factors in this equation are appropriate at this juncture. As the worked examples in 1.3.2 have illustrated, most practical engineering applications are concerned with turbulent flow in pipes. Under such conditions α is only marginally greater than unity ($\alpha \doteq 1.05$) and its omission will therefore introduce no significant errors into the calculations. On the other hand, for laminar flow conditions $\alpha = 2.0$ and it would thus appear necessary to include the correction factors when such flows are being considered. However, the existence of laminar flow is generally associated with very low velocity motions (the exception being flows of high-viscosity fluids–Example 1.2) and in these circumstances the velocity heads are likely to be negligibly small compared with the other terms in the equation. The multiplication of these velocity heads by a factor of 2.0 is consequently a matter of little importance.

It is obvious from the foregoing discussion that in most pipe-flow problems the kinetic energy correction factors may be omitted without incurring any significant inaccuracy. Neglecting α, equation 1.14 simplifies to

$$\frac{p_1}{\omega} + \frac{V_1^2}{2g} + z_1 = \frac{p_2}{\omega} + \frac{V_2^2}{2g} + z_2 + h_f \qquad (1.15)$$

If a pump is positioned between (1) and (2), the above equation must be further modified to take account of this supply of energy to the system. We write

$$\frac{p_1}{\omega} + \frac{V_1^2}{2g} + z_1 + H = \frac{p_2}{\omega} + \frac{V_2^2}{2g} + z_2 + h_f \qquad (1.16)$$

Here H represents the mechanical energy addition per unit weight of fluid and is known as the 'pump head'. Multiplying equation 1.16 by the specific weight, ω, gives another useful modified relationship

$$p_1 + \tfrac{1}{2}\rho V_1^2 + \rho g z_1 + P = p_2 + \tfrac{1}{2}\rho V_2^2 + \rho g z_2 + p_f \qquad (1.17)$$

where $P = \rho g H$ is the 'pump pressure' and $p_f = \rho g h_f$ is the 'pressure loss due to friction'. The term $\tfrac{1}{2}\rho V^2$ is referred to as the velocity pressure.

1.5 FLUID FLOW IN PIPES

1.5.1 The Darcy equation

Consider the steady flow of an incompressible fluid through a long horizontal pipe of internal diameter D (Figure 1.16). The average velocity of flow is V, the fluid density is ρ and the viscosity is μ. In this flow the head loss due to friction will appear simply as a drop in the static pressure along the length of the pipe. This reduction in pressure is related directly to the frictional shear stress τ_o, which acts at the boundary of the flow and gives rise to a force which opposes the motion of the fluid.

Let us examine the forces acting on the cylinder of fluid contained within the length L of the pipe. This fluid experiences a retarding force $\tau_o \pi D L$. In order to maintain the motion this force must be balanced by an equal and opposite force resulting from the difference in the pressures at sections (1) and (2). Thus

$$(p_1 - p_2)\frac{\pi}{4}D = \tau_o \pi D L$$

and the pressure drop over the length L is given by

$$(p_1 - p_2) = p_f = \tau_o \frac{4L}{D} \tag{1.18}$$

Figure 1.16

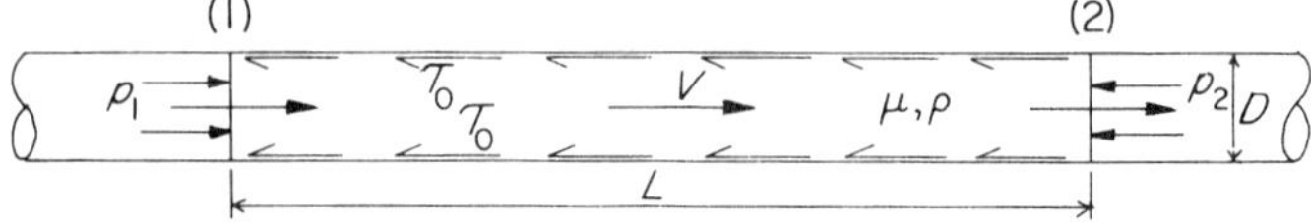

As $p_f = \rho g h_f$, we can now write a relation between the resistance stress τ_o and the head loss due to the action of this resistance

$$h_f = \tau_o \frac{4L}{\rho g D} \tag{1.19}$$

To proceed further with this investigation, the method of dimensional analysis is now applied to the boundary shear stress τ_o. A description of the theoretical basis and practical significance of dimensional analysis is contained in the next chapter. The analysis of the present pipe friction problem will therefore be given here without comment on these aspects of the procedure. From an examination of the problem (Figure 1.17), it appears

Figure 1.17

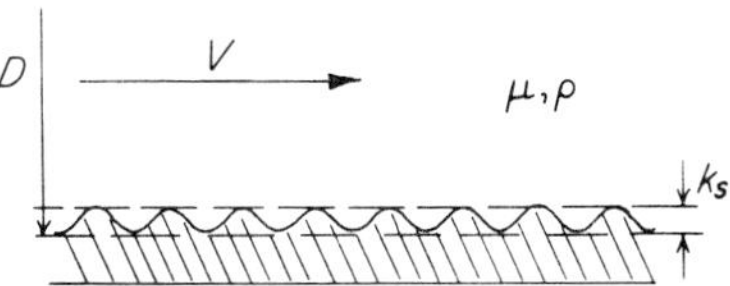

that τ_o will depend on D, V, ρ and μ. Clearly another relevant parameter is the roughness of the pipe surface. Assuming that the surface protuberances are of the same shape and distribution pattern and denoting the height of the protuberances as k_s, we can now write

$$\tau_o = \phi\,(D, V, \rho, \mu, k_s),$$

$$\text{or } \tau_o = \psi D^a V^b \rho^c \mu^x k_s^y \tag{1.20}$$

Here ϕ and ψ denote, respectively, an unknown function and an unknown dimensionless factor.

Expressing the terms of equation 1.20 in terms of the primary dimensions of mass M, length L and time T gives

$$\frac{M}{LT^2} = (L)^a \left(\frac{L}{T}\right)^b \left(\frac{M}{L^3}\right)^c \left(\frac{M}{LT}\right)^x (L)^y$$

The indices of M, L and T must be the same on both sides of this equation if the function for τ_o is to be dimensionally correct.

Thus

$$M:\quad 1 = c + x,$$

$$L: -1 = a + b - 3c - x + y,$$

$$T: -2 = -b - x$$

From the above equations we find that

$$c = 1 - x; b = 2 - x; a = -y - x$$

Substituting these results into equation 1.20 yields the following relationships for the boundary shear stress

$$\tau_o = \psi \rho V^2 \left(\frac{\mu}{DV\rho}\right)^x \left(\frac{k_s}{D}\right)^y$$

$$\text{or } \tau_o = \rho V^2 \phi [(Re)(k_s/D)] \qquad (1.21)$$

The ratio k_s/D is known as the *relative roughness* of the pipe.

If we now substitute for τ_o in equation 1.19 we obtain the relationship

$$h_f = \rho V^2 \phi [(Re)(k_s/D)] \frac{4L}{\rho g D},$$

which may be written in terms of some other unknown function ϕ' as

$$h_f = \frac{4LV^2}{2gD} \phi' [(Re)(k_s/D)]$$

This equation is alternatively expressed as

$$h_f = \frac{4fLV^2}{2gD} \qquad (1.22)$$

where $f = \phi' [(Re)(k_s/D)]$ is a dimensionless coefficient, usually referred to as the *friction factor*. Equation 1.22 is known as Darcy's equation. It is the basic equation for the loss of head due to friction in straight, uniform pipes. The form of the equation means that the problem of evaluating the head loss in a given pipe flow situation becomes, in effect, one of evaluating the appropriate value of the friction factor.

1.5.2 Friction factors in laminar flow

Laminar flow in a pipe can be visualised as a large number of infinitesimally thin concentric cylinders sliding over one another like the tubes of a telescope.

Consider the cylinder of fluid, of length L and surface area S, which moves within the hollow cylinder of inside radius r and thickness dr (Figure 1.18). The viscous drag on this fluid is

$$\tau S = -(2\pi r L)\mu \frac{dv}{dr}$$

Figure 1.18

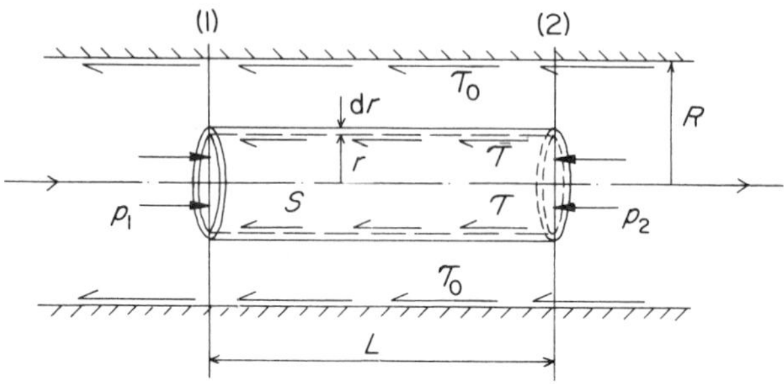

the minus sign being included here because the velocity gradient dv/dr is negative. This drag force must be balanced by the pressure difference $(p_1 - p_2)$ acting over the area πr^2. Thus

$$(p_1 - p_2)\pi r^2 = -(2\pi r L)\mu \frac{dv}{dr},$$

from which

$$dv = -\frac{(p_1 - p_2)}{2\mu L} r dr$$

The integration of this differential relationship yields

$$v = -\frac{(p_1 - p_2)}{4\mu L} r^2 + C \qquad (1.24)$$

where the constant of integration C is evaluated from the following boundary condition: at $r = R, v = 0$.

Therefore

$$C = \frac{(p_1 - p_2)}{4\mu L} R^2$$

and substituting this into equation 1.24 gives the expression for the velocity distribution in laminar pipe flow

$$v = \frac{(p_1 - p_2)}{4\mu L} R^2 \left(1 - \frac{r^2}{R^2}\right) \qquad (1.25)$$

This equation defines a parabolic velocity profile. The maximum velocity, which occurs at the pipe axis ($r = 0$) is

$$v_m = \frac{(p_1 - p_2)}{4\mu L} R^2 \qquad (1.26)$$

$(p_1 - p_2)/L$ is the pressure drop per unit length of pipe. Denoting this as Δp, equations 1.25 and 1.26 may be rewritten as

$$v = \frac{\Delta p R}{4\mu} \left(1 - \frac{r^2}{R^2}\right)$$

and $v_m = \dfrac{\Delta p R}{4\mu}$

Consider now the hollow cylinder of radius r and thickness dr. The differential volume flow rate associated with this cylinder is

$$d\dot{G} = v dA = \frac{(p_1 - p_2)}{4\mu L} R^2 \left(1 - \frac{r^2}{R^2}\right) 2\pi r dr$$

The total volume flow rate in the pipe is determined by integrating this equation between the limits $r = 0$, $r = R$

$$\dot{G} = \frac{\pi(p_1 - p_2)}{2\mu L} R^2 \int_0^R \left(r - \frac{r^3}{R^2}\right) dr$$

$$= \frac{\pi(p_1 - p_2)}{2\mu L} R^2 \left[\frac{r^2}{2} - \frac{r^4}{4r^2}\right]_0^R$$

Evaluating this expression we obtain the Hagen-Poiseuille formula

$$\dot{G} = \frac{(p_1 - p_2)\pi R^4}{8\mu L} \tag{1.27}$$

which in terms of the pipe diameter D is

$$\dot{G} = \frac{(p_1 - p_2)\pi D^4}{128\mu L} \tag{1.28}$$

From equation 1.28 the average velocity of flow V may be evaluated as

$$V = \frac{\dot{G}}{A} = \frac{(p_1 - p_2)\pi D^4}{128\mu L} \frac{4}{\pi D^2}$$

$$= \frac{(p_1 - p_2) D^2}{32\mu L} \tag{1.29}$$

Comparing this result with equation 1.26 we see that

$$V = \tfrac{1}{2} v_m \tag{1.30}$$

Now Darcy's equation is

$$h_f = \frac{(p_1 - p_2)}{\rho g} = \frac{4 f L V^2}{2 g D}$$

and from equation 1.29 it is also apparent that

$$\frac{(p_1 - p_2)}{\rho g} = \frac{32 L V^2 \mu}{D^2 \rho g}$$

Combining these two expressions, the friction factor for laminar flow is found to be

$$f = \frac{16}{(\rho VD/\mu)} = \frac{16}{Re} \tag{1.31}$$

The friction factor would thus appear to be independent of the surface roughness. As will be shown in 1.5.3, this fact has been fully borne out by experiment.

Consider for a moment the distribution of shear stress in the pipe. The differentiation of equation 1.25 expresses the velocity gradient as

$$\frac{dv}{dr} = -\frac{\Delta pr}{2\mu}$$

and the profile of shear stress is therefore defined by

$$\tau = -\mu\frac{dv}{dr} = \frac{\Delta pr}{2} \tag{1.32}$$

Figure 1.19

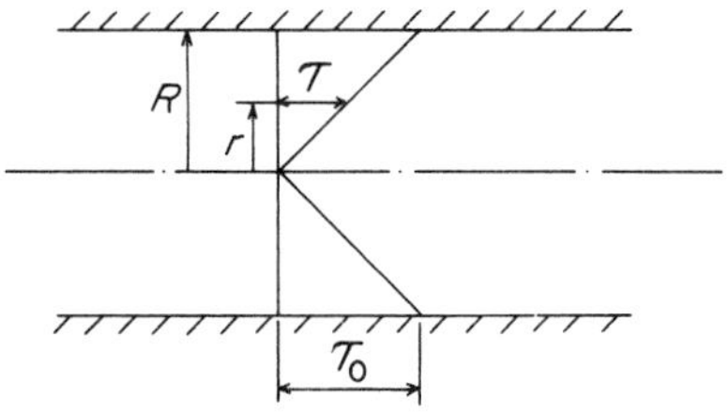

Consequently, the shear stress must vary linearly (Figure 1.19) from a value of zero at the pipe axis ($r = 0$) to a maximum at the pipe wall ($r = R$) where

$$\tau_o = \frac{\Delta pR}{2}$$

1.5.3 Friction factors in turbulent flow

By its nature, turbulent flow is not amenable to the type of treatment applied to laminar flow. Friction factors for turbulent flow have therefore had to be determined experimentally. In accordance with the analysis presented in 1.5.1, experimental data on pipe friction is correlated in terms of the friction factor, the Reynolds number and the relative roughness. The nature of the relationship between f, Re and k_s/D is clearly demonstrated by Figure 1.20, which is based on the results of experiments conducted by Nikuradse[(2)]. (Sir Thomas Stanton was one of the first to present experimental results in this way and this figure is thus frequently referred to as a Stanton diagram). Nikuradse used smooth circular pipes which had been artificially roughened by fixing coatings of uniform sand grains to the pipe walls. This technique provided easily measured values of relative roughness, k_s being taken simply as the sand grain diameter.

Figure 1.20

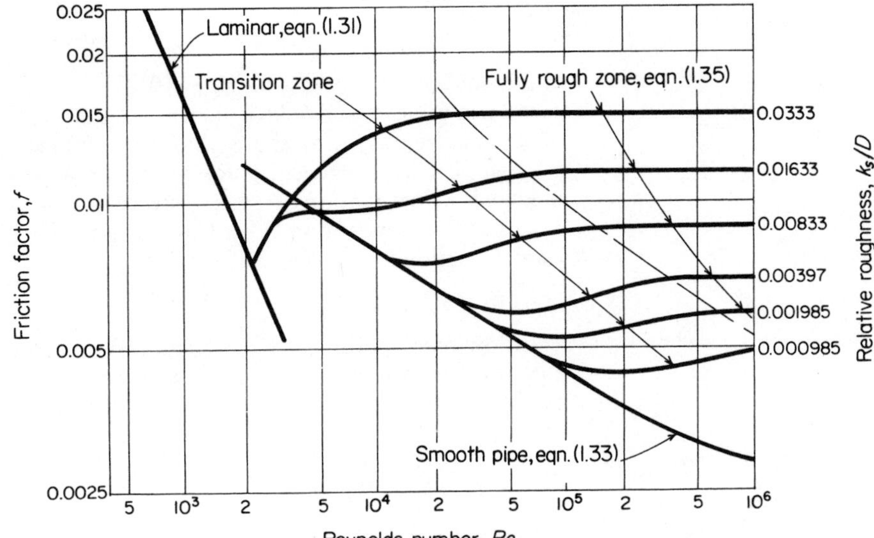

Several interesting points follow from an examination of Figure 1.20. With regard to the laminar flow regime, we see that this (as predicted in 1.5.2) is characterised by the single line $f = 16/Re$. This law applies to pipes of all surface roughnesses and confirms the independence of f from k_s/D. It is also of interest to note that the transition from laminar flow to turbulence is likewise unaffected by surface roughness.

With regard to the turbulent flow regime, two important observations can be made.

(i) For each rough pipe there is a Reynolds number, the value of which depends on the relative roughness, below which the pipe has the same resistance as a smooth pipe. The region within which the friction-factor curves for a rough pipe and a smooth pipe are coincident is known as the *smooth zone* of flow. From his own data and that of other investigators, Nikuradse proposed the following relationship for the smooth-pipe curve in Figure 1.20.

$$\frac{1}{\sqrt{f}} = 4.0 \log (Re\sqrt{f}) - 0.40 \qquad (1.33)$$

This equation is applicable over the whole range of Reynolds numbers investigated (from 4000 to 3.2 × 10^6). Up to a Reynolds number of 100 000, the curve is closely approximated by the Blasius formula

$$f = 0.079\,(Re)^{-1/4} \qquad (1.34)$$

Substituting this expression into Darcy's equation, it can be concluded that

$$h_f \propto V^{1.75},$$

in the smooth zone of flow for $Re < 10^5$.

(ii) There is a value of Reynolds number, again depending on the relative roughness, above which the resistance of a rough pipe is independent of *Re*. In this range, which is known as the *fully rough zone*, f becomes constant and totally dependent on k_s/D. It is evident from the Darcy equation that in this case

$$h_f \propto V^2$$

The horizontal portions of the friction-factor curves corresponding to the fully rough zone are characterised by the equation

$$\frac{1}{\sqrt{f}} = 4 \log\left(\frac{D}{k_s}\right) + 2.28 \tag{1.35}$$

Between the smooth zone and the fully rough zone lies the so-called *transition zone*.

The behaviour described by these observations may be explained (albeit in a somewhat idealised form) by considering briefly the flow structure immediately adjacent to a pipe surface. Although the flow in the pipe may be turbulent, there will exist at the surface an extremely thin film of fluid within which the flow is laminar. This film is known as the *laminar sub-layer*. As long as the roughness projections of the surface are fully submerged in the laminar sub-layer, they will not influence the main flow and in consequence the pipe behaves as a smooth pipe (Figure 1.21a). We say in this situation that the rough pipe is 'hydraulically smooth'. As the Reynolds number of the flow is increased so the thickness of the sub-layer decreases. Eventually the roughness projections will penetrate the sub-layer and the friction-factor curve for the rough pipe will diverge from the smooth-pipe curve. Obviously for rougher pipes this will happen at a lower value of *Re*. As the Reynolds number is further increased, a situation will finally develop in which the laminar sub-layer is completely disrupted, with viscous effects being consequently rendered negligible. This is the condition which corresponds to the fully rough zone. The turbulent flow around the surface protuberances will now lead to a continuous generation of eddies, the extent to

Figure 1.21

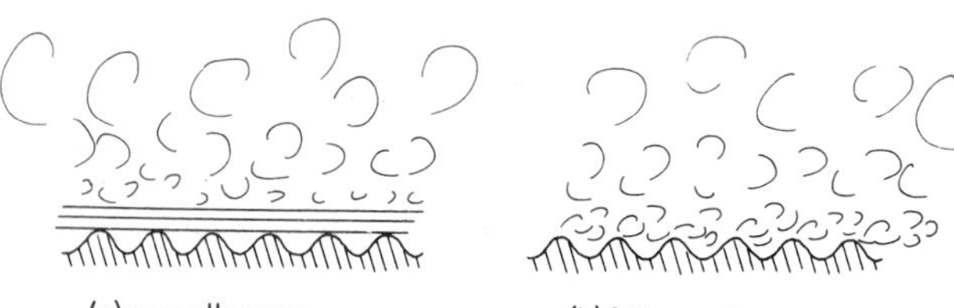

which these affect the main flow being dependent on the size of the roughness pattern, f increasing as k_s/D increases (Figure 1.21b). In the transition zone the surface protuberances are partly covered by the sub-layer and thus both viscous and roughness effects will be of significance.

Commercial pipes

Although the results of Nikuradse's experiments provide us with an important insight into the mechanics of pipe flow, they are unfortunately not representative of the behaviour of commercial pipes encountered in engineering practice. This is because commercial pipes generally contain protuberances of a varying size and shape and as a result have roughness patterns which are much more variable (and hence much less definable) than the uniform roughness patterns used by Nikuradse. Even if the *average* height of the roughness projections on a commercial pipe surface were to be measured, Figure 1.20 cannot be used to predict a value of friction factor. The largest projections cause an early departure from the smooth-pipe value of f and because the surface protuberances have different heights they penetrate the laminar sub-layer at different values of Re. In consequence, Nikuradse's transition curves do not agree at all well with those for commercial pipes. However, Nikuradse's results have been used to establish a quantitative measure of commercial pipe roughness. This was first done by C. F. Colebrook[3], whose experiments into the friction losses in commercial piping are of great importance.

Before looking at Colebrook's work let us consider Nikuradse's equations 1.33 and 1.35. Subtracting 4 log $(3.7D/k_s)$ from both sides of each equation yields

$$\text{'Smooth'} \quad : \frac{1}{\sqrt{f}} - 4\log\left(\frac{3.7D}{k_s}\right) = 4\log\left(Re\sqrt{f}\,\frac{k_s}{D}\right) - 2.68$$

$$\text{'Fully rough'} : \frac{1}{\sqrt{f}} - 4\log\left(\frac{3.7D}{k_s}\right) = 0$$

If these modified equations are plotted using $1/\sqrt{f} - 4\log(3.7D/k_s)$ and $\log(Re\sqrt{f}\,k_s/D)$ as co-ordinates (Figure 1.22), we obtain two straight lines characterising the smooth and fully rough zones of flow. Nikuradse's data for the transition zone plotted on the same basis gives a curve which connects the two straight lines. The family of friction-factor curves of Figure 1.20 has thus been reduced to the single curve of Figure 1.22.

Colebrook found that commercial pipes, when tested to a sufficiently high Reynolds number, exhibited friction factors which were independent of Re. The comparison of the appropriate constant value of f for a particular commercial pipe with equation 1.35 thus permits the evaluation of an *equivalent* sand grain size k_s for the pipe. Obtaining the equivalent roughness in

Figure 1.22

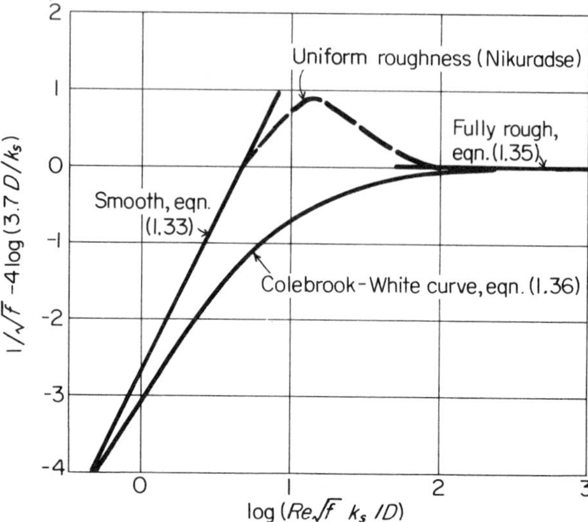

this manner and correlating the results of extensive tests on various types of commercial piping, Colebrook (in collaboration with C. M. White) proposed the following equation for the friction factor

$$\frac{1}{\sqrt{f}} - 4\log\left(\frac{D}{k_s}\right) = 2.28 - 4\log\left(1 + \frac{4.64}{Re\sqrt{f}\,(k_s/D)}\right)$$

which in a rearranged form is

$$\frac{1}{\sqrt{f}} = -4\log\left(\frac{k_s}{3.7D} + \frac{1.255}{Re\sqrt{f}}\right) \tag{1.36}$$

It is readily seen that the curve defined by this equation is asymptotic to the 'smooth' and 'fully rough' lines in Figure 1.22; as $k_s \to 0$, the above expression reduces to

$$\frac{1}{\sqrt{f}} = -4\log\left(\frac{1.255}{Re\sqrt{f}}\right)$$

i.e. $\frac{1}{\sqrt{f}} = 4\log(Re\sqrt{f}) - 0.4,$ . . . 'smooth' law

and as $Re \to \infty$ it becomes

$$\frac{1}{\sqrt{f}} = -4\log\left(\frac{k_s}{3.7D}\right)$$

i.e. $\frac{1}{\sqrt{f}} = 4\log\left(\frac{D}{k_s}\right) + 2.28$ . . . 'fully rough' law

Although the equation of Colebrook and White accurately summarises the experimental data on commercial pipes, it is (like the smooth-pipe equation 1.33) a transcendental function. It cannot be solved for f directly and is therefore hardly suitable for practical engineering purposes. However, just as Nikuradse's results reduce to a single curve, so the inverse process may be carried out by expanding the Colebrook-White curve (Figure 1.22) into a plot corresponding to Figure 1.20. This task was performed by L. F. Moody [4] and the result is the family of curves shown

Figure 1.23

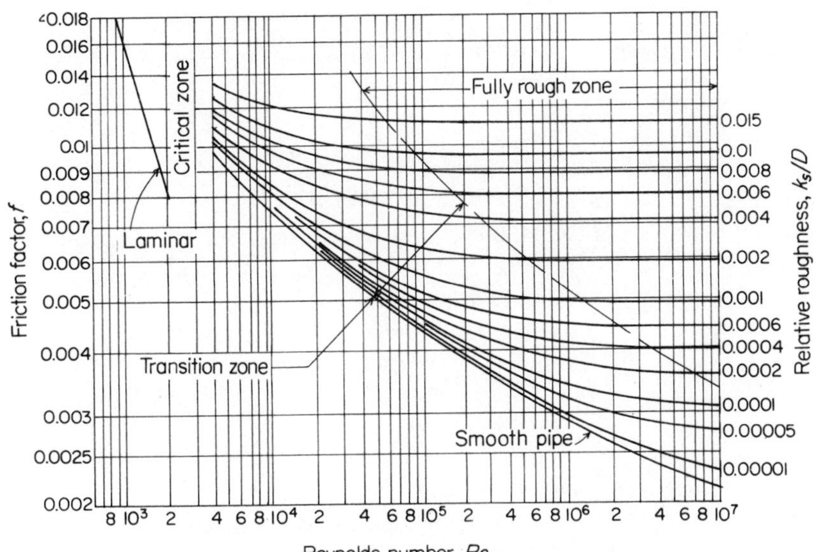

in Figure 1.23 (now known as the Moody diagram). Moody produced these curves by firstly solving equation 1.36 for Re and by subsequently plotting the computed value of Re against equal increments of f, k_s/D being held constant for each curve. Remember that the equivalent roughness of a commercial pipe cannot be inferred directly from roughness measurements, but only by determining from tests the limiting value to which f tends at a high Reynolds number.

Some representative values of k_s are:

cast iron	0.25 mm
galvanised steel	0.15 mm
uncoated steel	0.046 mm
drawn tubing	0.0015 mm
concrete	0.3–3 mm
riveted steel	0.9–9 mm

Comprehensive information on k_s and k_s/D for the various types of piping commonly used in heating services applications is given in Reference (5).

The use of the Moody chart is illustrated by the following worked examples.

Example 1.3

Water at a temperature of 75 °C flows in a 100 mm diameter

steel pipe at a rate of 15 l/s. Determine the pressure drop due to friction in a 100 m length of pipe.

$\rho = 975\ \text{kg/m}^3$, $\mu = 3.78 \times 10^{-4}\ \text{Ns/m}^2$

The average velocity of flow, $V = \dfrac{15 \times 10^{-3}}{\dfrac{\pi}{4} \times 0.1^2} = 1.91\ \text{m/s}$

$$Re = \frac{975 \times 1.91 \times 0.1 \times 10^4}{3.78} = 4.93 \times 10^5$$

$k_s/D = 0.046/100 = 0.00046$

From the Moody chart (Figure 1.23)

$f = 0.0044$

$$p_f = \rho g h_f = \frac{4 \rho f L V^2}{2D}$$

$$= \frac{4 \times 975 \times 0.0044 \times 100 \times 1.91^2}{2 \times 0.1}$$

$$= 3.13 \times 10^4\ \text{N/m}^2$$

Example 1.4

For the same pipe and volume flow rate, calculate the new pressure drop if the water temperature is reduced to 10 °C.

$\rho = 1000\ \text{kg/m}^3$, $\mu = 1.306 \times 10^{-3}\ \text{Ns/m}^2$

$$Re = \frac{1000 \times 1.91 \times 0.1 \times 10^3}{1.306} = 1.46 \times 10^5$$

In this case, $f = 0.0048$

$$p_f = \frac{4 \times 1000 \times 0.0048 \times 100 \times 1.91^2}{2 \times 0.1}$$

$$= 3.5 \times 10^4\ \text{N/m}^2$$

These examples illustrate two points of interest:

(a) At both temperatures the flow is in the transition zone. Indeed, because of the design limits imposed on velocity, this is the zone of flow generally associated with heating services systems.

(b) The reduction in the fluid temperature is accompanied by a corresponding rise in the density and the friction factor. Both these effects contribute to the increase in the pressure loss. This dependence of pressure loss on temperature has important ramifications in the design situation and is a matter which will receive further comment in Chapter 9.

1.6 ADDITIONAL LOSSES IN PIPES

1.6.1 Introduction

Besides the loss of head due to friction in uniform straight pipes, additional losses are caused by bends, valves and all other types of fitting. Such losses are commonly called *minor losses.* (Although it is convenient to refer to them collectively in this way, it should be noted that the term 'minor losses' is really a misnomer as many situations will be encountered (e.g. plant rooms), in which they may be at least as important as the straight pipe losses considered in the preceding section).

Minor losses are associated generally with sudden changes in velocity which result in the generation of large-scale turbulence. The useful energy extracted from the flow in the creation of this turbulence is dissipated to heat as the eddies decay. Although the fittings which constitute the source of a loss usually occupy only a relatively short length of pipe, the additional turbulence produced by them may persist for some considerable distance downstream and it is in this section of piping that the minor loss is in effect realised. The superposition of large-scale turbulence on the general turbulence pattern means that the nature of the flow and the associated pipe friction processes downstream of a turbulence-generating device are exceedingly complex. However, for general applications we assume that the effects of normal pipe friction and of the additional large-scale turbulence can be treated separately and that any loss caused by a fitting is concentrated at the device itself. This is an extremely convenient assumption for engineering calculations, since the total head loss in a pipeline may be determined by simply adding the friction loss for the length of pipe considered to the minor losses, without giving any consideration to the complex combination of the two which may exist in some parts of the line.

In almost all cases minor losses are evaluated experimentally as a theoretical approach to the problem is seldom possible. Since early experimentation indicated that the losses vary with the square of the average velocity of flow, they are generally expressed as

$$\text{Head loss, } h_l = K\frac{V^2}{2g} \tag{1.37}$$

or as,

$$\text{Pressure loss, } p_l = K\tfrac{1}{2}\rho V^2 \tag{1.38}$$

where K is the *loss factor* (it is also commonly referred to as the *velocity pressure factor* or simply as the *'K' factor*). For a given flow geometry (i.e. a given turbulence-generating device), the loss factor is practically constant at high Reynolds numbers.

1.6.2 Abrupt enlargement

A minor loss which can be treated analytically (with the aid of certain simplifying assumptions) is that due to an abrupt enlargement of the flow cross-section (Figure 1.24). An inspection of the flow pattern shows that the fluid leaving the smaller pipe does not follow the sudden deviation of the boundary and consequently pockets of turbulent eddies are formed at the corners of the enlargement.

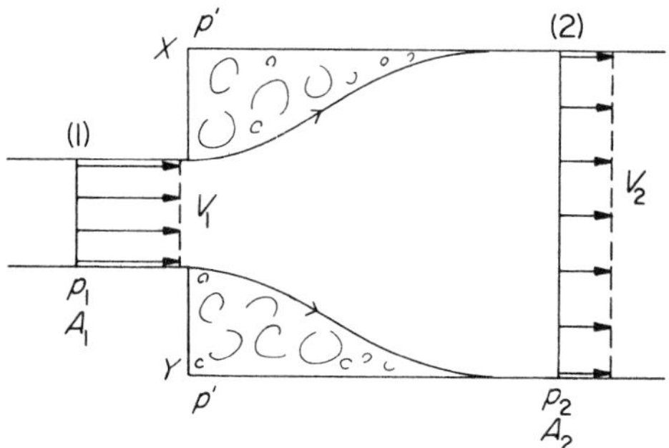

Figure 1.24
Abrupt enlargement

Consider a section (1) in the smaller pipe, across which it is assumed that the velocity is uniform. (It can be seen from 1.3.3 that, for the high Reynolds numbers generally encountered in practice, this assumption is reasonably valid). Consider also section (2) in the larger pipe, which is taken as being sufficiently far downstream for the velocity to be again assumed uniform. Due to the velocity change from (1) to (2), there is a corresponding change in momentum and this requires the application of a net force on the fluid between the sections. If we neglect the boundary shear stress acting over the short distance between (1) and (2), the net force is

$$p_1 A_1 + p'(A_2 - A_1) - p_2 A_2$$

If it is now assumed that the pressure p' in the plane of the enlargement is equal to p_1, this becomes

$$(p_1 - p_2) A_2$$

Equating the net force to the rate of change of momentum gives

$$(p_1 - p_2) A_2 = \rho \dot{G} (V_2 - V_1)$$

where $\dot{G}$ is the volume flow rate, and therefore

$$p_1 - p_2 = \rho \frac{\dot{G}}{A_2} (V_2 - V_1) = \rho V_2 (V_2 - V_1) \qquad (1.39)$$

If the modified Bernoulli equation 1.15 is applied to the flow between (1) and (2) we have

$$\frac{p_1}{\omega} + \frac{V_1^2}{2g} = \frac{p_2}{\omega} + \frac{V_2^2}{2g} + h_l$$

which yields

$$h_l = \frac{V_1^2 - V_2^2}{2g} + \frac{p_1 - p_2}{\omega}$$

Comparing this equation with equation 1.39,

$$h_l = \frac{(V_1 - V_2)^2}{2g} \tag{1.40}$$

$$\text{or } p_l = \tfrac{1}{2}\rho(V_1 - V_2)^2 \tag{1.41}$$

It is seen that the loss at an abrupt enlargement is proportional to the square of the velocity difference.

By continuity, $A_1\ V_1 = A_2\ V_2$ and equations 1.40 and 1.41 may be expressed alternatively as

$$h_l = \frac{V_1^2}{2g}\left(1 - \frac{A_1}{A_2}\right)^2 = \frac{V_2^2}{2g}\left(\frac{A_2}{A_1} - 1\right)^2 \tag{1.42}$$

$$p_l = \tfrac{1}{2}\rho V_1^2\left(1 - \frac{A_1}{A_2}\right)^2 = \tfrac{1}{2}\rho V_2^2\left(\frac{A_2}{A_1} - 1\right)^2 \tag{1.43}$$

The loss specified by these equations is sometimes called the *Borda-Carnot* loss. From the equations we see that the loss factor for an abrupt enlargement may be referred to either V_1 or V_2

$$K(V_1) = \left(1 - \frac{A_1}{A_2}\right)^2 ;K(V_2) = \left(\frac{A_2}{A_1} - 1\right)^2 \tag{1.44}$$

Due to the simplifying assumptions made in the analysis, equations 1.40 to 1.44 are subject to some inaccuracy. However, this inaccuracy is small and the equations are regarded as suitable for engineering applications.

1.6.3 Exit loss

A special case of an abrupt enlargement occurs when a submerged pipe discharges fluid into a large vessel such as a calorifier (Figure 1.25). Here $A_2 \gg A_1$, $V \rightarrow 0$ and equations 1.41 and 1.43 indicate that the pressure loss (now known as the *exit loss*) is equal to $\frac{1}{2}\rho V_1^2$, the velocity pressure in the pipe.

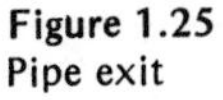

Figure 1.25
Pipe exit

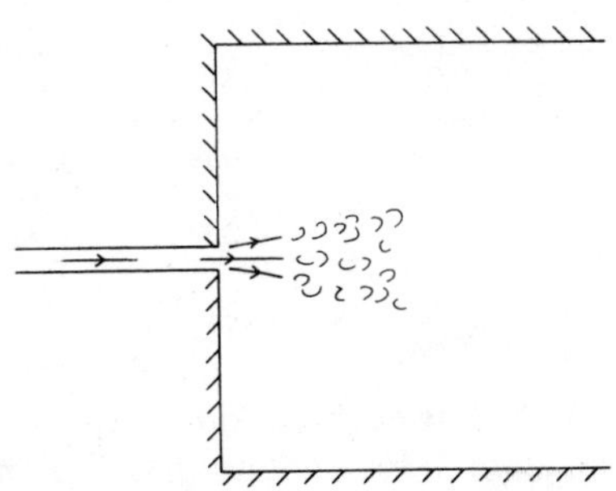

1.6.4 Abrupt contraction

The characteristics of the flow at an abrupt contraction are shown in Figure 1.26. At some distance upstream of the junction the flow streamlines start to converge towards the entrance to the smaller

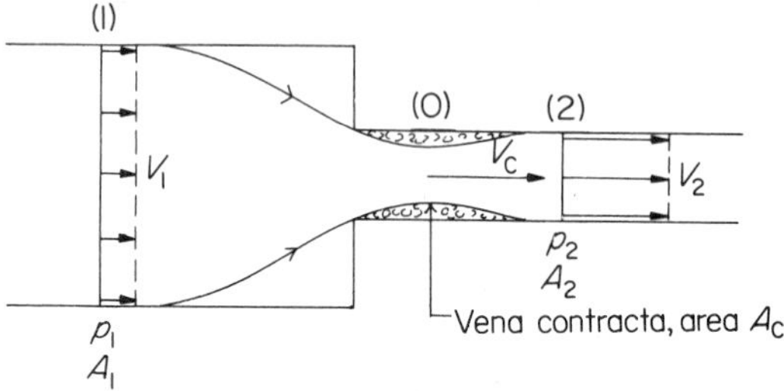

Figure 1.26
Abrupt contraction

pipe. This convergence continues after the junction and a *vena contracta* is formed, after which the fluid stream again expands to fill the pipe. Between the vena contracta and the pipe wall a zone of turbulent eddying is formed. Section (2) is again taken as being far enough downstream to enable the assumption of a uniform velocity profile to be made.

The process of converting pressure energy into kinetic energy is inherently more efficient than the reverse process. Hence the pressure loss from section (1) to the vena contracta is small compared with the loss from section (0) to section (2). If we neglect this small loss associated with the acceleration of the fluid and apply equation 1.43 to the expansion downstream of the vena contracta, the approximate pressure loss for the abrupt contraction is found to be

$$p_l = K\tfrac{1}{2}\rho V_2^2 = \tfrac{1}{2}\rho V_2^2\left(\frac{A_2}{A_c} - 1\right)^2 = \tfrac{1}{2}\rho V_2^2\left(\frac{1}{C_c} - 1\right)^2 \qquad (1.45)$$

where A_c represents the cross-sectional area of the vena contracta and C_c is the coefficient of contraction, A_c/A_2.

The coefficient of contraction C_c is a function of the ratio A_2/A_1. Early measurements by Weisbach[(6)] established accurately the values of C_c. These are given in Table 1.2 along with the corresponding computed values of K.

Table 1.2

A_2/A_1	0.2	0.4	0.6	0.8	1.0
C_c	0.632	0.659	0.712	0.813	1.0
K	0.34	0.27	0.16	0.05	0

1.6.5 Entry loss

As $A_2/A_1 \rightarrow 0$, it has been found experimentally that the value of K in equation 1.45 tends to 0.5. This limiting case corresponds to the flow from a large vessel into a pipe (Figure 1.27a), and here $0.25\rho V_2^2$ is known as the *entry loss*. The entry loss varies as the entry arrangement is varied. For well-rounded entrances, the loss is of the order of $0.01\rho V_2^2$ and is thus usually neglected (Figure 1.27b). A tapered entry also gives a much lower loss than the abrupt entry (Figure 1.27c). With re-entrant openings (where the pipe extends into the vessel), K may or may not exceed 0.5, depending upon the re-entry arrangement. Reference (5) contains information on various layouts.

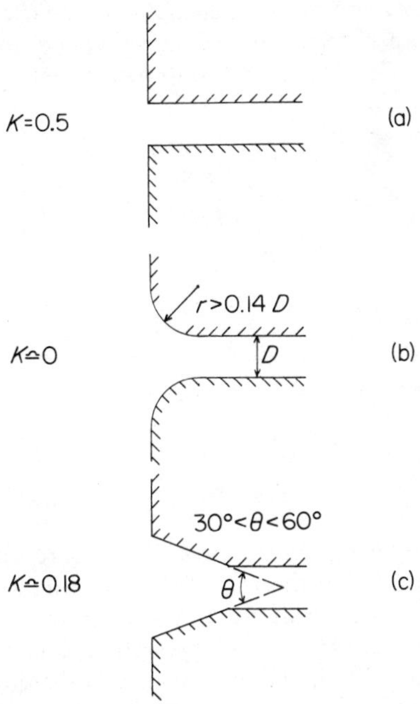

Figure 1.27
Pipe entry arrangements (From Massey, B. S., *Mechanics of Fluids*, 2nd ed., Van Nostrand Reinhold)

1.6.6 Pipe fittings

The loss of pressure in commercial pipe fittings is usually specified by equation 1.38, where V is the average velocity in the pipe. The magnitude of K depends upon the shape and construction of the fitting, the more tortuous the passage followed by the fluid the greater the pressure loss. A few typical values of K are given in Table 1.3 and more information is given in Chapter 9.

Table 1.3 Some Typical 'K' Factors

Fitting	*K*
Globe valve, fully open	10.0
Gate valve, fully open	0.2
¾ open	1.15
half open	5.6
¼ open	24.0
Strainer	2.0
Angle valve, fully open	5.0

References (5) and (7) contain sufficient information on loss factors for most design purposes, although where necessary the designer should consult the relevant manufacturer's data.

1.6.7 Bends

The loss of pressure in bends is greater than that experienced in the same length of straight piping. The additional loss is normally due to the combined effects of *separation* and *secondary flow*. These effects are produced due to the changes which occur in the pressure distribution at a bend in order to balance the centrifugal forces. Consider the bend shown in Figure 1.28.

Figure 1.28

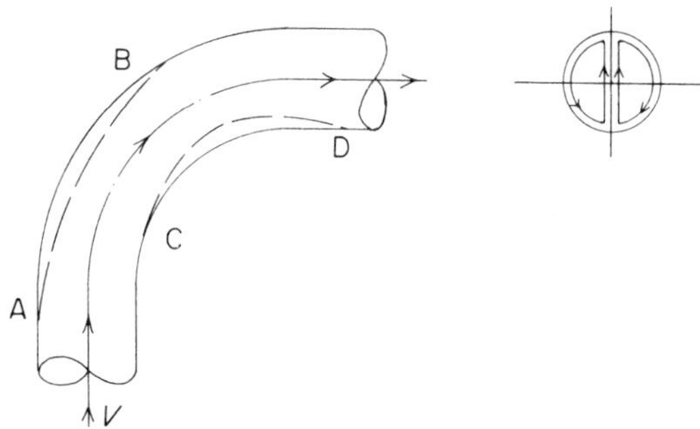

The change in the flow direction results in an increase in the pressure along the outside of the bend and a decrease along the inside. As a result fluid moving from A to B and from C to D experiences an adverse pressure gradient. Unless the radius of curvature of the bend is sufficiently large, separation of the flow from the boundary will occur between these points, giving rise to regions of turbulent eddying. The extent of the separations and the magnitude of the resulting losses in turbulence depend principally on the curvature of the bend. Now, due to the effects of fluid friction, the pressure gradient across the pipe is greater at the pipe centre than near the top and bottom of the section shown. The consequence of this is the creation of a secondary flow superimposed on the main flow, in which there is an outward movement of fluid at the centre and a corresponding inward movement of fluid at the top and bottom. The fluid leaving the bend thus possesses a double spiralling motion, the energy of which is dissipated as the motion is slowly destroyed by viscous action. Spiralling may persist for some considerable distance downstream of a bend—as much as 50 to 75 times the pipe diameter.

Reference (5) contains a comprehensive list of 'K' factors for elbows and bends, which has been compiled with reference to many sources of information.

1.6.8 Equivalent lengths

In piping design applications a convenient method of dealing with minor losses is to express them in terms of an *equivalent length* of straight pipe in which an equal pressure loss would occur. For the minor loss produced by a fitting in a pipe of diameter D, we can write

$$K\tfrac{1}{2}\rho V^2 = \frac{4\rho f L_e V^2}{2D}$$

where L_e is the equivalent length for the fitting. From the above equation we see that this is given by

$$L_e = \frac{KD}{4f} \tag{1.46}$$

The use of the equivalent length procedure in pipe sizing problems is fully covered in Chapter 9. (It will be seen that in these problems it is usual to work in terms of a length equivalent to the loss of one velocity pressure, i.e. EL = $L_e/K = D/4f$).

REFERENCES

1. REYNOLDS, O. 'An experimental investigation of the circumstances which determine whether the motion of water shall be direct or sinuous, and the law of resistance in parallel channels', *Phil. Trans. R. Soc.* Vol. 174, (1883).
2. NIKURADSE, J. 'Strömungsgesetze in rauhen Rohren', *VDI–Forschungsheft*, Vol. 361, (1933). (English translation in *N.A.C.A. Tech. Memo.*, 1292).
3. COLEBROOK, C. F. 'Turbulent flow in pipes with particular reference to the transition region between the smooth and rough pipe laws', *J. Inst. Civ. Engrs.*, Vol. II, (1939).
4. MOODY, L. F. 'Friction factors for pipe flow', *Trans. Amer. Soc. Mech. Engrs.*, Vol. 66, (1944).
5. C.I.B.S. Guide, Section C4.
6. WEISBACH, J. 'Die Experimental-Hydraulik', (Englehardt, Freiberg, 1855).
7. A.S.H.R.A.E. Handbook of Fundamentals.

SYMBOLS USED IN CHAPTER 1

A cross-sectional area
C_c coefficient of contraction
D pipe diameter
f friction factor
g acceleration due to gravity
h_f head loss due to friction
h_l head loss at fitting
K loss factor

K_E bulk modulus
k_s surface roughness
L length
L_e equivalent length
p pressure
Δp pressure difference
R pipe radius
Re Reynolds number
t time
V average velocity of flow
v local velocity; v' velocity fluctuation
$\bar{v}$ temporal mean velocity
V' volume
$\dot{G}$ volume flow rate
ω specific weight
z potential head
μ coefficient of viscosity
ρ density
ν kinematic viscosity
α kinetic energy correction factor
τ shear stress
τ_o boundary shear stress
τ_t turbulent shear stress
η eddy viscosity
ϕ function of dimensional analysis
ψ factor of dimensional analysis

2 The Fundamentals of Heat and Mass Transfer and Thermodynamics

2.1 INTRODUCTION

To the services engineer dealing with thermal environmental problems a thorough understanding of the basic principles of heat and mass transfer and elementary thermodynamics is of vital importance. This chapter is intended as a suitable introduction to the fundamental theory of these subjects. The application of this theory to practical design problems is covered extensively in later chapters.

2.2 CONDUCTION

2.2.1 Nature of conduction

Conduction is the term given to the process within a medium whereby heat is transmitted from a region of higher temperature to a region of lower temperature, without appreciable displacement of the particles of the medium. Conduction involves the direct transfer of kinetic energy at a molecular level and may have several distinct operating mechanisms associated with it, e.g. the elastic collision of molecules in a fluid medium or the motion of free electrons in metals. Regardless of the exact mechanism, a knowledge of which is unimportant anyway when conduction is considered as an engineering problem, the observable effect of the transfer between neighbouring regions will be the elimination of the temperature difference between them. Alternatively, if the temperatures of these regions are maintained constant by the addition and removal of heat at appropriate points, the result will be a continuous flow of heat from the higher temperature region to the lower temperature one.

Conduction is the only mode of heat transfer which occurs in opaque solids. It is also an important process in fluids but in these media it is usually combined with convection, and in some cases with radiation also. Conduction is the dominant heat transfer mechanism in fluid regions where laminar flow conditions exist, e.g. in the laminar sub-layer immediately adjacent to a solid boundary.

Fourier's Equation

The fundamental relationship for conduction in one dimension is the Fourier equation. This states that the rate of heat flow in a given direction is proportional to the temperature gradient in that direction and to the area normal to the flow. Considering, for example, heat flow in the x direction (Figure 2.1), Fourier's equation is

$$\dot{Q}_{(x)} = -kA \frac{d\theta}{dx} \tag{2.1a}$$

or

$$\dot{q}_{(x)} = \frac{\dot{Q}_{(x)}}{A} = -k \frac{d\theta}{dx} \tag{2.1b}$$

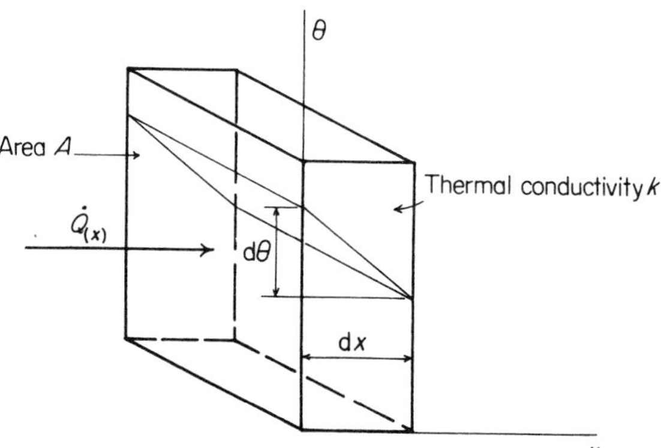

Figure 2.1
One-dimensional thermal conduction

where $\dot{Q}_{(x)}$ is the rate of heat flow (W) along the x coordinate axis through the cross-sectional area $A(m^2)$ and $d\theta/dx$ is the temperature gradient (°C/m) in this direction. The rate of heat flow per unit area (W/m^2), denoted here by $\dot{q}_{(x)}$, is referred to as the *heat flux*. The proportionality factor k is called the *thermal conductivity* and has units of W/m °C. The minus signs in equations 2.1a and 2.1b indicate the adopted convention that $\dot{Q}_{(x)}$ and $\dot{q}_{(x)}$ are taken as positive quantities when the heat flow is in the positive x direction.

Thermal Conductivity

The thermal conductivity k is a physical property of the heat-conducting material. The variation in the thermal conductivity of materials encountered in engineering problems is exceedingly large. The highest values of conductivity correspond to pure metals and the lowest to gases and vapours; building and insulating materials and liquids have values which lie between these upper and lower limits. Some idea of the orders of magnitude involved can be obtained from the following list of some typical values of k:

Copper	388 W/m °C
Concrete	1.4 W/m °C
Water	0.6 W/m °C
Asbestos	0.15 W/m °C
Air	0.026 W/m °C

The thermal conductivities of most substances vary with temperature, although, except at very low temperatures, this variation is generally not great. Over a moderate range of temperature, the relationship between k and the temperature θ may be assumed linear and we can write

$$k = k_0 \,[1 + b(\theta - \theta_0)] \tag{2.2}$$

where k_0 is the conductivity at the reference temperature θ_0 °C, k is the conductivity at θ °C and b is a constant known as the *temperature coefficient of thermal conductivity*. The temperature coefficient of metals and non-metallic crystalline materials is generally negative; for most other substances b is positive.

In heating services design applications, the complication of a varying conductivity with temperature is one which can be ignored and it is sufficient to consider mean values of k which are taken as constant over a given thickness of material. With regard to these applications, some comments on the conductivities of porous building and insulating materials seem appropriate at this juncture. In porous materials, heat flows by conduction through the solid parts while in the air spaces all three modes of heat transfer may be involved to varying degrees. For such materials quoted values of thermal conductivity are really 'apparent' conductivities, which are dependent upon the porosity or 'apparent' density of the material. An increase in porosity is generally associated with a decreasing k. If water penetrates the pores of a material, then a substantial increase in the apparent conductivity is exhibited, because water has a conductivity some twenty-three times greater than that of air. The most obvious manifestation of this phenomenon is the difference in conductivity between the outer and inner layers of a cavity brick wall. In selecting the

Table 2.1 Thermal Conductivities of Metals (k = W/m °C)

Metal	*Temperature* (°C) 0	100	200	300	400
Cast iron	55.4	51.9	48.5	45	43.3
Mild steel	45.4	45	44.8	43.3	39.8
Copper	387.6	377.2	372	366.8	363.4
Aluminium	202.5	206	214.6	230.1	249.2
Lead	34.6	32.9	31.1	31.1	—
Brass	96.9	103.8	109	114.2	115.9

design thermal conductivity of a porous material, it is therefore necessary to consider firstly the degree to which the material is likely to be subjected to water penetration [1].

Table 2.2 Thermal Conductivities of some Building and Insulating Materials

Material	*Density* (kg/m³)	*Temperature* (°C)	*k* (W/m °C)
Structural materials			
Asbestos-cement boards	1922	20	0.744
Asbestos sheets	890	50	0.166
Building bricks	1600	20	0.69
Concrete	1900–2300	20	0.81–1.4
Aerated concrete	320	10	0.087
	480		0.1
	640		0.14
	800		0.21
Plaster	1440	21	0.485
Wood	640	24	0.147
Insulating materials			
Expanded polystyrene	30	24	0.037
Urea formaldehyde foam	12	10	0.036
Fibre insulating board	237	21	0.0485
Kapok	14	20	0.035
Glass wool quilt	50	25	0.036
Wood felt	330	30	0.0519
Cork	160	30	0.0433

The thermal conductivities of various metals over a range of temperatures are given in Table 2.1. Table 2.2 gives values of conductivity for some common building and insulating materials. More detailed and comprehensive information on thermal conductivities is to be found in References (1), (2) and (3).

2.2.2 Differential equation of thermal conduction

The Fourier relationship of equation 2.1 was defined in terms of the heat flux and temperature gradient in one direction only. In general, however, conduction is a three-dimensional problem and the temperature is considered as a function of the three co-ordinates x, y and z and time t:

$$\theta = \theta\,(x, y, z, t)$$

The computation of the heat flux components $\dot{q}_{(x)}$, $\dot{q}_{(y)}$ and $\dot{q}_{(z)}$ requires the specification of the temperature gradients in the x, y and z directions and these gradients can be determined only if the temperature distribution in the medium is known. The temperature distribution is obtained from the solution of the differential equation of thermal conduction subject to the appropriate boundary conditions. A fundamental assumption used in the derivation of this differential relationship is that the heat-flow medium is isotropic, i.e. the conductivity k at any point in the medium does not vary with direction. This is true for most materials encountered in engineering applications. Some exceptions do however occur, a common example being wood, where k along the grain differs from that across the grain.

Besides dealing extensively with conduction in rectangular-shaped solids, the heating services designer also encounters problems of conduction in pipes and other cylindrical objects. To enable the analysis of both these cases to be carried out, the differential equation of thermal conduction will now be derived in both rectangular and cylindrical co-ordinates.

(a) Rectangular Co-ordinate System

Consider the infinitesimal rectangular element of material having sides dx, dy and dz as shown in Figure 2.2

Figure 2.2

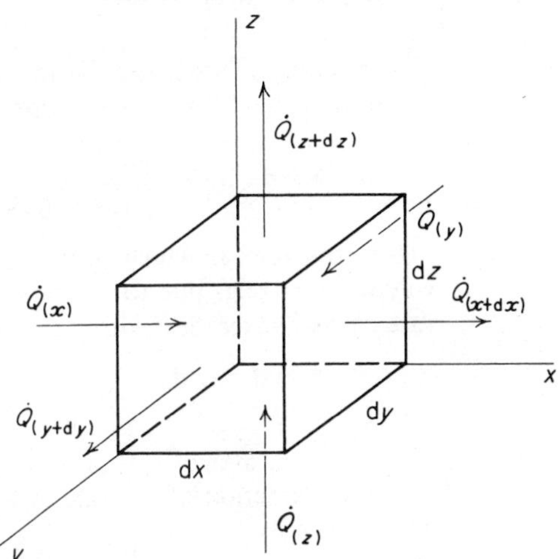

An energy balance applied to this element gives

$$\begin{pmatrix}\text{net rate of heat}\\ \text{entering element}\\ \text{by conduction}\end{pmatrix} + \begin{pmatrix}\text{rate of heat}\\ \text{generation}\\ \text{within element}\end{pmatrix} = \begin{pmatrix}\text{rate of heat}\\ \text{storage within}\\ \text{element}\end{pmatrix}$$

A B C

(2.3)

These three terms, denoted as shown by the symbols A, B and C , will now be evaluated.

A. The rate of heat flow into the element in the y direction is

$$\dot{Q}_{(y)} = -k \, \mathrm{d}x \, \mathrm{d}z \frac{\partial \theta}{\partial y}$$

The rate of heat flow out of the element in the y direction is

$$\dot{Q}_{(y+\mathrm{d}y)} = -k \, \mathrm{d}x \, \mathrm{d}z \frac{\partial}{\partial y}\left(\theta + \frac{\partial \theta}{\partial y} \mathrm{d}y\right)$$

$$= -k \, \mathrm{d}x \, \mathrm{d}z \frac{\partial \theta}{\partial y} - k \, \mathrm{d}x \, \mathrm{d}z \frac{\partial^2 \theta}{\partial y^2} \mathrm{d}y$$

The net rate at which heat enters the element by conduction in the y direction is therefore

$$\dot{Q}_{(y)} - \dot{Q}_{(y+\mathrm{d}y)} = k \, \mathrm{d}x \, \mathrm{d}y \, \mathrm{d}z \frac{\partial^2 \theta}{\partial y^2} \tag{2.4}$$

Similarly in the x and z directions

$$\dot{Q}_{(x)} - \dot{Q}_{(x+\mathrm{d}x)} = k \, \mathrm{d}x \, \mathrm{d}y \, \mathrm{d}z \frac{\partial^2 \theta}{\partial x^2} \tag{2.5}$$

$$\dot{Q}_{(z)} - \dot{Q}_{(z+\mathrm{d}z)} = k \, \mathrm{d}x \, \mathrm{d}y \, \mathrm{d}z \frac{\partial^2 \theta}{\partial z^2} \tag{2.6}$$

The net rate of heat entering the element by conduction is the sum of these three components

$$\mathrm{A} \equiv k \, \mathrm{d}x \, \mathrm{d}y \, \mathrm{d}z \left(\frac{\partial^2 \theta}{\partial x^2} + \frac{\partial^2 \theta}{\partial y^2} + \frac{\partial^2 \theta}{\partial z^2} \right) \tag{2.7}$$

B. If the rate at which heat is being generated per unit volume is $\dot{H}$ (e.g. due to the flow of electricity), then the rate of heat generation within the element is given by

$$\mathrm{B} \equiv \dot{\mathrm{H}} \, \mathrm{d}x \, \mathrm{d}y \, \mathrm{d}z \tag{2.8}$$

C. The rate at which heat is being stored within the element is a function of the rate of temperature change $\partial\theta/\partial t$ and is given by

$$\mathrm{C} \equiv \rho \, \mathrm{d}x \, \mathrm{d}y \, \mathrm{d}z C_{\mathrm{p}} \frac{\partial \theta}{\partial t} \tag{2.9}$$

where C_{p} is the specific heat and ρ is the density of the medium.

Substituting equations 2.7, 2.8 and 2.9 into equation 2.3, the energy balance becomes

$$\rho C_{\mathrm{p}} \frac{\partial \theta}{\partial t} = k \left(\frac{\partial^2 \theta}{\partial x^2} + \frac{\partial^2 \theta}{\partial y^2} + \frac{\partial^2 \theta}{\partial z^2} \right) + \dot{H}$$

and therefore

$$\frac{\partial \theta}{\partial t} = \alpha \left(\frac{\partial^2 \theta}{\partial x^2} + \frac{\partial^2 \theta}{\partial y^2} + \frac{\partial^2 \theta}{\partial z^2} \right) + \frac{\dot{H}}{\rho C_{\mathrm{p}}} \tag{2.10}$$

where $\alpha = k/\rho C_{\mathrm{p}}$ is known as the *thermal diffusivity*. Equation 2.10 is the general form of the differential equation of thermal conduction in a rectangular co-ordinate system. It may be simplified to suit particular situations. For example,

(i) The equation for unsteady heat conduction in one dimension (say x direction) without heat generation is

$$\frac{\partial \theta}{\partial t} = \alpha\left(\frac{\partial^2 \theta}{\partial x^2}\right) \tag{2.11}$$

(ii) In the steady state there is no variation of temperature with time and hence $\partial\theta/\partial t = 0$. The equations for steady one- and two-dimensional conduction with heat generation are respectively

$$\frac{\dot{H}}{\rho C_p} + \alpha \frac{d^2 \theta}{dx^2} = 0 \tag{2.12}$$

$$\frac{\dot{H}}{\rho C_p} + \alpha\left(\frac{\partial^2 \theta}{\partial x^2} + \frac{\partial^2 \theta}{\partial y^2}\right) = 0 \tag{2.13}$$

Figure 2.3

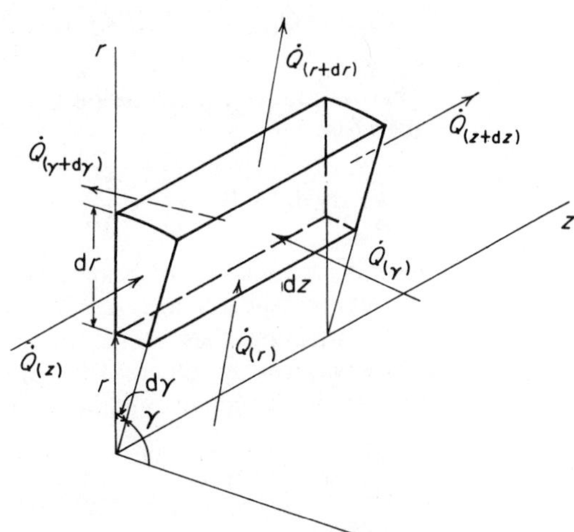

(b) Cylindrical Co-ordinate System

Consider the element of material shown in Figure 2.3. The procedure adopted here is the same as for rectangular co-ordinates except that in this case we consider heat flow in the radial, circumferential and axial directions.

The rate of heat flow into the element in the radial direction is

$$\dot{Q}_{(r)} = -k \, dz \, r d\gamma \frac{\partial \theta}{\partial r}$$

The rate of heat flow out of the element in the same direction is

$$\dot{Q}_{(r+dr)} = -k\,dz\,d\gamma\,(r+dr)\frac{\partial}{\partial r}\left(\theta + \frac{\partial\theta}{\partial r}dr\right)$$

$$= -k\,dz\,r d\gamma\frac{\partial\theta}{\partial r} - k\,dz\,d\gamma\,r\,dr\frac{\partial^2\theta}{\partial r^2}$$

$$-k\,dz\,d\gamma\,dr\frac{\partial\theta}{\partial r}$$

neglecting a term of higher order.

The net rate at which heat enters the element in the radial direction is

$$\dot{Q}_{(r)} - \dot{Q}_{(r+dr)} = k\,d\gamma\,dz\,dr\frac{\partial\theta}{\partial r} + k\,d\gamma\,dz\,r\frac{\partial^2\theta}{\partial r^2}dr \quad (2.14)$$

Similarly in the circumferential and axial directions

$$\dot{Q}_{(\gamma)} - \dot{Q}_{(\gamma+d\gamma)} = k\,r\,dr\,dz\frac{\partial^2\theta}{r^2\partial\gamma^2}d\gamma \quad (2.15)$$

$$\dot{Q}_{(z)} - \dot{Q}_{(z+dz)} = k\,dr\,d\gamma\,dz\,r\frac{\partial^2\theta}{\partial z^2} \quad (2.16)$$

The net rate of heat entering the element by conduction is therefore

$$k\,r\,d\gamma\,dz\,dr\left(\frac{1}{r}\frac{\partial\theta}{\partial r} + \frac{\partial^2\theta}{\partial r^2} + \frac{1}{r^2}\frac{\partial^2\theta}{\partial\gamma^2} + \frac{\partial^2\theta}{\partial z^2}\right)$$

In this case the rate of heat generation is $\dot{H}\,rd\gamma\,dr\,dz$ and the rate of heat storage is $rd\gamma\,dr\,dz\rho C_p\,\partial\theta/\partial t$.

Substituting these three items into the heat balance equation gives the general form of the differential equation of heat conduction in cylindrical co-ordinates

$$\frac{\partial\theta}{\partial t} = \alpha\left(\frac{\partial^2\theta}{\partial r^2} + \frac{1}{r}\frac{\partial\theta}{\partial r} + \frac{1}{r^2}\frac{\partial^2\theta}{\partial\gamma^2} + \frac{\partial^2\theta}{\partial z^2}\right) + \frac{\dot{H}}{\rho C_p} \quad (2.17)$$

As before, this equation may be simplified to suit particular situations.

2.2.3 One-dimensional steady-state conduction

Because of edge effects, steady-state conduction problems dealt with in engineering practice are generally never of a precisely one-dimensional nature. However, many simple problems approximate to this condition and may be solved satisfactorily by assuming heat transfer in one direction only. The conduction of heat through the wall of a building is one such important example and presages the use of one-dimensional conduction theory in heat loss calculations.

Conduction in Plane Slabs

The form of equation 2.10 for conduction in one dimension under steady state conditions with no heat generation is

$$\frac{d^2\theta}{dx^2} = 0 \tag{2.18}$$

Integrating this equation yields

$$\theta = C_1 x + C_2 \tag{2.19}$$

where the integration constants C_1 and C_2 are determined by the application of the appropriate boundary conditions. We note that equation 2.19 defines a linear variation of temperature with x.

Consider now the plane slab of thickness L and conductivity k illustrated in Figure 2.4. For the boundary conditions

$$\theta = \theta_1 \text{ at } x = 0$$

$$\theta = \theta_2 \text{ at } x = L$$

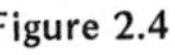
Figure 2.4

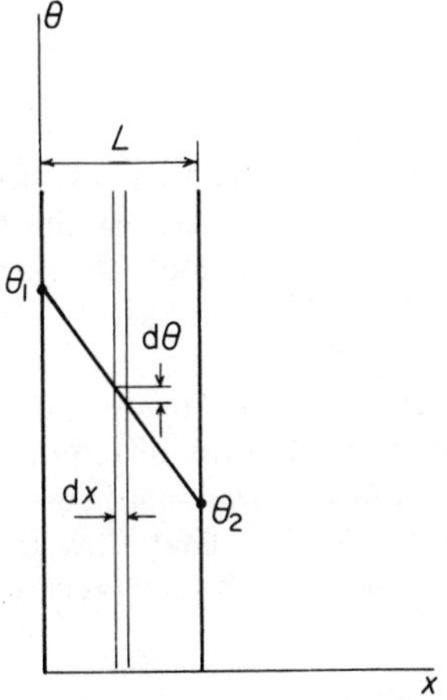

the temperature distribution is evaluated as

$$\frac{\theta - \theta_1}{\theta_2 - \theta_1} = \frac{x}{L} \tag{2.20}$$

The heat flux $\dot{q}_{(x)}$ at an arbitrary point in the material is given by Fourier's equation

$$\dot{q}_{(x)} = -k\,\frac{d\theta}{dx} \tag{2.21}$$

and as the temperature profile through the slab is linear this equation becomes

$$\dot{q}_{(x)} = k \frac{\theta_1 - \theta_2}{L} \tag{2.22}$$

The rate of heat flow $\dot{Q}_{(x)}$ through an area A of the slab is

$$\dot{Q}_{(x)} = Ak \frac{\theta_1 - \theta_2}{L} \tag{2.23}$$

From this analysis we see that in problems of one-dimensional steady-state conduction with no heat generation, the heat flux flowing through a plane slab may be written generally as

$$\dot{q} = \frac{\Delta\theta}{R} \tag{2.24}$$

where $\Delta\theta$ = difference in temperature between the boundary surfaces (°C)

and R = L/k, is the thermal resistance of the slab, (°C m²/W)

It is observed from equation 2.24 that the flow of heat by conduction is analogous to the flow of electricity as defined by Ohm's law

$$I = \frac{\Delta V}{R_E}$$

where I is the current (quantity of electricity per unit time) and ΔV is the voltage drop across the electrical resistance R_E. Clearly, heat flux is analogous to current while temperature difference, which provides the driving potential for heat flow, is analogous to voltage drop.

Let us now examine the steady state heat transfer through a series of plane slabs having different thermal conductivities. Consider the composite wall consisting of three parallel layers shown in Figure 2.5. The heat flow through unit area of this structure is the same through each layer and is given by

Figure 2.5

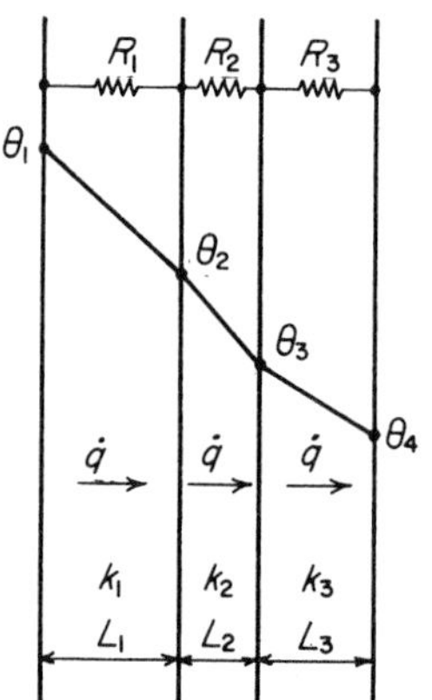

$$\dot{q} = \frac{k_1(\theta_1 - \theta_2)}{L_1} = \frac{k_2(\theta_2 - \theta_3)}{L_2} = \frac{k_3(\theta_3 - \theta_4)}{L_3} \tag{2.25}$$

This equation may be written in terms of the thermal resistances of the layers

$$\dot{q} = \frac{\theta_1 - \theta_2}{R_1} = \frac{\theta_2 - \theta_3}{R_2} = \frac{\theta_3 - \theta_4}{R_3} \tag{2.26}$$

where the various resistances are defined as

$$R_1 = \frac{L_1}{k_1};\ R_2 = \frac{L_2}{k_2};\ R_3 = \frac{L_3}{k_3} \tag{2.27}$$

By eliminating the interface temperatures θ_2 and θ_3 from equation 2.25, the heat flux through the composite structure is obtained in the following simple form

$$\dot{q} = \frac{\theta_1 - \theta_4}{R_t} \tag{2.28}$$

where $R_t = R_1 + R_2 + R_3$ is the total thermal resistance of the composite structure.

The temperature at either interface within the structure is readily determined from equations 2.26 and 2.28. For example, the temperature θ_3 is found to be

$$\theta_3 = \frac{R_3}{R_t}(\theta_1 - \theta_4) + \theta_4 \tag{2.29}$$

Radial Conduction in Cylindrical Layers

Steady-state conduction through pipe walls is a common heat transfer problem. It may be treated one-dimensionally if the surface temperatures are uniform, in which case heat flow is considered in the radial direction only. Figure 2.6 illustrates the situation for a single cylindrical layer of thermal conductivity k.

The differential equation for unidirectional radial conduction under steady-state conditions with no heat generation is

$$\frac{d^2\theta}{dr^2} + \frac{1}{r} \cdot \frac{d\theta}{dr} = 0 \tag{2.30}$$

which upon integration gives

$$\theta = C_1 \ln r + C_2 \tag{2.31}$$

Again the constants of integration C_1 and C_2 are determined from the appropriate boundary conditions, which in this case are

$$\theta = \theta_i \text{ at } r = r_i$$

$$\theta = \theta_o \text{ at } r = r_o$$

Figure 2.6

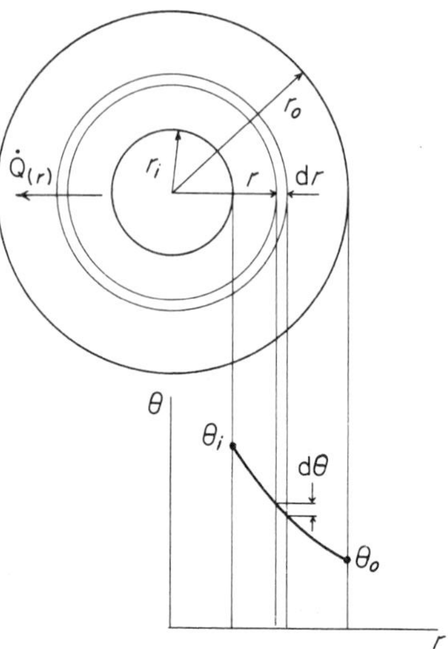

Evaluating C_1 and C_2 from these conditions gives the following equation for the temperature distribution through the layer:

$$\theta = \frac{\theta_o - \theta_i}{\ell n \frac{r_o}{r_i}} \ell n \frac{r}{r_i} + \theta_i \tag{2.32}$$

At the radius r, Fourier's equation may be applied to the cylindrical element of thickness dr

$$\dot{Q}_{(r)} = -k A \frac{d\theta}{dr} \tag{2.33}$$

Here A is the surface area at radius r. As this surface area depends upon r and thus varies between the inner and outer radii, it is convenient in this case to consider the conduction heat flow *per unit length* of cylinder

$$\dot{Q} = -k (2\pi r) \frac{d\theta}{dr} \tag{2.34}$$

Differentiating equation 2.32 to obtain the general expression for the temperature gradient and substituting this into equation 2.34 gives

$$\dot{Q} = \frac{2\pi k (\theta_i - \theta_o)}{\ell n \frac{r_o}{r_i}} \tag{2.35}$$

This equation can also be written as

$$\dot{Q} = \frac{\theta_i - \theta_o}{R} \tag{2.36}$$

Figure 2.7

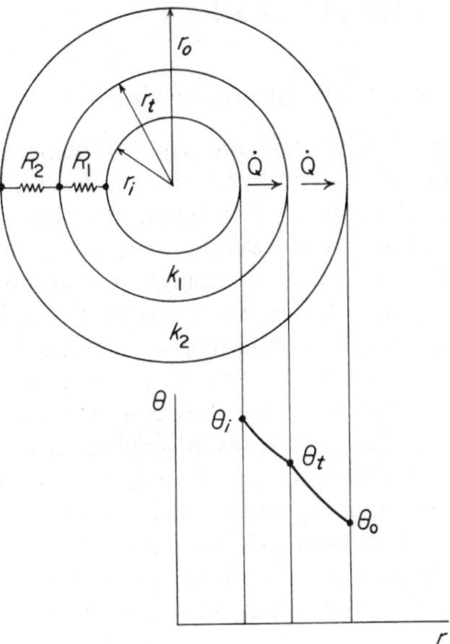

where $R = \ln (r_o/r_i)/2\pi k$ is the thermal resistance per unit length of cylinder, °Cm/W.

Consider now the composite cylindrical structure consisting of two co-axial layers having different thermal conductivities as shown in Figure 2.7. Insulated hot water or steam pipes are familiar examples of this type of multiple layer problem. The same quantity of heat must be conducted through both layers, hence

$$\dot{Q} = \frac{\theta_i - \theta_t}{R_1} = \frac{\theta_t - \theta_o}{R_2} \tag{2.37}$$

where the thermal resistances R_1 and R_2 are given by

$$R_1 = \frac{1}{2\pi k_1} \ln \frac{r_t}{r_i} ; R_2 = \frac{1}{2\pi k_2} \ln \frac{r_o}{r_t} \tag{2.38}$$

By eliminating the interface temperature θ_t from equation 2.37, the equation obtained for the heat flow rate per unit length of cylinder is

$$\dot{Q} = \frac{\theta_i - \theta_o}{R_t} \tag{2.39}$$

where $R_t = R_1 + R_2$.

Again, the interface temperature is determined readily from equations 2.37 and 2.39, thus

$$\theta_t = \theta_i - (\theta_i - \theta_o) \frac{R_1}{R_t} \tag{2.40}$$

2.3 CONVECTION

2.3.1 Nature of convection

Convection is the name given to the overall process by which heat is transferred between a moving fluid and an adjoining solid surface which is at a different temperature. The phenomenon of convection is a complex one and will generally combine the effects of different transport mechanisms, including that of conduction. By its nature convection is influenced greatly by the behaviour of the fluid flow occurring adjacent to the solid surface. In order to understand the mechanics of the heat transfer process, the mechanics of the flow process must be known. It is therefore obvious that any study of convective heat transfer must be based on a knowledge of the characteristics of fluid motion.

The basic elements of the convection process may be summarised conveniently by considering the flow structure formed by the passage of a turbulent fluid over a smooth solid surface as shown in Figure 2.8. In this example the surface is taken to be

Figure 2.8

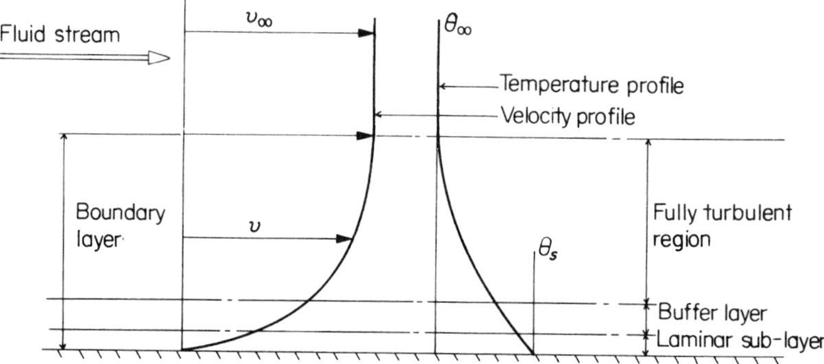

at the higher temperature. In the region of flow adjacent to the solid surface a boundary layer will form within which both the temperature and the velocity can be taken as varying from the values appropriate to the bulk fluid (i.e. θ_∞, v_∞) to those appropriate to the surface (i.e. θ_s, $v_s = 0$). Within this boundary layer, two sub-layers are identifiable. It has been shown in Chapter 1 that in the immediate vicinity of the surface there will exist an extremely thin layer of laminar flow. Next to this laminar sub-layer there is the so-called buffer layer. This acts as a transitional zone over which the flow changes from a fully laminar state to a fully turbulent one. The transfer of heat from the solid surface to the adjacent fluid particles and subsequently across the laminar sub-layer can only take place by conduction. In the buffer layer the effects of both conduction and turbulent mixing will be present and of significance. Beyond the buffer layer lies the fully turbulent part of the boundary layer. Here turbulent mixing, which we take as including energy storage and mass transport effects, will provide the dominant heat transfer mechanism.

Some idea of the complexities involved in the study of convection may be appreciated by noting that significant variations to this described flow structure, and consequently to the associated heat transfer processes, will result from changes in the characteristics of the flow system. For example, in certain low-velocity situations the entire flow may be laminar, in which case no buffer layer or fully turbulent region exists and the only heat transfer mechanism in the boundary layer is that of conduction. On the other hand, we have seen (1.5.3) that changes in the fluid velocity and surface roughness may lead to a partial or complete disruption of the laminar sub-layer, which in turn will lead to significant changes in the importance of conduction effects.

2.3.2 Natural and forced convection

Convection heat transfer is generally sub-divided into two different kinds of process, known as natural convection and forced convection. The classification of these sub-divisions is based upon the way in which the fluid motion is generated (Figure 2.9). Heat transfer by *natural convection* occurs when the main mass

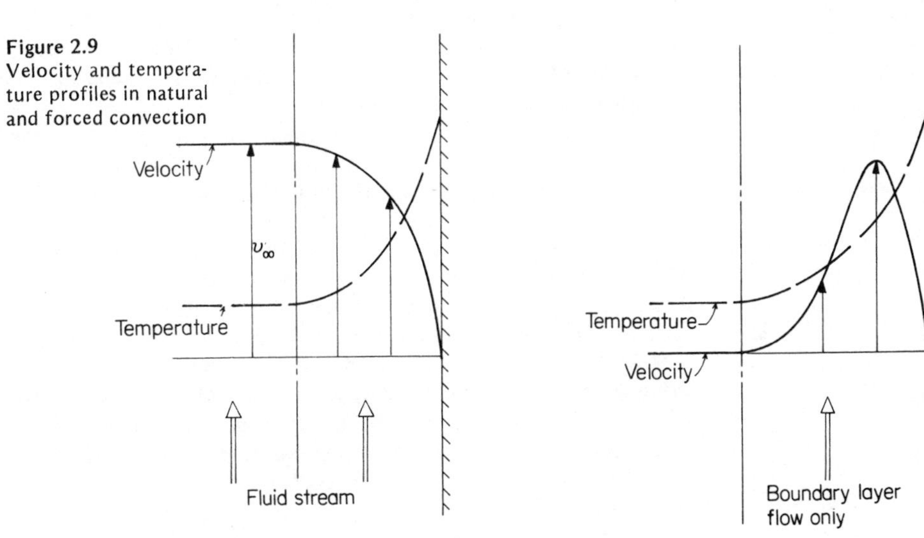

Figure 2.9 Velocity and temperature profiles in natural and forced convection

of the fluid is undisturbed and the motion which does occur is due entirely to buoyancy forces arising from density variations within the fluid in the vicinity of the surface. For example, a high temperature plate positioned vertically in cool stagnant air will set up a motion in the air layer adjacent to the plate as the temperature gradient in the air gives rise to a density gradient which in turn induces the motion. When the motion of a fluid is produced as a result of some external agency, such as a pump or a fan, the resulting heat transfer process is called *forced convection.*

2.3.3 Convection heat transfer coefficient

The rate of heat transfer by convection between a fluid and an adjoining solid surface may be computed by the following equation, which dates back to Newton (1701):

$$\dot{Q}_c = h_c A (\theta_s - \theta_\infty) \tag{2.41}$$

where $\dot{Q}_c$ = rate of heat transfer by convection (W),
A = area of surface (m^2),
θ_s = surface temperature (°C),
θ_∞ = bulk temperature of fluid (°C),
h_c = convection heat transfer coefficient (W/m^2 °C)

Several comments on this equation require to be made. Firstly, the equation has a general application and is used to describe all forms of convective heat transfer, including those involving a phase change. Secondly, although sometimes referred to as 'Newton's Law of Cooling', the equation does not in fact represent a phenomenological law of convection but is simply a definition of the convection heat transfer coefficient h_c. Thirdly, the form of the equation means that the calculation of convective heat transfer rates reduces basically to the evaluation of h_c for the particular situation being considered. The central object in the study of convection therefore becomes the prediction of numerical values of h_c for design purposes.

As convection presents an extremely complex problem, the simplicity of equation 2.41 can be misleading. Unlike the thermal conductivity in the problem of conduction, h_c is not simply a property of the fluid involved in the convection process. On the contrary, h_c must take account of all the factors in the problem which influence the transport of heat across the boundary layer. Accordingly, the numerical value of h_c is a function of many variables, such as the system geometry, the flow velocity and the specific heat, viscosity and other physical properties of the fluid. In view of the fact that these quantities are not necessarily constant over a given surface, the convection heat transfer coefficient may also vary over the surface. For this reason a distinction should be made between a *local* coefficient and the *average* coefficient as defined in equation 2.41. The local coefficient h'_c is such that

$$d\dot{Q}_c = h'_c \, dA (\theta_s - \theta_\infty) \tag{2.42}$$

while the average coefficient is given in terms of the local coefficient by

$$h_c = \frac{1}{A} \iint_A h'_c \, dA \tag{2.43}$$

In most engineering applications, including heating services design, interest generally centres on average values of the convection heat transfer coefficient.

2.3.4 Prediction of h_c

There are three general methods available for the evaluation of convection heat transfer coefficients:

(i) mathematical analysis of the boundary layer equations,
(ii) the Reynolds analogy, and
(iii) dimensional analysis and experimentation.

Although the first two methods have been used with reasonable success, the limitations associated with them mean that in practice heat transfer coefficients are generally evaluated by method (iii), the correlation of experimental data with the aid of dimensional analysis. Methods (i) and (ii) will not be considered further here but for information on these methods the reader should consult the excellent treatment given in Reference (4).

Dimensional Analysis

Dimensional analysis is a method which, although not producing an analytical solution to a given problem, does yield information about the form of the mathematical relationship connecting the pertinent variables. It does this by enabling an equation to be derived in which the variables are combined into dimensionless groups. The precise functional relationship between these groups can be determined only by experimentation, but by working in terms of such dimensionless parameters the number of quantities to be studied is significantly reduced, the number of experiments is minimised and the interpretation of experimental data is greatly aided.

The basis of dimensional analysis is the simple premise that any equation characterising a phenomenon must be dimensionally correct. The first step in the procedure is to select a system of primary dimensions and to express the variables entering the physical situation in terms of these dimensions. In applying dimensional analysis to heat transfer it is customary to use five dimensions: Mass M, time T, length L, temperature θ and heat H. The dimensions of some important physical quantities are therefore as listed

quantity	*symbol*	*dimensions*
velocity	v	L/T
viscosity	μ	M/LT
conductivity	k	$H/LT\theta$
density	ρ	M/L^3

Forced convection

From the description of the convective heat transfer process, it is reasonable to expect that the factors affecting the heat transfer coefficient h_c are:

a characteristic velocity v, a characteristic length ℓ, and the fluid properties of conductivity k, viscosity μ, specific heat C_p and density ρ. We therefore write

$$h_c = \phi\,(v, \mu, k, \rho, C_p, \ell)$$

where ϕ represents some unknown function. For convenience this can be replaced by

$$h_c = \psi\, v^a\, \mu^b\, k^c\, \rho^d\, C_p{}^e\, \ell^f \tag{2.44}$$

where ψ is a dimensionless factor.
Expressing each term in the primary dimensions gives

$$\frac{H}{L^2 T\theta} = \left[\frac{L}{T}\right]^a \left[\frac{M}{LT}\right]^b \left[\frac{H}{LT\theta}\right]^c \left[\frac{M}{L^3}\right]^d \left[\frac{H}{M\theta}\right]^e \left[L\right]^f$$

Summating the exponents of like dimensions gives the equations

$$\begin{aligned}
H&:\quad 1 = c + e\\
L&:\quad -2 = a - b - c - 3d + f\\
T&:\quad -1 = -a - b - c\\
M&:\quad 0 = b + d - e\\
\theta&:\quad -1 = -c - e
\end{aligned}$$

The simultaneous solution of these equations gives c = 1 – e, b = e – d, a = d, f = d – 1. Substituting in equation 2.44 yields the result

$$\frac{h_c \ell}{k} = \psi \left[\frac{\rho v \ell}{\mu}\right]^d \left[\frac{\mu C_p}{k}\right]^e \tag{2.45}$$

which can be expressed alternatively as

$$\frac{h_c \ell}{k} = \phi \left[\frac{\rho v \ell}{\mu}\right] \left[\frac{\mu C_p}{k}\right] \tag{2.46}$$

The groups in this relationship are dimensionless.

$\dfrac{h_c \ell}{k}$ is the *Nusselt number, Nu.*

$\dfrac{\rho v \ell}{\mu}$ is the *Reynolds number, Re.*

$\dfrac{\mu C_p}{k}$ is the *Prandtl number, Pr.*

Therefore the result of dimensional analysis applied to forced convection can be written as

$$Nu = \phi\,(Re, Pr) \tag{2.47}$$

and the experimental data can now be correlated in terms of these three parameters instead of the original seven.

Natural Convection

If ρ_o is the density of the cold undisturbed fluid, ρ is the density of the warmer fluid at the heated surface and $\Delta\theta$ is the corresponding temperature difference, then the buoyancy force per unit mass of fluid is

$$\frac{g\,(\rho_o - \rho)}{\rho}$$

or $\quad \beta g \Delta\theta$

where β is the *coefficient of cubic expansion*. To the variables considered in the forced convection problem we must therefore add the factors (βg) and $\Delta\theta$.

We write

$$h_c = \phi\,(\Delta\theta, k, \rho, \mu, C_p, (\beta g), \ell)$$

or $\quad h_c = \psi\,\Delta\theta^a\, k^b\, \rho^c\, \mu^d\, C_p{}^e\, (\beta g)^f\, \ell^h$

In terms of the primary dimensions this gives

$$\frac{H}{L^2 T\theta} = \theta^a \left[\frac{H}{LT\theta}\right]^b \left[\frac{M}{L^3}\right]^c \left[\frac{M}{LT}\right]^d \left[\frac{H}{M\theta}\right]^e \left[\frac{L}{\theta T^2}\right]^f \left[L\right]^h$$

Summating the exponents as before we find b = 1 – e, c = 2f, a = f, d = e – 2f, h = 3f – 1 and hence

$$\frac{h_c \ell}{k} = \psi \left[\frac{\Delta\theta(\beta g)\,\ell^3\,\rho^2}{\mu^2}\right]^f \left[\frac{C_p \mu}{k}\right]^e \tag{2.48}$$

The first dimensionless grouping on the right hand side of this equation is known as the *Grashof number, Gr.* Therefore for natural convection the result of the dimensional analysis can be written as

$$Nu = \phi\,(Gr, Pr) \tag{2.49}$$

and the experimental data obtained for natural convection is correlated in terms of these three parameters.

Empirical formulae

Extensive information is available on the precise forms of equations 2.47 and 2.49 for a wide range of heat transfer situations. A comprehensive selection of this data is given in References (3,5,8). Some of the principal relations of natural and forced convection are:

Forced convection

(i) Turbulent flow in pipes $\quad Nu = 0.0225 Re^{0.8}\, Pr^{0.33}$ (2.47a)

(ii) Laminar flow in pipes $\quad Nu = 4.1$ (2.47b)

(characteristic length : internal diameter of pipe)

(iii) Flow across a cylinder $Nu = 0.26Re^{0.6}\ Pr^{0.3}$ (2.47c)

(Characteristic length : diameter of cylinder)

Natural convection

(i) Horizontal cylinders

laminar range $10^3 < GrPr < 10^9$, $Nu = 0.53\,(GrPr)^{0.25}$ ... (2.49a)

(Characteristic length : diameter of cylinder)

(ii) Vertical flat plates and cylinders

laminar range $10^4 < GrPr < 10^9$, $Nu = 0.13\,(GrPr)^{0.33}$ ... (2.49b)

turbulent range $10^9 < GrPr < 10^{12}$, $Nu = 0.13(GrPr)^{0.33}$ ... (2.49c)

(Characteristic length : height of vertical surface)

(iii) Horizontal flat plates

Heated surface facing up (or cooled surface facing down)

laminar range $10^5 < GrPr < 2 \times 10^7$, $Nu = 0.54\,(GrPr)^{0.25}$ (2.49d)

turbulent range $2 \times 10^7 < GrPr < 3 \times 10^{10}$,

$Nu = 0.14\,(GrPr)^{0.33}$ (2.49e)

Heated surface facing down (or cooled surface facing up)

laminar range $3 \times 10^5 < GrPr < 3 \times 10^{10}$,

$Nu = 0.27\,(GrPr)^{0.25}$ (2.49f)

(Characteristic length : mean length of side)

The problems of natural convection encountered by the heating services engineer normally involve air as the fluid medium. Over the temperature range associated with these problems, the corresponding variations in the physical properties of air are relatively small. It is therefore possible to assign approximately constant values to the properties which appear in the Nusselt, Prandtl and Grashof numbers.

In this way the general formulae of natural convection may be modified to yield simplified expressions for the convection heat transfer coefficient. For example, applying this procedure to equations 2.49a and 2.49e gives

$$h_c = 1.35\left(\frac{\Delta\theta}{\ell}\right)^{0.25} \quad (2.49g)$$

$$h_c = 1.7\Delta\theta^{0.33} \quad (2.49h)$$

The natural convection heat transfer from plane and cylindrical surfaces will be considered further in a later chapter.

2.4 THERMAL RADIATION

2.4.1 Nature of radiation

All bodies continuously emit energy in the form of electromagnetic waves which we know as radiation. These waves travel with the speed of light, c, and do not require an intervening medium for their propagation. Indeed, radiation occurs most freely in a vacuum. The frequency of radiation depends solely on the nature of the emitting source. For example, a metal bombarded by high-frequency electrons emits X-rays while the fission of nuclei or radioactive disintegration produces gamma radiation. The wavelength of radiation, λ, is related to the frequency f, by

$$\lambda = c/f \tag{2.50}$$

The unit of wavelength most commonly used is the *micron* ($1\ \mu m = 10^{-6}$ m). In a vacuum $c = 2.9979 \times 10^8$ m/s.

Thermal radiation, by which we mean radiation emitted by a body in consequence of its temperature, occupies in effect only a small part of the total electromagnetic spectrum, in the wavelength range 0.1–100 μm. This includes all of the visible wavelengths (0.4–0.7 μm) and parts of the infra-red and ultra-violet spectra (Figure 2.10).

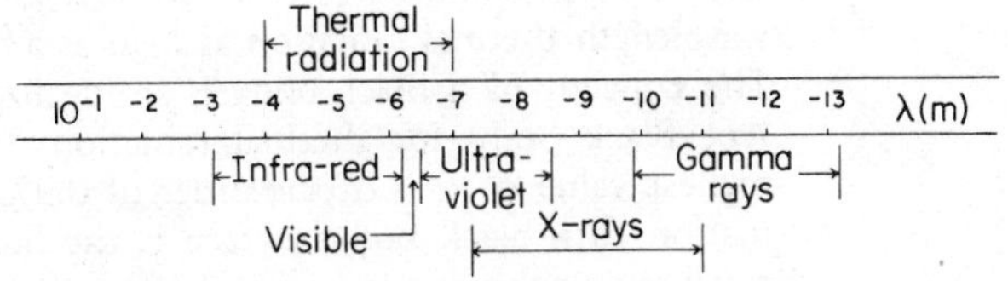

Figure 2.10 Electromagnetic spectrum

Absorption and Reflection

Figure 2.11 shows the possibilities which exist following the incidence of thermal radiation on a body surface.

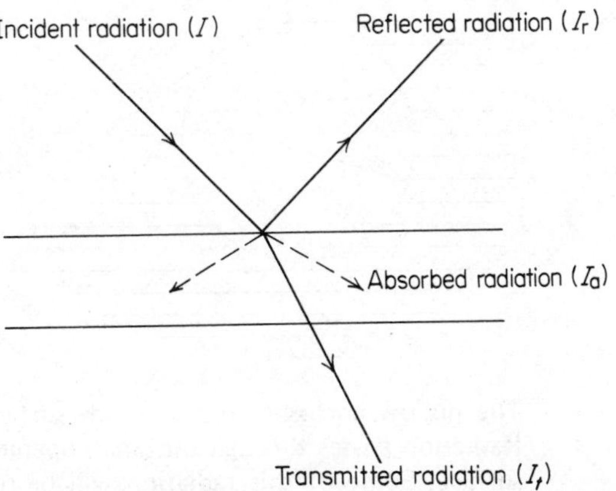

Figure 2.11 Reflection, absorption and transmission of radiation

Part of the incident radiation may be reflected (I_r) and part may be transmitted (I_t). The remainder will be absorbed (I_a) by the material and manifested as heat. We define the *reflectivity* as $\rho = I_r/I$, the *absorptivity* as $\alpha = I_a/I$ and the *transmissivity* as $\tau = I_t/I$. Thus

$$\rho + \alpha + \tau = 1 \tag{2.51}$$

This equation applies only to diathermanous materials such as glass. For *opaque* materials which do not transmit thermal radiation, $\tau = 0$ and

$$\rho + \alpha = 1 \tag{2.52}$$

A common problem in heating design applications involves the radiant exchange between surfaces with air as a separating medium. Radiation is freely transmitted in air (although not in some other gases) and we take the absorptivity and reflectivity of air as being zero, i.e. $\tau = 1$.

Black Body

An important concept in thermal radiation theory is that of the *black body*. A black body or black surface is taken as one which absorbs all incident radiation. The term 'black' in this context should not be confused with the actual colour of the surface, since a white-coloured surface can have an absorptivity for long-wavelength thermal radiation as high as a black-coloured surface. The concept of a black body is an idealization—there is no perfect black body for thermal radiation. For real materials the highest value of α is of the order of 0.97. An accurate approximation to a black body surface is the hollow enclosure with a small opening shown in Figure 2.12.

Figure 2.12

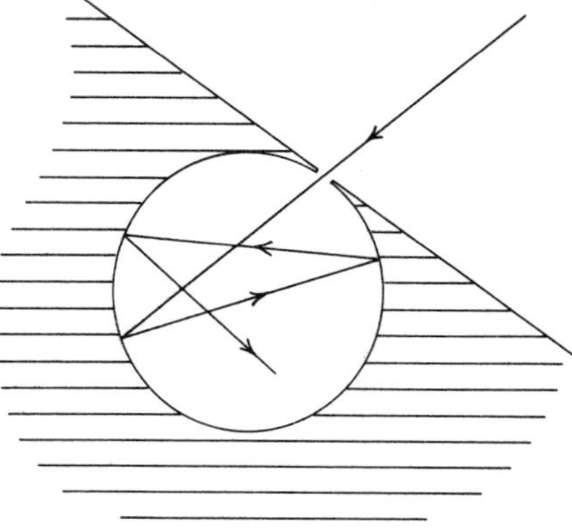

The hollow enclosure has an inside surface of high absorptivity. Radiation passes through the small opening and strikes the inside surface. Some of this radiation will be reflected but because of the high surface absorptivity most will be absorbed. Of the portion reflected most will be absorbed when it strikes the surface for the second time. After a number of such incidences the amount unabsorbed is exceedingly small and very little of the original radiation is reflected back out through the opening. The area of the opening may thus be regarded as black.

The *emissive power E* of a surface at a given temperature is the amount of heat radiated by unit area of the surface per unit time. The *emissivity* ϵ of a surface is defined as

$$\epsilon = \frac{E}{E_b} \tag{2.53}$$

where E_b is the emissive power of a black surface at the same temperature. The emissivity of a given surface is found to vary with temperature.

2.4.2 Laws of Radiation

Three important laws for the emission of black body radiation exist—the laws of Stefan-Boltzmann, Planck and Wien. The *Stefan-Boltzmann* law states that for a black body the emissive power is given by

$$E_b = \sigma T^4 \tag{2.54}$$

where T is the absolute temperature, K, and σ is the Stefan-Boltzmann constant, 5.663×10^{-8} W/m^2 K^4. The other two black body laws concern the distribution of this radiation throughout the wavelength spectrum and the shift in the distribution with changes in temperature. If $E_{b\lambda}$ is the *monochromatic emissive power* at wavelength λ, the relationship between $E_{b\lambda}$, λ and T is given by *Planck's Law*

$$E_{b\lambda} = \frac{C_1 \lambda^{-5}}{e^{C_2/\lambda T} - 1}, \tag{2.55}$$

where λ = wavelength, μm
T = absolute temperature, K
$C_1 = 3.743 \times 10^8$ W μm^4/m^2
$C_2 = 1.4387 \times 10^4$ μm K

Wien's displacement law states that the product of the wavelength at which $E_{b\lambda}$ is a maximum and the absolute temperature is a constant

$$\lambda_{max} T = C_3 \tag{2.56}$$

where $C_3 = 2897.6\,\mu$mK.

A schematic illustration of these three black body radiation

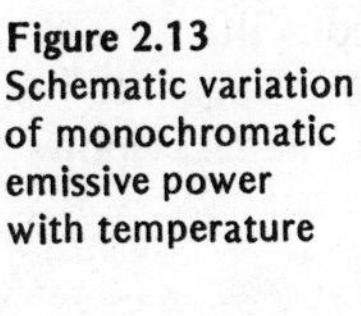
Figure 2.13 Schematic variation of monochromatic emissive power with temperature

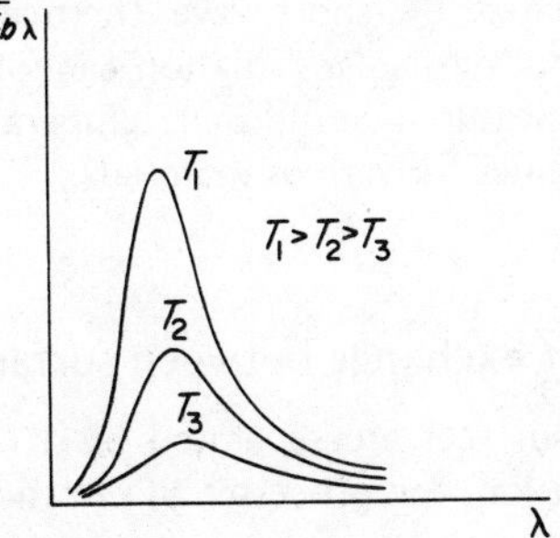

laws is given by Figure 2.13. The curves may be plotted using equation 2.55. As the temperature increases, the maximum monochromatic emissive power shifts to a shorter wavelength as required by equation 2.56. The area under any curve represents the total emissive power which is equal to the value calculated from equation 2.54.

$$E_b = \int_0^\infty E_{b\lambda}\, d\lambda = \sigma T^4 \tag{2.57}$$

Kirchhoff's Law

Kirchhoff's law provides a relation between the emissivity and the absorptivity and states that the emissivity of a surface at some temperature T is equal to the absorptivity which the surface exhibits for radiation from a black body at the same temperature. It is important to note that the law is derived on the basis of thermal equilibrium in an isothermal enclosure and therefore is subject to these restrictions. The generalization of Kirchhoff's law to say that

$$\epsilon = \alpha \tag{2.58}$$

for other conditions requires caution. Equation 2.58 is valid in ordinary thermal radiation problems where the incident and emitted radiation have the same spectral distribution. It is not valid for a surface emitting long-wave thermal radiation while

Table 2.3 Emissivity of Various Materials

Material	*Temperature* (°C)	*Emissivity*
Aluminium: polished	38	0.04
rough plate	38	0.0625
roofing	38	0.216
Zinc: galvanised sheet iron	24	0.25
Lead	24	0.28
Asbestos	90	0.933
Brickwork	21	0.93
Wood	16	0.9
Water	20	0.953
Plaster	10–80	0.91

being irradiated by short-wave thermal radiation—the absorptivity of a body to solar radiation cannot be taken as equal to its emissivity measured at normal temperatures. Table 2.3 shows emissivity values for various materials.

2.4.3 Heat exchange between surfaces

When two surfaces are separated by a medium which does not absorb radiation, the processes of emission and absorption result in an energy exchange between the surfaces. If the surfaces are at the same temperature, the net exchange is zero. If the surfaces are at different temperatures, there is a net transfer of heat from the hotter surface to the colder one.

Consider two surfaces which form an enclosure, one of area A_1, absolute temperature T_1 and emissivity ϵ_1; the other of area A_2, absolute temperature T_2 and emissivity ϵ_2. The net rate of radiant heat transfer between these surfaces may be calculated from the relationship

$$\dot{Q}_R = \sigma A_1 \mathcal{F}_{12} (T_1^4 - T_2^4) \tag{2.59}$$

$\mathcal{F}_{12}$ is a dimensionless factor, introduced by Hottel[6], which takes into account the emissivities of the surfaces and the geometric configuration of the surfaces as specified by the appropriate view factor

$$\mathcal{F}_{12} = \frac{1}{\left(\frac{1}{\epsilon_1} - 1\right) + \frac{1}{F_{12}} + \frac{A_1}{A_2}\left(\frac{1}{\epsilon_2} - 1\right)} \tag{2.60}$$

The view factor F_{12} is defined as the fraction of radiation leaving surface A_1 which is intercepted by A_2 (the reader is referred to References (4,5,7) for detailed information on the computation of view factors).

Heat Transfer Coefficient

When combined modes of heat transfer are being examined, it is desirable to express the rate of radiant heat transfer in the same form as the rate of convective heat transfer is expressed in equation 2.41. We therefore write

$$\dot{Q}_R = \epsilon_1 A_1 h_R (\theta_1 - \theta_2) \tag{2.61}$$

where θ_1 and θ_2 are the temperatures of the surfaces in degrees Celsius and h_R is the radiation heat transfer coefficient, W/m^2 °C. Comparing equations 2.59 and 2.61 yields

$$h_R = \frac{\sigma \mathcal{F}_{12} (T_1^4 - T_2^4)}{\epsilon_1 (\theta_1 - \theta_2)}$$

or $$h_R = \frac{\sigma \mathcal{F}_{12}}{\epsilon_1} (T_1 + T_2)(T_1^2 + T_2^2) \tag{2.62}$$

We see that h_R is a function of the Hottel factor $\mathcal{F}_{12}$, the temperatures of the surfaces and the emissivity ϵ_1.

Reference (1) quotes two values of h_R for heat loss calculation purposes—5.7 W/m^2 °C for a mean surface temperature of 20 °C (the inside surface coefficient) and 4.6 W/m^2 °C for a mean surface temperature of 0 °C (the outside surface coefficient). It is interesting to investigate how such values can be arrived at from equation 2.62.

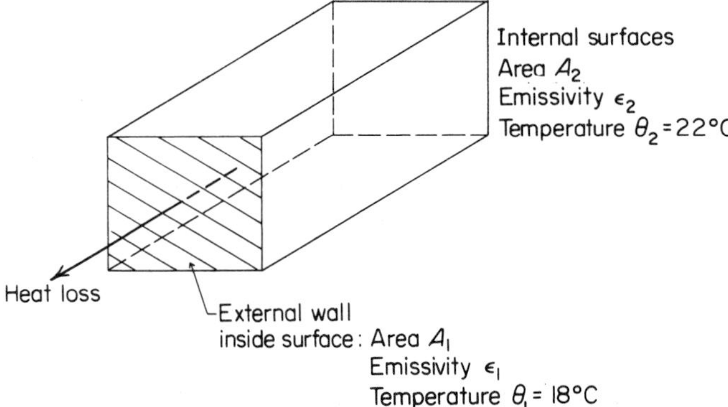

Figure 2.14 Enclosure for evaluation of h_R

Consider the enclosure shown in Figure 2.14. There is one plane surface through which a heat loss to the outside is being experienced. The temperatures of the surfaces of the enclosure are as indicated. For this situation, let us assume that A_2 is large compared with A_1. Thus $A_1/A_2 \ll 1$ and

$$\frac{A_1}{A_2}\left(\frac{1}{\epsilon_2} - 1\right) \simeq 0$$

(The same result is obtained from the alternative assumption that the surface A_2 is black, i.e. $\epsilon_2 = 1$[(2)]). We note also that for the configuration of the surfaces in the enclosure, $F_{12} = 1$. Therefore, equation 2.60 reduces to $\mathcal{F}_{12} = \epsilon_1$ and equation 2.62 becomes

$$h_R = \sigma(T_1 + T_2)(T_1^2 + T_2^2)$$

Substituting for T_1 and T_2, h_R evaluates to

$$h_R = \frac{5.663}{10^8}(586)(87025 + 84681)$$

$$= 5.7 \text{ W/m}^2\ \text{°C}$$

2.5 OVERALL TRANSFER OF HEAT

2.5.1 Introduction

Most practical problems in heating services design concern the transfer of heat between two fluids separated by some solid medium. Generally all three modes of heat transfer are involved in the overall process. We will now derive equations defining the overall transfer of heat in both rectangular and cylindrical co-ordinates.

2.5.2 Plane surfaces

Consider two fluids of bulk temperatures θ_{f1} and θ_{f2} which are separated by the composite wall of area A shown in Figure 2.15.

Figure 2.15

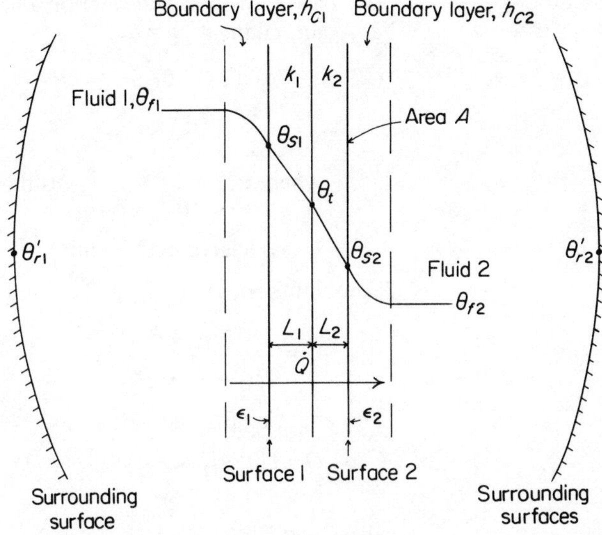

We will assume $\theta_{f1} > \theta_{f2}$. If the fluids concerned are gases such as air then radiation effects as well as convection effects will require to be included in the analysis.

We can consider the overall heat transfer process in three parts:

(i) Heat transfer by convection and radiation to Surface 1. Using equations 2.41 and 2.61 the total heat flow $\dot{Q}$ can be written as

$$\dot{Q} = \dot{Q}_c + \dot{Q}_R = A\, h_{c1}\,(\theta_{f1} - \theta_{s1}) + A\,\epsilon_1\, h_{R1}\,(\theta'_{r1} - \theta_{s1})$$

For the present let us assume that $\theta'_{r1} = \theta_{f1}$ (the validity of this assumption will be considered in detail in Chapter 6). Therefore

$$\dot{Q} = A\,(h_{c1} + \epsilon_1\, h_{R1})\,(\theta_{f1} - \theta_{s1})$$

or $$\dot{Q} = \frac{A\,(\theta_{f1} - \theta_{s1})}{R_{s1}} \qquad (2.63)$$

where $R_{s1} = \dfrac{1}{h_{c1} + \epsilon_1\, h_{R1}}$ is the surface resistance associated with Fluid 1.

(ii) Heat transfer by conduction through the wall. Using equation 2.28,

$$\dot{Q} = \frac{A(\theta_{s1} - \theta_{s2})}{R_t} \tag{2.64}$$

where $R_t = \dfrac{L_1}{k_1} + \dfrac{L_2}{k_2}$ is the total thermal resistance of the wall.

(iii) Heat transfer by convection and radiation from Surface 2. Again, taking $\theta'_{r2} = \theta_{f2}$,

$$\dot{Q} = \frac{A(\theta_{s2} - \theta_{f2})}{R_{s2}} \tag{2.65}$$

where $R_{s2} = \dfrac{1}{h_{c2} + \epsilon_2 h_{R2}}$ is the surface resistance associated with Fluid 2.

Combining equations 2.63, 2.64 and 2.65 gives

$$\dot{Q} = \frac{A(\theta_{f1} - \theta_{f2})}{R_{s1} + R_t + R_{s2}} \tag{2.66}$$

This equation may be written as

$$\dot{Q} = UA(\theta_{f1} - \theta_{f2}) \tag{2.67}$$

where $\dfrac{1}{U} = \Sigma R = \dfrac{1}{h_{c1} + \epsilon_1 h_{R1}} + \sum \dfrac{L}{K} + \dfrac{1}{h_{c2} + \epsilon_2 h_{R2}}$

$1/U$ is the overall thermal resistance in the path of the heat flow between the fluids, m^2 °C/W. U is known as the *overall heat transfer coefficient* (or the *overall thermal transmittance*), W/m^2 °C.

Air Spaces

If the layers of the composite wall considered in the foregoing analysis are separated by an air space, then the equations require to be modified to take account of the increased thermal resistance. The total resistance of the wall becomes

$$R_t = \frac{L_1}{k_1} + \frac{L_2}{k_2} + R_a$$

where R_a is the thermal resistance of the air space, m^2°C/W. The overall heat transfer coefficient is accordingly given by

$$\frac{1}{U} = \Sigma R = \frac{1}{h_{c1} + \epsilon_1 h_{R1}} + \sum \frac{L}{k} + R_a + \frac{1}{h_{c2} + \epsilon_2 h_{R2}}$$

Conduction effects in building air spaces are generally negligible and the heat transfer across a cavity results principally from

Figure 2.16

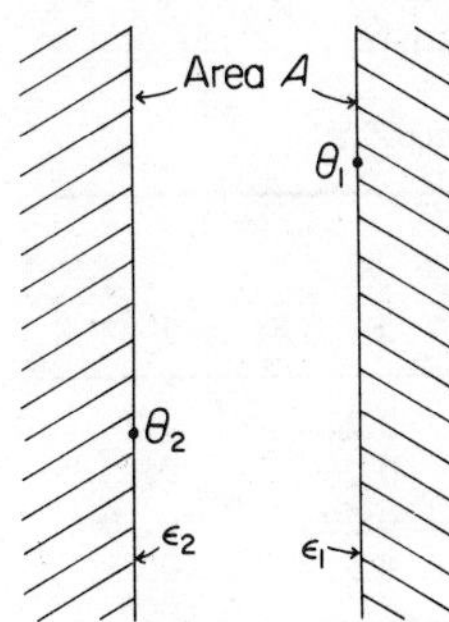

the radiation exchange between the bounding surfaces and the movement of convection air currents. The thermal resistance of an air space is consequently dependent on the physical factors which influence these two processes. These factors may be considered with reference to the typical air space shown in Figure 2.16. Expressing the radiation and convection transfers in terms of equations 2.61 and 2.41, we obtain

$$\dot{Q} = \dot{Q}_R + \dot{Q}_c = \epsilon_1 h_R A\,(\theta_1 - \theta_2) + \frac{h_c A}{2}(\theta_1 - \theta_2)$$

$$= \frac{A\,(\theta_1 - \theta_2)}{R_a}$$

The air space resistance is given therefore by

$$R_a = \frac{1}{h_c/2 + \epsilon_1 h_R} \tag{2.68}$$

The radiation component of the heat transfer may be approximated by the radiation exchange between infinite parallel planes. For such a configuration

$$\mathcal{F}_{12} = \frac{1}{\dfrac{1}{\epsilon_1} + \dfrac{1}{\epsilon_2} - 1}$$

The radiation term in equation 2.68 can now be written as

$$\epsilon_1 h_R = \frac{\sigma\,\mathcal{F}_{12}\,(T_1^4 - T_2^4)}{(\theta_1 - \theta_2)}$$

$$= \frac{5.663\,\epsilon_1\,\epsilon_2}{(\epsilon_1 + \epsilon_2 - \epsilon_1\,\epsilon_2)}\,(T_1 + T_2)\left[\left(\frac{T_1}{10^4}\right)^2 + \left(\frac{T_2}{10^4}\right)^2\right]$$

and the influence of surface emissivity and mean surface temperature on the air space resistance is clearly demonstrated.

The value of the convection term in equation 2.68 is a function of several factors, namely, the width and orientation of the air space and the temperature difference across it, the direction of the heat flow and the nature of the air space ventilation. A ventilated space will of course offer a lower thermal resistance than a similar unventilated one. The air space width determines the extent of the interaction between the convection processes occurring at the two bounding surfaces. Experiments on vertical air spaces have indicated a rapid decrease in h_c as the width is increased from zero to about 20 mm. Above 20 mm there is little further change in the convection coefficient. It would thus appear that for widths in excess of 20 mm the convection processes are virtually independent, but that for decreasing widths below this value the associated air streams increasingly interfere with one another, with resulting increases in the heat transfer.

Table 2.4 Thermal Resistance of a 90 mm Air Space

Mean Temperature (°C)	*Temperature Difference* (°C)	R_a, (m² °C/W) $\mathcal{F}$ = 0.05	$\mathcal{F}$ = 0.2	$\mathcal{F}$ = 0.5	$\mathcal{F}$ = 0.8
−18	11	0.48	0.38	0.27	0.2
10	6	0.6	0.4	0.25	0.18
10	17	0.45	0.33	0.22	0.16
32	6	0.6	0.38	0.22	0.15

Adapted from A.S.H.R.A.E., *Book of Fundamentals*

Table 2.5 Thermal Resistance of an Air Space Bounded by High-emissivity Surfaces

Mean Temperature (°C)	*Width* (mm) 6	12	18	25	40
	R_a(m² °C/W)				
−1	0.12	0.15	0.16	0.164	0.165
4	0.118	0.148	0.156	0.158	0.159
10	0.115	0.14	0.15	0.152	0.153
16	0.11	0.136	0.146	0.147	0.148

Adapted from N. S. Billington, *Thermal Properties of Buildings*, (Cleaver-Hume, London, 1952)

Some of the important characteristics of air spaces are clearly demonstrated by Tables 2.4 and 2.5, which display some experimentally determined resistance values for closed cavities. Table 2.4 gives the resistance of 90 mm air space for different surface temperature and emissivity conditions. Table 2.5 illustrates the effect of a varying cavity width on the thermal resistance of an air space bounded by high-emissivity surfaces. For detailed design information on the resistances of a wide range of air spaces the reader should consult References (1,2).

2.5.3 Cylindrical surfaces

Figure 2.17 shows two fluids separated by two cylindrical layers. Only convective effects need be considered for the hot fluid. If the cold fluid is air then both convection and radiation effects must be considered. This type of situation corresponds to an insulated hot water or steam pipe losing heat to its surroundings.

As before with cylindrical problems we will consider the radial heat transfer per unit length of pipe. The overall heat transfer process can be considered in three parts. At the inside surface, the heat transfer rate is

Figure 2.17

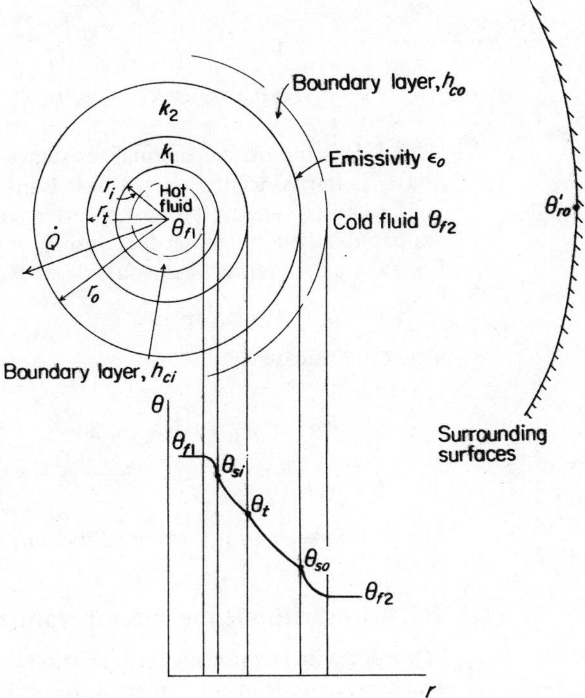

$$\dot{Q} = 2\pi r_i h_{ci} (\theta_{f1} - \theta_{si}) \tag{2.69}$$

The same quantity is conducted through the layers and transmitted to the surroundings by convection and radiation. Thus

$$\dot{Q} = \frac{2\pi k_1 (\theta_{si} - \theta_t)}{\ln \frac{r_t}{r_i}} = \frac{2\pi k_2 (\theta_t - \theta_{so})}{\ln \frac{r_o}{r_t}} \tag{2.70}$$

$$\dot{Q} = 2\pi r_o (h_{co} + \epsilon_o h_{Ro}) (\theta_{so} - \theta_{f2}) \tag{2.71}$$

where, as before we assume $\theta'_{ro} = \theta_{f2}$.
Combining these equations gives

$$\dot{Q} = \frac{(\theta_{f1} - \theta_{f2})}{\frac{1}{2\pi r_i h_{ci}} + \frac{\ln \frac{r_t}{r_i}}{2\pi k_1} + \frac{\ln \frac{r_o}{r_t}}{2\pi k_2} + \frac{1}{2\pi r_o (h_{co} + \epsilon_o h_{Ro})}}$$

Again an overall heat transfer coefficient can be introduced to give

$$\dot{Q} = U(\theta_{f1} - \theta_{f2}) \tag{2.72}$$

$$\text{where } \frac{1}{U} = \frac{1}{2\pi r_i h_{ci}} + \frac{\ell n \frac{r_t}{r_i}}{2\pi k_1} + \frac{\ell n \frac{r_o}{r_t}}{2\pi k_2} + \frac{1}{2\pi r_o (h_{co} + \epsilon_o h_{Ro})}$$

Here $1/U$ is the overall thermal resistance *per unit length of pipe* and U is the associated overall heat transfer coefficient, W/m°C.

If required, an overall heat transfer coefficient may easily be evaluated based on either the inside or outside surface areas. For example, in terms of the outside surface area

$$\dot{Q} = U A_o (\theta_{f1} - \theta_{f2}) \tag{2.73}$$

where in this case

$$\frac{1}{U} = \frac{r_o}{r_i h_{ci}} + \frac{r_o \ \ell n \frac{r_t}{r_i}}{k_1} + \frac{r_o \ \ell n \frac{r_o}{r_t}}{k_2} + \frac{1}{h_{co} + \epsilon_o h_{Ro}}$$

The group $(h_{co} + \epsilon h_{Ro})$ is termed the *outside surface heat transfer coefficient*, h_{so}.

2.5.4 Logarithmic mean temperature difference

In the preceding section no consideration was given to any changes which may occur in the temperature of the fluids involved in the heat transfer process. In a system, such as a heat exchanger, where these fluids undergo a continuous variation in temperature, equations 2.72 and 2.73 can be used only to specify the rate of heat transfer at local sections of the system. In order to calculate the total heat exchange between the fluids, the local temperature difference in these equations must be replaced by an effective mean temperature difference $\Delta\theta_m$. In terms of the outside surface area separating the two fluids, we write the total rate of heat transfer as

$$\dot{Q} = U A_o \Delta\theta_m \tag{2.74}$$

Let us consider the evaluation of $\Delta\theta_m$ with reference to the typical counter-flow heat exchange process shown schematically in Figure 2.18. The following assumptions are made

(i) The overall heat transfer coefficient is constant throughout the exchanger.

(ii) The mass flow rate and specific heat of each fluid is constant.

(iii) The heat loss to the surroundings is negligible.

Consider an elemental area dA_o. The rate of heat transfer associated with this area is

$$d\dot{Q} = U dA_o (\theta_{f1} - \theta_{f2}) = U dA_o \Delta\theta \tag{2.75}$$

If $\dot{m}_1$ and $\dot{m}_2$ are the mass flow rates of the hot and cold fluids and C_{p1} and C_{p2} are the respective specific heats, then

$$d\dot{Q} = -\dot{m}_1 C_{p1} d\theta_{f1} = -\dot{m}_2 C_{p2} d\theta_{f2}$$

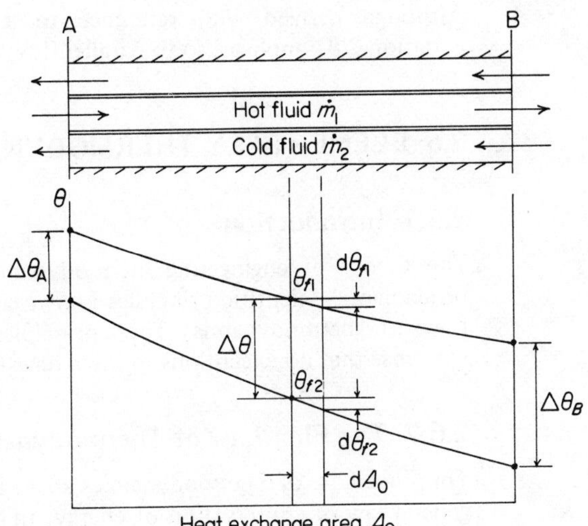

Figure 2.18 Counterflow heat exchange process

Hence,

$$d(\Delta\theta) = d\theta_{f1} - d\theta_{f2} = -\left(\frac{1}{\dot{m}_1 C_{p1}} - \frac{1}{\dot{m}_2 C_{p2}}\right) d\dot{Q} \quad (2.76)$$

The integration of this equation between A and B yields

$$\Delta\theta_B - \Delta\theta_A = -\left(\frac{1}{\dot{m}_1 C_{p1}} - \frac{1}{\dot{m}_2 C_{p2}}\right) \dot{Q} \quad (2.77)$$

Substituting for $d\dot{Q}$ from equation 2.75 into equation 2.76 gives

$$\frac{d(\Delta\theta)}{\Delta\theta} = -\left(\frac{1}{\dot{m}_1 C_{p1}} - \frac{1}{\dot{m}_2 C_{p2}}\right) U dA_o \quad (2.78)$$

Integrating this equation between A and B, we obtain

$$\ln\frac{\Delta\theta_B}{\Delta\theta_A} = -\left(\frac{1}{\dot{m}_1 C_{p1}} - \frac{1}{\dot{m}_2 C_{p2}}\right) U A_o \quad (2.79)$$

Dividing equation 2.77 by 2.79,

$$\dot{Q} = UA_o \frac{\Delta\theta_B - \Delta\theta_A}{\ln\dfrac{\Delta\theta_B}{\Delta\theta_A}} \quad (2.80)$$

Comparing equation 2.80 with 2.74, it is clear that the required temperature difference is

$$\Delta\theta_m = \frac{\Delta\theta_B - \Delta\theta_A}{\ln(\Delta\theta_B/\Delta\theta_A)} \quad (2.81)$$

This is known as the *logarithmic mean temperature difference*.

Although derived with reference to a counter-flow process, equation 2.81 applies also to parallel flow arrangements.

2.6 ELEMENTARY THERMODYNAMICS

2.6.1 Introduction

The theory of engineering thermodynamics is based upon two fundamental scientific principles known as the First and Second Laws of Thermodynamics. These principles can be proved only in the sense that no exceptions to them have ever been found.

2.6.2 The First Law of Thermodynamics

The First Law of Thermodynamics arises as a consequence of the general law of conservation of energy. In its fundamental form it states that when a system executes a cyclic process the net work delivered to the surroundings is proportional to the net heat taken from the surroundings.

$$\Sigma W \; \alpha \; \Sigma Q \tag{2.82}$$

An important corollary follows from the First Law—when a system undergoes a change of state, the difference between the heat supplied and the work done is equal to the change in the energy of the system.

For a *non-flow process* in a closed system this corollary is expressed mathematically by the *non-flow energy equation*

$$Q - W = \Delta U \tag{2.83}$$

where U is the *internal energy*. The internal energy of a substance is the energy associated with the motion and configuration of the molecular particles which compose the substance. It contains both the kinetic energy of the molecules due to their translation, rotation and vibration and the potential energy due to intermolecular forces. For a *flow process* in an open system, the system energy must also include the macroscopic mechanical energy forms, both kinetic and potential

$$Q - W = \Delta E \tag{2.84}$$

where $E = U + (KE) + (PE)$. Some non-simple thermodynamic systems may additionally involve other energy forms due to electric, magnetic or capillary effects. For such a case, terms representing the appropriate additional energy components are incorporated into E.

Equation 2.84 forms the basis of the *steady-flow energy equation*, which will be considered later. A fluid property which is of importance in flow processes is that of *enthalpy*. In specific terms, this composite property is defined by

$$h = u + pv' \tag{2.85}$$

where h = specific enthalpy, J/kg
u = specific internal energy, J/kg
p = absolute pressure, N/m^2
v' = specific volume, m^3/kg

2.6.3 The Second Law of Thermodynamics

The Second Law of Thermodynamics may be stated in the following form—it is impossible to construct a system which, operating in a cycle, will extract heat from a single reservoir and do an equivalent amount of work on the surroundings. This form of the law is known as the *Kelvin-Planck* statement. A number of important corollaries of the Second Law exist. One of these, known as the *Clausius* statement, states that heat cannot pass from a cooler to a hotter body without the assistance of some external agency.

Following from the Second Law the concept of entropy is established. Entropy, S, is a property of a system such that a change in its value is equal to

$$S_2 - S_1 = \int_1^2 \frac{dQ}{T} \tag{2.86}$$

for any reversible process undergone by the system between states 1 and 2. The introduction of entropy simplifies considerably the analyses of reversible adiabatic and isothermal processes. Another use of entropy in more advanced thermodynamic problems is as a criterion of equilibrium and reversibility.

2.6.4 The Steady-Flow energy equation

Most of the processes encountered in environmental engineering work are of a flow nature. Usually these processes may be considered as steady with respect to time and the basic relationship specifying them is the *steady-flow energy equation*. This will now

Figure 2.19

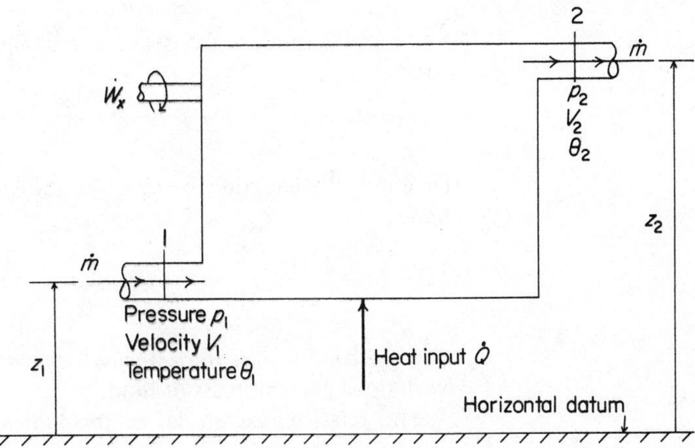

be derived. Consider the steady flow of a fluid through the open system shown in Figure 2.19.

Fluid enters the system at Section 1 at a constant flow rate of $\dot{m}$ kg/s and with conditions p_1, V_1, θ_1. The fluid leaves the system at Section 2 at the same rate and with conditions p_2, V_2, θ_2. The rate of heat input is $\dot{Q}$ and the shaft work output $\dot{W}_x$. Although the properties of a fluid element will change as it moves from point to point through the system, the mechanical and thermodynamic state of the fluid at any point in the system is constant with respect to time.

Equation 2.84 may be written as

$$\dot{Q} - \dot{W} = \dot{m}\,\Delta e \tag{2.87}$$

The heat input $\dot{Q}$ requires no comment. The evaluation of the work $\dot{W}$ does however require more attention because in addition to the shaft work $\dot{W}_x$, $\dot{W}$ must incorporate the flow work done by moving parts of the system boundary at Sections 1 and 2. If v_1' and v_2' are the specific volumes of the fluid at entry and exit, this flow work done is given by

$$\dot{W}_f = \dot{m}\,(p_2 v_2' - p_1 v_1')$$

and therefore

$$\dot{W} = \dot{W}_x - \dot{m} p_1 v_1' + \dot{m} p_2 v_2'$$

In the absence of electrical, magnetic or other complex effects, Δe is defined by the changes in the internal, potential and kinetic energy between Sections 1 and 2.

$$\Delta e = [\,(u_2 - u_1) + \tfrac{1}{2}(V_2^2 - V_1^2) + g(z_2 - z_1)\,]$$

Substituting for $\dot{W}$, Δe and $\dot{Q}$ in equation 2.87 gives

$$\dot{m}u_1 + \dot{m}\frac{V_1^2}{2} + \dot{m}gz_1 + \dot{m}p_1 v_1' + \dot{Q}$$

$$= \dot{m}u_2 + \dot{m}\frac{V_2^2}{2} + \dot{m}gz_2 + \dot{m}p_2 v_2' + \dot{W}_x$$

Since by definition the specific enthalpy $h = u + pv'$ we can write

$$\dot{m}h_1 + \dot{m}\frac{V_1^2}{2} + \dot{m}gz_1 + \dot{Q} = \dot{m}h_2 + \dot{m}\frac{V_2^2}{2} + \dot{m}gz_2 + \dot{W}_x$$

Dividing this equation by $\dot{m}$ gives the *steady-flow energy equation*

$$h_1 + \frac{V_1^2}{2} + gz_1 + Q' = h_2 + \frac{V_2^2}{2} + gz_2 + W_x' \tag{2.88}$$

where Q' and W_x' are respectively the heat added and the external work done per unit mass of fluid.

The relative magnitudes of the individual terms in equation

2.88 depend upon the type of flow process being considered. In most processes some of the terms will be either zero or negligible. For example, in heat transfer devices such as boilers, heater batteries and calorifiers, the potential and kinetic energy terms are usually negligible and the work term is zero. In such cases the equation thus reduces to

$$h_1 + Q' = h_2$$

2.7 PSYCHROMETRY AND MOISTURE TRANSMISSION

2.7.1 Air-vapour mixtures

Moist atmospheric air is a mixture of a gas, air, and a vapour, water vapour. The air itself is a mixture of several gases, the principal constituents being oxygen and nitrogen. It is a matter of common knowledge that at a certain temperature there is a limit to the amount of water vapour that can be held in the air. At this limit the air is said to be *saturated* with water vapour. As the air temperature is increased so the quantity of water which can be held in the air also increases. If the temperature of saturated air is decreased then water vapour is shed from the air, or in other words, condensation occurs.

The key to comprehension of this phenomenon is Dalton's Law of Partial Pressures. According to this Law the air and the water vapour each occupy any given space exerting their own individual partial pressures, each behaving independently of the other. Consider for a moment a p–v' diagram for the vapour as shown in Figure 2.20a.

The process line 1234 may be taken as representing the three stages of a process in which water enters the air to become water vapour, the process taking place at a constant pressure as follows:

1 to 2 the water receives heat and is raised to its saturation (boiling) temperature,

2 to 3 the water evaporates into the air as it continues to receive heat, until it becomes a dry saturated vapour at its saturation temperature,

Figure 2.20 Air/water vapour processes

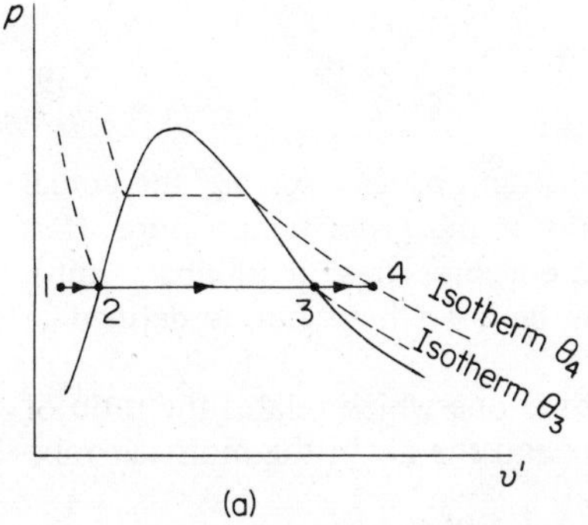

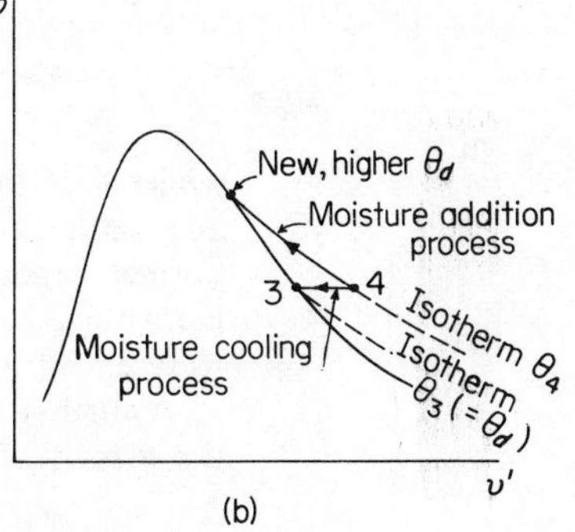

3 to 4 the dry saturated vapour continues to receive heat until it becomes a superheated vapour in the air at the same temperature as the air.

Thus generally we can say that moisture in the air is in fact a superheated vapour.

It is interesting now to consider how condensation may occur from the point 4.

(a) Cooling of the air: Suppose the moist air passes across a cool surface such as a window and the moisture is cooled (desuperheated) from 4 back to 3. At this point condensation begins. θ_d is known as the *dew point temperature* for the moist air containing vapour with a partial pressure of p_{1234}.

(b) Adding more moisture to the air: Suppose further quantities of moisture are added to the air in processes similar to process 1234. In this case the moisture would remain at temperature θ_4 (the same as the air temperature) but the partial pressure of the water vapour would increase until once again a saturation condition occurred. This time a higher dew point temperature (in this case $\theta_d = \theta_4$) occurs, corresponding to the increased saturation vapour pressure. These two possible processes are illustrated in Figure 2.20b.

The two processes illustrated are the condensation processes which are likely to arise in practice either as a result of localised cooling (emphasis on designer) or as a result of an excessive addition of moisture (tenant or occupier beware).

2.7.2 Psychrometric Properties

Clearly, a useful concept is one which determines how close the condition of mosit air is to saturation. Relative humidity or percentage saturation are the appropriate properties for making such a judgement. Relative humidity is defined as the ratio of the partial pressure of the water vapour in moist air at a given temperature to the partial pressure of the water vapour in saturated air at that temperature. Thus

$$\phi = 100 \frac{p_s}{p_{ss}} \% \tag{2.89}$$

where ϕ is the relative humidity and p_s and p_{ss} are the actual and saturation vapour pressures at the given temperature. It is normal practice to record vapour pressures in millibars (mb). Percentage saturation will not be used here but is defined in Reference (9).

A further important concept is one which relates the ratio of the masses of the water vapour and the air in the moist air mixture. This is available under a variety of guises, such as moisture content, humidity ratio and specific humidity. It will be sufficient for our purpose to use the definition:

$$\text{moisture content} = \frac{\text{mass of water vapour}}{\text{unit mass of dry air}} \text{ kg/kg} \qquad (2.90)$$

In order to assess the relative humidity of moist air a wet bulb thermometer may be used. A current of unsaturated moist air passing across an ordinary mercury in glass thermometer having its bulb moistened by a wetted wick of muslin or other suitable material causes a depression of the reading below that of the air dry bulb temperature. The drier the air stream, the greater the depression, and in the particular situation where the airstream is saturated there is no depression at all. The sling psychrometer (discussed in Chapter 4) combines a dry bulb thermometer and a wet bulb thermometer in the one instrument. For such an instrument the following empirical equation is appropriate:

$$p_s = p'_{ss} - p_{at} A\,(\theta - \theta')$$

where p_s is the vapour pressure of the air under test and θ and θ' are the dry and wet bulb temperatures read from the instrument. p'_{ss} is the saturation vapour pressure corresponding to the wet bulb temperature and may be obtained from hygrometric tables and p_{at} is the atmospheric pressure obtained from a barometer. A is a constant for the particular instrument. Normally the instrument is purchased with a slide rule and the slide rule is used to obtain a relative humidity of the air directly from the dry and wet bulb readings, assuming a standard value for the atmospheric pressure. Other psychrometric properties may be obtained but are of no relevance to this discussion.

The Psychrometric Chart

Psychrometric data are available in the form of tables(9) or in the form of a chart. Many varieties of chart are available but the type presented by the B.R.E. in their Digest 110 ‘Condensation’, is perhaps the most useful for our purposes and is presented here as Figure 2.21. The two condensation processes discussed earlier in relation to the p–v' diagram are shown superimposed for comparison and to aid understanding, starting from the quite arbitrary point 4. Notice in the case of the cooling process how the relative humidity gradually increases (without any change in the moisture content) until it reaches 100% at the dew point temperature, any further cooling then causing moisture to be shed. Again, in the case of the moisture addition process, notice that the moisture content and the relative humidity gradually increase (without any change in the dry bulb temperature) until the humidity reaches 100%. Any further addition of moisture would cause deposition of moisture onto surfaces.

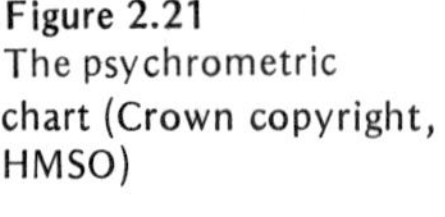

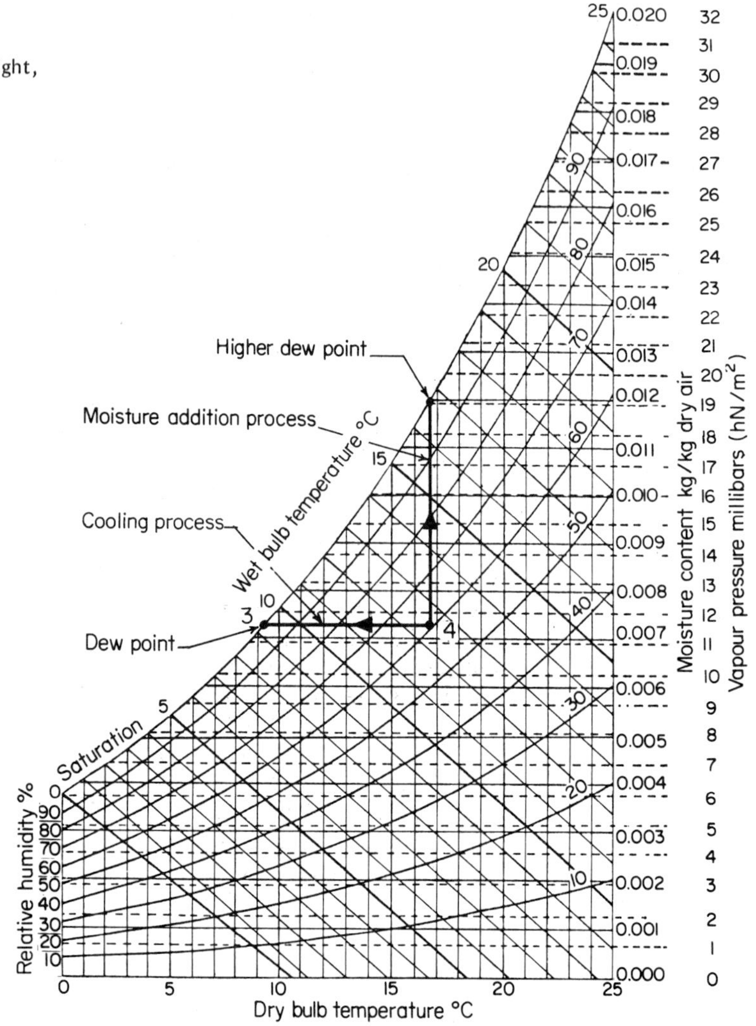

Figure 2.21
The psychrometric chart (Crown copyright, HMSO)

2.7.3 Fick's Law of Diffusion

The molecular diffusion of water vapour into a stagnant gas is defined by Fick's Law, which states that the mass rate of vapour flow per unit area (the mass flux) is proportional to the concentration gradient. For diffusion in the *x*-direction this is expressed mathematically as:

$$\frac{\dot{m}_s}{A} = -D\,\frac{\partial \rho_s}{\partial x} \tag{2.91}$$

where $\dot{m}_s$ is the mass flow rate of vapour (kg/s) along the *x* coordinate axis through the cross-sectional area A (m^2) and $\partial\rho_s/\partial x$ is the vapour density gradient (kg/m^3 per m) in this direction. The proportionality factor D is termed the *diffusion coeffi-*

cient. For the diffusion of water vapour in air D has been experimentally evaluated at approximately 0.000026 m^2/s.

The obvious formal similarity between equation 2.91 and the Fourier Law as given by equation 2.1b identifies the process of water vapour diffusion as directly analogous to theprocess of conduction heat transfer. It is generally convenient to re-write equation 2.91 in terms of the vapour pressure gradient

$$\frac{\dot{m}_s}{A} = -\frac{D}{R_s T}\frac{\partial p_s}{\partial x} \tag{2.92}$$

where the vapour pressure $p_s = \rho_s R_s T$ and R_s is the gas constant for water vapour.

2.7.4 Vapour transmission through materials

Generally speaking, most building and insulating materials allow the transmission of water vapour through them. In the presence of a vapour pressure difference across these materials a corresponding movement of water vapour will occur. Although this process is in fact more complex than the process of simple gaseous diffusion, it is nevertheless treated in the same manner. The equation used to describe the vapour transmission through a material of thickness L is the following integrated form of Fick's Law

$$\frac{\dot{m}_s}{A} = \delta\frac{\Delta p_s}{L} \tag{2.93}$$

where Δp_s is the difference in vapour pressure between the surfaces of the material and δ is the material *permeability.* With Δp_s expressed in N/m^2 and L in metres, the unit of permeability is kgm/Ns. An indication of the permeability of various building and insulating materials is given in Table 2.6. More comprehensive data will be found in References (2), (10).

Table 2.6 Permeability of some Building and Insulating Materials

Material	*Permeability* (kgm/Ns × 10^{11})
Concrete	0.5 -3.8
Brickwork	0.55-4.0
Plaster	1.6 -2.7
Plasterboard	1.7 -2.9
Wood	0.1 -1.00
Wood wool slab	2.4 -7.7
Corkboard	0.3 -1.6
Plywood	0.1 -0.7
Polyurethane foam	0.1 -3.5
Expanded polystyrene	0.17-0.90

Following the analysis of steady-state conduction in plane slabs, the vapour mass flux may alternatively be expressed as

$$\frac{\dot{m}_s}{A} = \frac{\Delta p_s}{R_v} \tag{2.94}$$

where $R_v = L/\delta$ is the *vapour resistance* of the material, Ns/kg. The reciprocal of vapour resistance is termed the *permeance, P.* This coefficient is used in situations where it is inconvenient or unnecessary to specify δ and L separately, e.g. when considering vapour barriers. Thus

$$\frac{\dot{m}_s}{A} = P\Delta p_s \tag{2.95}$$

The permeance of some common materials is listed in Table 2.7. Again, more comprehensive information is to be found in References 2,10.

Table 2.7 Permeance of Various Materials

Material	*Permeance* (kg/Ns X 10^{11})
Aluminium foil	0.01–0.03
Polythene	0.4 –0.6
Roofing felt	1 –23
Kraft paper	160 –480
Paint–2 coats:	
Flat paint on plaster	9 –17
Gloss paint on plaster	3 –13

Figure 2.22

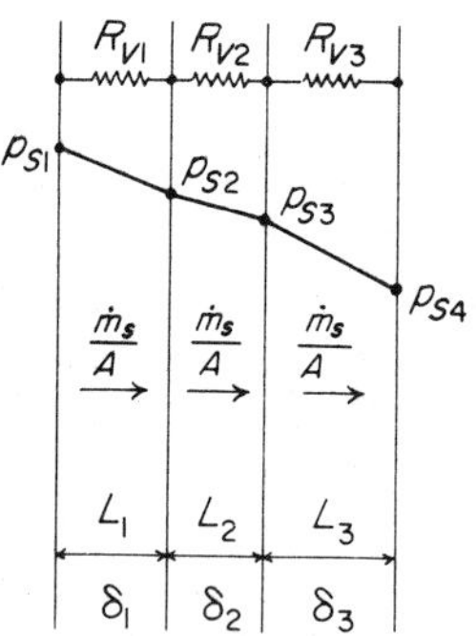

The transmission of water vapour through a composite plane structure (Figure 2.22) is treated in the same way as the heat conduction problem. Thus

$$\frac{\dot{m}_s}{A} = \frac{p_{s1} - p_{s2}}{R_{v1}} = \frac{p_{s2} - p_{s3}}{R_{v2}} = \frac{p_{s3} - p_{s4}}{R_{v3}} \tag{2.96}$$

where $R_{v1} = L_1/\delta_1$; $R_{v2} = L_2/\delta_2$; $R_{v3} = L_3/\delta_3$. By eliminating the interface vapour pressures p_{s2} and p_{s3} the mass flux through the composite structure is obtained in the form

$$\frac{\dot{m}_s}{A} = \frac{p_{s1} - p_{s4}}{R_{vt}} \tag{2.97}$$

where $R_{vt} = R_{v1} + R_{v2} + R_{v3}$ is the total vapour resistance of the wall. The interface vapour pressures are readily determined from equations 2.96 and 2.97. For example

$$p_{s3} = \frac{R_{v3}}{R_{vt}}(p_{s1} - p_{s4}) + p_{s4} \tag{2.98}$$

REFERENCES

1. C.I.B.S. GUIDE, Section A3.
2. A.S.H.R.A.E., Handbook of Fundamentals.
3. McADAM, W. H., *Heat Transmission,* 3rd ed, (McGraw-Hill Book Company, New York, 1954).
4. KREITH, F., *Principles of Heat Transfer,* 2nd ed, (International Textbook Company, 1969).
5. C.I.B.S. GUIDE, Section C3.
6. HOTTEL, H. C., *Notes on Radiant Heat Transmission,* (Chem. Eng. Dept, M.I.T., 1951).
7. HOTTEL, H. C. and SARAFIN, A. F., *Radiative Transfer,* (McGraw-Hill Book Company, New York, 1967).
8. FISHENDEN, M. and SAUNDERS, O. A., *An Introduction to Heat Transfer,* (Oxford University Press, 1950).
9. C.I.B.S. GUIDE, Section C1.
10. C.I.B.S. GUIDE, Section A10.

SYMBOLS USED IN CHAPTER 2

A	cross-sectional area; psychrometric constant
b	temperature coefficient of thermal conductivity
C_p	specific heat at constant pressure
c	speed of light
D	diffusion coefficient
E	emissive power; energy
e	specific energy
F	view factor
f	frequency
g	acceleration due to gravity
h	specific enthalpy
h_c	convection heat transfer coefficient
h_R	radiation heat transfer coefficient
h_s	surface heat transfer coefficient
k	thermal conductivity
L	length; material thickness
ℓ	characteristic dimension

$\dot{m}$ rate of mass flow
$\dot{m}_s$ rate of vapour transmission
P permeance
p pressure
p_s vapour pressure
Δp_s vapour pressure difference
p_{ss} saturation vapour pressure
$\dot{Q}$ rate of heat transfer; $\dot{Q}_c$ by convection; $\dot{Q}_R$ by radiation
$\dot{q}$ heat flux
Q heat quantity
R thermal resistance
R_a thermal resistance of air space
R_v vapour resistance
r radius
S entropy
t time
T absolute temperature
U overall heat transfer coefficient; internal energy
u specific internal energy
v velocity
V average velocity of flow
v' specific volume
$\dot{W}$ rate of work done
W work quantity
α thermal diffusivity; absorptivity
τ transmissivity
ϵ emissivity
θ temperature; θ' wet bulb reading
$\Delta\theta$ temperature difference
$\Delta\theta_m$ logarithmic mean temperature difference
μ viscosity
ρ density; reflectivity
λ wavelength
ψ factor of dimensional analysis
ϕ function of dimensional analysis; relative humidity
δ permeability
σ Stefan-Boltzmann constant

3 Fundamentals of Automatic Control

3.1 INTRODUCTION

Automatic control of heating services is a branch of process control engineering. Heating services processes can be summarised as:

(i) Comfort processes which consist mainly of ensuring adequate control of air and radiant temperatures (see also Chapter 4).
(ii) Domestic hot and cold water supply.
(iii) Process hot water supply (e.g. swimming pools).

Each of the above primary processes usually involve some secondary or sub-processes such as boiler water temperature control, calorifier temperature control, water level control, etc.

Unlike most other branches of process control engineering, much of the application of automatic control to heating services appears to comprise selecting more or less standard purpose-made controllers from the range offered by various manufacturers without any prior system analysis or system simulation to check whether satisfactory performance will be achieved. In contrast, chemical process engineering controls are often designed into the plant on the drawing board to ensure that optimum plant performance will be obtained [1,2,3]. As mentioned elsewhere, it is to be hoped that the increasing use of computers in building services design will soon begin to remedy this situation.

Despite the fact that it is generally difficult (and often impossible due to lack of data) to predict the performance of a proposed control system, it is nevertheless useful to set down the main functions of the control system for reference:

(i) safety of personnel and plant,
(ii) optimum control over the controlled condition(s),
(iii) minimal overall energy consumption,
(iv) convenience (and improved performance compared with direct human control).

In order to achieve 'satisfactory' (if not 'optimum') control of a particular plant the designer will use a variety of control equipment comprising detectors, controllers, regulators (e.g. control valves), time switches, etc. Space restrictions permit only a minimum of detailed discussion of particular applications giving control and wiring diagrams (Section 3.9). Fortunately, these are readily available in many control manufacturers' catalogues and also in References (8) and (12).

The main purpose of this Chapter, therefore, is to concentrate on the broad principles rather than to attempt to detail a topic which in itself would warrant a rather hefty tome.

3.1.1 Control definitions and terminology

It is hoped that readers will find many of the terms used to be largely self-explanatory and formal definitions are omitted here. They can be found in References (4,5,6,7).

3.2 FEEDFORWARD AND FEEDBACK CONTROL

Figure 3.1 shows a feedforward (open loop) control system and Figure 3.2 shows a feedback (closed loop) control system.

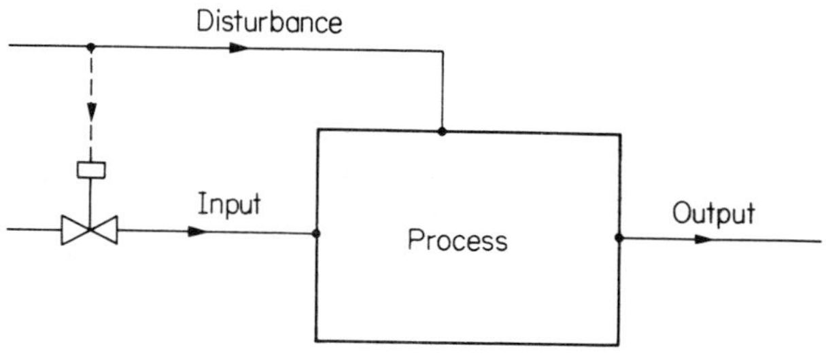

Figure 3.1 Feedforward (open loop) control

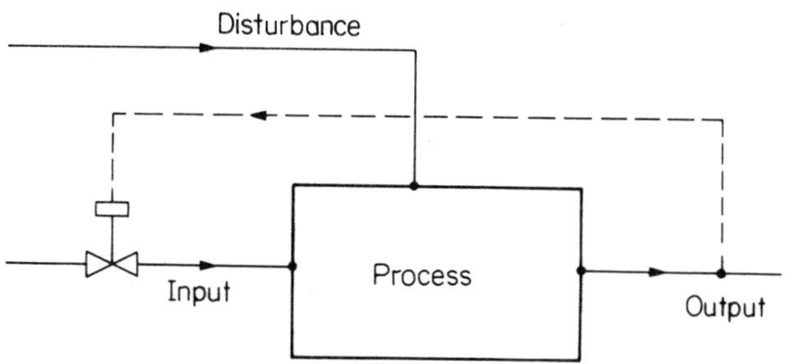

Figure 3.2 Feedback (closed loop) control

In *feedforward* systems the disturbance is measured *before* it arrives at the process and the input is regulated in advance to counteract the disturbance effect. It is thus (theoretically) possible to achieve a constant desired value of the controlled condition.

In *feedback* systems the *resultant effect* of the disturbance is measured and used to regulate the input. Notice that the controlled condition *must* change in order to produce control action, i.e. the system is 'error-actuated'. Feedback systems are much more common for heating services because:

(i) Although Figure 3.1 shows only one disturbance, in practice there are usually a number. To counteract one (or more) disturbance requires a knowledge of the disturbance effect(s) and would probably involve complicated controller tuning.

(ii) Feedback systems are quite satisfactory for most heating services as the process itself is then used as a disturbance integrator and thus the need for complicated controller tuning is reduced or eliminated.

(iii) Feedback error can generally be reduced to an acceptable value.

(iv) Feedback systems generally provide a larger factor of safety by taking into account unforeseen system disturbances.

Compound control systems (feedforward + feedback) are also used (e.g. inside/outside temperature control of LPHW heating systems), but again it is common practice to use an extra feedback loop (TRV or hand radiator valve) for final (overriding) control.

3.3 PROCESS REACTION RATE

Before discussing the various modes of control, an important term 'process reaction rate' (p.r.r.) will be introduced.

Consider Figure 3.3 which depicts a water tank being filled at a rate $\dot{Q}_I = 0.3\ m^3/s$ while supplying at a rate of $\dot{Q}_o = 0.1\ m^3/s$ (considered here to be constant, i.e. independent of the water level in the tank; this could be achieved by a constant displacement pump).

Figure 3.3

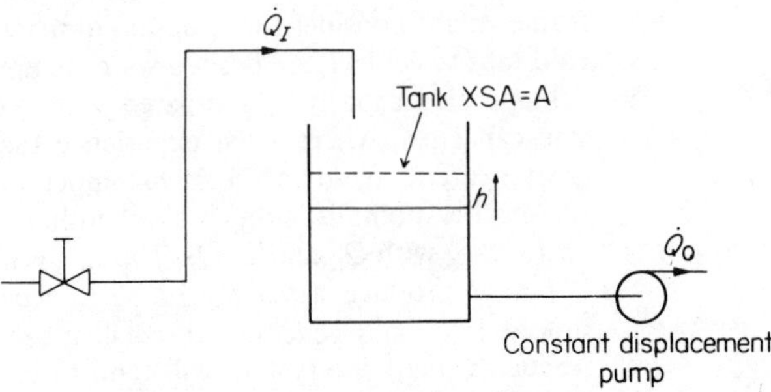

For the situation shown, the net rate of inflow is $(\dot{Q}_I - \dot{Q}_o) = 0.2\ m^3/s$. If $A = 4\ m^2$ then the p.r.r. $= (\dot{Q}_I - \dot{Q}_o)/A = +0.05$ m/s. If, however, $A = 0.1\ m^2$ then the p.r.r. $= +2$ m/s. Figure 3.4 shows these graphically.

During emptying, if $\dot{Q}_I = 0$ the net outflow is $\dot{Q}_o = 0.1\ m^3/s$ and the corresponding p.r.r.s are -0.025 m/s ($A = 4\ m^2$) and -1 m/s ($A = 0.1\ m^2$). These are shown in Figure 3.5.

If we consider manual control of the water level in order to

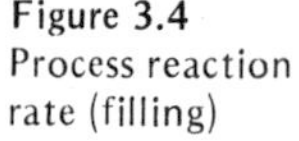

Figure 3.4 Process reaction rate (filling)

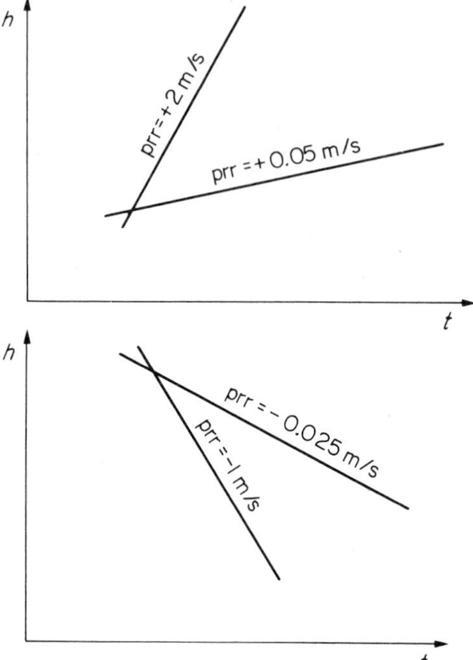

Figure 3.5 Process reaction rate (emptying)

prevent both tank overflow and pump starvation at the outflow, clearly the tank with the larger cross-section is more controllable. The cross-sectional area (A) represents the system demand side capacity in this case.. Thus we can generalise by saying that systems with 'large', (a relative term), demand side capacities have smaller p.r.r.s and are easier to control than those with 'small' demand side capacities. As an illustration of the importance of the demand side capacity on the control problem the reader might consider the situation of manually adjusting hot and cold taps to achieve the desired water temperature in a bath (large demand side capacity), compared with a shower (small demand side capacity), where from experience the latter requires much greater care to avoid an excessive temperature.

Another point to note is that both the filling and emptying p.r.r.s vary with $\dot{Q}_o$ and/or $\dot{Q}_I$. Thus a given demand side capacity (A) may produce a variety of p.r.r. depending on the load or supply rates and generally this will mean that as $\dot{Q}_o$ decreases (reducing load) the system will tend to become more difficult to control.

There are other important features of p.r.r.s and demand side capacities which will be discussed further later but the main point to be emphasised here is that most heating services are inherently easy to control at *design* load conditions. At *light* loads, however, this inherent controllability may be reduced to the extent that a serious lack of control may occur or that a more sophisticated mode of control may be required than that indicated by the 'design' p.r.r.s.

3.3.1 Inherent regulation

If the constant displacement pump in Figure 3.3 were replaced by an adjustable valve to represent varying load then the outflow rate would become dependent on the water level in the tank (i.e. $\dot{Q}_o = C_o\sqrt{h}$ where C_o varies with valve stem position). In this situation if the tank were sufficiently deep it would be possible for the system to operate without supervision as the outflow rate would eventually increase to match the inflow rate. This may, of course, be undesirable both from the point of view of excessive water level in the tank and too high a flow rate $\dot{Q}_o$. However, in general, this self-regulating tendency called inherent regulation will be found in most physical systems and can often be considered as an aid to control.

3.4 SYSTEM TIME LAGS

Consider the feedback system shown in Figure 3.6. The main elements of the system are:

(i) the detector sensing outlet water temperature,
(ii) the controller,
(iii) power unit (pneumatic, electric or electromagnetic motor),
(iv) control valve,
(v) primary coil,
(vi) calorifier.

Figure 3.6

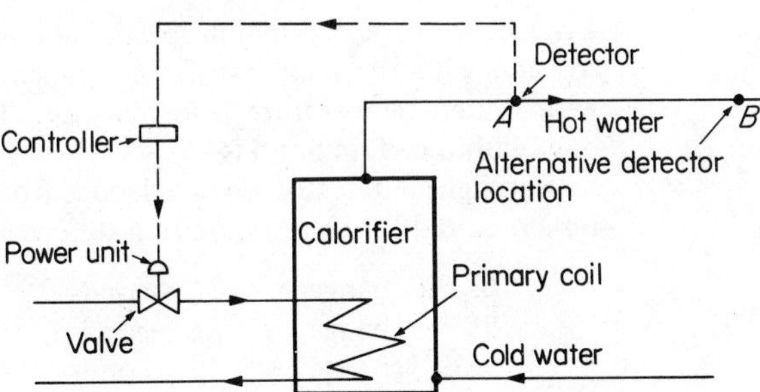

'Dead Time'

Consider a situation in which there is a change in water temperature leaving the calorifier. If we consider the outflow pipe to have negligible thermal capacity and to be perfectly insulated, this change in temperature will appear *unaltered* at some time later at A and some time after that at B. The actual time required for the change in outlet temperature to travel to the detector is found by dividing the length of outflow pipe by the water velocity. The name given to this delay is 'dead time' or 'transport lag'. (In practice, such changes will be attenuated by the thermal capacity and heat losses/gains from the pipe.)

Detector Lags

By considering the detector as a first-order system it can be shown that its response to a sudden or step change in temperature is

$$\theta_{\mathrm{d}} = \theta_{\mathrm{o}} + (\theta_{\mathrm{f}} - \theta_{\mathrm{o}})\,(1 - \mathrm{e}^{t/L}) \tag{3.1}$$

This is shown graphically in Figure 3.7.

Figure 3.7

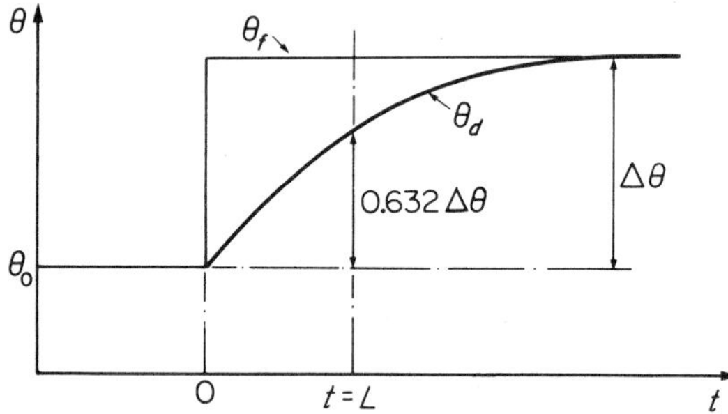

The important characteristic here is that the response of a first order detector to a step change is exponential, (i.e. its p.r.r. changes as the potential $(\theta_{\infty} - \theta_{\mathrm{t}})$ changes). The detector characteristic which governs this response is called its time constant (L) and is given by

$$L = \frac{\rho c d}{4h} \tag{3.2}$$

L varies with the thermal properties and dimensions of the detector and with the heat transfer coefficient (h) which itself varies with water temperature and velocity. Thus L is a 'constant' only within certain bounds.

For a given detector we can deduce from equation 3.1 that the time required for its response to a step change is as follows:

63.2% change $=$ 1L seconds
95% change $\triangleq$ 3L seconds
98% change $\triangleq$ 4L seconds
100% change $=$ ∞

Generally, for good control, detector time constants should be small; this is a matter of good detector design, and *equally*, of careful detector positioning.

Controllers

Time delays associated with *electronic* and *pneumatic* controllers are small and are generally negligible compared with other elements in the control loop although some signal distortion may occur (i.e. the controller output may not correspond exactly to the theoretical output given by the equations described later). The reasons for this signal distortion are too numerous to detail in the space available. It is sufficient to note that the effect is generally most marked at high and low loads giving close to ideal response between say, 10% to 90% load.

Most *electrical* controllers also behave as described above with the exception of room thermostats, where the detector and controller are integrated. Many factors influence room thermostat performance and the reader is referred to other works for guidance[9,10].

Power Unit

Electromagnetic motors (including on/off solenoids) are usually very fast acting as are most pneumatic motors and delays caused by them can generally be ignored. Most electrical motors are also fast but *some* are relatively slow and may introduce significant errors in systems controlling fast acting processes.

A more serious defect of many power units is that they incorrectly position the valve stem, i.e. often the valve stem does not correspond exactly to the controller output. Two factors cause this:

(i) most power units operate against a spring, which positions the valve stem in a 'fail-safe' position in the event of power failure or when the control system is de-energised, and

(ii) most valves operate under pressure, which requires gland packing (which exerts a friction force) around the valve stem to prevent leakage.

The combined effect of spring and friction forces can lead to the need for a valve stem positioner which is simply a feedback control sub-system on the power unit/valve linkage which ensures that the valve stem is correctly positioned. Valves not incorporating valve positioners may give rise to serious 'hunting' if the overall response of the control loop is too slow for the process being controlled.

Valves

Valve characteristics are very important in process control and are discussed in detail in Section 3.5.3.

Primary Coil

If the primary coil is considered to be a first-order system then by analogy with the detector it will have a time constant governed by its thermal inertia and its heat transfer rate. In Figure 3.7 control of secondary flow temperature is achieved by reducing the flow of primary water to the coil. For a given coil it may be deduced that its time constant is largely a function of the coil internal heat transfer coefficient (which varies with primary flow rate) as all the other factors are nearly constant. Thus the coil time constant will tend to increase at low loads.

There is little published data on primary coil time constants, however it is likely that they are of the same order as those for air heating and cooling coils[11].

Calorifier

The thermal capacity of the calorifier relative to the heat input power of the primary coil has a major influence on the process reaction rate and hence on the controllability of the system as a whole.

Figure 3.8

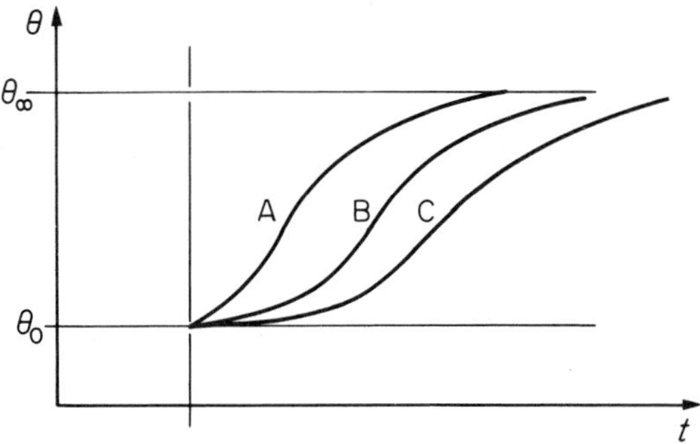

Figure 3.8 shows the response of the outlet temperature for three thermal capacity/heat input ratios in which A represents a 'non-storage' (low demand side capacity) calorifier and C represents a 'storage' (large demand side capacity) calorifier.

It should be noted that these curves represent the overall response of the heat input coil, the calorifier and the detector and therefore incorporate the time lags associated with each. As can be seen, each curve exhibits the shape which is characteristic of a multi-stage system. A common method of reducing higher-order systems of this type to an approximately equivalent first-order system is shown in Figure 3.9 where a tangent is drawn through the turning point at the foot of the curve which extends to cut

Figure 3.9

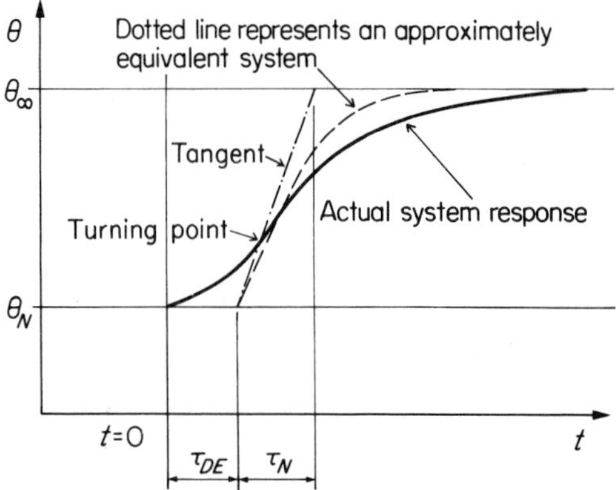

the abscissa giving an equivalent dead time (τ_{DE}) and to the asymptotic value to give an equivalent or 'notional' (single) system time constant (τ_N). This representation is not exact as it provides an 'equivalent' system whose response is that shown by the dotted curve. Further details can be found in Reference (8).

Most storage calorifiers have notional time constants of about 0.5 to 0.75 *h*. This makes the calorifier the dominant item in the control loop and most of the time delays etc., of the other elements in the control loop generally become unimportant.

For DHWS storage calorifiers the control specification on outflow water temperature is generally not strict (e.g. 65 °C ± 5 °C) and, because of the large time constants usually involved, this specification can normally be achieved using simple on/off control. For process hot water supply, however, a much tighter control of temperature may be necessary (e.g. 45 °C ± 1 °C) and proportional or integral control may be required to ensure that this is achieved, even when a large demand side capacity exists.

Non-storage calorifiers may have much smaller notional time constants (say 0.5 to 5 min). In such a case the time delays, etc., of the other elements in the control loop will play a significant part in determining control system performance. This is often coupled with the need for a tight control specification (e.g. 90 °C ± 1 °C) on outflow water temperature with a primary heat supply from steam or MPHW and hence the inherent danger of producing steam in the LPHW side of the calorifier in the event of control failure. The combination of an inherently less controllable system and a tighter control specification often leads to the application of a modulating form of control such as proportional (*P*), integral (*I*), *P* + *I* or even *P* + *I* + *D* in such cases.

3.5 CONTROL VALVES

3.5.1 Types of valve

The two basic types of valve used in heating services are the two-port and three-port. (Four-port valves are also used but are much less common and are not discussed here.) The complete valve assembly normally comprises the power unit, valve linkage and valve body. The valve body incorporates two main parts:

(i) one (two in the case of three-port valves) characterised plug(s) attached to the end of the valve stem,
(ii) one (or two) characterised valve seat(s) (orifice(s)) through which the controlled fluid flows.

3.5.2 Valve characteristics

The mode of control will determine the movement of the valve stem and hence the relative position of the plug(s) and seat(s) which will in turn determine the flow rate versus valve stem position characteristic. For on/off valves the plug/seat requirements are simply that tight shut-off is required during the 'off' (closed) phase and (normally) a minimal resistance to flow rate during the 'on' (open) phase. (Note: Control valves are seldom used to give multi-step control of liquid flow rates.)

For other modes of control the requirements of the plug/ seat characteristics are usually either:

(i) to provide an (ideal) linear characteristic (Figure (3.10)

or

(ii) to provide a non-linear or 'characterised' relationship.

Figure 3.10
Valve characteristics

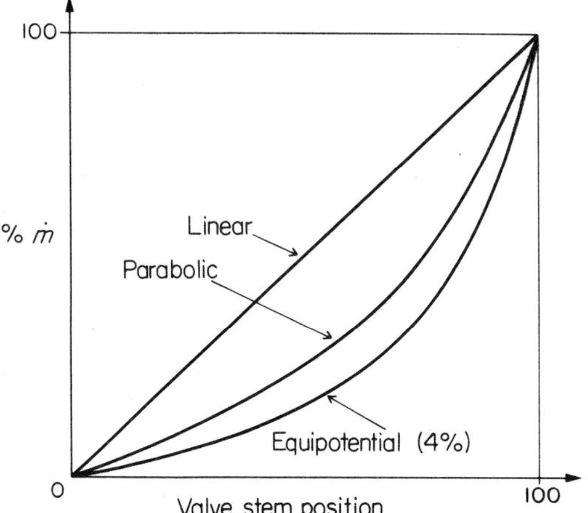

Figure 3.10 also shows some of the common non-linear plug/ seat characteristics available; these are required when a non-linear relationship exists between the fluid flow rate and the heat input to the process. By 'matching' the heat input and valve characteristics an approximate *overall linear* valve stem position versus heat input characteristic can often be achieved. Matching is a complex process; space does not permit this to be discussed and readers are referred to other works,[8,13], which outline the basis upon which valve manufacturers decide to characterise the control valves they offer. It will be seen from this that at present the system designer has little option but to choose from what is offered.

3.5.3 Valve authority

The valve characteristics shown in Figure 3.10 generally apply only when the pressure drop across the valve seat is held *constant* at all flow rates. In most cases, however, the pressure drop across the control valve seat will change (generally increases) as the valve closes and this increased pressure drop tends to cause an increased flow rate counteracting the throttling action of the valve. A valve which under such conditions maintains a flow rate versus valve stem position characteristic close to its design characteristic (i.e. that shown in Figure 3.10), is said to have a good valve 'authority' (N). The concept of valve authority is discussed in some detail in Reference (6), and from this it can be shown that in order to ensure good valve authority it is necessary to incur a fairly high pressure drop at its design flow rate (when the valve is normally fully open). Figure 3.11 and 3.12 show how valve authortity,

Figure 3.11
Two-port valve

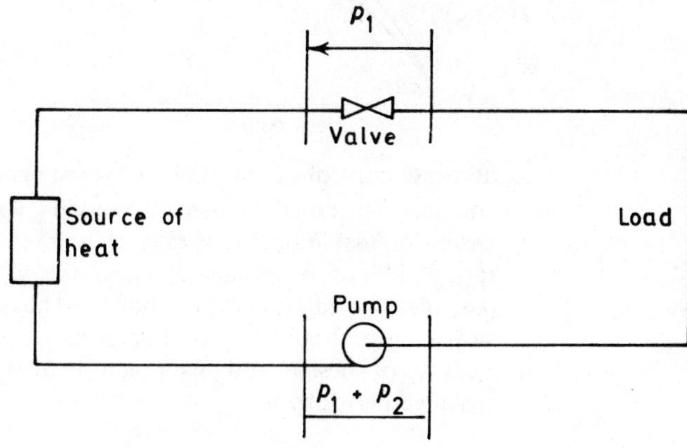

Figure 3.12
Three-port valve circuit

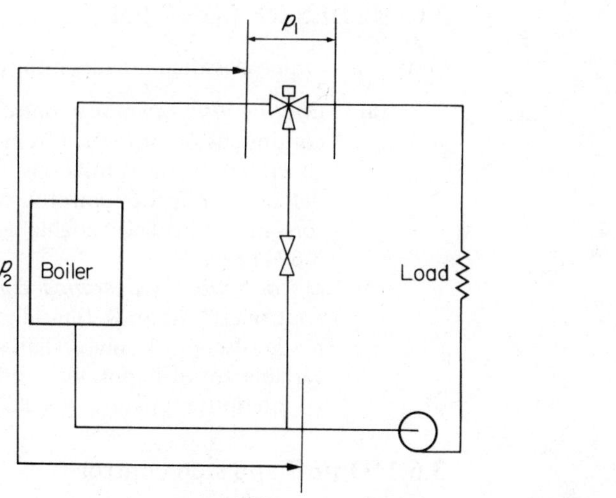

$N = p_1/(p_1 + p_2)$ in both cases. Figure 3.13 shows the overall flow rate versus valve stem position for a given two-port valve controlling an air heating coil for different values of N. Obviously a high valve authority is desirable but it is closely in terms of pump power and although it is difficult to generalise, high valve authority is less likely to be necessary when a two or three term controller is used than when proportional control alone is employed.

Two further terms which are used to describe characterised valves are 'turndown ratio' and 'rangeability'; their meanings are similar and are used to express the ability of a valve to maintain

Figure 3.13

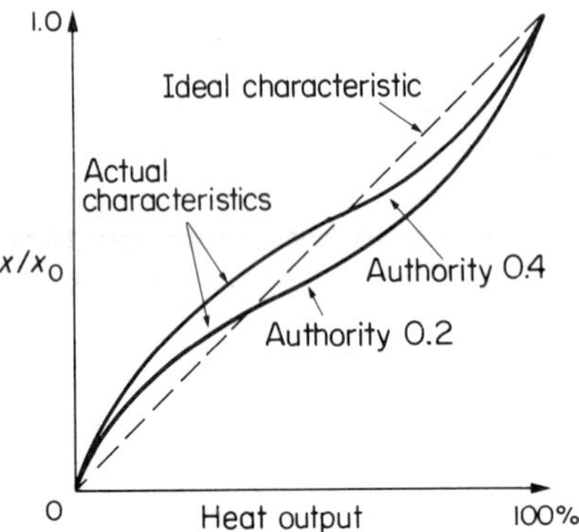

accurate control of the flow rate, (and hence of the heat input to the process), down to a small fraction (typically $\frac{1}{30}$ to $\frac{1}{50}$) of the design or maximum flow rate. It is important to note that this ratio will not be obtained *unless* the valve has a satisfactory (i.e. high) *installed* valve authority. Thus rangeability and turn down ratio should be considered more as an indication of the quality of design and manufacture of a valve rather than its installed performance.

3.6 MODES OF CONTROL

Controllers can be classified broadly into two groups.

(a) *Continuously operating* controllers in which there is a continuous output signal from the controller which is generated by a continuous input signal from the detector. Thus the control system *continuously* monitors the controlled variable as described in Sections 3.62 *et seq.*

(b) *Discontinuously operating controllers* in which the controller output is typically 'on' or 'off'. In this mode of control only certain values of the controlled variable are of importance and the 'monitoring' is of an intermittent nature as is described in Section 3.6.1.

3.6.1 On/off and step-control

On/off (sometimes called two-position) is perhaps the simplest form of control. Here only two magnitudes of supply are available, full and zero. In step control the supply is usually divided into two or more steps (normally equal). Thus on/off control is a form of step control and both forms exhibit the same features.

Consider the tank filling situation shown in Figure 3.14 in which the outflow rate ($\dot{Q}_o$) is constant, (achieved by means of a constant displacement pump). The liquid level (h) is the controlled variable which varies between h_1 and h_2 (provided $\dot{Q}_I > \dot{Q}_o$). When the level rises to h_2 the valve closes and $\dot{Q}_I = 0$. When the level falls to h_1 the valve opens and $\dot{Q}_I = \dot{Q}_I$. Figure

Figure 3.14

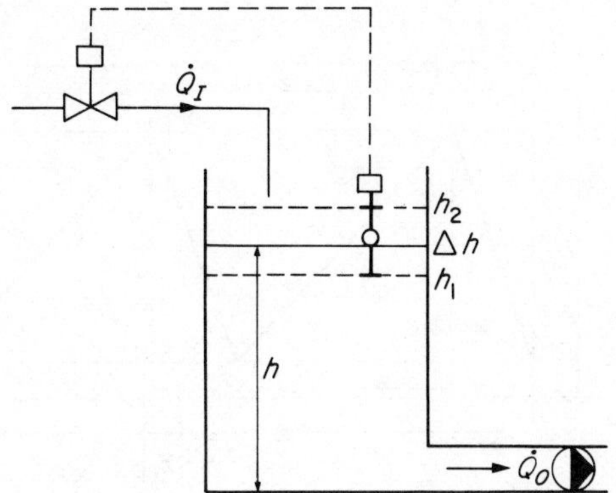

Figure 3.15
Ideal on/off control

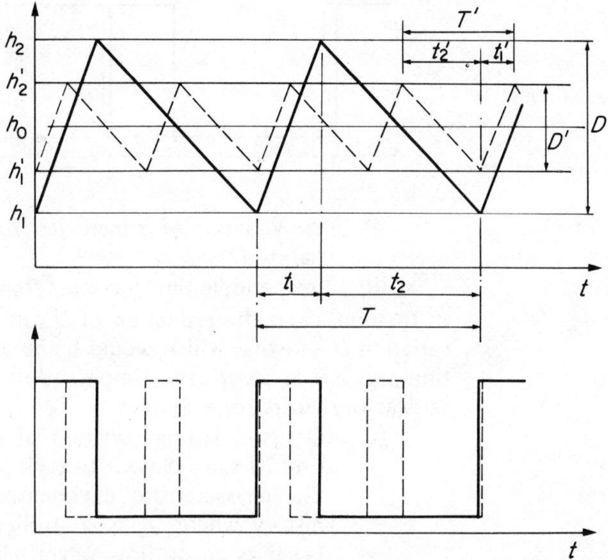

3.15 shows the valve position and liquid level histories (full lines) for an ideal case where no time delays occur.

Superimposed on Figure 3.15 are the valve position and liquid level histories (broken lines) for another ideal case (i.e. no time delays) but where the differential gap is reduced to D'. This reduces the periodic time to T' and gives closer control (i.e. less variation from the desired value h_0). Continuing this trend of reducing D in order to obtain more accurate control generally results in excessive valve wear and early system failure. Thus even in the absence of time delays the choice of a setting for D usually involves a compromise between accuracy and practicality.

As discussed previously, many factors cause time lags. Consider the system previously described in Figures 3.14 and 3.15 but for a case when simple dead times, (each with a value of t_D), occur in sensing liquid levels h_1 and h_2. Figure 3.16 illustrates that the consequences of these time lags are:

Figure 3.16

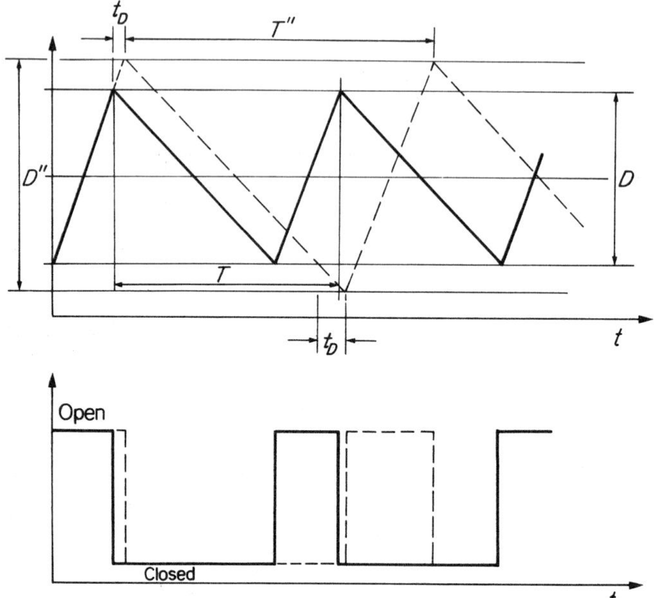

(i) the variation of h increases from D (the set differential) to D''.

(ii) the periodic time increases from T' to T''.

In some cases the reduction of D can reduce the liquid level variation D'' to that which would be obtained in the absence of time lags but in other cases simple on/off control may prove unsatisfactory where for example:

(i) $\dot{Q}_o$ is a strong function of h and when the 'off' (inflow valve closed) phase is long; Figure 3.17 shows the characteristic exponential curved liquid level history where $\dot{Q}_o$ is controlled by a fixed resistance (such as an outflow valve). Here the length of time (dwell time) of the controlled condition (h) in the neighbourhood of the lower limit (h_1) may be unacceptable. (This is particularly so with on/off control of room air temperature and leads to complaints from occupants and usually results in set points being raised.) This effect is aggravated by time lags.

Figure 3.17

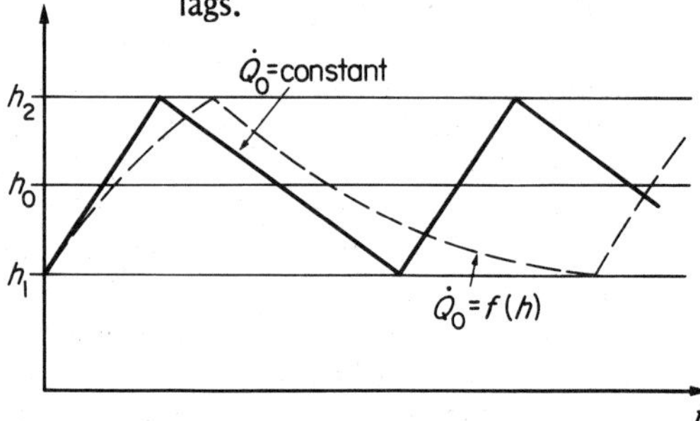

(ii) Where $\dot{Q}_o$ varies widely the 'on' phase p.r.r. may be so steep that dangerously high h_2 values occur at light loads. In this situation it may be impossible to reduce D enough to contain D''.

The inability of simple on/off control (due mainly to high 'on' phase p.r.r.s) to cater for light load situations, leads naturally to the consideration of step control. (Note: For those wishing to study on/off process control in greater depth Reference (14) provides further guidance).

In step control each supply step can be considered to act as an individually operated on/off control system with its own set differential. Figures 3.18, 3.19 and 3.20 show a two-step control system.

By increasing the number of steps the controllability of the process (i.e. reduced p.r.r.s) under light loads can be improved. As can be seen from Figures 3.19 and 3.20 however, it is necessary to introduce a dead band between $h_{2\,(A)}$ and $h_{1\,(B)}$ to prevent un-

Figure 3.18

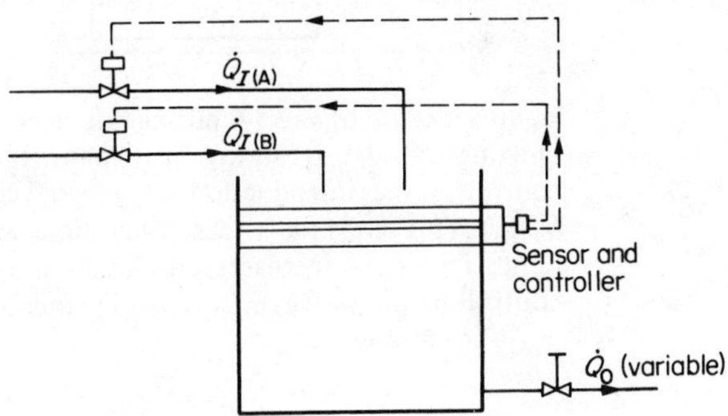

Figure 3.19
Full load
($\dot{Q}_o$ = max)

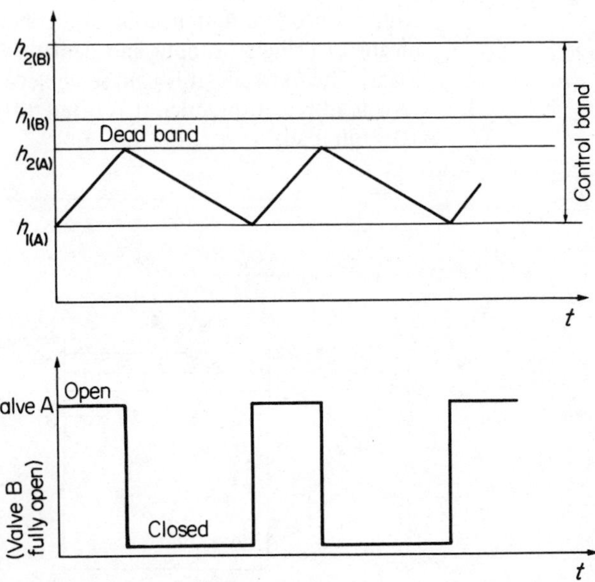

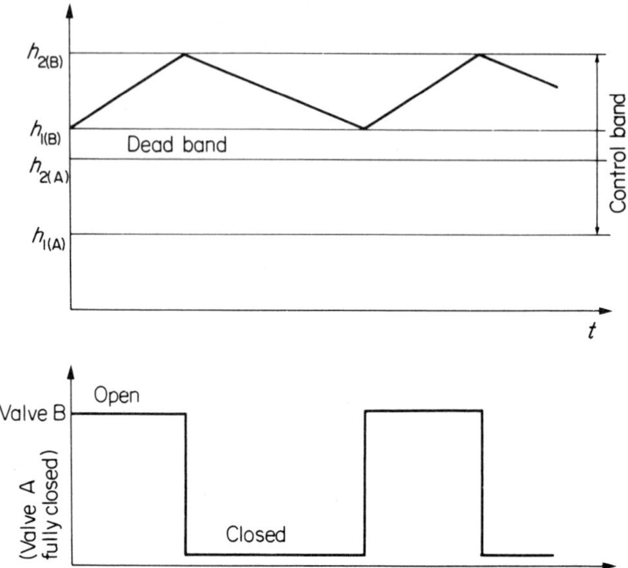

Figure 3.20 Light load ($\dot{Q}_0 \rightarrow 0$)

wanted closure of valve A during full or heavy loads and unwanted opening of valve B during light loads. The main feature of step control is the introduction of a wide control band ($h_{1(A)}$ to $h_{2(B)}$). This tends to increase when time lags occur and when the number of steps increases. In many control situations such wide control bands prove unacceptable and other modes of control must be employed.

3.6.2 Proportional control

Unlike on/off and step control, proportional control provides a means of bringing supply and demand *into balance* at all system loads. Theoretically this can be achieved from full load down to zero load but in practice it is often difficult to cater for less than 10% full load.

Figure 3.21

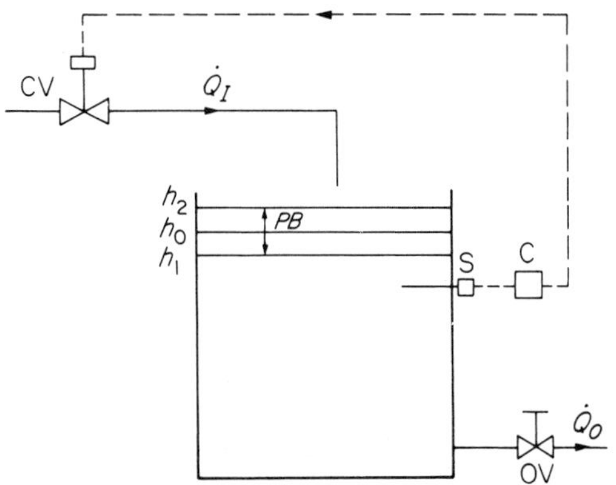

Consider the *ideal* control loop shown in Figure 3.21 in which all elements behave *linearly*. For purposes of analysing the system response it is convenient to consider the elements as follows:

(i) The outflow valve (OV) controls the load ($\dot{Q}_o$) which can be varied from full to zero. The actual flow rate through the valve is, however, influenced by the water level in the tank as well as the valve plug/seat position. The characteristic equation for the valve is

$$\dot{Q}_{o(t)} = C_o \sqrt{h_{(t)}} \tag{3.3}$$

where $\dot{Q}_{o(t)}$ = outflow rate (m/s) at any time t,
C_o = flow coefficient which varies with the outlet valve stem position x,
$h_{(t)}$ = head level in the tank at any time t.

Note: Provided the control loop is stable and $h_{(t)}$ lies within the proportional band ($P: h_1 \rightarrow h_2$) it may be possible to use a mean value of $h_{(t)} = (h_1 + h_2)/2$ if the proportional band is small for which $\dot{Q}_{o(t)}$ is then nominally constant.)

(ii) The sensor (S) and the controller (C) are considered together. They are assumed to produce a maximum signal to close the valve fully when the liquid level reaches h_2 and a minimum signal to open fully the valve when the liquid level reaches h_1. Assuming that the time delays in the controller are negligible (i.e. we assume a zero-order controller[(17)]), then this behaviour can be expressed in equation form,

$$P_{(t)} = (K_c h_{(t)} + C_c) \tag{3.4}$$

or

$$(P_{(t)} - P_o) = K_c (h_{(t)} - h_o) \tag{3.5}$$

where $P_{(t)}$ = controller output signal (e.g. N/m^2 or V) at time t,
P_o = controller output signal for liquid level at h_o,
$h_{(t)}$ = liquid level at time t,
h_o = set point of liquid level
K_c = controller gain
C_c = a constant (determines set point)

Thus from equation (3.4) the controller output versus liquid level relationship is linear and its response to changes is (assumed to be) instantaneous.

(iii) The power unit is also considered to position the valve stem position linearly as the controller signal strength varies, thus

$$P_{(t)} = -x_{(t)}K_m + C_m \tag{3.6}$$

where $x_{(t)}$ = valve stem position measured from fully closed at time t,

K_m = power unit (motor) constant,
C_m = a constant

(iv) The valve plug/seat position versus flow rate characteristic is considered initially to be linear, thus

$$\dot{Q}_{I(t)} = C_I x_{(t)} \tag{3.7}$$

where C_I = a constant

The overall response of these elements can now be studied graphically as shown in Figures 3.22 and 3.23 which show the

Figure 3.22

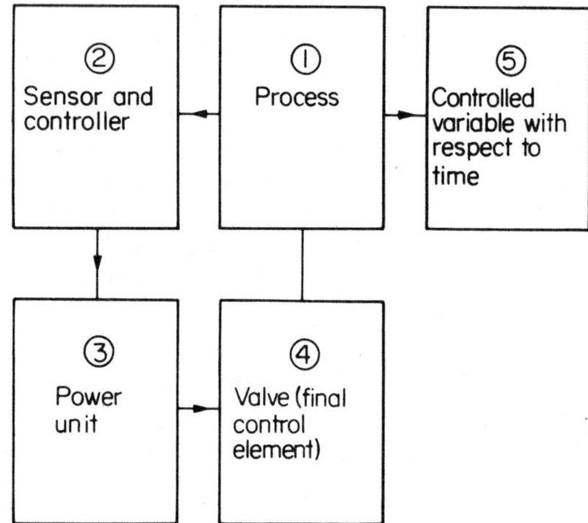

Figure 3.23

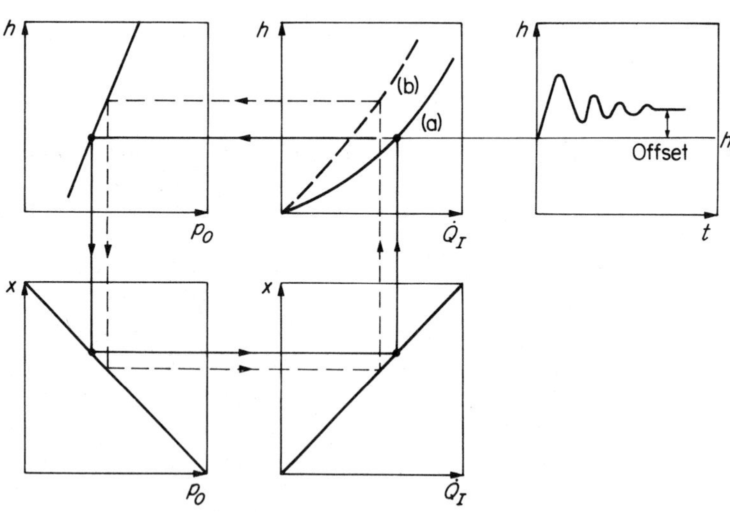

effect of a change in the outlet valve position from (a) to (b) to reduce $\dot{Q}_o$. (Note: For clarity the result shown is for a system having a small demand side capacity and a low gain; this does not correspond directly to Figure 3.21 which shows a small PB, high gain, situation.)

In a given control system, generally the only element characteristic which can be significantly modified is that of the controller. Altering K_c (the gain which varies inversely with PB) varies the controller sensitivity while altering C_c varies the set point h_o (Figure 3.24). Although tuning of controllers is discussed fully in Section 3.8 it is important to note here that the main feature of proportional control is *offset.* Offset can be reduced by increasing the controller gain but in certain cases this can cause instability. Instability is particularly likely when a 'small' demand side capacity is present. Figure 3.25 shows how the controlled variable history varies as the gain is increased.

Figure 3.24
(a) Altering gain,
(b) altering setpoint

Figure 3.25

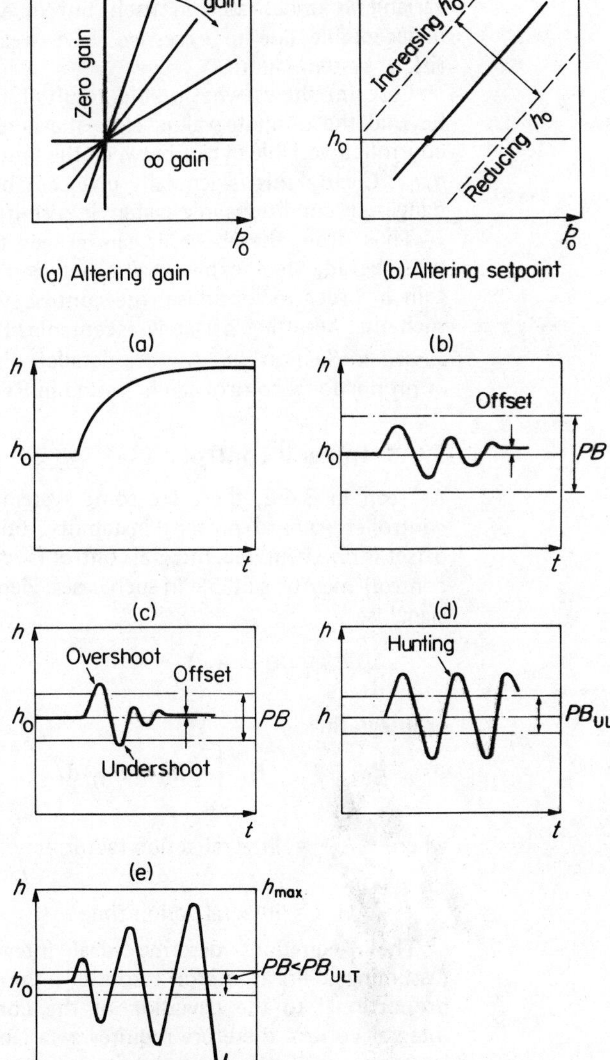

Case (a) shows the system response with zero gain, i.e. the CV position remains constant regardless of the level h. This is generally unacceptable as, for example, it could cause overflow as $\dot{Q}_o \rightarrow 0$.

Case (b) shows the response with a low controlled gain resulting in offset.

Case (c) shows an improvement (i.e. reduction in offset) as a result of increasing the gain. Notice, however, that over-shoot and under-shoot and considerable oscillation occur before the system settles down.

Case (d) shows the system response when the ultimate system gain is used, i.e. minimum PB. Again over-shoot and under-shoot are present but unlike case (c), the system does not settle down. Often this continuous oscillation ('hunting') of the controlled variable is undesirable in itself but in any event it is usually unacceptable due to excessive valve wear and the consequent risk of system failure.

Case (e) shows what would result if the gain were increased beyond the ultimate value. Here the amplitude grows until the controlled variable cycles between the limiting valves of h_{max} and h_{min}. Clearly this is generally unacceptable and will often cause dangerous conditions in a system if permitted.

Thus, from the above, it can be seen that the controller gain must be adjusted to have value between zero and the ultimate gain in order to 'optimise' the control system performance and such that resulting offset is acceptable. How this is done is discussed in Section 3.8. A more detailed discussion on the theory of proportional control can be found in Reference (14).

3.6.3 Integral control

As seen in 3.6.2, there are some systems which require a low controller gain to prevent instability, but where the attendant offset is unacceptable. Integral control (sometimes called 'floating' control) may be suitable in such cases. Here the controller output signal is:

$$\frac{dP}{dt} = K_I\,(h_o - h_{(t)}) \tag{3.8}$$

or integrating

$$P_{(t)} - P_o = K_I \int_o^t (h_o - h_{(t)})\,dt \tag{3.9}$$

where K_I = integral action factor = $\dfrac{1}{T_I}$

T_I = integral action time

These equations describe ideal integral control in which, (assuming a linear motor response), the valve stem velocity is proportional to the deviation of the controlled variable. Ideal integral control therefore requires a variable-speed motor, which is expensive. Single-speed and two-speed motors are often used as they are cheaper and this gives rise to single-speed and two-speed floating control. A further practical modification is to incorporate a small dead band (i.e. a range of values about the set point for which no controller output is produced), in order to reduce valve and motor wear and tear.

Although these practical and commercial considerations mean that pure integral control is often not provided by commercial 'integral' or 'floating' controllers, they all possess the essential feature which is that they maintain the desired value with zero (or very small) offset. A detailed discussion of the details of the above modifications is provided in Reference (12).

The formal analysis of integral control using the simple graphical methods used previously for on/off and proportional control is not possible. Investigations[15,16] of the stability of integral control systems have shown, however, that they are normally suitable for processes which have *fast* process reaction rates (e.g. a non-storage calorifier) *provided* sensing and other system lags are small. The three main applications in heating services are:

(i) control of mixed water flow temperature from a three-way valve,

(ii) control of outflow water temperature from a non-storage calorifier (provided sensing lags are small and a fast acting motor is used),

(iii) control of leaving air temperature from an air heating coil.

In these applications integral control can be (and often is) used with a master-sub-master (cascade) control loop. This is discussed in Section 3.7).

Integral control is also suitable for control of fluid flow rates and pressure levels provided (as is normally the case) sensing and distance velocity lags are small.
(Note: Integral control cannot be used on plants which have long time lags as over-correction will tend to occur with the result that self-generating 'hunting' will tend to occur.)

3.6.4 Proportional plus Integral (P + I) control

Systems which prove unsuitable for either proportional only or integral only control can often be controlled by combining both modes to give a P + I controller. Figure 3.26 (based on Figure 3.23) shows that by resetting proportional controller set points manually, (without altering the gain), offset can be removed. Thus a P + I controller *behaves as though* the set-point is reset automatically. The P + I controller output signal can be expressed:

$$P_{(t)} = (P_{\mathrm{TERM}})_{(t)} + (I_{\mathrm{TERM}})_{(t)} \qquad (3.10a)$$

or $$P_{(t)} = (K_c\, h_{(t)} + C_c) + \int_0^t \left(\frac{dP}{dt}\right) \qquad (3.10b)$$

or $$(P_{(t)} - P_o) = K_c\,(h_{(t)} - h_o) + K_I \int_o^t (h_o - h_{(t)})\,dt \quad (3.10c)$$

or $$(P_{(t)} - P_o) = K_c\,(h_{(t)} - h_o) + \frac{1}{T_I} \int_o^t (h_o - h_{(t)})\,dt \quad (3.10d)$$

The integral action generates an additional valve movement sufficient to remove the offset which would result from the use of proportional action alone. Both modes generate their responses simultaneously however, and the overall controlled variable response is as shown in Figure 3.27 and not as shown in Figure 3.26.

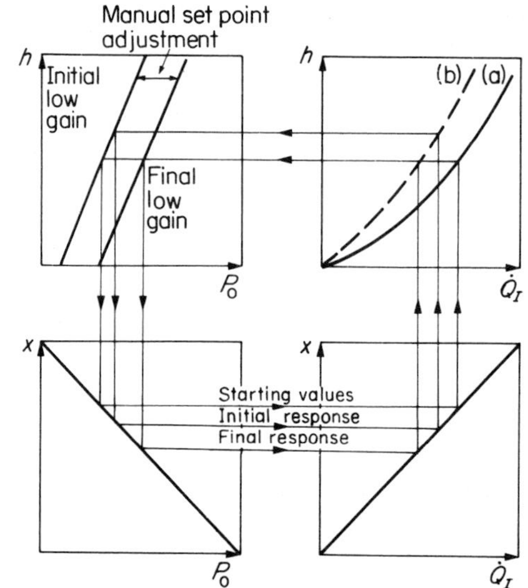

Figure 3.26
Equivalent P and I system

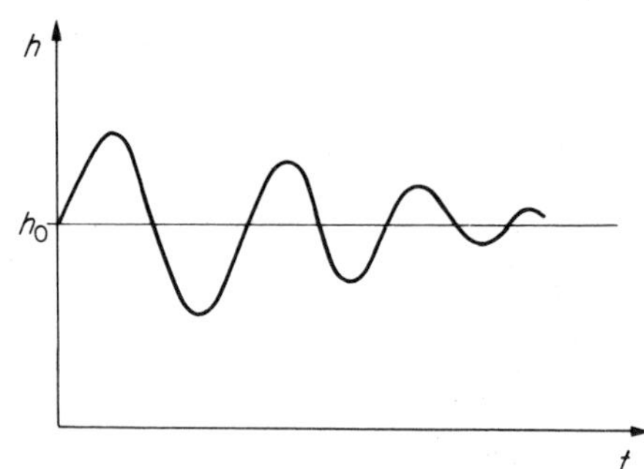

Figure 3.27
Actual P and I response

As with integral controls commercially available, P + I controllers seldom produce a pure P + I output. Despite this P + I control extends the benefits of integral control, viz. zero (or very small) offset, to a much wider range of applications than would be suitable for either mode used alone. In fact, there are very few applications in heating services where satisfactory performance

Figure 3.28

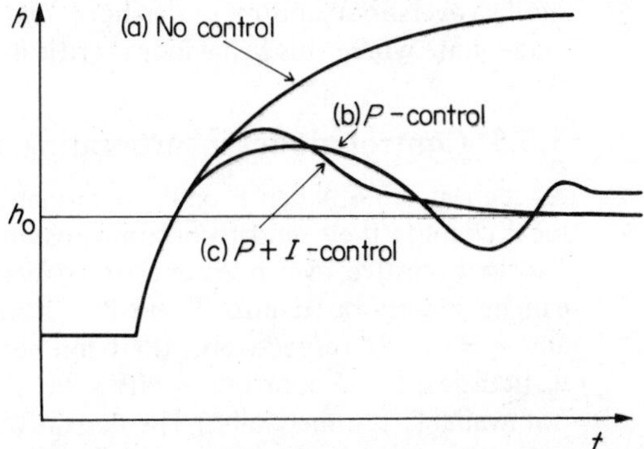

cannot be achieved using P + I control provided they are correctly tuned (see Section 3.9).

P + I control does, however, have one inherent weakness which is not immediately apparent. Figure 3.28 shows the controlled variable response to a sudden change in load for a case where the process exhibits a large exponential lag. A particular example might be the start up of a boiler plant following overnight or weekend shut down. Curve (a) shows the overshoot without control. Curve (b) shows the overshoot with P only control. (Note: I only control is unsuitable because the p.r.r. is slow.) Curve (c) shows that overshoot with P + I control is larger than with P only control. Up to a point this can be removed by increasing T_I but large T_I values can lead to other problems (see Section 3.9).

Figure 3.29

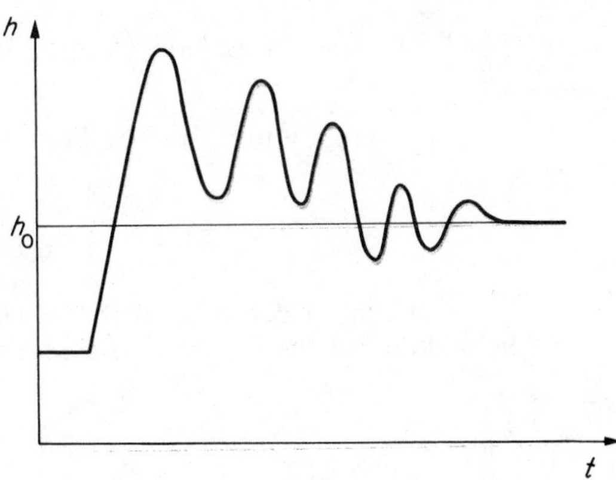

Figure 3.29 shows the response for a more reactive system when subjected to a sudden load change while operating under P + I control.

In cases where this overshoot (and/or undershoot) is known to be a problem, it may be necessary to start the plant under manual control and switch to automatic P + I control only when the controlled variable is close to the desired value.

It should also be noted that step changes in the desired value, (caused say, by manual adjustment of the set point), can cause similar overshoot and/or undershoot and that P + I may prove inadequate where this behaviour is critical.

3.6.5 Control modes incorporating derivative action

As seen in 3.6.4, when P or P + I modes are subjected to sudden local changes they tend to become unstable (i.e. oscillate) or give rise to excessive overshoot and/or undershoot. Derivative action can be incorporated into P and P + I controllers to give P + D and P + I + D respectively. (Perhaps because of manufacturing difficulties, I + D controllers, although theoretically possible, are not available commercially.) The derivative action term (note that derivative action cannot be used alone, as can be deduced from equation 3.11), improves control performance of other modes when sudden load changes occur. The derivative action term is defined by

$$(P_{(t)} - P_o) = T_D \frac{dh}{dt} \qquad (3.11)$$

where T_D = derivative action time

This leads to the defining equations for P + D and P + I + D controllers where

P + D: $$(P_{(t)} - P_o) = P_{TERM} + D_{TERM} \qquad (3.12a)$$

$$(P_{(t)} - P_o) = K_c(h_{(t)} - h_o) + T_D\left(\frac{d(h_{(t)} - h_o)}{dt}\right) \qquad \ldots (3.12b)$$

P + I + D: $$(P_{(t)} - P_o) = P_{TERM} + I_{TERM} + D_{TERM} \qquad (3.13a)$$

$$(P_{(t)} - P_o) = K_c(h_{(t)} - h_o) + \frac{1}{T_I}\int_o^t (h_{(t)} - h_o)dt + T_D\left(\frac{d(h_{(t)} - h_o)}{dt}\right) \qquad (3.13b)$$

P + D control does not, of course, eliminate offset although the addition of the D term does permit a higher K_c value to be

Figure 3.30

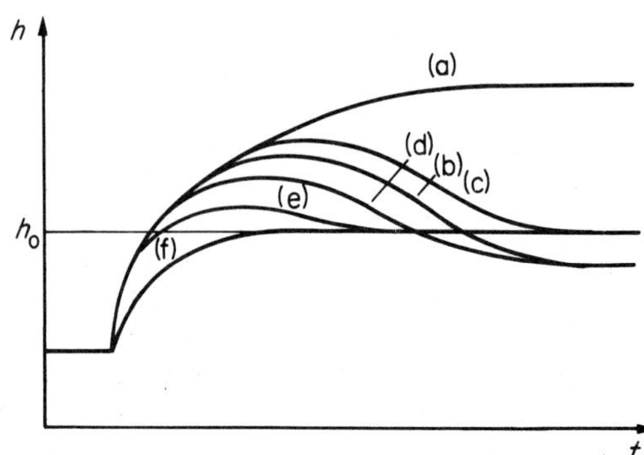

used without causing instability and thereby reducing offset compared with that produced by P only control.

P + I + D control represents the most sophisticated and comprehensive mode of control normally used in heating services. It caters for large and/or sudden load changes and ensures very small or zero offset. By careful tuning (see Section 3.8) of the various controller parameters, (h_o, K_c, T_I and T_D) good control can almost always be obtained (provided of course that the nature of the system lags and process behaviour does not make the system basically uncontrollable). Figure 3.30 compares P + D and P + I + D with other modes for the situation previously shown in Figure 3.28 where the curves are:

(a) no control (i.e. inherent regulation)
(b) P only
(c) P + I
(d) P + D
(e) possible P + I + D response
(f) possible P + I + D response

The only real disadvantages of P + I + D are:

(i) cost and availability. At present, pneumatic P + I + D controllers are not available commercially. Electronic P + I + D controls have been expensive until recently but perhaps costs will reduce significantly as electronic components become cheaper.
(ii) P + I + D control can be difficult to 'tune' (and to keep in 'tune') and hence less sophisticated modes are generally preferred provided they give sufficiently accurate control.

3.7 CASCADE AND MULTI-CONTROLLER SYSTEMS

Most systems are in fact controlled by a number of sub-systems each incorporating their own controls. An example is that of a typical heating system shown diagrammatically in Figure 3.31. From this we see that in an attempt to reduce the effects of inherent time lags and signal distortions, a variety of sub-systems are used. Sometimes these operate in *parallel*, e.g. the 'room sensor/controller/radiator valve (feedback) control circuit' operates in parallel with the 'outside detector/controller/flow detector/controller/burner controller circuit feedforward-feedback controller circuit', while both of these systems are overridden or assisted as required by boiler temperature limit controls, pump/boiler stop/start (perhaps by time clock) and, perhaps in addition to those shown, frost protection devices, and/or low-limit fabric (anti-condensation) control and/or optimum start control for energy saving purposes.

Considering the 'outside detector/controller/flow detector/controller/burner controller/feedforward-feedback control circuit'

Figure 3.31

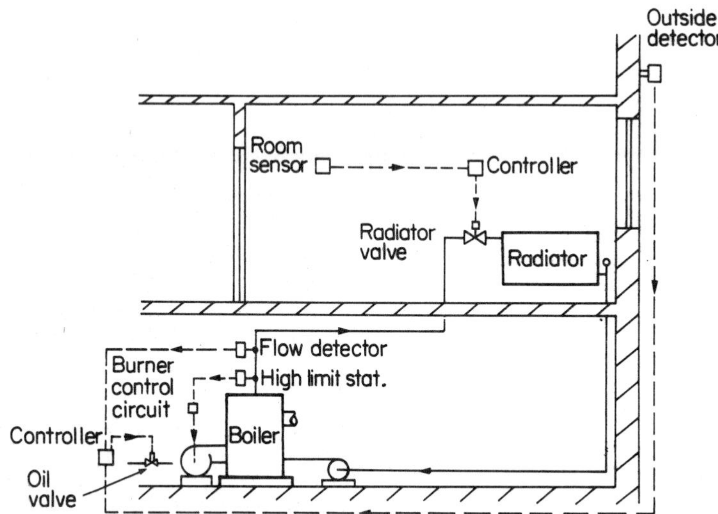

in more detail reveals that the control point of the flow detector sensing the water leaving the boiler is reset or arranged in 'cascade' with the controller on the basis of information received by the controller from the outside detector and in accordance with the desired water flow temperature schedule specified by the system designer. If in this case, the boiler serves only a small system, then on/off burner control would be appropriate while the normal 'mode' (if such a description is really valid), of the outside detector circuit is normally proportional only, the proportional band adjustment being implicitly selected as the water flow temperature schedule is adjusted at commissioning. Figure 3.32 shows a similar system incorporating a three-way mixing valve to adjust water flow temperature. Boiler control here may be simple, e.g. on/off or high/low/off in order to maintain 'constant' leaving water temperature, or more complex, e.g. where more than one boiler

Figure 3.32

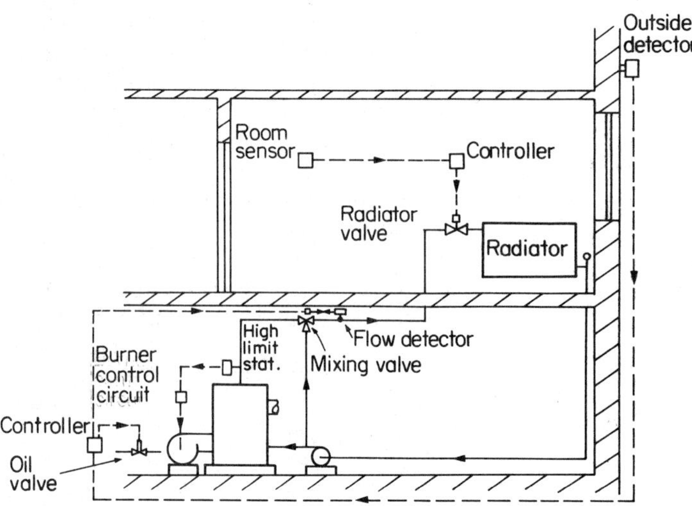

is installed boiler sequencing control (a form of step control) may be used to advantage.

From the above it will be obvious to the reader that all but the very simplest installations will employ cascade or multi-controller circuits in order to minimise the effects of time delays and thus ensure safe and economic operation. Some typical practical control arrangements are detailed in Section 3.9.

3.8 COMMISSIONING OF SYSTEMS AND THEIR CONTROLS

When commissioning systems, reference should be made to the Chartered Institution of Building Services Commissioning Codes which outline the various safety checks which must be carried out before a system is started. This section assumes that these preliminary steps have been carried out and that it is then required that the controls be adjusted ('tuned') to give satisfactory (or perhaps even 'optimum') control and various procedures for achieving this are discussed.

It must be pointed out that while methods presented here are in everyday use in the chemical and process control engineering field, they are *not* at present in common use for heating systems. Despite this the authors believe that the discussion of these techniques highlights an area which deserves greater attention during commissioning and, of equal importance, which provides an insight into the operation of systems and their controls.

3.8.1 General requirements

In setting out a control specification, it should be kept in mind that very often other (uncontrolled or indirectly controlled) variables may also significantly affect the process which it is desired to control. For example, although room air temperature may be the controlled variable for room heating, many other factors will influence the thermal comfort of the room occupants, (see Chapter 4).

Having decided on the most suitable parameter to use as the controlled variable a control specification should ideally provide the following information about its controlled behaviour:

(i) the *desired value* (usually although not always this can be used directly as the controller set point),

(ii) the *allowable range* of the controlled variable. In general the differential gap or proportional band will require to be *less* than this by an amount to suit expected overshoot/undershoot during normal operation,

(iii) although not common in heating services installations, cases do occur where it is necessary to limit the *rate of change* of the controlled variable and/or prevent oscillation at some or all frequencies. This may be a process heating control requirement,

(iv) although seldom stated in heating services control specifications, ideally the requirements on *control quality* to be achieved by the system should be set out. This is discussed in Section 3.8.2.

3.8.2 Control quality

As stated previously all feedback control systems are error actuated and few feedforward control systems are capable of providing error-free control. Thus in general, control response to changes produces deviation and various control quality criteria exist [17] which can be used in conjunction with the tuning methods discussed to 'optimise' the performance of a particular control loop. Space restrictions permit only a very brief discussion of these here.

Figure 3.33

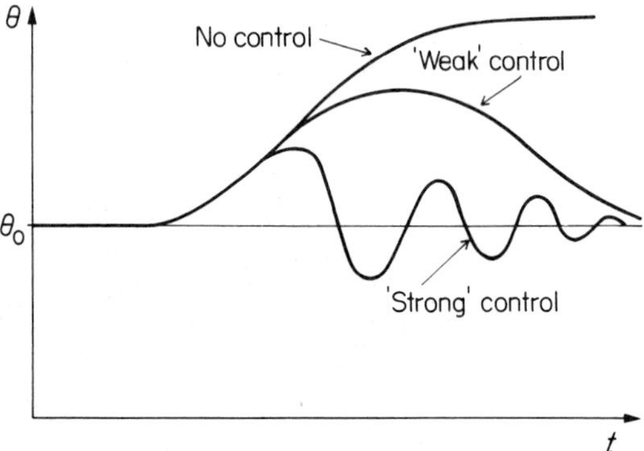

Figure 3.33 outlines the three general cases of system response to a step change in load. From this figure and from previous discussion the reader will be aware of the accuracy versus stability compromise inherent in all forms of process control.

The advantage of the aperiodic (limiting case) control is that despite the large initial deviation this 'weak' form of control can

Figure 3.34
Control area as criterion of quality

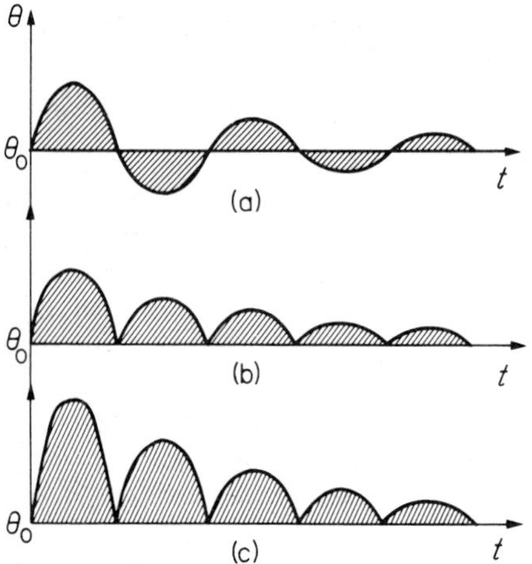

often (although not necessarily), result in a quicker return to the desired value (i.e. a shorter 'settling time') or, (in the case of proportional only control) to minimum offset, following a disturbance than will result from using 'strong' control. Furthermore if disturbances are known to be small and to occur slowly then aperiodic control may be satisfactory or even desirable. In heating services, however, 'strong' control is usually desirable or necessary to minimise deviations.

Figure 3.34 shows the four main criteria used for assessing control quality (shown here in relation to strong control response). These criteria are:

(a) *Linear control area*: This area is found by evaluating the equation from the results of a small step disturbance test (either in load or set value, the latter being almost always used as it is much easier to impose).

$$A_{\mathrm{LIN}} = \int_{0}^{\infty} \theta \, dt$$

This measures the *net* (+ or –) effect of the disturbance and when used as a means of assessing the performance of a temperature control loop it provides a useful means of optimising for minimum energy consumption, as negative and positive areas *cancel* to give a net area which corresponds to an energy quantity which should be minimised to give optimum control. Common sense indicates caution regarding the use of this criterion without a specification on the degree of damping as, for example, in the case of continuous undamped oscillation, the net area would be zero and the apparent optimum would appear to have been achieved.

(b) *Absolute control area*: This is found from equation

$$A_{\mathrm{ABS}} = \int_{0}^{\infty} \theta \; dt$$

This measures the absolute deviation effect of the disturbance and which it is also desirable to minimise for good control. When used in conjunction with method (a) this provides a useful check on the degree of damping inherent in the response.

(c) *Quadratic control area*: This is found from equation

$$A_{\mathrm{QUA}} = \int_{0}^{\infty} (\theta^2) \, dt$$

As with (b) this method provides a useful check on the amount of damping inherent in the response. It also readily distinguishes between the contrasting cases of aperiodic creep and that of very small damping. This criterion takes particular account of the first overshoot and it follows that controllers optimised to give minimum A_{QUA} will have smaller overshoot but take longer to settle than controllers optimised using method (a).

3.8.3 Tuning methods

Having decided on a control specification (which in itself normally involves compromise) a further compromise must be made between the accuracy and stability with which the systems can be controlled. Although there are a number of methods which can be used to tune controllers only a few of the main ones are described here.

In tuning controllers it is important to realise:

(i) Systems will often be tuned initially under conditions which are unrepresentative. When this occurs they should be re-tuned under the load condition at which they most frequently operate (e.g. for U.K. heating systems this will be at about 50% load). Care should be taken to ensure that such tuning does not give rise to unsafe behaviour at other loads.

(ii) Even when the system and its controls have been set to work satisfactorily, frequent re-tuning is essential to ensure safe and satisfactory performance. This is because valve wear, fouling of heat transfer surfaces, etc. tend to alter the behaviour of system and controls as time passes. Re-tuning should be carried out at not more than yearly intervals and in many cases six or even three-monthly intervals will be justified.

3.8.4 Tuning discontinuously operating controllers

Very often tuning on/off and step controllers is restricted to adjusting the set point(s) and choosing suitable differential gap(s) to ensure that too short a periodic time is avoided and, as this has already been discussed in some detail in Section 3.6.1, it is not discussed further here. On/off control of heating using room thermostats is perhaps an exception and much research work has been carried out on this problem. Space restrictions prohibit any meaningful discussion here and the reader is referred to References (9), (10).

3.8.5 Tuning continuously operating controllers

As discussed earlier, it will be often very difficult or impossible to make anything other than a *qualitative* assessment of the controller requirements for a particular system. Provided this assessment leads to the selection of a suitably flexible controller and that the other system components have been correctly sized, etc. it should be possible to tune the chosen controller to obtain optimum (or nearly so) performance.

A number of techniques are available for doing this but many show little improvement over those proposed by Ziegler and Nichols[18] who were the originators of such tuning methods. Z-N methods are restricted to tuning feedback controllers and as such apply to the majority of heating services installations. No such methods are yet available for tuning feedforward controllers but Reference (8) gives useful advice. No information is yet available about tuning cascade controller systems as a whole but provided each controller is tuned according to the following principles satisfactory performance can be expected.

Although it is likely that more than 90% of all heating controls are tuned using 'rules of thumb', the Z-N methods described below allow a more deductive approach. Z-N are time domain methods but Laplace and frequency domain methods are also available[3]. Two procedures are available both based on the same theory.

Procedure No. 1, Open Loop Test: In this procedure the Z-N method gets the required system dynamic information

Figure 3.35
Z-N open loop test

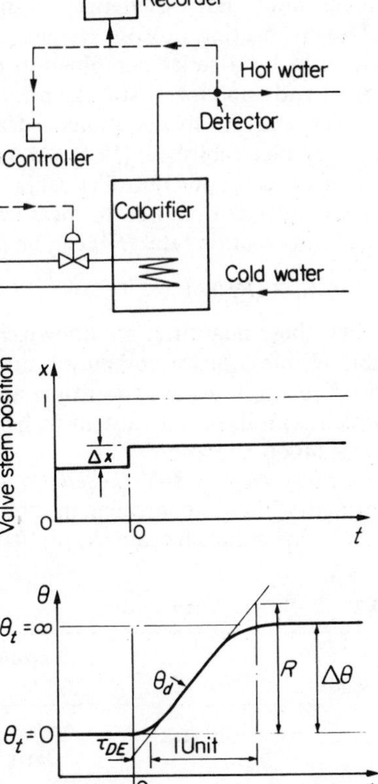

by means of open loop step-response test. The physical arrangement and typical test results are shown in Figure 3.35 and the test procedure is shown in Table 3.1. The

Table 3.1 Z-N Open-loop Test Procedure

1. The loop is opened at a point between the controller and the final element.
2. A recorder is set up to record the measured value (θ_d) of the controlled variable.
3. The final control element (e.g. valve) is positioned manually so that the controlled variable is in equilibrium *slightly below the desired value*
4. A *small* step change (Δx) is imposed manually at the FCE.
5. This upsets the process and causes the controlled variable to undergo a transient before coming to a new equilibrium.
6. The record of this response contains the information needed to extract the dynamic characteristics of the controlled system and the measuring means.

response curve of θ_d versus time is due actually to complex physical processes occurring within the particular system under study, but experience has shown that the behaviour of many heating process systems can be represented satisfactorily by a series combination of equivalent dead time (τ_{DE}) and a nominal first-order process. From measurements of Δx and θ_d the equivalent dead time τ_{DE} and p.r.r. (R) can be established. (Note: Any unit of time appropriate to the process and controller calibration (i.e. hours, minutes or seconds) can be used to measure R). When R is known, the 'unit reaction rate' (R') can be found from

$$R' = (R/\Delta x)$$

When these quantities are known simple formulae are available (Table 3.2) for computing suitable settings of proportional band, integral action time and derivative action time which will allow the system to be started up safely under closed loop operation.

Procedure No. 2, Z-N Closed Loop Test: This test is performed with the controller in circuit (as shown in Figure 3.36). The controller gain (K_c) is first set to a very low value

Table 3.2 Z-N Open Loop Settings

Controller Setting	P	*Controller Type* P + I	P + I + D
K_c	$1/(R'\tau_{DE})$	$0.9/(R'\tau_{DE})$	$1.2/(R'\tau_{DE})$
T_I	—	$3.3\tau_{DE}$	$2\tau_{DE}$
T_D	—	—	$0.5\tau_{DE}$

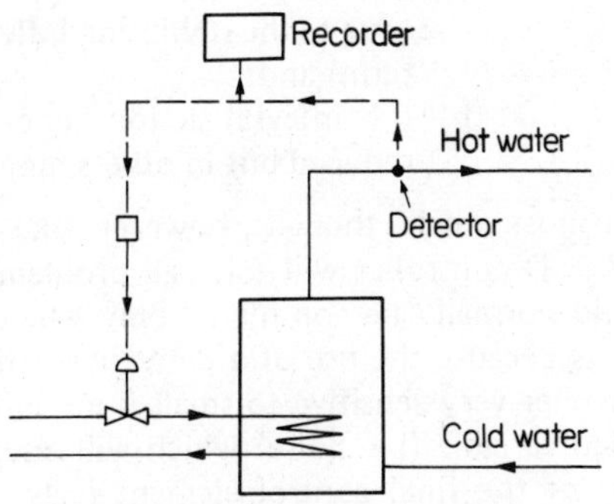

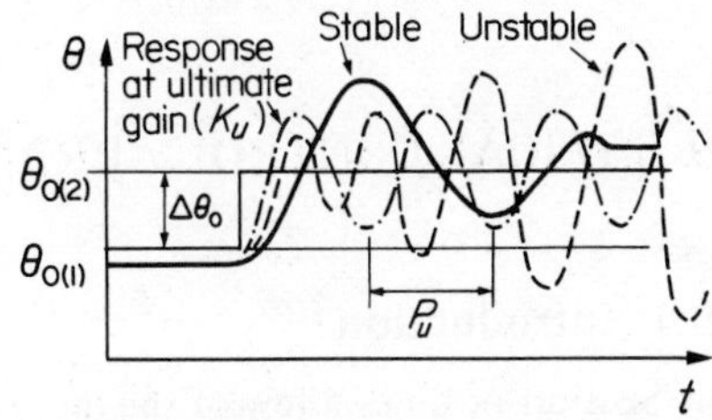

Figure 3.36
Z-N closed loop test

and integral and derivative actions are removed entirely or set to have minimum effect.

A small step change is applied (either a load or set point change will suffice) and the response of the system is noted. (If under these circumstances the process is unstable the open loop test should be used instead.) Normally it will be possible to increase the gain gradually until the system starts to exhibit a continuous oscillation. The gain at which this just starts to occur is known as the ultimate gain (K_u) and the corresponding periodic time is the ultimate periodic time (P_u). Z-N optimum controller settings can be computed by substituting K_u and P_u in values, (Table 3.3).

Table 3.3 Z-N Closed Loop Settings for Optimum Performance

Controller Setting	P	*Controller Type* P + I	P + I + D
K_c	0.5 K_u	0.4 K_u	0.6 K_u
T_I	—	0.33 P_u	0.5 P_u
T_D	—	—	0.125 P_u

Conclusions

From the above Tables it can be seen that

(i) When an integral term by itself is added the proportional gain must be reduced slightly in order to avoid initial overshoot.

(ii) When a differential term is added to a P + I controller (a) the proportional gain can usually be increased (due to the stabilising influence of the differential term) and;

(b) the integral action time may in some cases be reduced but in others may require to be increased.

It must not be thought, however, that the use of a P + I + D or P + D controller will solve all problems. Indeed, its adoption should normally be considered only when there is a proven need. This is because the use of a derivative action term can make the controller very sensitive to small (random) variations in the controlled variable (i.e. noise) which will result in continuous adjustment of the final control element (valve) resulting in excessive valve and motor wear and early control system breakdown.

3.9 TYPICAL CONTROL ARRANGEMENTS

3.9.1 Introduction

This Section outlines a few of the more common control arrangements. The emphasis is on controls for medium and large sized installations. Domestic and small commercial installations are discussed in Section 3.9.7.

Most control systems can usefully be considered as comprising 'local or terminal controls' and 'central plant controls' each operating in cascade. At the present state of the art, control system behaviour can be discussed only in a qualitative manner and hence an outline of what has proved to be suitable in the past is likely to be of benefit in designing new control systems.

3.9.2 Local control of heat emission

Table 3.4 provides a brief outline of some of the more common methods of regulating typical LPHW emitters and each is discussed briefly below:

Radiators

(a)(i) Although *manual radiator valves* (MRVs) are still widely used, generally their regulating ability is poor. This is because most MRVs have a poor valve characteristic and, further, are installed with poor valve authority. MRVs provide what is in effect virtually on/off control and do not in practice give proportional control as is commonly believed. Improved control can be obtained by dividing the heating surface into a number of radiators (thus providing a form of step control) but this may be less effective and more costly than one radiator fitted with a TRV.

(a)(ii) The two basic types of *thermostatic radiator valves* (TRV) are shown in Figure 3.37(a) and (b). Most TRVs

Table 3.4 Method of Local Control

Emitter Type		*Method of Control*	*Description of Operation*
(a) LPHW radiators (all types)	(i)	manual radiator (MRV) valve on flow or return	virtually on/off due to poor regulating characteristics of most valves
	(ii)	thermostatic radiator valve (TRV)	roughly proportional (but see text and Section 3.9.3)
	(iii)	room thermostat (RT) controlling zone via a two- or three-port valve	most two-port valves give on/off control (i.e. they are solenoid type); three-port may be on/off or proportional (diverting)
	(iv)	RT as (a) (iii) but incorporating a time switch; occasionally this arrangement will also incorporate an elevated pre-heat boost arrangement and/or a depressed night set-back arrangement (see text)	as (a) (iii)
	(v)	weather compensated zone control (see text and Section 3.9.3)	general proportional-reset-integral or rarely proportional-reset (proportional + integral)
(b) LPHW natural convectors including skirting heating, cill line heating, pipe coils and similar	(i)	the principal control method is by means of a manual damper or baffle plate used alone or with (b) (ii) below	very poor proportional control virtually equivalent to high/low (see text)
	(ii)	generally speaking any of the methods (a) (i) to (a) (v) can be applied together or alone with (b) (i) to give the required control	as (a) (i) to (a) (v)
(c) LPHW fan convectors (fully recirculating type)	(i)	generally individual units have high/low/off fan speed control which gives a form of manual control	effectively this gives (manual) high/medium/low heat control of heat output (see text) which is thus a form of step-control
	(ii)	a common addition to (c) (i) is to time switch the electrical supply to the fans to obtain a measure of time and zone control	as (c) (i)
	(iii)	as an alternative to (c) (i) a thermostat located behind the recirculating air grille is used to give high/low/off control of fan motor speed	as (c) (i) but control is thermostatic rather than manual
(d) LPHW fan convectors (fresh air type)	(i)	as (c) (i) to (c) (iii) with the addition of an immersion thermostat in the flow water pipeline at entry to unit to hold off fan operation until the flow water temperature rises above say, 60 °C	an on/off frost protection device
(e) LPHW unit heaters	(i)	generally fans are quite large and are commonly arranged for manual on/off fan control from push button starters	effectively this gives on/off control of heat supply into the occupied zone
	(ii)	an alternative is to have a room thermostat to control one or more unit fans	as (e) (i)

Emitter Type	*Method of Control*	*Description of Operation*
	(iii) time switching may be applied as for fan convectors and fresh air types may employ a pipeline frost protection thermostat as (d) (i)	
(f) LPHW radiant panels (low temperature sources)	(i) methods similar to those described for radiators (a) (i) to (a) (v) can generally be used provided design checks show that size and distribution of panels do not give noticeable vector radiant temperature effects	as (a) (i) to (a) (v)
(g) LPHW heated floors	(i) again, design checks should prevent v.r.t. discomfort but in addition the floor surface temperature must be controlled (usually at about 25 °to 30 °C) to prevent local foot discomfort; this is done usually by means of a three-port mixing valve arrangement and pipeline thermostat (set at about 40 °C); this is a functional necessity as most boilers will provide a flow temperature well above this. Room temperature control can be difficult to achieve; methods (ii) and (iii) describe arrangements normally used but these generally give only partial success (see text)	flow water temperature control usually by integral action (sometimes proportional plus integral)
	(ii) one method is to provide a manual (should be a globe type) valve with a good regulating characteristic to allow control of water flow rate	roughly proportional
	(iii) an alternative is to use a room or zone thermostat (RT) to reset the water flow detector in g (i)	usually proportional reset integral, but sometimes proportional reset (proportional + integral)
(h) LPHW heated walls and embedded ceilings	(i) methods usually similar to those for floors but not the same restriction on surface temperature provided v.r.t. effects cause no discomfort. Again control is inherently difficult due to the slow response of the heated element to changes in load	as (g) (i)
(i) LPHW heated suspended ceilings	(i) again methods similar to h, v.r.t. discomfort to be avoided but control considerably easier than (g) and (h) due to faster system response	as (g) (i)

Figure 3.37
Thermostatic radiator valves. (a) Integral sensor, (b) remote bulb

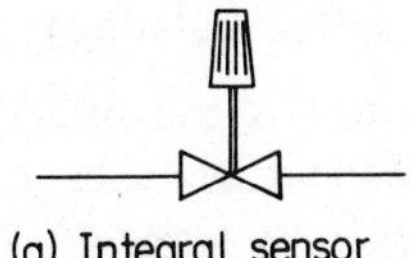

(b) Remote bulb

have adjustable set-points and most modern valves can give good proportional control provided they are correctly installed. See also Section 3.9.3 (Control of heat emission). Some valves incorporate a desirable feature to give a minimum temperature setting (say 10 °C) when at the nominal 'off' position for use with frost/fabric protection controls. TRVs have now recieved widespread acceptance and provide a good terminal control device to tune emitter output thereby utilizing casual heat gains. They give best service when used in conjunction with 'weather compensated control' (Sections 3.9.3 and 3.9.7.).

(a)(iii) The use of a room thermostat and a two- or three-port valve arrangement to control a few radiators generally proves more expensive and less effective than TRVs. It may prove viable for a large room or a number of rooms in a zone but control quality would be inferior to a TRV.

(a)(iv) This arrangement can be simplified to give useful zone time control using a simple on/off (solenoid) valve.

(a)(v) Rooms must have similar solar and/or casual gains. This arrangement can and probably will receive wider application as considerable energy savings are possible in larger buildings.

Natural Convectors

(b)(i) Most units have a hand operated damper. As with MRVs, regulation is virtually on/off for air flow but, because units radiate and convect 10 to 20% from the casing with damper 'off', heat output control is high/low.

(b)(ii) In addition to (b)(i) or to replace it most of the methods described for radiators can be applied to natural convectors. If TRVs are used then natural convectors and radiators can be mixed provided a suitable water flow schedule (Section 3.9.3) is chosen to suit the natural convectors. Pipework savings can often be used to finance the additional TRVs and give better overall control.

Fan Convectors (fully recirculating type)

(c)(i) Speed control is the principal means of varying heat output. Manual switches to give high/low speed and power on/off are common. With the fan off natural convection and casing emission give about 10 to 15% output provided hot water is circulated to the units. This can prove useful for frost/fabric protection. Thus overall heat output control is normally high/low/minimum.

(c)(ii) It is often useful to provide one or more separate power circuits for fan convectors with central timeclock control for time zoning purposes.

(c)(iii) Units can be individually thermostatically controlled, the thermostat switching the fan to high or low speed. The sensor is normally located inside the unit behind the return air grille. Some units have a boost/normal switch by means of which the occupant can override the thermostat and set the fan to high speed. Again local on/off switch control and central time clocking are common.

(Note: It is commonly believed that the heat index of fan convectors is $n = 1.5$. This is not so, as $n = 1.0$ and hence fan convectors can be fitted to weather compensated circuits without danger of underheating. When this is done the fans will, of course, run for longer periods in mild weather but this may give better heat distribution than with more intermittent fan operation when connected to a constant temperature circuit.)

Fan Convectors (fresh air type)

(d)(i) A low water temperature sensor is an essential requirement in addition to methods (c)(i) to (c)(iii).

Unit Heaters

(e)(i) Normally each unit has an individual room thermostat in the occupied zone to give fan on/off control. With fan off, 10 to 20% of heat output at high level is generally wasteful and individual shut off valves (sited at an accessible height) should be provided to give positive shut down. When shutting off units in spring and autumn this should be staggered to prevent uneven heat distribution.

(e)(ii) Group control from one thermostat is generally unsatisfactory and seldom employed.

(e)(iii) As for fan convectors but fresh air inlet units should have some means of shutting off air inlet when the fan is off in order to prevent unwanted stack effect ventilation. Pneumatic or cable operated dampers can be used.

Radiant Panels

(f)(i) As for radiators. (Note: High temperature heaters should generally have manual controls as it is often difficult to sense 'comfort' when heating by this method.)

Heated Floors, Walls and Ceilings

Floor and wall heating generally employs embedded pipe coils or electric heating cables. This method of heating has given rise to serious problems of over or underheating due to the slow response of the system to changes in heating load, because of high thermal inertia. This is particularly so with electric off-peak systems. For these reasons such floor and wall heating systems are seldom employed in modern buildings but when they are used a combination of weather compensated water flow schedule and a manual (globe) control valve used by an experienced occupant to anticipate the lag is often as good if not better than a room thermostat. Room thermostats should be fitted with accelerated heaters calibrated for the installation.

Embedded ceiling heating suffers from the above difficulties but suspended (lightweight) panels are much easier to control. A simple manual (globe) is often sufficient. A TRV (remote bulb type) can also be used or a room thermostat as (a) (iii).

3.9.3 Central plant control

Controls for domestic and other small installations are discussed in Section 3.9.7; this Section deals with medium and large installations. The layouts refer basically to LPHW systems but they can also be applied to pressurised water installations provided the additional safety interlocks are fitted (where required) to cater for de-pressurisation failures. Controls for the pressurisation system are normally a specialised package and are not discussed here, nor are the special requirements of steam boiler plants.

For reasons given in Chapter 5, most modern boiler plants employ two boilers, although some older plants may incorporate three, four or more. Controls for modular boiler installations (with 3, 4, 5, 6 or more boilers) are not discussed; modular boiler manufacturers have developed special sequencing control systems to suit such layouts.

Two-boiler installations employing simple manual control

For a large part of the United Kingdom heating season one boiler (rated at 60% design load) will often prove sufficient for requirements and thus manual boiler operation as per Figure 3.38 may be suitable. This layout shows only the basic logic and does not include the various safety controls such as flame failure devices, purge timers, etc. usually provided as part of the boiler package.

The operation of the control equipment is described below.

(a) Time controls

The timeswitch (TS) dictates when power is available to the pump and to both boiler on/off switches A and B. Normally

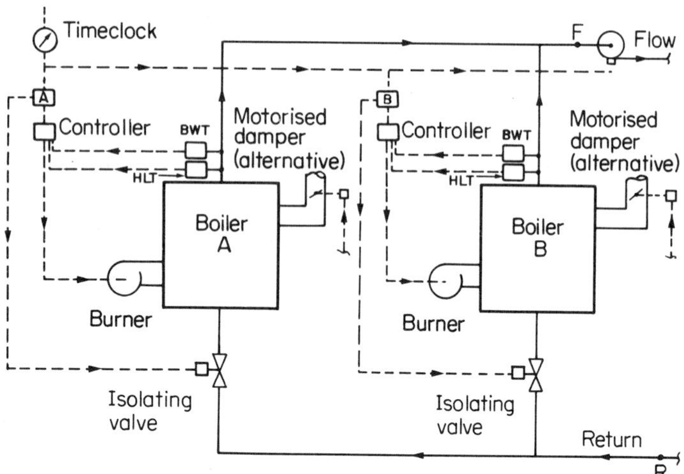

Figure 3.38
Two boiler installation with simple manual control

two or three parallel pipework circuits are required with dual pumps (see Chapter 9) and automatic changeover arrangements, but only one circuit is shown here for clarity. A relay is usually fitted to shut off power to the boilers in the event of pump(s) failure.

(b) Boiler on/off switches
With power available to A and B the boilers will now operate as determined by the switch positions and under the thermostatic control system.

(c) Boiler isolating valves and flue dampers
Isolating valves (IV) are a very necessary addition to prevent hot water circulating through the idle boiler which would give rise to wasteful flue and casing losses. Although this is often done manually, spring return normally-closed solenoid valves held open only when the corresponding boiler switch is 'on' are convenient and safer. An alternative method is to fit dampers on the flue gas outlets from each boiler which are closed when the boiler is not in use. This cuts down flue heat losses but casing losses persist. As with the isolating valves, dampers may be manual or motorised for convenience and safety.

(d) Boiler safety thermostats
The boiler safety or high limit thermostats (HLT) are designed to shut off power to the burner and fuel supply solenoid valve which closes, cutting off fuel in the event of a fault (e.g. failure of pumps or working thermostats, system blockage or leakage etc.), thereby preventing excessive temperatures which could lead to steam generation and consequent explosion risk in the plant. The HLT is nearly always mounted on the boiler and is usually pre-wired as part of the boiler/burner safety controls mentioned prreviously. It is almost invariably a manual reset type thus preventing automatic re-occurrence of the fault condition.

(e) Boiler working thermostats

(i) Parallel operation: When both boilers are in operation it is common for both boiler working thermostats (BWT) to have the same set point. This has the advantage that both boilers share the load about evenly.

(ii) Lead-lag arrangement: An alternative is to reduce the set point of one boiler by a significant amount (say 5 °C) below the other. The boiler with the higher set point then becomes the 'lead' boiler and the other becomes the 'lag' boiler. Because the return water temperature (at R) to each boiler is nominally equal the lead boiler will operate for a greater proportion of the time. Unless the boiler chosen as the lead boiler is known to have a higher operating efficiency, then there is little advantage in this arrangement. Furthermore the mixed outflow temperature (at F) to the heating system will vary about a value which is the mean of the two set points. As can be seen, this arrangement does not 'rest' one boiler (until temperature at R rises to the lag boiler set point). In order to 'rest' the lag boiler both working thermostats could be repositioned in the common outflow at F. Notice, however, that the lag boiler is nominally 'on' while resting and incurring casing and flue heat losses as a result. One method of preventing such losses is to place the boiler isolating valves directly under BWT control but this should not normally be attempted as the residual heat in the boiler following BWT operation and IV closure would probably give rise to frequent lock-out on HLT which apart from the risk of boiling would prove very inconvenient in practice. Motorised flue dampers should only be placed under direct BWT control when this arrangement can be incorporated into the various safety interlocks, purge checks etc., required to prevent explosion on start-up. Thus the best method is normally to adopt parallel operation with equal set-points and to shut down and isolate one boiler entirely when the other can cope with the load alone.

(f) High/low and modulating burners

The above discussion has been based implicitly on boilers which operate in on/off mode only and many boilers, including those of quite large output, operate in this way for reasons discussed in Chapter 5.

High/low/off burners employ a form of step control and produce a low flame setting of around 40% full input in addition to high flame and off. *Modulating* burners provide a limited form of proportional control and vary the flame size continuously down to about 40% full load and thereafter operate on on/off mode. In both cases the lower limit of about 40% is caused by problems of flame stability, size and shape to prevent local overheating, and poor combustion efficiencies. Such burners are useful in preventing excessive cycling and consequent flow temperature fluctuations which tend to occur in large on/off boiler control and which can impose a difficult control problem on subsequent control loops operating in series.

Multi-boiler return water temperature sequence control.

As shown in Figure 3.39, the return water temperature is the controlled variable. Only two boilers are shown for simplicity

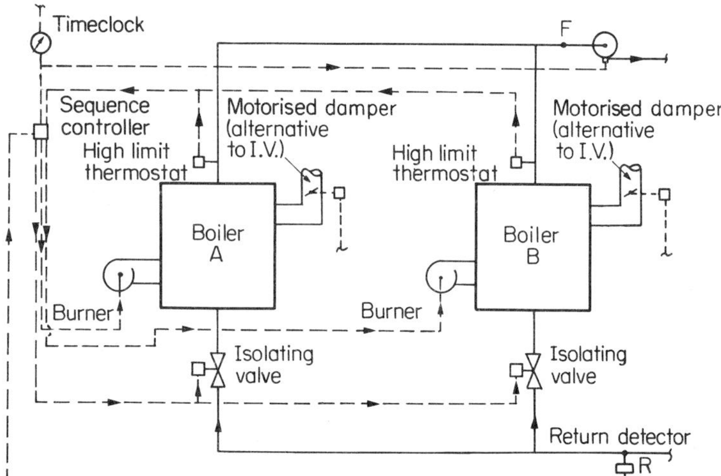

Figure 3.39 Multi-boiler return water sequence control

but more can be controlled if required. A common return water temperature detector at R operates a proportional motor which in turn operates the switches on a step controller which in turn determimines the number of boilers 'on stream' in accordance with a pre-selected sequence. The sequence is generally variable to ensure that all boilers share the load evenly. In the boiler 'off' position the boiler isolating valves (or flue dampers) should close to prevent losses. In this system each boiler (when 'on' under the sequence control) will fire under normal BWT control and flow temperature fluctuation will occur but in most systems these fluctuations are sufficiently damped out and delayed that they seldom cause any unstable behaviour via the sequence control. (This contrasts sharply with what would be likely to occur were the sequences detector sited at F as discussed previously under 'boiler working thermostats'.)

Weather-compensated systems

Figure 3.40 gives a general outline of a typical system, which comprises an outside detector (OD), a controller (C), a temperature sensor (T), a three-port mixing valve and a time switch.

The control circuit monitors outside conditions (feedforward control) and adjusts the flow temperature detector set-point in

Figure 3.40
Weather compensated control

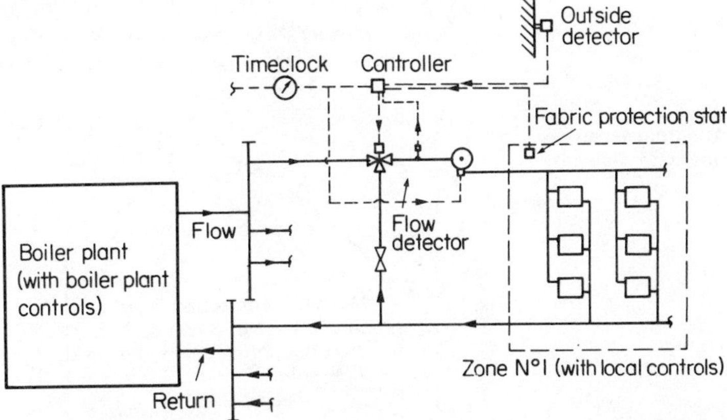

accordance with a pre-determined (proportional) outside/flow water schedule. The flow detector monitors the flow temperature (integral control) and smooths out fluctuations in boiler water and return water temperatures. The former may be quite rapid but the latter are normally smoothed and damped by the heating system/building fabric-interaction.

(a) Outside Detector
In its simplest form the OD is designed to measure external air temperature but some detectors are available which generate a signal which takes account of solar radiation and wind speed, thus providing a 'weather' signal.

(b) Zoning
Probably more important than the type of detector is its position which must be representative of the zone which it serves. When a single weather compensated system is used for the whole building it is common practice to choose a north (i.e. coldest) wall thereby ensuring that the flow temperature is scheduled to meet the worst conditions. The decision to use more than one zone depends largely on the expected solar gain and other heat gains and upon the quality of local control provided. Most United Kingdom buildings having TRVs or thermostatically controlled fan convectors, for example, would require only one zone whereas buildings with only MRVs or flap controlled natural convectors would probably benefit from more than one zone.

(c) Control of heat emission
The need for a weather-compensated system to supplement good local control may not be immediately apparent. It arises from the characteristics of the heat emitters. As discussed in Chapter 8 the heat output of most LPHW emitters can be expressed

$$\dot{Q} \simeq K(\Delta\theta)^n$$

where $n \simeq 1.3$ for radiators
$n = 1.4$ for natural convectors, skirting heating, etc.
$n = 1.0$ for forced convectors, ie. fan convectors and unit heaters.

Consider a LPHW radiator heating system with a design flow temperature (θ_1) of 80 °C and a return temperature (θ_2) of 70 °C. Figure 3.41 shows the variation in heat

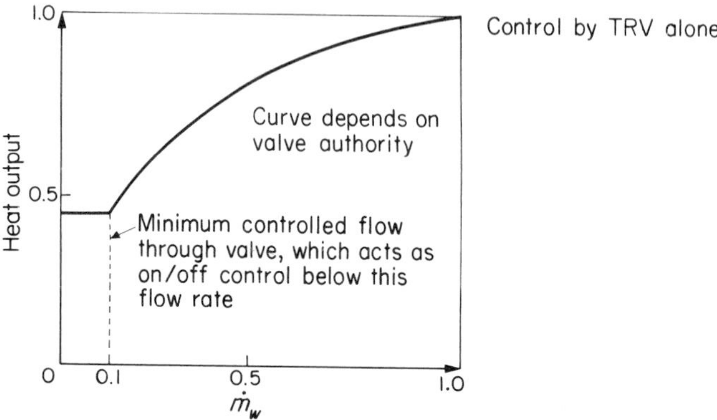

Figure 3.41
Heat output versus emitter mass flow rate

emitted with variation of water flow rate ($\dot{m}_w$) assuming θ_1 = 80 °C is constant for a constant room temperature (θ_r) of 20 °C. This is how a TRV operates in a system without weather-compensated flow temperature adjustment, from

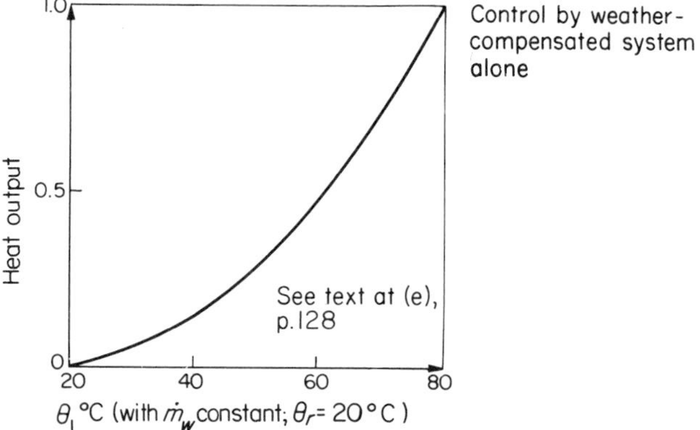

Figure 3.42
Heat output versus flow temperature

which we can deduce that the TRV gives on/off control below about 50% load (i.e. about 85% of a normal heating season). Figure 3.42 shows the variation in heat output with variation in θ_1 where θ_2 is assumed to equal (θ_1 – 10), $\dot{m}_w$ is constant and θ_r = 20 °C is constant. Although scheduling θ_1 does exert a strong influence on heat output, it lacks the essential feedback mechanism of the TRV. By combining both influences as shown in Figure 3.43, weather-compenated (feedforward) flow temperature schedule supplements the water flow rate TRV (feedback) control. Indeed where TRVs are employed there the flow water schedule chosen need not be precise, provided it errs on the high side as the TRVs will adjust the flow rate to suit.

(d) Choosing the flow temperature schedule

It has been common practice to set the water flow tempera-

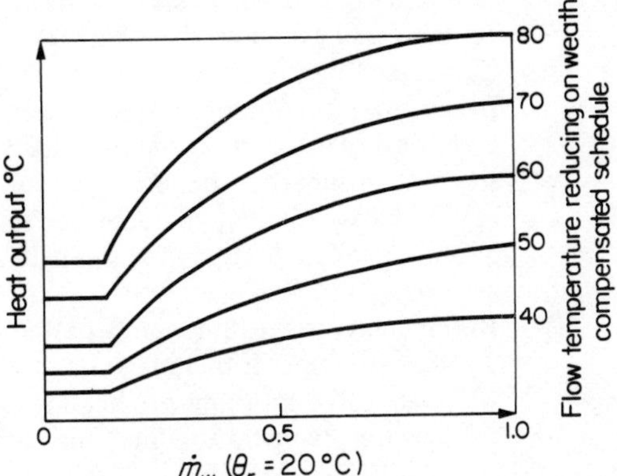

Figure 3.43 Combined TRV and weather compensated control

ture at about 80 °C at an outside design temperature of −1 °C and a minimum water flow temperature of 40 °C when the outside temperature is 18 °C; (above this outside temperature the building being assumed to be self-heating due to solar and casual gains and the lower flow temperature limit of 40 °C being set to avoid thermal stress corrosion in the boilers.) Adjustment of the schedule is normally provided at the controller and this allows the designer to trim the schedule to a given building or zone with known solar and casual gains. Figure 3.44 indicates a typical sche-

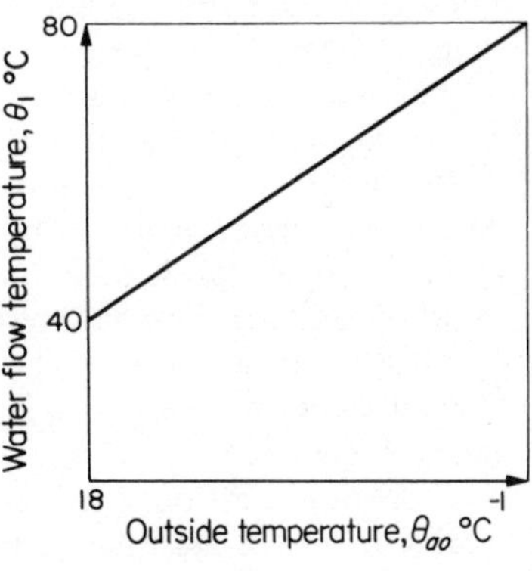

Figure 3.44 Typical weather compensated flow schedule

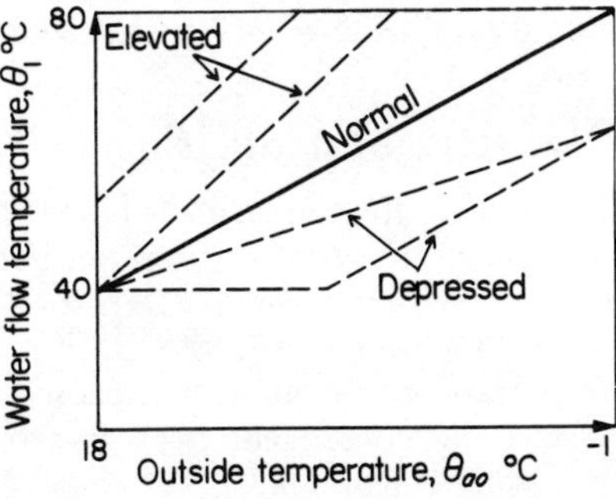

Figure 3.45 Elevated and depressed schedules

dule and Figure 3.45 shows how the schedule can be elevated or depressed as required.

(e) Characterising the mixing valve ports
Because weather-compensated systems are so widely used a great deal of effort has been expended by controls manufacturers to ensure that the mixing valves offered for this application have suitably characterised ports. Space does not allow a detailed discussion of this here and interested readers are referred to (13).

(f) Refinements to weather compensated control systems
Typical refinements include
 (i) an early morning pre-heating period when the flow water temperature may be set at a fixed value of 80 °C.
 (ii) a night set-back period for which a fixed (depressed) value of (say) 50 °C, chosen to maintain an internal temperature of about 10 °C for fabric protection (see also below).
 (iii) day omission, manual override, etc.

(g) Fabric protection
An alternative method of providing fabric protection is to install a fabric protection thermostat (FPT) in a chosen index room which overrides the normal time controls to start the plant in order to maintain about 10 °C. The aim is to prevent condensation (both surface and interstitial) but it also increases the base temperature and reduces preheat times. It also reduces the likelihood of cold spots and frost damage might occur to pipes and may involve one or more of the following measures:
 (i) an index room thermostat but with a lower set point of about 5 °C to start the whole system, and/or
 (ii) an external detector set to bring on the circulating pump at a temperature of around 3 °C, and/or
 (iii) as (ii) but to start the whole system, and/or
 (iv) a return water detector set at 3 °C to start the boiler plant and controls often used in conjunction with (ii).

(i) Optimum start control
This is a development of weather-compensated control systems which seeks to start the plant at the optimum time to minimise the pre-heat period. It is closely related to intermittent heating and is discussed in Chapter 6.

3.9.4 Non-storage calorifiers

An initial classification can be made based on the heat source
 (i) steam (not considered in this book)
 (ii) HPHW/LPHW (source $>$ 100 °C)
 (iii) LPHW/LPHW (source $<$ 95 °C)

In all cases the control problem is caused by low demand side capacity and the consequent high process reaction rates. The problem is worst at light loads.

HPHW/LPHW calorifiers

This arrangement can be used to provide either:

(i) a nominally *constant* LPHW flow temperature, or
(ii) a *variable* (weather-compensated) LPHW flow temperature.

(a) Constant temperature circuits
Figure 3.46 outlines a typical arrangement. The controller (C) produces a variable speed floating action (i.e. approximately true integral control) in response to the detector

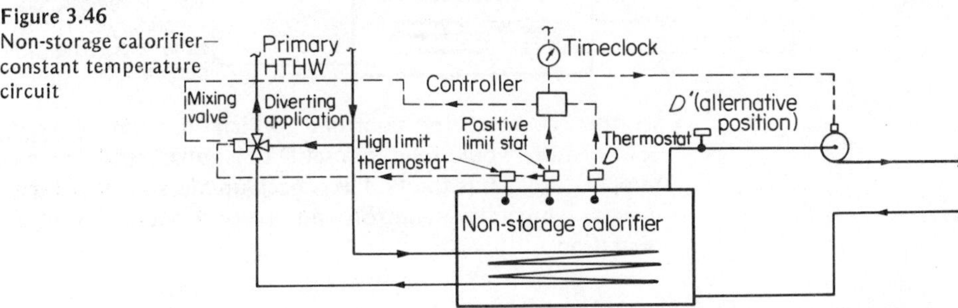

Figure 3.46
Non-storage calorifier—constant temperature circuit

signal (D); (D′ is an alternative detector location sometimes used). Occasionally two-port valves or three-port diverting valves are used but best results are obtained with the three-port mixing valve arrangement shown. The valve motor must be capable of producing a rapid valve stem movement in response to the controller signal. This is essential to counter the rapid changes in process reaction rates which occur and thus prevent overheating (and consequent boiling), especially at start-up and/or under light loads. Because of the fast response of the calorifier it is also essential to instal the high limit thermostat HLT (set at say 90 °C). This operates directly on the valve motor and drives it to full by-pass in the event of overheating (caused, say, by secondary system pump failure or valve closure). The HLT is normally a manually reset device (by removal of the thermostat cover). The use of a positive temperature limiter (PTL) for a constant temperature circuit is optional (see variable temperature circuits).

(b) Variable temperature circuits
Figure 3.47 shows a typical arrangement used to provide a weather-compensated LPHW circuit from a HPHW source. The operation of the weather compensator is similar to that discussed in Section 3.9.3. As before a proportional-reset-integral control mode is used but with a smaller integral action time to give a much higher valve stem velocity to ensure fast response and prevent boiling. As before the best performance is achieved using the three-way mixing valve/pump arrangement and again a manually reset HLT is essential. A PTL must also be used. This device is normally set at about 85 °C and provides override features to the normal

Figure 3.47
Non-storage calorifier—variable temperature circuit

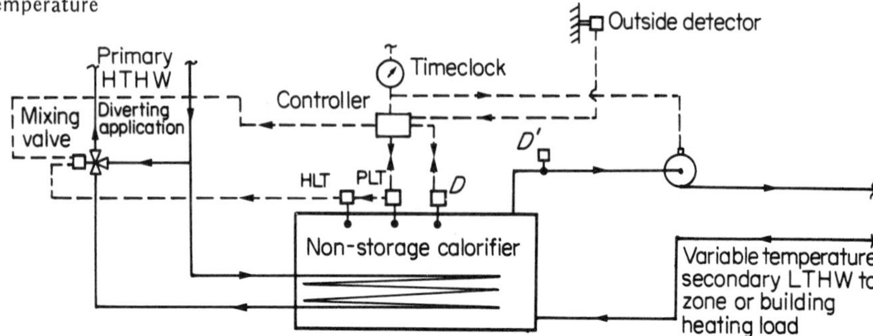

weather compensator controller which in event of very cold weather would tend to reset D to an unacceptably high temperature. Thus the PLT is a back-up measure to aid the normal temperature controls and prevent unnecessary HLT operation.

LPHW/LPHW Calorifiers

Non-calorifiers of this type are generally required only for process requirements (e.g. pool water heating, heat reclaim, etc.). The control specification on water flow temperature and the load requirements will determine the actual control requirements but in all cases there is no danger of boiling and generally speaking because of the lower heat flux the control problem is easier.

3.9.5 Storage calorifiers

Most storage calorifiers are used to provide DHWS and the discussion here focusses on this. DHWS calorifiers can be classified according to their heat source

(i) steam (not discussed in this book),
(ii) HPHW ($>100\,^{\circ}C$),
(iii) LPHW ($<95\,^{\circ}C$),

Figure 3.48 shows a typical arrangement. The design basis of

Figure 3.48
Direct hot water storage calorifier

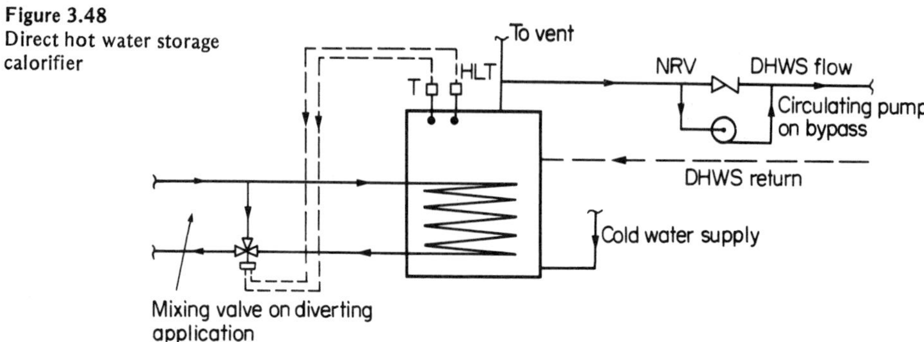

storing DHWS at around 65 °C is discussed in Section 3.9.6. The controls operate in the same way for both HPHW and LPHW sources with the HLT being set at 70 °C to drive the diverting valve to full by-pass.

Compared with a non-storage calorifier, control is much easier due to the larger demand side capacity. The working thermostat (T) set at 65 °C normally provides a rudimentary form of on/off control with a differential gap of 2 or 3 °C. At the upper limit T opens and a spring return mechanism drives the three-way valve to full by-pass and vice versa. The slow response of the calorifier means that the HLT will only require to override T in the event of a thermostat failure.

In Figure 3.48 the three-port valve is shown in a by-pass arrangement in the return line. However, because only on/off control is required and because questions of valve authority and characterisation do not arise, the valve could also be placed in the flow line to divert the flow (but check for suitability with the valve manufacturer). Further, a two-port valve can also be used but if it must shut against a high pump head then leakage may give rise to overheating (or even boiling with HPHW). This sort of leakage is most unlikely with three-port valves due to the much smaller operating head differences.

3.9.6 Control of DHWS temperature

If domestic hot water is supplied at less than 35 °C it is normally considered to be 'cold'. At 40 °C it is lukewarm, at 45 °C it is 'hot' and just bearable at 50 °C. Temperatures above 50 °C cannot normally be tolerated even for short periods unless rubber gloves are worn, e.g. dishwashing.

Chapter 4 describes the physiological basis of the above range but a few simple facts are noted here for guidance;

(i) Normally, skin surface temperatures are around 30 °C to 35 °C which sets the lower acceptable limit for 'hot' water.

(ii) The body deep core temperature is close to 37 °C and as the body's ability to store heat is limited, temperatures above this cannot be tolerated over a long period.

(iii) 'Pain' is sensed at skin temperatures above about 50 °C although generally great discomfort (due to overheating as in (ii)), is normally experienced above about 45 °C in baths and showers where a large area of contact is involved.

Upon this basis we can set down a set of nominal ('process') hot water temperatures for different requirements:

Washbasins	40 °C to 50 °C
Baths and showers	40 °C to 45 °C
Handwashing of clothes and dishes	40 °C to 50 °C (up to 55 °C using rubber gloves)
Machine washing of clothes and dishes	up to 60 °C

As stated in Section 3.9.5, it is conventional practice to store domestic hot water at around 65 °C, (reasons are given in Chapter 10). Thus for most uses (wash basins, sinks and most baths), the stored water temperature must be reduced, usually by simple mixing with cold water (at around 10 °C). A more positive means of control may be required for:

(i) most showers,
(ii) washbasins and baths in schools, hospitals, etc., which are to be used by young children, geriatrics and hospital patients.

Non-thermostatic mixing valves

Non-thermostatic control is achieved by simple manual adjustment of the cross-sectional area of the hot and cold water inlet ports leading to a common outlet port and affords a cheap, maintenance-free from of control. Unfortunately, this generally provides a poor form of control because the flow rate through the ports is affected by pressure fluctuations in the hot and cold supply lines and this in turn generates temperature fluctuations in the mixed outflow. In a large system such pressure fluctuations will be random and frequent and may preclude the use of non-thermostatic control except for some single outlet applications; it is generally unsuitable for supplying a number of outlets. It finds its main application in domestic installations where the pressure fluctuations are usually less of a problem and its low first cost proves attractive.

Thermostatic mixing valves

As before the technique is to vary the inlet port areas to obtain the desired outflow temperature but in this type of valve variations are minimised by means of a detector/actuator element placed in the mixed outflow stream which varies the relative areas automatically to cater for inlet pressure and temperature variations. The detector/actuator element is normally wax-filled and is self-acting. It also provides quick shut off in the event of cold water failure. The control is proportional with a fairly narrow control band.

For single output valves set point adjustment is normally provided which can be preset to give any desired maximum temperature (schools, hospitals, etc.). For multi-outlet valves normally one fixed set-point is provided and adjustment is by special key. Thermostatic valves are quite expensive and often require frequent maintenance especially in hard water areas. To reduce capital and maintenance costs it is common practice to use one larger valve to serve a number of common outlets, (e.g. a group of basins or showers). A common problem with this arrangement is that at low flow rates (e.g. only one or two outlets in operation), the reduction in valve authority causes high outflow temperatures. This is particularly noticeable with groups of showers and in practice it is often best to use say, three groups of five showers with three mixing valves rather than one large valve serving all fifteen.

3.9.7 Domestic heating controls

It is not possible in the limited space available to do more than describe in brief detail the layout and operation of one of the very many heating system/control arrangements which are to be found in practice.

Basic Temperature Controls

A typical arrangement is shown in Figure 3.49 employing a LPHW gas- or oil-fired boiler serving radiators and an indirect

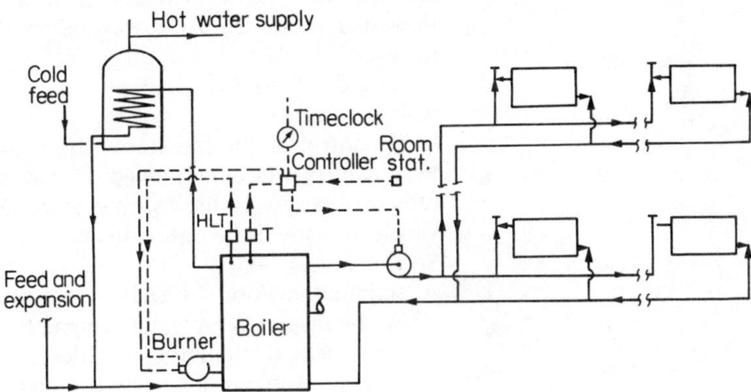

Figure 3.49 Domestic hot water and heating supply installation

DHWS storage cylinder. Though basically a domestic arrangement, a similar layout is often employed for small commercial buildings.

Control is achieved as follows:

(a) BWT is an on/off boiler working thermostat which cycles the burner on and off controlling heat input to the boiler. Normally BWT has an adjustable set-point, (but with a fixed differential gap), which allows the occupier to 'schedule' the flow temperature to suit outside conditions thereby limiting radiator emission and achieving a measure of central control.

(b) BLT is a 'boiler limit thermostat', a safety device normally factory preset (at say 95 °C) to lock out the burner in the event of BWT or system failure, and having a manual reset action.

(c) RT is a 'room thermostat' which, (when employed), normally exercises start/stop control over the heating system circulating pump. When the pump is stopped the rate of natural circulation around the system is much reduced and the building/system cools until RT starts the pump to renew the water circulation. This provides a fairly coarse form of on/off 'feedback' control which assists the 'feed-forward' BWT setting fixed by the occupier.

Choosing a suitable location (i.e. index room), for RT is also something of a compromise. If it is located in the main living room, (which generally experiences the largest 'casual' gains), this can lead to underheating of the hall and bedrooms and may give rise to condensation problems. For this reason it is often located in the entrance hall and set to a 'background' temperature level of around 18 °C. As the set point of RT is adjustable the occupier will discover by experience the most suitable temperature level for his particular building/heating system/usage pattern.

An alternative method of employing the RT is to use it to give direct control over the burner. This can give rise to underheating of the DHWS cylinder during light loads. It can also give rise to problems of wide variations of water flow and room temperatures due to the large time lags incurred.

(d) BWT and RT (if fitted) provide a course form of control which can then be 'tuned' locally by the occupier using local control at the radiators. The quality of and ease of local control will be improved if good quality thermostatic radiators valves are used rather than simple manual valves which provide little more than on/off control (see also Section 3.9.3).

(e) A common method of DHWS temperature control is to allow the mean stored water temperature to be determined by the thermal equilibrium achieved by the gravity circulating pressure generated by the boiler/cylinder temperature difference supplying boiler water to offset the cylinder demand pattern and cylinder and pipework heat losses. With this arrangement a perfectly insulated system could give rise to excessively high temperatures during prolonged spells of light or zero draw-off but generally this does not give rise to a serious problem in practice because the combination of cylinder and pipework losses, intermittency of DHWS demand, heating system demand on the boiler and the scheduling of the BWT setting during light heating load conditions are generally sufficient to ensure satisfactory operation and limit the stored water temperature to 65 °C in practice.

(A more positive form of control is sometimes achieved using either a cylinder thermostat/solenoid valve arrangement or a self-actuating sensor/valve arrangement.)

Time clock (programmer) controls

As can be deduced from the above, the basic temperature controls are quite simple, generally employing only a series on simple on/off loops arranged in cascade with a fair amount of occupier adjustment required for fine tuning. The application of time switch control, however, has already grown quite complex (and this trend will probably increase with the application of microprocessors). A typical simple 'programmer' allows the user to pre-select during the normal 24 hour period:

(a) one or two 'hot water only' periods (i.e. boiler 'on' under BWT/BLT control with pump 'off'), and/or
(b) one or two 'hot water and heating periods' (i.e. boiler 'on' under BWT/BLT control with pump 'on' under RT control (if fitted).

Normally a manual override feature is provided to bring hot water and heating (under BWT/BLT and RT control) when required at times out with the normal pre-select on periods.

This type of programmer is fairly common and allows adequate flexibility for most domestic installations. More sophisticated arrangements (similar to those for larger installations) provide the above facilities but on a weekly basis and some 'domestic' programmers also incorporate day omission, weather compensation, zone control, etc.

3.9.8 Future trends for control systems

The application of microprocessors and sophisticated and accurate electronic temperature sensors will almost certainly have a great impact on all control systems.

For larger systems greater zoning combined with 'active' rather than 'passive' supervisory control, (trimming heat output and boiler capacity more finely to building usage patterns and weather conditions), seem quite likely.

For smaller systems (domestic and small commercial), greater application of weather-compensated control and perhaps timed boiler off peiods seems likely, in an attempt to reduce boiler standing losses and hence improve overall boiler efficiencies. This is desirable and may soon be possible at reasonable cost.

Perhaps also the increased use of more accurate control valves, (e.g. the electromagnetic valves now becoming available), may also give improved control performance.

LIST OF SYMBOLS USED IN CHAPTER 3

A	cross-sectional area
A_{ABS}	absolute control area
A_{LIN}	linear control area
A_{QUA}	quadratic control area
c	specific heat
C_{c}	controller constant
C_{I}	valve plug/seat constant
C_{m}	motor constant
C_{o}	valve flow coefficient
D	differential gap (theoretical)

D', D''	effective differential gaps; detector location
d	diameter
h	detector heat transfer coefficient; liquid level
h_1, h_2, h_o	liquid levels
K	heat emitter constant
K_c	controller gain
K_I	integral action factor
K_m	motor gain
K_u	ultimate controller gain
L	detector time constant
N	valve authority
n	heat output index
P	controller output
P_u	ultimate periodic time
$\dot{Q}$	heat output
$\dot{Q}_I$	inflow rate
$\dot{Q}_o$	outflow rate
R	process reaction rate (p.r.r.)
R'	unit process reaction rate
t	time
t_D	simple dead time
T	periodic time (theoretical)
T', T''	effective periodic time
T_I	integral action time
T_D	derivative action time
x	valve stem position
Δx	change in valve stem position
θ	temperature
ρ	density
τ_{DE}	equivalent dead time
τ_N	notional time constant

other subscripts

d	detector
o	original
f	fluid
t	at time t
∞	as $t \to \infty$

REFERENCES

1. HIMMELBLAU, D. M. and BISCHOFF, K. B., *Process Analysis and Simulation,* (Wiley, 1968).
2. CROWE, HAMIELEC, HOFFMAN, JOHNSON, SHANNON AND WOODS, *Chemical Plant Simulation,* (Prentice-Hall, 1971).
3. LUYBEN, W. L., *Process Modelling, Simulation and Control for Chemical Engineers,* (McGraw-Hill, 1973).
4. B.S. 1523: Part 1: 1967 'Glossary of Terms used in Automatic Controlling and Regulating Systems: Part 1: Process and Kinetic Control'.
5. C.I.B.S. GUIDE, 'Section B11 Automatic Control'.
6. JONES, W. P., *Air Conditioning Engineering,* 3rd edn., (Arnold, 1978).

7. FABER & KELL (Revised by P. L. Martin), *Heating and Air Conditioning of Buildings,* 6th edn., (Architectural Press, 1979).
8. WOLSEY, W. H., *Basic Principles of Automatic Controls with special reference to heating and air conditioning systems,* (Hutchinson Educational, 1975).
9. STANLEY, E. E., SHORTER, D. N., and BADGER, P. A. (1964) 'A combined theoretical and analogue computer study of on-off thermostats', *H.V.R.A. Lab. Report No. 20.*
10. FITZGERALD, D. (1968), 'Room thermostats--choice and performance', *H.V.R.A. Lab. Report No. 42.*
11. ADAMS, S. and HOLMES, M. (1977), 'Determining time constants for heating and cooling coils' *B.S.R.I.A. Technical Note TB6/77.*
12. HAINES, R. W., *Control systems for heating, ventilating and air conditioning,* 2nd edn., (Van Nostrand Reinhold Environmental Engineering Series, 1977).
13. WOLSEY, W. H., 'A theory of 3-way valves with some practical conclusions', *J. Ind. Heating and Ventilating Engrs.* Vol. 39, May 1971.
14. HADLEY, W. A. and LONGOBARDO, G., *Automatic Process Control,* (Pitman, 1963).
15. JUNKER, B., 'Die Regeltechnisen Grundlagen der Anwendung selbstatinger regler in der Heizung- und Klimatechnik', *Heizung-Luftung-Haustechnik,* 7, 1956, No. 11, pp 177-86 (An English version is avaialble as Sauter Bulletin No. 12-Re.)
16. HECK, E., 'Regelkreise in der Klimatechnik und Ihre Stabilisierung', 1810, 15, April, 1964, pp 125-29.
17. CERBE, G., 'Optimising continuously operating controllers in heating and air conditioning', *B.S.E.* Vol. 41, August 1973.
18. ZEIGLER, J. G. and NICHOLS, N. B., 'Optimum settings for automatic controllers', *Trans. A.S.M.E.,* 1942.

4 Thermal Comfort

4.1 INTRODUCTION

The primary function of most building heating systems is the provision of a comfortable environment for the occupants of the building. The study of the thermal environment and its effect upon people and their feelings of subjective comfort is therefore of central importance in heating services design.

There have been many definitions of comfort given by many authorities. Perhaps a definition which is most meaningful for the heating services designer is that 'thermal comfort exists when the subject is thermally unaware of his surroundings'.

This Chapter opens with an examination of the physical and physiological processes which allow the human body to maintain a thermal equilibrium with its surroundings and subsequently the identification and measurement of those environmental parameters which influence the comfort of the individual. The most important thermal indices are then reviewed, after which follows a detailed discussion of the methods, illustrated by worked examples, of employing these indices in the assessment of comfort at the design stage.

4.2 BODY PHYSIOLOGY AND TRANSFER PROCESSES

4.2.1 Body internal heat production

In examining the heat interchanges between an individual and his surroundings, the human body may be considered analogous to a heat engine. Food is converted into energy through the process of oxidation by air breathed into the lungs. From an engineering point of view the thermal efficiency of the body is low, rarely rising above 20% and being nominally zero in many situations, and as such it must continuously reject to the surrounding environment as heat the greater portion of its energy intake. This process of heat rejection occurs in the main from the body surface.

The rate at which heat is produced by the body is of course dependent on the degree of muscular activity being undertaken and may well vary from about 75 W for an adult sleeping to some 700 W for sustained heavy work. Table 4.1 is given as an indication of the rates of heat production over the full range of physical

Table 4.1 Internal Heat Production for Various Levels of Activity (adult male)

Level of Activity	*Internal Heat Production* (W)
Sleeping	75
Seated at rest	115
Light seated work	140–160
Light bench work	160–190
Moderate work	190–400
Heavy work, intermittent	440–590
sustained	590–700

activity levels. More detailed information is readily available from a number of references (Fanger[1], A.S.H.R.A.E.[2]).

The human thermo-regulatory system is capable of maintaining the correct thermal balance with the surrounding environment (i.e. the internal heat production equal to the rate of heat dissipation with no heat storage within the body) over a wide range of environmental conditions, thus maintaining an essentially constant internal body temperature consistent with the well-being of the individual. This internal temperature is nominally 37.2 °C for deep tissue, although some departure from it does occur according to age and certain other factors. However, the range of environmental conditions within which an individual may feel *comfortable* is very much narrower than the range over which *thermal equilibrium* is possible and comfort may be thought of as being governed by the *ease* with which the thermal balance is maintained. As the necessary balance is dependent on the body losing heat to the environment, attention must be given to the ways in which this is achieved and to the physical parameters of the environment which influence the transfer processes.

4.2.2 Energy transfer processes

There are three principal modes of energy exchange between the human body and its surroundings (Figure 4.1) which are

(i) convection; (ii) radiation; (iii) evaporation.

Other small losses also occur, such as by conduction or by the rejection of excreta from the body, but these are of little significance and are normally ignored.

Mathematically, the thermal balance may be expressed as

$$\dot{H} = \pm \dot{q}_c \pm \dot{q}_R + \dot{q}_E \tag{4.1}$$

where $\dot{H}$ is the internal heat production,
$\dot{q}_c$ is the heat transfer by convection to the surrounding air at θ_a,
$\dot{q}_R$ is the heat transfer by radiation to the surrounding surfaces at θ_r and,
$\dot{q}_E$ is the heat loss by evaporation.

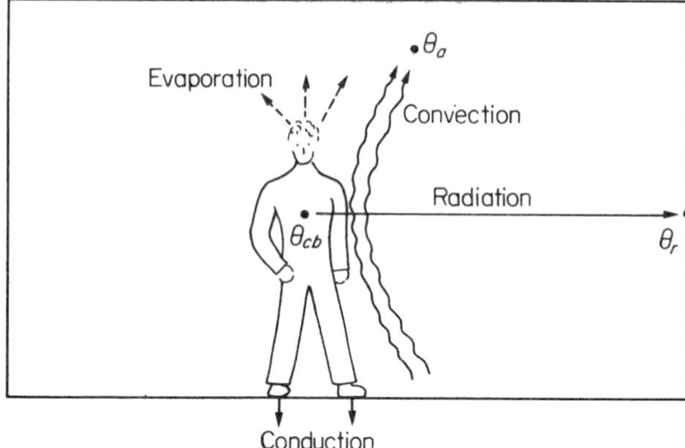

Figure 4.1
Energy exchanges between occupant and surroundings

(a) Convection and Radiation

The body loses heat by convection if the mean temperature of the outer surface of the clothed body, θ_{cb}, exceeds the air temperature, θ_a, and by radiation if this mean surface temperature exceeds the mean temperature of the surrounding enclosure, θ_r. These convection and radiation exchange processes are expressible in simple form as

$$\dot{q}_c = h_c A_{cb} (\theta_{cb} - \theta_a) \tag{4.2}$$

$$\dot{q}_R = h_R A_{cb} (\theta_{cb} - \theta_r) \tag{4.3}$$

where h_c and h_R are heat transfer coefficients by convection and radiation respectively, and A_{cb} is the clothed body external surface area. For a given body surface temperature the convection exchange is dependent upon the air temperature and the coefficient h_c. For low air velocities (natural convection) h_c is a function of the temperature difference $(\theta_{cb} - \theta_a)$, while for higher velocities (forced convection) h_c is a function of the velocity itself. The radiation loss is dependent primarily on the mean radiant temperature, θ_r. The coefficient h_R as given in equation 4.3 will incorporate a view factor between the body and the surrounding enclosure, the temperatures θ_{cb} and θ_r, and an emissivity factor for the outer surface of the clothed body (2.4.3).

With a given set of environmental conditions, equations 4.2 and 4.3 show that the sensible heat loss from the body may be increased or decreased by changes in the mean body surface temperature. This the thermo-regulatory system of the body can achieve by virtue of a shunt mechanism which varies the supply of blood to the skin (Figure 4.2).

When the body is subjected to a high ambient temperature a sympathetic stimulation of the autonomic nervous system produces a constriction of the smooth muscle of the main blood vessels. The blood supply to the skin is increased, which in turn increases the skin surface temperature, thus

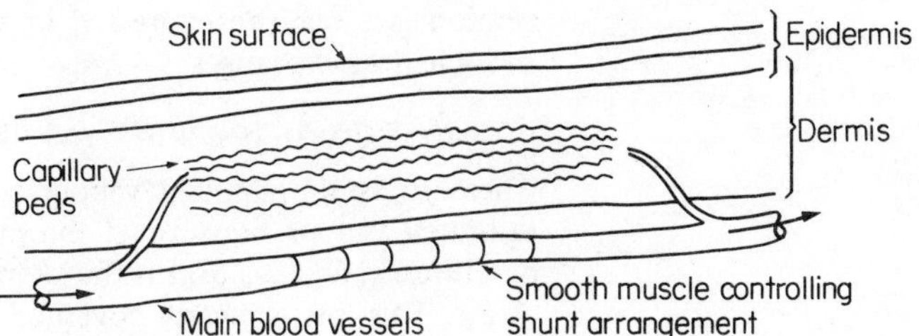

Figure 4.2 Skin temperature control mechanism

attempting to compensate for the high ambient temperature. Conversely, when subjected to a low ambient temperature the supply of blood to the skin is reduced by a dilation of the main blood vessels. This lowers the skin surface temperature which leads to a subsequent reduction in the rate of heat loss.

(b) Evaporation

The evaporation loss from the body can be considered as composed of three component parts

(i) exhalation,
(ii) insensible perspiration,
(iii) sweating.

Air inhaled into the lungs eventually becomes saturated with water vapour at the deep tissue body temperature. Although some cooling of the outward moving air from the lungs does occur, the exhaled air is generally at a higher moisture content and temperature than the ambient air. This component of the evaporation loss therefore contains a degree of sensible loss in addition to the latent respiration loss, although this sensible part is relatively small. The latent loss depends principally upon the pulmonary ventilation rate and therefore on the activity level of the individual and, to a lesser extent, upon the moisture content and temperature of the surrounding air. The functions of components (ii) and (iii) of the evaporation loss, although seemingly similar, are in fact quite distinct from each other. The principal difference is that the insensible perspiration process is not under thermo-regulatory control. In this process body fluid continuously diffuses through the skin to form microscopic droplets on the surface. The subsequent evaporation of these droplets is so rapid as to prevent any build up of moisture on the body surface.

Sweating, on the other hand, is a thermo-regulatory function of the body which comes into operation under a sympathetic stimulation when the normal exchange processes are unable to provide the necessary heat loss. In order therefore to increase the heat loss by evaporation, the sweat glands flood certain regions of the skin surface with fluid.

For an unclothed subject (of surface area A_b) the evaporation loss may be expressed in a manner similar to the sensible heat exchanges

$$\dot{q}_E = h_D A_b [p_{ss}(\theta_b) - p_s] h_{fg}(\theta_b) \tag{4.4}$$

where p_s is the ambient vapour pressure, $p_{ss}(\theta_b)$ is the saturated vapour pressure at the mean body surface temperature θ_b, and $h_{fg}(\theta_b)$ is the latent energy of evaporation at θ_b. This loss is clearly governed by the vapour pressure of the surrounding air and the mass transfer coefficient h_D, which like the convection coefficient h_c, is highly sensitive to air velocity.

Another thermo-regulatory function of the body which should also be mentioned is that of shivering. This process involves an involuntary twitching of the muscles and is an attempt by the body to increase its internal heat production rate in situations where there is a tendency for the body temperature to fall.

4.2.3 Factors affecting comfort

By the processes and mechanisms just described, the human body is capable of maintaining a thermal balance with the surrounding environment. The relation between the sensible and latent components of the energy loss varies with air temperature and is shown in Figure 4.3 for different internal heat production rates.

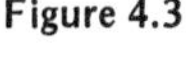
Figure 4.3

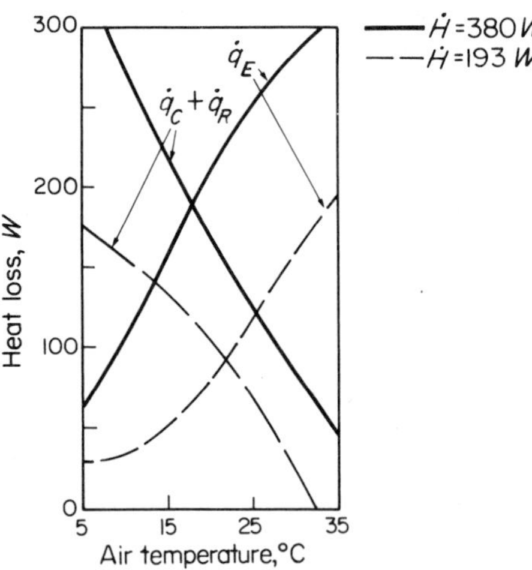

Although, as has been previously stated, the thermo-regulatory system of the body is capable of maintaining a heat balance over a wide range of environmental conditions, there is only a comparatively narrow band within which feelings of comfort may be attained. Clearly however, the parameters which control the heat transfer processes from the body to the surroundings will also be the parameters which influence the attainment of thermal comfort.

The heat transfer processes are themselves described mathematically in equations 4.1 to 4.4 and inspection of these equations shows that four environmental variables influence the body heat balance, viz:

air temperature,
air velocity,
mean radiant temperature,
air humidity.

It is worth noting at this time that generally the heating services designer has no direct control over two of these variables, namely air humidity and air velocity.

However, it is known that, over a wide range of values, air humidity has little influence on thermal comfort. This range, expressed in terms of relative humidity, is effectively 35% to 70%. It should also be noted that values of relative humidity outwith this range can lead to more far-reaching effects than purely an increased influence on thermal comfort. Low values of relative humidity are associated frequently with such unpleasant symptoms as sore throats and a build up of static electricity, while high humidities are commonly associated with condensation problems.

The energy exchange equations also include certain personal variables. By recognising that the rate of internal heat production, $\dot{H}$, is a function of activity and that the external mean body temperature, θ_{cb}, is a function of the thermal resistance of the clothing worn, the personal variables are identified as:

(i) activity level,
(ii) clothing.

These are difficult to assess accurately at the design stage and the designer must use his judgement to select 'appropriate design values'.

4.3 MEASUREMENT OF THE ENVIRONMENTAL VARIABLES

4.3.1 Introduction

This section deals briefly with the methods available for measurement of the environmental variables discussed previously. These variables may require to be measured during the commissioning process. It is also important to note that in the final analysis, thermal comfort can be assessed accurately only by direct measurement and in some cases such measurements will be required to verify the assessments (discussed in Section 4.5) made at the design stage.

4.3.2 Air temperature

A wide variety of thermometric devices are available but the mercury-in-glass thermometer is probably still the most convenient instrument. In cases where there is likely to be a significant difference between the air and mean radiant temperatures, a radiation shield (which may simply be a piece of aluminium foil on a frame) can be placed around the bulb of the instrument.

4.3.3 Air speed

A visual appreciation of air movement within the space may be obtained by means of a smoke trail. In order to quantify thermal comfort at a point in the space, the non-directional magnitude, or air speed, requires to be established over a period of time.

This average air speed may be measured by means of the Kata thermometer used in conjunction with air temperature measurement. This thermometer has a large bulb filled with coloured spirit. In making an observation the instrument is heated by immersion in hot water and then dried. It is then suspended in the space and the time taken for the meniscus of the spirit to fall between the two marks on the stem is noted. The rate of cooling is affected by radiation and convection heat transfer exchanges with the surroundings, and for this reason it was used formerly as a comfort index. The convection effect is sensitive to changes in air speed past the bulb and for this reason it makes a good anemometer. The normal Kata thermometer has a cooling range from 38 °C to 35 °C, three degrees around body temperature. Having noted the cooling time and the air temperature, the usual technique is to use a nomogram to obtain an indication of the average air speed. In situations where the mean radiant temperature is substantially different from the air temperature, a Kata thermometer with a silvered bulb should be used. A comprehensive description of the Kata instrument is available[3].

Another instrument coming into more common use is the hot-wire thermo-anemometer. It is energised by a dry cell and has a hot-wire probe which normally consists of two fine nickel resistance wires, one heated, the other at ambient temperature. These resistors form arms of a Wheatstone bridge circuit. Air passing across the wires cools the hot wire thereby lowering its resistance. A moving coil galvanometer connected across the bridge gives a visual display of the air speed in m/s. Air temperature readings may also be recorded using this instrument by simply disconnecting the heated wire from the bridge circuit and using the unheated wire as a temperature-sensitive resistance element in the bridge circuit. The hot-wire instrument is compact and gives instantaneous readings of air speed and temperature at the throw of a switch. The accuracy is good, and it is more suitable for field use than the Kata thermometer with its need for hot water, etc.

4.3.4 Vapour pressure

The vapour pressure tends to be reasonably uniform throughout the space and for this reason it is necessary to take readings at only one location. The normal method of measurement is to use an aspirated instrument such as the sling or Assman psychrometers. Each of these instruments contains a dry and wet bulb mercury-in-glass thermometer, the wet bulb being covered with a moist muslin wick. In order to keep radiation influences as small as possible, air speeds of 4.5 m/s or higher are necessary, across the wet bulb. This is accomplished manually with the sling instrument by whirling it around at the point under consideration while it is achieved with the Assmann by movement of a sample of air through the instrument by means of a small fan within the instrument. The sling psychrometer is the likely instrument to be used in field tests and it should be whirled briskly for about thirty seconds before noting the temperatures. The wet bulb reading should always be read first, as soon as rotation is stopped. The instrument is normally supplied with a humidity slide rule from which the relative humidity is obtained from the readings of dry and wet bulbs. Alternatively, the humidity may be obtained from this information by using psychrometric tables or charts.

Thermo-hygrometers are now available, energised by dry cells, and incorporating Wheatstone bridge circuits. Like the thermo-anemometer previously described, these instruments have a temperature-sensitive resistance wire. The humidity sensor is a fine wire enveloped with glass wool impregnated with lithium chloride solution. Lithium chloride is hygroscopic and on absorption of water, changes in electrical resistance occur. Direct readings of air temperature and humidity are obtained with such instruments but unfortunately the instruments require very careful maintenance, otherwise inaccuracies arise of an order seldom found with the aspirated instruments.

4.3.5 Mean radiant temperature

There are two basic methods of measuring mean radiant temperature. These are:

(a) use of integrating instruments giving a direct assessment of the mean radiant temperature,

(b) measurement of individual surface temperatures and solid angles, prior to a mathematical integration.

The former represents the method for site investigation and the latter is useful when surface temperatures have been assessed as part of the design process.

The traditional integrating instrument for direct measurement of mean radiant temperature is the Globe thermometer which consists of a 150 mm diameter hollow blackened sphere, normally of copper. The temperature sensor is usually a mercury-in-glass thermometer with its bulb placed at the centre of the sphere. The equation which relates the Globe temperature (θ_g) to the mean radiant temperature (θ_r), the air temperature (θ_a) and the air speed (v_a) is

$$\theta_g = \frac{\theta_r + 2.35\theta_a\sqrt{v_a}}{1 + 2.35\sqrt{v_a}} \tag{4.5}$$

Thus in order to assess the mean radiant temperature at a point in space, simultaneous readings of Globe and air temperatures and air speed are necessary. The mean radiant temperature is then calculated from the preceding equation. The likely alternative to this instrument, for field work, is the similar blackened globe of 100 mm diameter. This instrument gives a direct reading of the dry resultant temperature (θ_{res}), the appropriate equation being

$$\theta_{res} = \frac{\theta_r + \theta_a\sqrt{10v_a}}{1 + \sqrt{10v_a}} \quad (4.6)$$

A point to note is that the numerical value of the mean radiant temperature obtained from these equations will differ in each case. This is so, since the values are obtained as a result of an equilibrium condition being attained, due to radiation exchanges between the sphere and the room surfaces and this will vary slightly with the size of the globe.

Where mathematical integration techniques are to be used it is most likely that a surface pyrometer will be used to obtain the temperature of all room surfaces. It is then possible to define the mean radiant temperature at the point under consideration in terms of the individual surface temperatures (θ_s) and their associated solid angle factors with relation to the point, hence

$$\theta_r = \Sigma(F_a\,\theta_s)\ (^\circ\mathrm{C}) \quad (4.7)$$

Apart from the fact that the field collection of surface temperatures is a tedious business there is the further problem of evaluating solid angles. This will be dealt with later in this Chapter. As an alternative to this expression and accepting a little loss of accuracy, the following definition of mean radiant temperature may be used at the centre of reasonably cubical rooms:

$$\theta_{ri} = \frac{\Sigma(\theta_s A_s)}{\Sigma A_s}\ (^\circ\mathrm{C}) \quad (4.8)$$

where A_s represents the area of room surfaces.

From the foregoing it is evident that there are numerous ways of defining the mean radiant temperature. So far the definition has been restricted in relation to a point in space. A more meaningful definition would relate to body posture and type of clothing. Fanger, in his book[1] presents 'angle factor diagrams' from which angle factors may be evaluated between the person and the various room surfaces (F_b would represent the angle factor between the person and surface). He proposes that the mean radiant temperature be evaluated from the expression

$$T_r^4 = \Sigma(T_s^4\,F_b)\ (\mathrm{K}) \quad (4.9)$$

Fanger suggests that if only small differences in temperature exist between all room surfaces then only a small loss of accuracy arises out of the use of this equation in linearised form, thus

$$\theta_r = \Sigma(\theta_s\,F_b)\ (^\circ\mathrm{C}) \quad (4.10)$$

4.3.6 The Field kit

The foregoing discussion has indicated a variety of instruments available for measuring the environmental variables. Some of these are more suited to measurements in the laboratory rather than in the field. A suitable field kit might consist of:

shielded mercury-in-glass thermometer (air dry bulb temperature)
hot-wire anemometer (air speed/dry bulb temperature)
100 mm dia./150 mm dia. globe (mean radiant temperature)
sling psychrometer (humidity)

Furthermore, it is useful to have a box which will contain the group of instruments. Figure 4.4 shows a range of instruments.

Figure 4.4

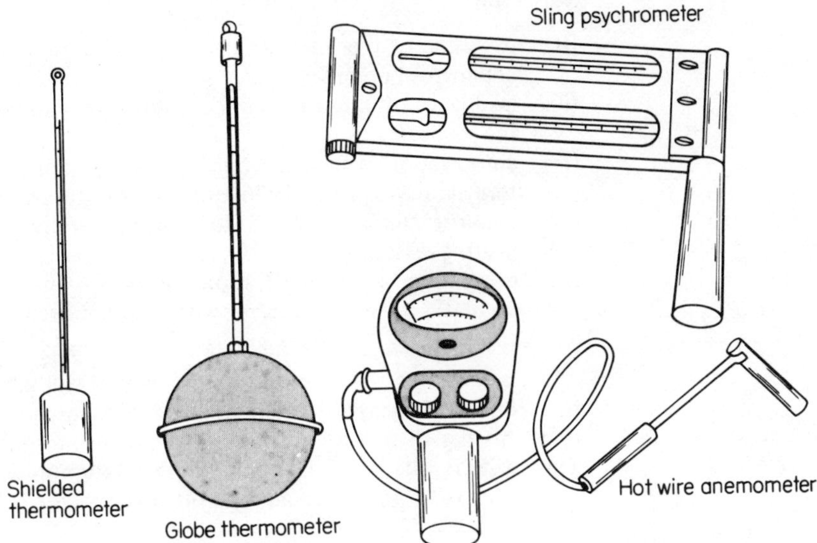

4.4 THERMAL INDICES

Thermal indices are used to express thermal comfort in terms of a simple number which is often an index temperature. Examples are air temperature (a poor thermal index when used alone), globe temperature, equivalent temperature and resultant temperature. Most indices of this kind are convenient to use when designing heating systems and for assessing their subsequent performance. Such indices are often a weighted combination of two or more of the six major factors influencing thermal comfort, viz. metabolic rate, thermal resistance of clothing (clo-value), air dry bulb temperature (θ_a), mean radiant temperature, (θ_r), air velocity (v_a), and humidity. Some indices have been correlated with subjective assessments of thermal comfort in laboratory experiments while others have not. For convenience, thermal indices may be considered as being:

(i) Statutory Indices,
(ii) Direct Indices,
(iii) Fanger's Comfort Criteria.

4.4.1 Statutory Indices

Various forms of legislation (i.e. 1976 Energy Act; Building Regulations; Factories Act; Health and Safety at Work Act, etc.) provide a number of statutory thermal indices of which the designer must take cognisance. Such indices (e.g. 'air' temperature, 'ambient' temperature, and the like) are generally ill-defined and of little direct value to the designer who must decide which of the following main requirements are intended by the relevant legislation:

(i) provision of thermal comfort particularly for domestic and commercial premises;
(ii) avoidance of undue heat or cold stress; (this topic is outside the scope of this book);
(iii) conservation of fuel, generally by setting statutory temperature limits (e.g. the 1976 Energy Act sets an upper 'temperature' limit of 20°C for office premises during the heating season);
(iv) prevention of condensation problems, particularly in domestic premises and/or where intermittent heating is employed.

As a first step the heating system designer will generally aim to meet requirements (i) (thermal comfort), subsequently modifying his design to take account of requirements (iii) and (iv). This often results in a final compromise solution in which other factors such as cost, space and controls are considered.

4.4.2 Direct indices

Thermal indices in this category are generally measured on so-called 'temperature' scales by which the influence of two or more of the six major comfort parameters are represented, either implicitly or explicity. Examples of this type are:

effective temperature,
corrected effective temperature,
equivalent temperature,
globe temperature,
dry resultant temperature.

In effect, many of these are very similar and of the above list, dry resultant temperature (θ_{res}), (a 'wet' form is also defined but hereafter 'resultant temperature' should be taken to mean dry resultant temperature), is currently recommended by the Chartered Institution of Building Services as being appropriate for most winter heating situations. The majority of rooms heated by means of LPHW heating systems will have reasonably uniform thermal environments and for these cases θ_{res} will be found to be an acceptable index of thermal comfort. Unlike some other direct indices θ_{res} has been experimentally correlated with subjective comfort responses from which the comfort criteria outlined in Section 4.5 were proposed.

As previously discussed the (dry) resultant temperature can be measured directly using a 100 mm blackened globe in a manner similar to the globe temperature or it can be found from the expression

$$\theta_{res} = \frac{\theta_r + \theta_a \sqrt{10 v_a}}{1 + \sqrt{10 v_a}} \quad (°C) \tag{4.11}$$

At a velocity of 0.1 m/s this gives

$$\theta_{res} = 0.5\,\theta_r + 0.5\,\theta_a, \quad (°C) \tag{4.12}$$

a result which is easy to remember. In fact, most direct indices reduce eventually to a form similar to equation 4.12 and this reveals the desire of most experimenters and designers to deal with the thermal comfort problems in a perhaps over-simplified manner.

4.4.3 Fanger's Comfort Criteria

Fanger's method takes account of the influence of the six major parameters influencing thermal comfort. As stated previously, these are:

- (i) metabolic rate (which is a function of activity level) } personal variables
- (ii) thermal resistance of clothing (clo-value) } personal variables
- (iii) air dry bulb temperature } environmental variables
- (iv) mean radiant temperature } environmental variables
- (v) air velocity } environmental variables
- (vi) humidity } environmental variables

By monitoring these variables during extensive laboratory tests, Fanger developed a heat balance equation (known as Fanger's Comfort Equation) which describes the steady-state heat balance between man and his environment for the condition of zero heat storage. By statistical methods the subjective comfort 'votes' of the experimental subjects were correlated to the imbalance in the comfort equation for a wide variety of conditions. This led to the development of procedures which allow comfort levels to be quantified for a wide variety of practical situations. Although the comfort equation itself is rather unwieldy, the comfort diagrams and charts which have been developed from it are easy to use. The theoretical basis and development of Fanger's Comfort Equation and diagrams are fully explained in his excellent book[1] and further discussion is excluded. The scope and application of his methods of assessing thermal environments are perhaps conveyed most readily by means of the design procedures and worked examples which are to follow. The principle terms employed in these examples are defined as follows for reference:

(i) *PMV* is the Predicted Mean Vote. It is the thermal sensation index recorded by subjects in the climate chamber tests used to correlate subjective response to different thermal environments.

(ii) *PPD* is the Predicted Percentage Dissatisfied. This is applied to a particular room location. PPD is a quantitative measure of the thermal comfort of a group of people at a particular point in the room. If the environment is thermally uniform this PPD value will apply to the room as a whole. Normally this should not exceed a design target of 7.5% at any point. (As Fanger discovered with thermal comfort as in other things 'you cannot please all of the people all of the time' and at best the PPD would have a minimum value of 5%.) McIntyre(4) at E.C.R.C. has proposed a similar 'comfort equation' to Fanger's employing 'subjective temperature' (θ_{sub}). This may be used as an alternative to Fanger's methods. It is simpler to apply than Fanger's methods although it provides less information. Lack of space precludes its detailed discussion and the reader is referred to McIntyre's paper for further information.

(iii) *LPPD* is the Lowest Possible Percentage Dissatisfied. It is a quantitative measure of the thermal comfort of the room as a whole for a group of people in a thermally *non*-uniform environment. It is more useful for large rooms than for small ones. It is found by modifying the design, if possible, to ensure that the average PPD of the room is minimised. Normally LPPD should not exceed 6% for design purposes.

4.5 'WHOLE BODY' COMFORT CHECKS

As has been stated previously, the surest way of ascertaining whether comfort criteria have been met is by direct measurement, but at the design stage a procedure is required which allows a comfort 'check' to be carried out by calculation and it is these procedures which are now described.

Two basic methods are:

(i) Method 1—Resultant Temperature (θ_{res}) comfort criteria for which

Procedure A provides a simple method of checking comfort at the room centre point

Procedure B provides a method of checking comfort at other room locations

(ii) Method 2—Fanger's comfort criteria for which

Procedure C describes Fanger's method for checking the room centre point

Procedure D describes the method for checking at other room locations.

These procedures describe the traditional approach in which comfort criteria are discussed in terms of 'whole body' criteria. It should be noted, however, that due to asymmetric radiation, 'localised discomfort' may occur in an environment in which 'whole body' criteria indicate that comfort exists. Thus in situations where asymmetric radiation is suspected further 'localised discomfort' checks must be performed. These are described in Section 4.7.

For reasons discussed more fully in Chapters 6 and 7, comfort criteria are seldom used as the starting point in the design process. Heat requirements and emitter sizes are generally calculated first, after which a comfort 'check' is performed. (In this context it is perhaps unfortunate that internal environment temperature (θ_{ei}), which is a useful and accurate heat transfer index temperature, has been considered by some designers to serve also as a comfort index–which it is not.) Resultant temperature, θ_{res}, is the comfort index currently recommended by C.I.B.S. and the resultant temperature methods described in this section should be used when checking comfort in accordance with the C.I.B.S.[5] as an appropriate index of 'whole body' thermal more complex and time consuming, they do lead to a greater understanding of, and give more quantitative information about, thermal environments at the design stage. Compared with the simpler direct indices such as θ_{res}, Fanger's methods require a greater effort both to learn and to apply and busy designers have been understandably conservative in adopting them. It is hoped that by the comparison of the two methods now outlined, designers will gain a fuller understanding of their relative merits.

4.5.1 Method 1: Resultant temperature (θ_{res}) comfort criteria

The (dry) resultant temperature is currently recommended by C.I.B.S. (5) as an appropriate index of 'whole body' thermal comfort for most winter heating system design calculations. The following criteria apply to 'normally' clothed individuals in 'office type' environments engaged in sedentary and low activity occupations where the body heat loss rate by convection and radiation is in the range 65 W to 83 W. The θ_{res} criteria under these constraints are:

(i) $19\,°C \leqslant \theta_{res} \leqslant 23\,°C$,
(ii) air velocity $\leqslant 0.1$ m/s,
(iii) air humidity is in the range 40% to 70%.

Notes:

(i) The wide range of θ_{res} values (compared with Fanger's recommendations) appear to suggest that in any given situation occupants will in the long term adapt to the temperature level provided by adjustment of clothing levels. Humphrey's work[6] tends to confirm this but obviously heating systems must not give rise to rapid variations in θ_{res} within the permitted range. In any case the 1976 Energy Act prohibits 'temperatures' above 20 °C in offices during the heating season which would seem to narrow the permitted range.

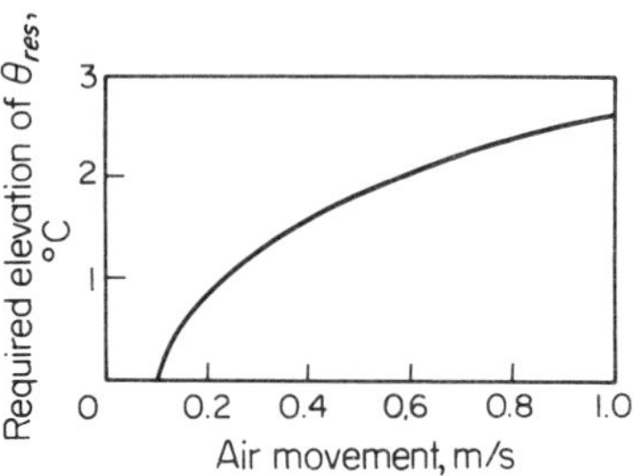

Figure 4.5
Corrections to the dry resultant to take account of air movement

(ii) Figure 4.5 (reproduced from [5]) shows the amount by which θ_{res} values must be increased to compensate for higher air velocities where it is assumed that air temperature is held constant. Thus even higher θ_{res} values are required to offset cold draughts when air temperatures are reduced (as per equation 4.11).

(iii) As will be discussed later, solar penetration (particularly in spring and autumn) can lead to significant reductions in the required θ_{res} value (see Section 4.7) but to some extent this effect is offset by occupants adopting 'summer' clothing levels.

The degree of accuracy with which the θ_{res} method can be applied depends largely on a detailed knowledge of air and surface temperatures. In general, air temperature is assumed to be nominally uniform in the occupied zone and if surface temperatures are known, Procedure B can be employed at any room location. Where surface temperatures are not fully known, or when a lower degree of accuracy is acceptable, Procedure A will give reasonable results by checking for the room centre point only.

(a) Procedure A: θ_{res} criteria room centre comfort check
This is the simplest comfort check procedure to be detailed and is the basic method recommended by the C.I.B.S. It is based on a procedure first proposed by Cornell and Fortey [9]. In this procedure the θ_{res} criteria listed previously are checked for the room centre position only. The basic heat loss calculation using internal environmental temperature (θ_{ei}) provides corresponding air (θ_{ai}) and mean radiant (θ_{ri}) temperatures at the room centre point. The steps in the procedure are:

(i) determine the room centre θ_{ai} } from heat loss
(ii) determine the room centre θ_{ri} } calculation (Chapter 7)
(iii) if air speed is unknown assume a value of 0.1 m/s
(iv) if humidity is unknown assume a value of 50%
(v) calculate θ_{res} and check θ_{res} criteria as previously described.

Obviously this simple comfort check does not consider other room locations (e.g. close to hot or cold surfaces) and such cases can be dealt with using Procedure B.

(b) Procedure B: θ_{res} comfort criteria check for other room locations

Although this procedure is perhaps best understood by means of the worked examples given in Section 4.6, the various steps are given here for completeness. This is an approximate method intended primarily to check comfort after heat losses have been computed by the environmental temperature method and when heat emitters have been sized (see Chapter 8). The steps are:

(i) carry out basic room centre comfort check (Procedure A) and, if this is satisfactory,
(ii) choose one or more representative room locations at, say, 0.6 m (the mean height of a sitting occupant) or 1.0 m (standing) above floor.
(iii) Determine the temperature (θ_s) of any internal surfaces which can be calculated or fixed by the designer (e.g. radiator, heated ceilings, etc.) or any exposed surfaces such as windows, or external walls. Determine also the areas of these surfaces.
(iv) Determine the equivalent radiant temperature (θ'_r) of the surfaces for which temperatures are unknown at the centre of the room using *either* (a) an area weighting method:

$$\sum_1^m (A_s\theta_{ri}) = \sum_1^n (A_s\theta_s) + (\sum_1^m A_s - \sum_1^n A_s)\,\theta'_r$$

from which

$$\theta'_r = \frac{\sum_1^m (A_s\theta_{ri}) - \sum_1^n (A_s\theta_s)}{(\sum_1^m A_s - \sum_1^n A_s)} \quad (°C) \qquad (4.13)$$

where m = total number of surfaces 'seen' by the room centre point
n = number of surfaces 'seen' by the room centre point whose temperatures are known

or

(b) by calculating the solid angle factors (F_{ai}) for

each of the surfaces whose temperatures are known:

$$\theta_{ri} = \sum_{1}^{n} (F_{ai}\theta_s) + (1.0 - \sum_{1}^{n} F_{ai})\, \theta'_r$$

$$\therefore \qquad \theta'_r = \frac{\theta_{ri} - \sum_{1}^{n} (F_{ai}\theta_s)}{(1.0 - \sum_{1}^{n} F_{ai})} \quad (°C) \qquad (4.14)$$

where 1.0 = total solid angle factor for m surfaces

Notes:

(i) For method of calculating solid angle factors see Appendix 4.1.

(ii) The above expressions yield approximately equal results for most rooms and method (a) is recommended as it is generally much quicker.

(iii) It is important to note, however, that for L-shaped rooms, etc., certain surfaces will not be 'seen' by the room centre point. Provided such hidden surfaces are at about the average surface temperatures (i.e. do not include hot or cold surfaces) then the calculated θ'_r value will be reasonably accurate. In other cases θ'_r will vary with the location(s) chosen for the comfort check. In both cases, however, it may be of some help to the designer to think in terms of one or more 'equivalent rectangular box' room(s); thus an L-shaped room could be considered as two 'rectangular box' rooms as shown in Fig. 4.6)

Figure 4.6

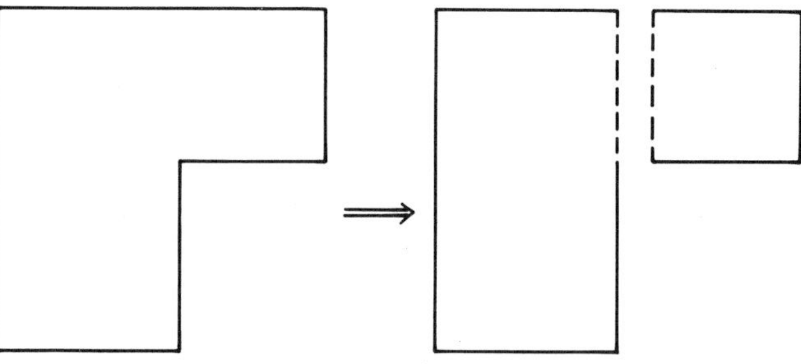

(v) For each of the other chosen locations consider a point in space and determine the solid angle (as Appendix 4.1) for each of the surfaces whose temperature is known. The total solid angle for the remaining surfaces (all of which are assumed to be at θ'_r) can then be found by subtraction, thus

$$F_{rem} = 1.0 - \sum_{1}^{n} (F_{ap})$$

(vi) Calculate the mean radiant temperature for the chosen location using the approximate formula

$$\theta_{rp} = \sum_{1}^{n} (F_{ap}\,\theta_s) + (F_{rem}\,\theta'_r)\ (^\circ C) \qquad (4.15)$$

where the subscript 'p' denotes the point under consideration.

(vii) If variations in room air temperature, velocity and humidity are unknown, assume them to be negligible and use the room centre values for air temperature and assume that the air speed is 0.1 m/s.

(viii) Calculate θ_{res} from $\theta_{res} = 0.5\ \theta_{ap} + 0.5\ \theta_{rp}$ and check θ_{res} criteria.

Additional steps if significant thermal *non-uniformity* is suspected are:

(ix) Repeat steps (ii) to (vii) at different room locations using a grid of 4 or more points and check θ_{res} criteria.

(x) C.I.B.S. recommend that provided thermal non-uniformity of the room does not produce a variation greater than $\pm$ 1.5 °C about the chosen design θ_{res} value then variations of activity and clothing levels are such that no significant dissatisfaction will arise. (Compared with Fanger's recommendations on LPPD this is an optimistic view and rooms which satisfy this C.I.B.S. criterion would, at the limit, prove quite unsatisfactory by Fanger's standards.)

4.5.2 Method 2: Fanger's Comfort Criteria

As with the resultant temperature method the main difficulties arise in determining surface temperatures and view factors. Fanger's method recommends that a steady state heat balance be struck at all room surfaces taking account of conductive, convective and radiative heat transfers between room and heating equipment surfaces. Almost invariably this requires a computer without which the designer must estimate the values of mean radiant temperature, etc., by the methods described previously, (i.e. either step A(ii) or steps B(i) to B(vi) for other room locations). Having obtained the necessary information concerning the major environmental variables for the room centre point, the designer can proceed to make a quantitative estimate of the room thermal comfort level as summarised below.

(a) Procedure C: Fanger's criteria check for room centre point
As with the θ_{res} method a simple comfort check can be carried out for the room centre point for which θ_a ($= \theta_{ai}$) and θ_r ($= \theta_{ri}$) are known for the heat loss calculation using the environmental temperature method. The steps are:

(i) Determine room centre values for θ_a, θ_r, v_a and r.h. as steps A(i) to (iv) for the θ_{res} method. If r.h. is in

Figure 4.7
Comfort lines (air temperature versus mean radiation temperature with relative air velocity as parameter) for persons with medium clothing (Icl = 1.0 clo, $f$cl = 1.15) at three different activity levels (rh = 50%)

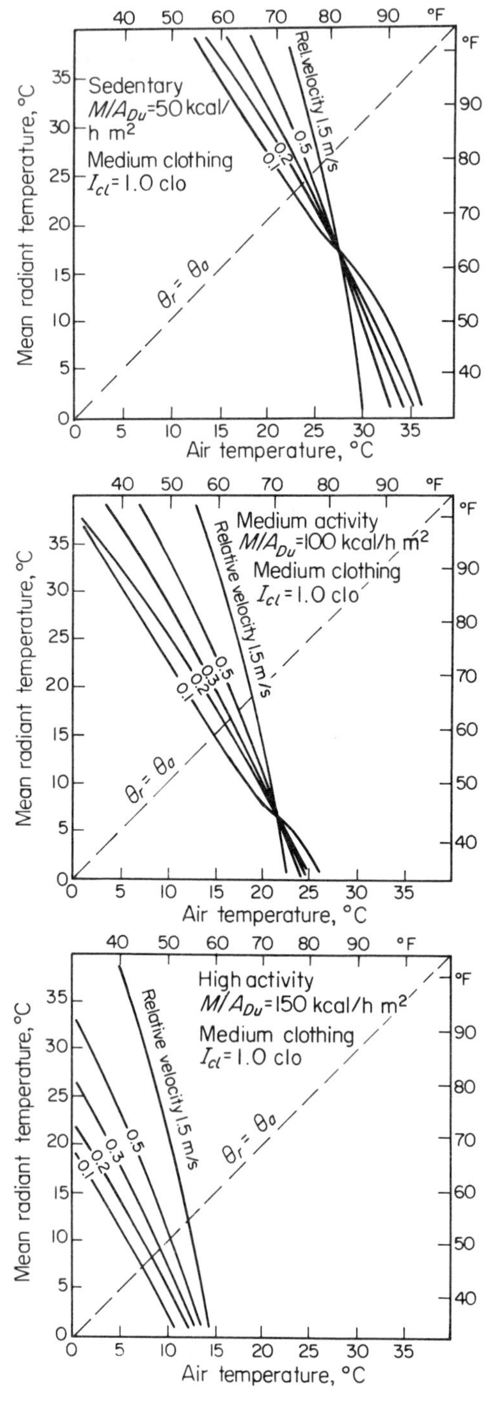

Figure 4.8
Comfort lines (air temperature versus mean radiant temperature with relative air velocity as parameter) for persons with heavy clothing (Icl = 1.5 clo, $f$cl = 1.2) at three different activity levels (rh = 50%)

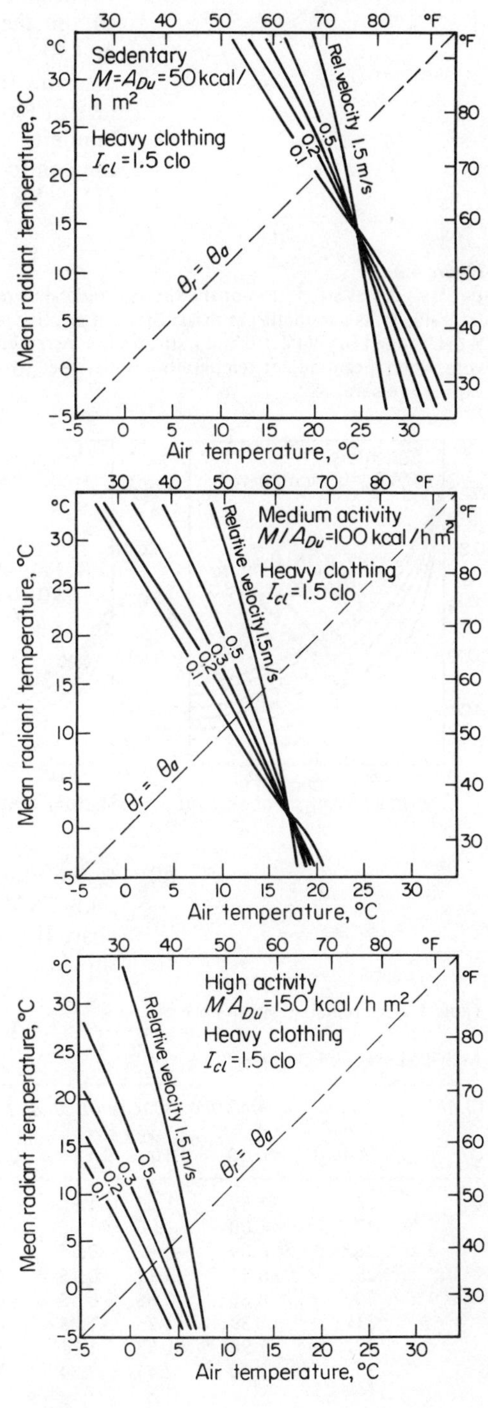

the range 35% to 70% assume a nominal value of 50%; as discussed previously, humidities outside this range will require corrective action.

(ii) Estimate the clothing and activity levels and hence choose the appropriate comfort diagram (Figure 4.7 or 4.8). (Fanger[1] provides a complete set of such diagrams to suit a wide range of activity and clothing levels.) Plot θ_a versus θ_r against v_a (assuming relative humidity = 50%). If the plot lies on the comfort line the comfort equation has been satisfied, PMV = 0 and PPD = 5% for the location.

Figure 4.9
∂ (PMV) $\partial\theta_r$ as a function of the thermal resistance of the clothing with rel. velocity as parameter, at three different activity levels. ∂ (PMV) /$\partial\theta_r$ is determined for PMV = 0 and indicates the increment of predicted mean vote, when mean radiant temperature is increased by 1 °C (constant vapour pressure)

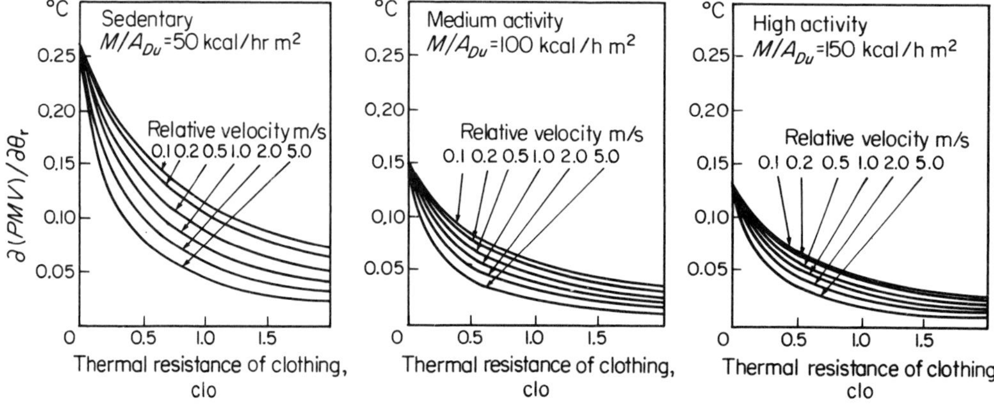

(iii) If the plot does not lie on the comfort line, determine PMV from Table 4.2 and hence the PPD from chart (Figure 4.10); (Fanger suggests that the design should be modified if necessary to ensure PPD ⩽ 7.5).

Table 4.2 Predicted Mean Vote†

Activity Level 50 kcal/m² hr (58 W/m²)

Clothing clo	*Ambient Temp.* °C (θ_a)	*Relative Velocity (m/s)* v_a <0.10	0.10	0.15	0.20	0.30	0.40	0.50	1.00	1.50
0	26.	−1.62	−1.62	−1.96	−2.34					
	27.	−1.00	−1.00	−1.36	−1.69					
	28.	−0.39	−0.42	−0.76	−1.05					
	29.	0.21	0.13	−0.15	−0.39					
	30.	0.80	0.68	0.45	0.26					
	31.	1.39	1.25	1.08	0.94					
	32.	1.96	1.83	1.71	1.71					
	33.	2.50	2.41	2.34	2.29					

Activity Level 50 kcal/m² hr (58 W/m²)

Clothing clo	*Ambient Temp.* °C (θ_a)	*Relative Velocity (m/s)* v_a <0.10	0.10	0.15	0.20	0.30	0.40	0.50	1.00	1.50
0.25	24.	−1.52	−1.52	−1.80	−2.06	−2.47				
	25.	−1.05	−1.05	−1.33	−1.57	−1.94	−2.24	−2.48		
	26.	−0.58	−0.61	−0.87	−1.08	−1.41	−1.67	−1.89	−2.66	
	27.	−0.12	−0.17	−0.40	−0.58	−0.87	−1.10	−1.29	−1.97	−2.41
	28.	0.34	0.27	0.07	−0.09	−0.34	−0.53	−0.70	−1.28	−1.66
	29.	0.80	0.71	0.54	0.41	0.20	0.04	−0.10	−0.58	−0.90
	30.	1.25	1.15	1.02	0.91	0.74	0.61	0.50	0.11	−0.14
	31.	1.71	1.61	1.51	1.43	1.30	1.20	1.12	0.83	0.63
0.50	23.	−1.10	−1.10	−1.33	−1.51	−1.78	−1.99	−2.16		
	24.	−0.72	−0.74	−0.95	−1.11	−1.36	−1.55	−1.70	−2.22	
	25.	−0.34	−0.38	−0.56	−0.71	−0.94	−1.11	−1.25	−1.71	−1.99
	26.	0.04	−0.01	−0.18	−0.31	−0.51	−0.66	−0.79	−1.19	−1.44
	27.	0.42	0.35	0.20	0.09	−0.08	−0.22	−0.33	−0.68	−0.90
	28.	0.80	0.72	0.59	0.49	0.34	0.23	0.14	−0.17	−0.36
	29.	1.17	1.08	0.98	0.90	0.77	0.68	0.60	0.34	0.19
	30.	1.54	1.45	1.37	1.30	1.20	1.13	1.06	0.86	0.73
0.75	21.	−1.11	−1.11	−1.30	−1.44	−1.66	−1.82	−1.95	−2.36	−2.60
	22.	−0.79	−0.81	−0.98	−1.11	−1.31	−1.46	−1.58	−1.95	−2.17
	23.	−0.47	−0.50	−0.66	−0.78	−0.96	−1.09	−1.20	−1.55	−1.75
	24.	−0.15	−0.19	−0.33	−0.44	−0.61	−0.73	−0.83	−1.14	−1.33
	25.	0.17	0.12	−0.01	−0.11	−0.26	−0.37	−0.46	−0.74	−0.90
	26.	0.49	0.43	0.31	0.23	0.09	0.00	−0.08	−0.33	−0.48
	27.	0.81	0.74	0.64	0.56	0.45	0.36	0.29	0.08	−0.05
	28.	1.12	1.05	0.96	0.90	0.80	0.73	0.67	0.48	0.37
1.00	20.	−0.85	−0.87	−1.02	−1.13	−1.29	−1.41	−1.51	−1.81	−1.98
	21.	−0.57	−0.60	−0.74	−0.84	−0.99	−1.11	−1.19	−1.47	−1.63
	22.	−0.30	−0.33	−0.46	−0.55	−0.69	−0.80	−0.88	−1.13	−1.28
	23.	−0.02	−0.07	−0.18	−0.27	−0.39	−0.49	−0.56	−0.79	−0.93
	24.	0.26	0.20	0.10	0.02	−0.09	−0.18	−0.25	−0.46	−0.58
	25.	0.53	0.48	0.38	0.31	0.21	0.13	0.07	−0.12	−0.23
	26.	0.81	0.75	0.66	0.60	0.51	0.44	0.39	0.22	0.13
	27.	1.08	1.02	0.95	0.89	0.81	0.75	0.71	0.56	0.48
1.25	16.	−1.37	−1.37	−1.51	−1.62	−1.78	−1.89	−1.98	−2.26	−2.41
	18.	−0.89	−0.91	−1.04	−1.14	−1.28	−1.38	−1.46	−1.70	−1.84
	20.	−0.42	−0.46	−0.57	−0.65	−0.77	−0.86	−0.93	−1.14	−1.26
	22.	0.07	0.02	−0.07	−0.14	−0.25	−0.32	−0.38	−0.56	−0.66
	24.	0.56	0.50	0.43	0.37	0.28	0.22	0.17	0.02	−0.06
	26.	1.04	0.99	0.93	0.88	0.81	0.76	0.72	0.61	0.54
	28.	1.53	1.48	1.43	1.40	1.34	1.31	1.28	1.19	1.14
	30.	2.01	1.97	1.93	1.91	1.88	1.85	1.83	1.77	1.74
1.50	14.	−1.36	−1.36	−1.49	−1.58	−1.72	−1.82	−1.89	−2.12	−2.25
	16.	−0.94	−0.95	−1.07	−1.15	−1.27	−1.36	−1.43	−1.63	−1.75
	18.	−0.52	−0.54	−0.64	−0.72	−0.82	−0.90	−0.96	−1.14	−1.24
	20.	−0.09	−0.13	−0.22	−0.28	−0.37	−0.44	−0.49	−0.65	−0.74
	22.	0.35	0.30	0.23	0.18	0.10	0.04	0.00	−0.14	−0.21
	24.	0.79	0.74	0.68	0.63	0.57	0.52	0.49	0.37	0.31
	26.	1.23	1.18	1.13	1.09	1.04	1.01	0.98	0.89	0.84
	28.	1.67	1.62	1.58	1.56	1.52	1.49	1.47	1.40	1.37

†Fanger[1] provides a set of such tables to suit a wide range of activity levels.

Figure 4.10
Predicted percentage of dissatisfied (PPD) as a function of Predicted Mean Vote (PMV).

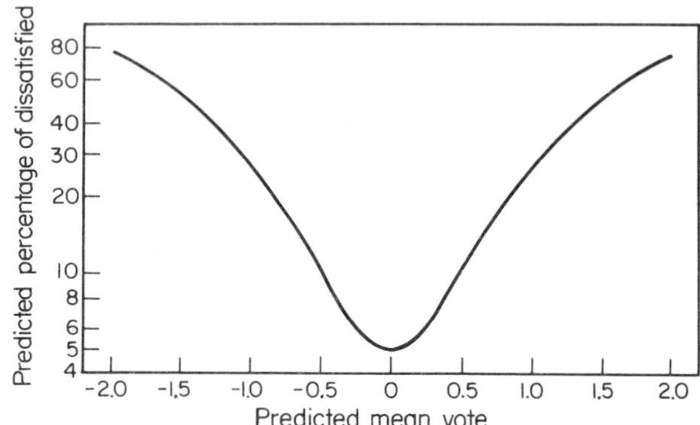

(b) Procedure D: Fanger's comfort criteria check for other room locations
As for Procedure B of the θ_{res} method the main problem is to estimate the mean radiant temperature at the chosen location. The steps are:

(i) Follow steps B(i) to B(vii) of θ_{res} method to determine θ_a, θ_r, v_a. Again use a nominal r.h. value of 50% as discussed previously.
(ii) Determine PMV and PPD for each location as described for steps C(ii) to C(iii).

Additional steps if significant thermal *non-uniformity* is suspected, are:
(iii) Choose a grid of not less than four equally spaced points as for procedure B of θ_{res} method and determine PMV and PPD as described previously.
(iv) (a) the mean of the grid PPD values gives a preliminary 'figure of merit' for the room;
(b) if the mean of the grid PMV values is not equal to zero the room is 'hot' or 'cold'; if possible the designer should alter the proposed heating system design and repeat the previous steps to obtain a mean PMV equal to or close to zero.
(v) Now subtract the mean PMV for the whole room from *each* grid PMV value and use the PMV differences to get new PPD values at each grid point. The mean of these new PPD values is the LPPD for the room as a whole; (Fanger suggests that the design should be modified if necessary to ensure that the LPPD $\leqslant$ 6%).
(vi) The final 'figure of merit' for the 'thermal non-uniformity' of the room is found by subtracting 5% (the minimum possible PPD for the room) from the LPPD value found in (e).

4.5.3 Summary

Procedure A (θ_{res}) and Procedure C (Fanger) allow the designer to make a fairly elementary check on comfort at the room centre point.

By Procedure B (θ_{res}) and Procedure D (Fanger) the designer can check comfort levels at other more representative room locations (e.g. known work locations).

Procedure D can be extended, using LPPD, to provide an estimate of the overall effect of thermal non-uniformity; this is of much greater significance in a large enclosure (e.g. a large open-plan office) than in a small office with only a few occupants.

By the above method, 'whole body' comfort checks can be made and by and large such checks will prove sufficient in most circumstances. In cases where 'localised discomfort' due to asymmetric radiation effects are suspected, further check(s) should be made as described in Section 4.7.

4.6 WORKED EXAMPLES ON 'WHOLE BODY' COMFORT CHECKS

4.6.1 Introduction

In order to check comfort by the methods outlined in Section 4.5, surface temperatures must first be established. In the design process this would follow the computation of heat losses and the sizing of emitters and, although these topics are discussed in detail in later chapters, the steps are included here to illustrate the method of establishing surface temperatures.

In the worked examples which follow, a fairly typical office situation is considered heated in each of three ways:

Proposal A — Fully convective method (e.g. fan convector, sill-line etc., where the radiant effect of any warm surface is neglected).

Proposal B — A LPHW radiator sited below the window.

Proposal C — A LPHW radiator sited on a sidewall near to the window.

4.6.2 Preliminary Steps

Consider the example of a private office on the fourth floor of a six-storey building for which it is proposed to provide thermally comfortable conditions for sedentary occupation. Details of the room are shown in Figures 4.11 and 4.12.

Preliminary step No. 1: Computation of heat losses

This has been carried out using the environmental temperature method (see Chapter 7) assuming a 100% convective heating system. Design environmental temperatures are:

Figure 4.11

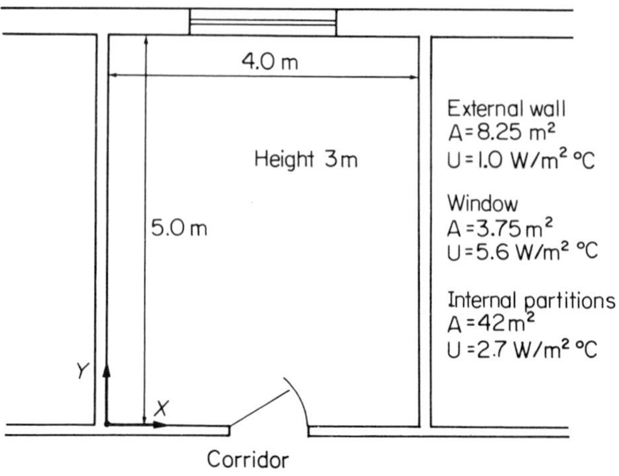

office $\theta_{ei} = 20\,°C$
corridor $\theta_{ei} = 16\,°C$
$\theta_{eo}\ (= \theta_{ao}) = -1\,°C$

All other adjacent rooms are assumed to be at $\theta_{ei} = 20\,°C$. Fabric heat losses are summarised in Table 4.3.
Air infiltration rate (n) = 1/h
Ventilation allowance $\dot{q}_v = 0.33$ W/m³ °C
From the analysis of heat flow paths in 7.4.3,

$$(\theta_{ai} - \theta_{ei}) = \frac{\Sigma \dot{Q}_f}{4.8 \sum\limits_{1}^{m} A_s} = \frac{744}{4.8 \times 94} = 1.7\,°C$$

$\therefore\ \theta_{ai} = 21.7\,°C$

Figure 4.12

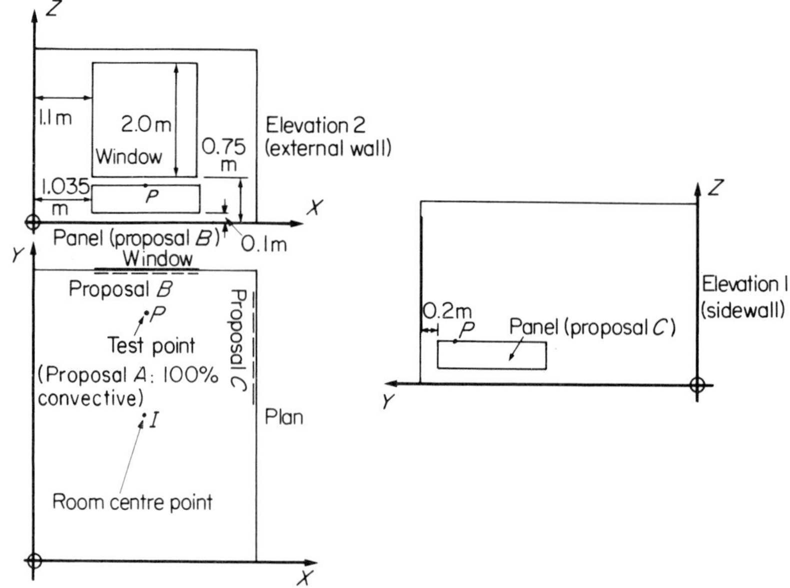

Table 4.3

Item	*Area* (m^2)	*U Value* ($W/m^2\ °C$)	*Temp. Diff.* (°C)	*Heat Loss* (W)
External wall	8.25	1.00	21	173
Window	3.75	5.60	21	441
Corridor	12.00	2.70	4	130
Floor	20.00	—	—	—
Ceiling	20.00	—	—	—
Other partitions	30.00	—	—	—
	$\sum_{1}^{m} A_s = 94.00$ m			$\Sigma \dot{Q}_f = 744$ W

$$\dot{Q}_v = \dot{q}_v \ V' (\theta_{ai} - \theta_{ao})$$

$$= 0.33 \times 60 \times 22.7$$

$$= 450\,\text{W}$$

Hence $\dot{Q}_t = \Sigma \dot{Q}_f + \dot{Q}_v = 1194$ W

$$\theta_{ri} = \frac{3}{2}\theta_{ei} - \frac{1}{2}\theta_{ai} = 30 - 10.9 = 19.1\,°\text{C}$$

Preliminary step No. 2: Emitter sizing

Proposal A — A suitable unit would be chosen noting that the heat output will refer to a room air temperature of 21.7°C

Proposals B and C — A particular manufacturer's data allows the following selection to meet this requirement; single panel radiator 2.0 m long x 0.460 m high, assumed panel surface temperature = 70°C

4.6.3 Proposal A (100% convective heating system)

Procedure A: θ_{res} comfort criteria check at the room centre point (I)

step (i) $\theta_{ri} = 19.1\,°C$ } from heat loss calculations
step (ii) $\theta_{ai} = 21.7\,°C$ }
step (iii) $v_{ai} = 0.1$ m/s (assumed)
step (iv) r.h. = 50% (assumed)
step (v) $\theta_{res} = 0.5\theta_{ai} + 0.5\theta_{ri} = (10.85 + 9.55) = 20.4\,°C$

Thus the θ_{res} criteria have been satisfied for the room centre point (I)

Procedure B: θ_{res} comfort criteria check at other room locations

step (i) Having carried out a satisfactory comfort check at (I) it is

step (ii) proposed to check the θ_{res} criteria at the point P; x = 2m, y = 4.5 m and z = 0.6 m (see Figure 4.12).

step (iii) Surface temperatures must now be established as follows:

External wall: $\dot{Q}_f/A_s = 1.0 \times 21 = 21$ W/m²

$$(\theta_{ei} - \theta_s) = (\dot{Q}_f/A_s)\, R_{si} = 21 \times 0.123$$
$$= 2.6\,°\mathrm{C}$$
$$\therefore \theta_s = 17.4\,°\mathrm{C}$$

Window: $\dot{Q}_f/A_s = 5.6 \times 21 = 117.6$ W/m²

$$(\theta_{ei} - \theta_s) = 117.6 \times 0.123 = 14.5\,°\mathrm{C}$$
$$\therefore \theta_s = 5.5\,°\mathrm{C}$$

Corridor partition:

$$\dot{Q}_f/A_s = 2.7 \times 4.0 = 10.8 \text{ W/m}^2$$
$$(\theta_{ei} - \theta_s) = 10.8 \times 0.123 = 1.33\,°\mathrm{C}$$
$$\therefore \theta_s = 18.7\,°\mathrm{C}$$

step (iv) the equivalent radiant temperature (θ'_r) of the surfaces whose temperatures are unknown can now be established by either of the two methods mentioned previously.

(a) Area weighting method

Surface	*Area* (m²)	θ_s (°C)	*Area* × θ_s
(1) External wall (net)	8.25	17.4	143.6
(2) Window	3.75	5.5	20.6
(3) Corridor partition	12.0	18.7	224.4
	$\sum_1^n = 24.0$		$\sum_1^n = 388.6$

The total room surface area $\sum_1^m A_s = 94$ m and hence

$$(\theta_{ri} \times \sum_1^m A_s) = (\sum_1^3 A_s\, \theta_s) + (\sum_1^m A_s - \sum_1^3 A_s)\, \theta'_r$$

$\therefore$ $19.1 \times 94 = 388.6 + (94 - 24)\theta'_r$

$\therefore$ $\theta'_r = 20.1\,°\mathrm{C}$

(b) Solid angle factor (F_{ai}) method (see Appendix 4.1)

Surface	F_{ai}	θ_s	$F_{ai}\theta_s$
(1) External wall (net)	0.063	17.4	1.096
(2) Window	0.041	5.5	0.226
(3) Corridor partition	0.104	18.7	1.945
	$\sum_1^3 = 0.208$		$\sum_1^3 = 3.267$

$$1.0 \times \theta_{ri} = \sum_1^3 (F_{ai}\theta_s) + (1.0 - \sum_1^3 F_{ai})\,\theta'_r$$

$$\therefore \quad \theta'_r = \left(\frac{19.1 - 3.267}{1.0 - 0.208}\right) = 19.99 \triangleq 20\,^\circ\mathrm{C}$$

Notes:

(i) That the value of $\theta'_r \triangleq \theta_{ei}$ is coincidental to this example and generally this will not be so.

(ii) Although, strictly, absolute temperatures should be used in the above calculations the error involved in using Celcius values is negligible provided that individual room surfaces do not differ by more than 20 °C.

(iii) As there is little difference between the values of θ'_r obtained by the two methods, the area weighting method being simpler and quicker is advantageous. At other room locations it must *not* be used as only solid angle factor methods will give meaningful results.

step (v) The remaining surfaces can thus be assigned a temperature value of 20.1 °C (= θ'_r) and for the point P solid angle factors of the various surfaces can be found from which the mean radiant temperature at P (θ_{rp}) can be found using Table 4.4.

Table 4.4

Surface	F_{ap}	θ_s	$F_{ap}\theta_s$
External wall (net)	0.210	17.4	3.65
Window	0.123	5.5	0.68
Corridor partition	0.039	18.7	0.73
Remaining surfaces (by subtraction)	0.628	20.1	12.62
	$\sum_1^m = 1.0$		$\therefore \theta_{rp} = 17.68 \triangleq 17.7\,^\circ\mathrm{C}$

step (vi) Thus the θ_{res} criteria can be checked for P making some assumptions:

$$\left.\begin{array}{lll} \theta_{ap} & \triangleq & \theta_{ai} = 21.7\,^\circ\mathrm{C} \\ v_{ap} & \leqslant & 0.1\ \mathrm{m/s} \\ \mathrm{r.h.} & \triangleq & 50\% \end{array}\right\} \text{(assumed)}$$

Thus $\theta_{res} = 0.5\ \theta_{ap} + 0.5\ \theta_{rp} = (10.85 + 8.85) = 19.7\,^\circ C$, from which we conclude that the θ_{res} criteria have been satisfied.

Notes:

(i) In the above calculations, values of θ_a and v_a have been assumed equal to those for the room centre. In fact a reduction in θ_a and an increase in ν_a are to be expected at P due to cold down draughts generated by the window. If, for example, the local air temperature were assumed to be 19 °C and the local air velocity 0.2 m/s then from equation 4.11 $\theta_{res} = 15.7\,^\circ C$ which perhaps raises doubts about the comfort level which might be expected in practice.

(ii) Whether or not 'whole body' comfort criteria have been satisfied at P it is to be expected that the proximity of the large cold window surface might give rise to 'localised discomfort' for which the checks outlined in Sections 4.7 and 4.8 would be required.

Procedure C: Fanger's comfort criteria check for the room centre point (I)

Fanger's method can be applied to the room centre point (I) in the same way as the θ_{res} Procedure A:

step (i) $\left.\begin{aligned}\theta_{ai} &= 21.7\,^\circ C\\ \theta_{ri} &= 19.1\,^\circ C\end{aligned}\right\}$ (as before)

$v_{ai} = 0.1$ m/s (assumed)

r.h. = 50% (assumed)

step (ii) clothing level = 1.0 clo (assumed)

activity level = M/A_{DU} = 50 kcal/hm^2 (58 W/m^2) (sedentary)

The above information can now be plotted on Figure 4.7. This point lies to the left of the comfort line indicating that the occupants will suffer some cold discomfort.

step (iii) The degree of cold discomfort can be estimated from Table 4.2 assuming an ambient temperature of 21.7 °C (= θ_{ai}). This gives a preliminary value of PMV = 0.41 (by interpolation). This must be corrected using Figure 4.9 for the difference between mean radiant and air temperatures, viz.

$$\frac{\partial\,(\text{PMV})}{\partial\theta_r} = 0.12$$ from the figure and thus the corrected PMV value is

$$\text{PMV} = -0.41 - (21.7 - 19.1)0.12 = 0.72$$

step (iv) The PPD for the room centre point can now be obtained from Figure 4.10, thus

PPD = 16%

This result indicates that this room would be considered by many people to be too cold to ensure sedentary comfort. It also demonstrates that by Fanger's standards designing for θ_{res} values to lie in the range 19 °C to 23 °C will not necessarily ensure comfort in office type environments. Thus referring again to Figures 4.7 and 4.8 the following combinations of θ_r and θ_a can be plotted which satisfy the comfort equation in this situation for two different clothing levels (all other parameters having the same values):

Clothing level = 1 clo (medium clothing level)

Required θ_{ai}/θ_{ri} combinations to produce optimum comfort		corresponding values	
θ_{ai}	θ_{ri}	θ_{res}	θ_{ei}
23.0	23.0	23.0	23.0
21.7	24	22.9	22.2
26.0	19.1	22.5	21.4

Clothing level = 1.5 clo (heavy clothing level)

Required θ_{ai}/θ_{ri} combinations to produce optimum comfort			
θ_{ai}	θ_{ri}	θ_{res}	θ_{ei}
20.5	20.5	20.5	20.5
21.7	18.0	19.9	19.2
21.0	19.1	20.1	19.7

Thus comfort can be obtained either by increasing the internal design temperature to θ_{ei} = 22.2 °C (say) or by requiring that the occupants wear higher clothing levels (= 1.5 clo). Although clothing levels are seldom under the direct control of the heating system designer, a recent study indicates that building users do in fact adapt to (deliberate) reductions in internal temperature levels over a period of time. Although all of the reasons for this adaptation are not yet known it would seem that the use of higher clothing levels is probably the most significant factor.

If the designer chooses to increase the design θ_{ei} temperature then the previous calculation method must be repeated before proceeding to the next step.

In this case, however, it will be assumed that the higher clothing level (= 1.5 clo) is acceptable and using the previous θ_{ai}, θ_{ri}, v_{ai} and humidity values the procedure can be readily repeated to show that

(a) uncorrected PMV = +0.24 (at $\theta_{ambient}$ = 21.7 °C)

(b) $\dfrac{\partial\,(\mathrm{PMV})}{\partial\theta_r} = 0.09$

(c) corrected PMV = +0.24 – (21.7 – 19.1) 0.09 i.e. PMV = 0.01 ≏ 0 (as expected) hence PPD ≏ 5%, which gives optimum comfort conditions.

Procedure D: Fanger's comfort criteria check for other room locations

From Procedure C we assume a clothing level of 1.5 clo and from Procedure B we have the following information about the point P:

step (i) θ_{rp} = 17.6 °C
θ_{ap} = 21.7 °C
v_{ap} = 0.1 m/s
(M/A_{DU}) = 50 kcal/hm² (58 W/m²)
step (ii) From Table 4.2
uncorrected PMV = +0.24 (as Procedure B)

and $\dfrac{\partial\,(\mathrm{PMV})}{\partial\theta} = 0.09$ (from Figure 4.10)

corrected PMV = +0.24 – (21.7 – 17.6) 0.09
= 0.129 ≏ 0

∴ PPD ≏ 5%

Thus the Fanger 'whole body' comfort check at P indicates that the PPD is close to optimum. Further, the small variation in PPD suggests that serious thermal non-uniformity is not suspected and that there is no need to make further checks at other room locations; in any case the size of this room would suggest one or two occupants and hence the use of the LPPD procedure would be inappropriate even if non-uniformity were suspected.

Notes:

(i) As with Procedure B the θ_a and v_a assumed at P are perhaps optimistic. Re-checking assuming θ_{ap} = 19 °C and v_{ap} = 0.2 m/s gives

uncorrected PMV = –0.90

and $\dfrac{\partial\,(\mathrm{PMV})}{\partial\theta_r} = 0.08$

corrected PMV = –0.90 – (19.0 – 17.6) 0.08 = –1.0 which gives PPD ≏ 27% which is clearly uncomfortable.
Here again we see that the local values assumed for θ_a and v_a have a significant effect on the calculation.

(ii) Again whether or not 'whole body' comfort criteria have been satisfied at P, it is to be expected that the proximity of the large cold window surface might give rise to 'localised discomfort', for which checks are outlined in Sections 4.7 and 4.8.

4.6.4 Proposal B (panel radiator below window)

Procedure A: θ_{res} comfort criteria check at the room centre point (I)

In the environmental temperature heat loss method (Chapter 7) LPHW radiator systems are assumed to be 100% convective and as a result the room centre values (θ_{ai} and θ_{ri}) used in 4.6.3 (Procedure A) can be assumed with little error. In fact the addition of the panel radiator (θ_s = 70 °C) which replaces an equivalent area of external wall surface (θ_s = 17.4 °C) will tend to raise the room centre θ_{ri} very slightly (using an area weighting method this increase can be shown to be about 0.3 °C). Thus neglecting the small increase in θ_{ri} caused by the panel radiator, the comfort level at I is deemed to be satisfactory as in Section 4.6.3 (Procedure A).

Procedure B: θ_{res} comfort criteria check at other room locations

Steps (i) to (iv) are carried out as per Section 4.6.3 (Procedure B) noting that in the use of the area weighting method to establish θ'_r = 20.1 °C we do *not* include the radiant component from the radiator in order to abide by the environmental temperature assumption of LPHW radiator systems being fully convective.

Step (v) Moving now from I to P, Table 4.5 can be set down (based on Table 4.4) which takes account of

Table 4.5

Surface	θ_s(°C)	T_s(K)	$T_s^4 \times 10^{-8}$	F_{ap}	$(F_{ap} \times T_s^4 \times 10^{-8})$
Radiator (under window)	70	343	138	0.096	13.2
External wall (net)	17.4	290.4	71	0.114	8.9
Window	5.5	278.5	60	0.123	7.4
Corridor partition	18.7	291.7	72	0.039	2.8
Remaining surfaces (by subtraction)	20.1	293.1	74	0.628	46.5
				$\sum_1^m = 1.0$	$\sum_1^m = 78.0$

the solid angle subtended by the radiator at the point P in order to establish the value of θ_{rp} but in this case it is advisable to use absolute temperatures in order to ensure accuracy, as the radiator is at a much higher temperature.

$$T_{rp}^4 = \sum_1^m (F_{ap} T_s^4)$$

$$\therefore \quad \theta_{rp} = \sqrt[4]{78.0 \times 10^8} - 273 = 297.2 - 273$$
$$= 24.2\,^\circ\text{C}$$

(Note: Using Celsius surface temperatures in this calculation yields an incorrect value of θ_{rp} = 22.7 °C)

step (vi) Again the θ_{res} criteria can be checked for P making some assumptions:

$$\left.\begin{aligned} \theta_{ap} &\triangleq \theta_{ai} = 21.7\,^\circ\text{C} \\ v_{ap} &\leqslant 0.1 \text{ m/s} \\ \text{r.h.} &= 50\% \end{aligned}\right\} \text{(assumed)}$$

from which $\theta_{res} = (0.5\theta_{ap} + 0.5\theta_{rp})$
$= (10.85 + 12.10) = 23.0\,^\circ\text{C}$

and comfort criteria have been satisfied.

Notes:

(i) Thus we see that the traditional location of the panel radiator has the benefit that the 'warm' radiator surface counteracts the 'cold' window surface and reduces the radiant field non-uniformity. Furthermore there is no need to consider the effect of cold downdraughts as the convective output of the radiator will counteract these.

(ii) As will be seen in Section 4.8, 'local cooling' effects are eliminated but at the expense of creating a perceptible radiant field asymmetry and probably some local overheating.

Procedure C: Fanger's comfort criteria check for the room centre point (I)

Again by neglecting the small increase in θ_{ri} caused by the radiator the same conclusions as for Section 4.6.3 (Procedure C) can be arrived at, viz. that a clothing level of 1.5 clo is required to ensure comfort.

Procedure D: Fanger's comfort criteria check for the other room locations

Again as outlined previously in Procedure B above we can determine a value of θ_{rp} = 24.2 °C as per Table 4.5.

step (i) Summarising the other data known or assumed:

$$\begin{aligned} \theta_{ap} \triangleq \theta_{ai} &= 21.7\,^\circ\text{C} \\ v_{ap} = v_{ai} &= 0.1 \text{ m/s} \\ \text{r.h.} &= 50\% \end{aligned}$$

step (ii) clothing level = 1.5 clo (assumed)
activity level = M/A_{DU} = 50 kcal/hm^2 (58 W/m^2) (sedentary).
This information can now be plotted on Figure 4.8 which shows that the point is slightly to the right of the comfort line (i.e. warm discomfort).

step (iii) The degree of warm discomfort can be assessed as before, viz.

uncorrected PMV = +0.24 (Table 4.2)

$$\frac{\partial\,(\text{PMV})}{\partial\theta_r} = 0.09 \text{ (Figure 4.9)}$$

corrected PMV = +0.24 + (24.2 – 21.7) 0.09

= +0.47

step (iv) Hence at the point P with the panel radiator located below the window

PPD = 9%

This represents warm discomfort with a 'high' clothing level of 1.5 clo; re-calculating at 1.0 clo indicates that approximately optimum conditions would be achieved. No 'localised discomfort' is to be expected but see also Section 4.8.

4.6.5 Proposal C (panel radiator on sidewall)

Procedure A: θ_{res} comfort criteria check at room centre point (I)

As per Sections 4.6.3 (Procedure A) and 4.6.4 (Procedure A) the θ_{res} criteria are satisfied for the room centre point.

Procedure B: θ_{res} comfort criteria check at other room locations
Steps (i) to (iv) as per Sections 4.6.3 (Procedure B) and 4.6.4 (Procedure B).

Table 4.6

Item	θ_s(°C)	T_s(K)	$T_s^4 \times 10^{-8}$	F_{ap}	($F_{ap} T_s^4 \times 10^{-8}$)
Radiator (on sidewall)	70	343	138	0.014	1.93
External wall (net)	17.4	290.4	71	0.210	14.91
Window	5.5	278.5	60	0.123	7.38
Corridor partition	18.7	291.7	72	0.039	2.81
Remaining surfaces (by subtraction)	20.1	293.1	74	0.614	45.44
					$\sum_1^m = 72.47$

step (v) Moving from I to P, Table 4.6 can now be set down as per Section 4.6.4 (Procedure B).

$$\therefore\ \theta_{rp} = \sqrt[4]{72.47 \times 10^8} - 273 = 18.8\,°\text{C}$$

(Note: Using Celsius surface temperature as per Table 4.6 gives θ_{rp} = 18.4 °C, thus we see that when F_{ap} of the 'hot' surface is small then the calculation can often be simplified without great loss of accuracy.)

step (vi) Assuming room centre point values of

$$\theta_{ap} \triangleq \theta_{ai} = 21.7\,^\circ\mathrm{C}$$

$$v_{ap} = v_{ai} = 0.1 \text{ m/s}$$

$$\text{r.h.} = 50\%$$

from which $\theta_{res} = (0.5\theta_{ap} + 0.5\theta_{rp})$

$$= (10.85 + 9.4) = 20.3\,^\circ\mathrm{C}$$

and we conclude that the comfort criteria are satisfied.

Notes:

(i) The point raised about Proposal A (viz. local changes in θ_a and v_a caused by the cold window), may give rise to discomfort again highlighting the importance of the values assumed for θ_a and v_a.

(ii) Again Sections 4.7 and 4.8 should be seen for localised discomfort checks.

Procedure C: Fanger's comfort criteria check at the room centre point (I)

As per Sections 4.6.3 (Procedure B) and 4.6.4 (Procedure C), Fanger's criteria can be satisfied with a clothing level = 1.5 clo.

Procedure D: Fanger's comfort criteria check at other room locations

As outlined in Procedure B of this Section, we can determine $\theta_{rp} = 18.8\,^\circ\mathrm{C}$ in this case.

step (i) Summarising other data known or assumed:

$$\theta_{ap} \triangleq \theta_{ai} = 21.7\,^\circ\mathrm{C}$$

$$v_{ap} \triangleq v_{ai} = 0.1 \text{ m/s}$$

$$\text{r.h.} = 50\%$$

step (ii) Clothing level = 1.5 clo (assumed)
activity level = M/A_{DU} = 50 kcal/hm^2, (58 W/m^2) (sedentary)
Plotting this on Figure 4.8 shows that the point lies almost exactly on the comfort line and hence

step (iii) PMV $\triangleq$ 0

step (iv) PPD $\triangleq$ 5%

As with Procedure B 'localised discomfort' is suspect and checks discussed in Sections 4.7 and 4.8 should be applied.

4.6.6 Summary

	Proposal A	Proposal B	Proposal C
Procedure A ($\theta_{res,i}$)	20.4°C	20.4°C	20.4°C
Procedure B ($\theta_{res,p}$)	19.7°C(a)	23.0°C	20.3°C(a)
Procedure C (PPD_i)	5% (b)	5% (b)	5% (b)
Procedure D (PPD_p)	5% (a,b)	9% (c)	5% (a,b)

Notes:

(a) Based assumptions $\theta_{ap} \triangleq \theta_{ai}$ and $v_{ap} \triangleq v_{ai}$; cold window may reduced θ_{ap} and increase v_{ap} and cause 'whole body' discomfort. 'Localised discomfort' is suspected.

(b) This optimum level requires that occupants adopt a 1.5 clo level which is quite heavy.

(c) This warm discomfort could be reduced easily by reducing clothing from 1.5 clo to 1.0 clo.

4.7 'LOCALISED DISCOMFORT' CHECKS

'Localised discomfort' may still occur in situations where 'whole body' criteria checks indicate that an approximate heat balance has been achieved which should produce thermal comfort.

A comic illustration is that of a person with one foot in a bucket of hot water at 40°C and the other in cold water at 0°C. Although the average water temperature of 20°C might give rise to an acceptable whole body heat balance the subject would be unlikely to admit to being comfortable.

In practice localised discomfort is usually caused by a marked degree of asymmetry in the radiant field which may be due to:

(i) Local cooling: caused by heat loss to adjacent cold surfaces such as single-glazed windows.

(ii) Radiant heating: care is required in the design of all types of radiant heating systems to ensure that localised overheating or unacceptable asymmetry does not occur. The methods described in this section can be used for long-wave radiant heating systems (surface temperatures less than about 300°C). For shortwave radiant heating systems see Section 4.9.

(iii) Heat from lighting: the radiant energy emitted by some lighting systems can produce thermal discomfort effects.

(iv) Solar penetration: as with high-temperature radiant heating systems, solar penetration via windows can give rise to localised discomfort.

4.7.1 Longwave radiant field parameters

For environments incorporating heating surfaces not above 300 °C and usually less than 200 °C the following parameters are defined. See also the discussion in Sections 2.4 and 4.9.1. These parameters were first used by McIntyre[8] and give a logical method of describing radiant fields in a manner similar to that used in lighting design.

(i) *Mean radiant temperature* (θ_r): As discussed previously a variety of definitions of θ_r exist. The most useful for design purposes is that defined for a small sphere for which solid angle factors are used to 'weight' the effect of the various surface temperatures in the enclosure.
In the absence of shortwave radiation the mean radiant temperature can be calculated from

$$T_r^4 = \frac{1}{4\pi} \int_0^{4\pi} T_s^4 \, d\psi \tag{4.16}$$

where T_r = (absolute) mean radiant temperature (K)
T_s = (absolute) surface temperatures (K)
ψ = solid angle subtended by surfaces (steradians)

From this $\theta_r = (T_r - 273)$ °C can be found easily, and introducing solid angle factors,

$$F_a = \frac{1}{4\pi} \int_0^{\psi} d\psi \tag{4.17}$$

where F_a = solid angle factor referred to an elemental sphere, equation 4.16 can be re-written as

$$T_{rp}^4 = (F_{ap} T_s^4) \tag{4.18}$$

and for cases where surface temperatures do not differ by more than about 20 °C (as in 100% convective systems), we can simplify this to

$$\theta_{rp} = (F_{ap}\,\theta_s) \quad (°C) \tag{4.19}$$

Although the small sphere does not ideally represent the human body, for most purposes it is a useful measure of the radiant field when checking whole body comfort. Unless all room surfaces are at the same temperature, θ_r will vary throughout the room. Most rooms do exhibit some asymmetry but generally this is small and in practice, checks need only be carried out where experience indicates such a problem may exist. Where problems are suspected the following parameters can often be used to quantify the radiant field effects. For methods of measurement see Section 4.3.5.

(ii) *Plane radiant temperature* (θ_p): The plane radiant temperature is defined as the surface temperature inside a uniform hemisphere which would produce the same radiation exchange with a small plane element, (at the test point in the centre of the basal plane of the hemisphere), as would exist between the actual room radiant field and the small plane element.
In the absence of short-wave radiation the plane radiant temperature is given by

$$T_p^4 = \Sigma\,(F_p\,T_s^4) + \Sigma\,(F_o\,T_s^4) \qquad (4.20)$$

where T_p = (absolute) plane radiant temperature (K)
F_p = the form factor of a surface *parallel* to an elemental plane
F_o = the form factor of a surface *orthogonal* to an elemental plane

and $\Sigma\,(F_p + F_o) = 1.0$

As with the mean radiant temperature where surface temperatures differ by less than about 20 °C we can simplify the calculation to

$$\theta_p = \Sigma\,(F_p\,\theta_s) + \Sigma\,(F_o\,\theta_s) \quad (°C)$$

If a room is considered to be divided by a plane through it, the plane radiant temperature can be visualised as the radiant temperature resulting from the surfaces to one side of the plane. Thus a normal cubical enclosure will have six

Figure 4.13
The plane thermometer

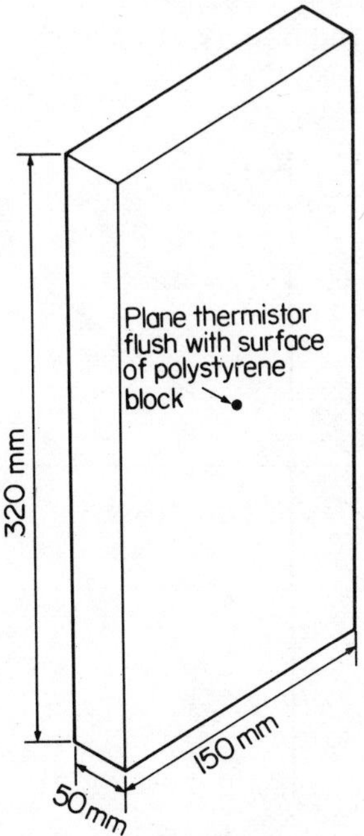

principal plane radiant temperatures measured or calculated parallel to each of the six principal surfaces (i.e. four walls, ceiling and floor). Direct measurement of θ_p can be made using a plane thermometer (see Figure 4.13). The thickness of the polystyrene block ensures that heat transfer between the sensor and the rear and sides is negligible. The 'sensor' temperature is a weighted average of the plane radiant temperature and the local air temperature. The actual weighting depends on the air velocity and on the radiant characteristics of the various surfaces which the sensor 'sees'. These are surface temperature, emissivity and view (form) factor. When measuring longwave radiant fields the reading obtained by the apparatus in Figure 4.13 will give a sensor reading weighted approximately as:

$$\theta_{\text{sensor}} = \frac{h_c\theta_a + h_R\theta_p}{h_c + h_R} \quad (°\text{C}) \tag{4.22}$$

where $h_c \triangleq 3.4$ W/m^2 °C

$h_R \triangleq 4.9$ W/m^2 °C (i.e. a linearised radiation coefficient)

Thus

$$\theta_{\text{sensor}} = 0.41\,\theta_a + 0.59\,\theta_p \quad (°\text{C}) \tag{4.23}$$

Thus the approximate value of θ_p can be measured simply with the equipment shown. More accurate measures can be obtained using a differential radiometer which is an expensive item and is really a laboratory technique. At the design

Figure 4.14

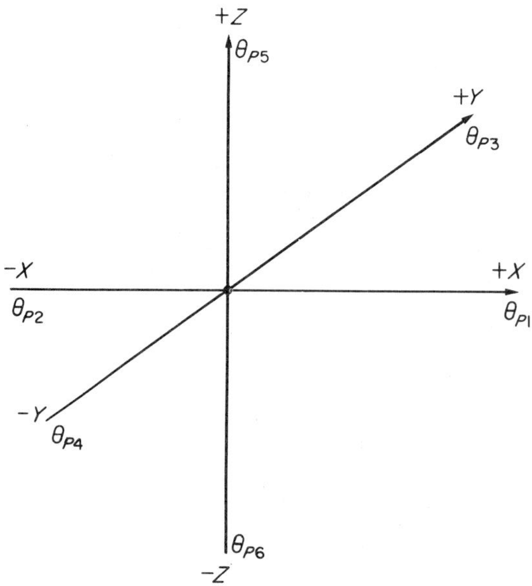

stage θ_p can be calculated as discussed in Appendix 4.2.

(iii) *Air radiant temperature* (θ_{ar}): The plane thermometer was developed originally to quantify the effect of cold window surfaces on comfort and for longwave radiant fields. The 'sensor' temperature is generally given the name of 'air-radiant temperature' (θ_{ar}). At the design stage this can often be estimated by inserting known or estimated values of θ_a and θ_p into equation 4.23.

(iv) *Vector radiant temperature* (θ_v); This is defined as the maximum difference between two opposite plane radiant temperatures. It is a quantity having both magnitude and direction and obeys the laws of vector addition. It can be visualised as the average radiant temperature of one side of a room minus the other half in a direction such that it is a maximum.

Figure 4.15

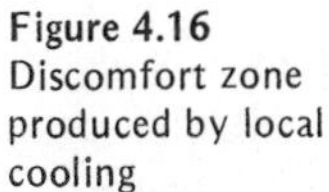

Direct measurement requires a differential radiometer. At the design stage estimates can be made by first calculating the six principal plane radiant temperatures (Figure 4.14) then computing the differences along each axis and adding the differences vectorially, (Figure 4.15), thus

$$\theta_v^2 = (\theta_{p1} - \theta_{p2})^2_X + (\theta_{p3} - \theta_{p4})^2_Y + (\theta_{p5} - \theta_{p6})^2_Z \tag{4.24}$$

where θ_v = vector radiant temperature (°C)

4.7.2 Localised discomfort criteria

Section A1 of the C.I.B.S. Guide[5] outlines the various criteria to avoid localised discomfort, as follows:

Local cooling (windows)

(a) C.I.B.S. recommends that provided θ_{res} criteria have been met then ($\theta_{res} - \theta_p$) should not exceed 8 °C. Figure 4.16 provides an indication of the likely extent of the discomfort zone adjacent to a single glazed window. Figure 4.17 gives

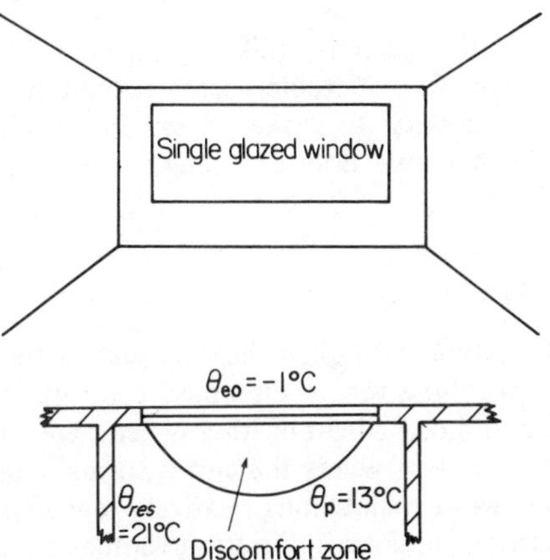

Figure 4.16 Discomfort zone produced by local cooling

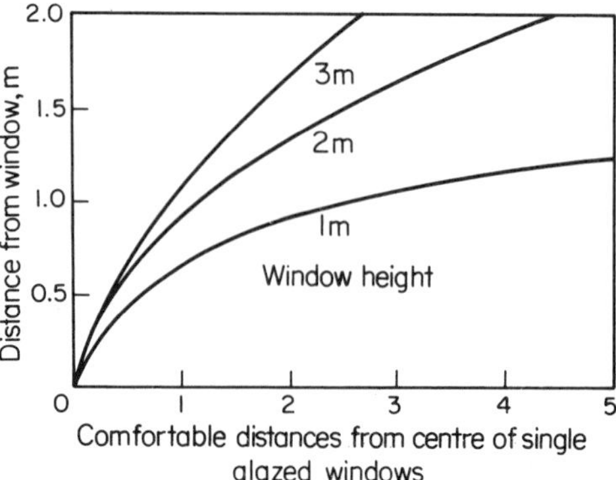

Figure 4.17 Comfortable distances from centre of single glazed windows

an indication of the comfortable distances from the centre of single glazed windows. These diagrams can be used to make an initial assessment of the problem. They do not take account of the effects of heated surfaces such as radiators placed below windows.

(b) As a result of experimental work using the plane thermometer, Anquez and Croiset[6] recommend that the air-radiant temperature (θ_{ar}) should not be less than about 17 °C adjacent to cold window surfaces. (Using equation 4.23 and referring to Figure 4.16, assuming θ_a = 21 °C, the profile indicated for θ_p = 13 °C corresponds to θ_{ar} = 16.3 °C.)

Notes:

(i) Double glazing effectively eliminates local cooling problems.

(ii) Plane thermometers give unreliable readings at distances less than 0.5 m from cold windows due to cold downdraughts causing increased air velocities and low air temperatures. Cognisance of these effects must also be taken when calculations are used to assess θ_{res} and θ_{ar}.

(iii) Tall single-glazed windows (> 2.0 m high) can give rise to noticeable discomfort by causing downdraughts which may penetrate into the room by as much as 1.5 m[7] at floor level. This can largely be offset by placing a convective heat source (e.g. a radiator) below the window.)

(c) Although it is unlikely that cold windows alone will give rise to noticeable vector radiant temperatures, it may be necessary to make checks on this (as described below), when wall, floor or ceiling heating is employed.

Radiant heating

For all types of radiant heating system the primary criteria to be met are those for 'whole body' comfort as discussed in Sections 4.5 and 4.6. Radiant heating systems are applied mainly to industrial situations where the aim is to maintain satisfactory comfort levels while maintaining relatively low air temperatures and thus incurring minimal infiltration/ventilation heat losses. As stated previously there are two main categories of radiant heating systems:

(a) In *longwave* radiant heating systems the heat is normally derived from LPHW, pressurised hot water or steam and surface temperatures are normally less than about 300 °C. Such surfaces do not emit light. 'Radiant' systems of this type are really 'mixed' systems, i.e. they have a significant convective component (see Chapter 8, example 8.3) emitted directly from the heated surface. Normally the radiant energy is emitted diffusely, i.e. there are no strongly directional effects. Such systems can be described in terms of the radiant field parameters previously discussed in Section 4.7.1. It is necessary to employ absolute temperatures in calculations in order to obtain the necessary accuracy. As yet no *upper* limits have been recommended for θ_p or θ_{ar} and the localised discomfort criteria are simply that

   (i) In *office-type* situations it is assumed that relatively low-temperature/large-area heating panels will be employed (e.g. wall, floor or ceiling heating). Although θ_v values up to 20 °C do not cause discomfort, values in excess of 10 °C are noticeable and should be avoided.

   (ii) in *industrial* situations the upper limit can be relaxed to $\theta_v \leqslant 20$ °C provided the irradiance criteria given below are not exceeded; this would be unlikely to arise in practice.

(b) Shortwave radiant heating systems (source temperatures above about 500 °C) are discussed in Section 4.9.

Heat from lighting

Fluorescent lighting is a relatively low temperature heat source and is unlikely to cause any thermal discomfort effects at normal lighting levels. Tungsten lighting can however cause serious problems and may cause θ_v values of around 10 °C at illuminances of 850 lux. Table 4.7 reproduced from the C.I.B.S. Guide(5) gives further guidance.

Table 4.7 Lamp data. (After C.I.B.S. Guide Table A1.4)

Lamp type	*Radiation output* (W/m² lux)	*Elevation of* θ_r (°C) *for 1 klux diffuse*	*Reduction in design air temp*/°C *for 1 klux diffuse*	*Illuminance for* θ_v *of* 10 °C/*lux*
Tungsten filament Spot and GS	0.07	6	2.6	850
Tungsten halogen	0.05	5	2.2	1200
Sodium I.P. SOX	0.006	0.5	0.2	10000
Sodium HP HPS/U	0.009	0.8	0.4	6500
Mercury fluorescent MBFU	0.015	1.4	0.6	4000
Fluorescent–MCFE	0.008	0.7	0.3	7500

Solar penetration

Although the problem of solar penetration is greatest in summer it is still of some interest to the heating system designer because spring and autumn sunshine can be surprisingly strong and can give rise to localised discomfort complaints. The approximate effect on θ_r and θ_{res} can be seen from Figure 4.18 which is reproduced from the C.I.B.S. Guide[5]. See also Section 4.9.

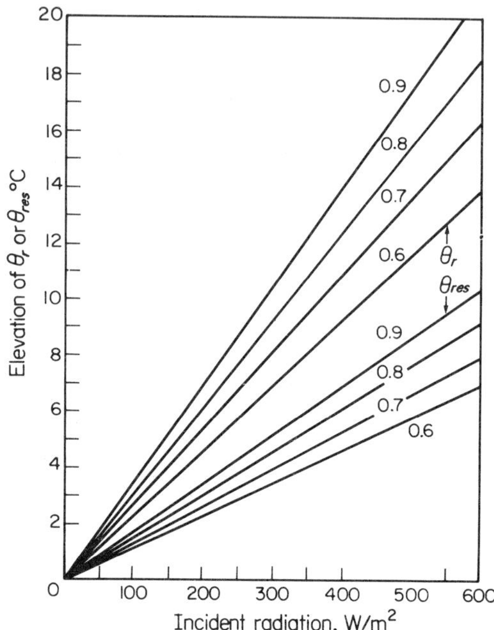

Figure 4.18
Effect of short-wave radiation on mean radiant and dry resultant temperatures

4.8 WORKED EXAMPLES ON 'LOCALISED DISCOMFORT' CHECKS

4.8.1 Introduction

The worked examples detailed below are based on the earlier examples used in Section 4.6, viz.

Proposal A—fully convective system
Proposal B—a LPHW radiator sited below the window
Proposal C—a LPHW radiator sited on a sidewall near to the window

When the 'whole body' checks have proved satisfactory but localised discomfort is suspected the preliminary steps to be carried out are:

(i) to check the 'local cooling' effect of the window, determine θ_p for the window wall or walls and then check the criteria in Section 4.7.

(ii) to check radiant field asymmetry, first determine θ_p for each of the six principal planes and then calculate θ_v as shown below.

The worked examples below detail these steps for each proposal although in practice serious radiant field asymmetry is unlikely with LPHW heating. The procedures can however be applied to all longwave radiant systems as discussed in Section 4.7. They cannot be used for shortwave radiant systems (for which see Section 4.9).

4.8.2 Proposal A (100% convective heating system)

(i) Local cooling check: Table 4.8 sets out the method of calculating the plane radiant temperatures based on the view (form) factors which can be determined as shown in

Table 4.8

Surface	θ_s	F_p	F_o	$F_p\theta_s, F_o\theta_s$
External wall (net)	17.4	0.5375	–	9.35
Window	5.5	0.2508	–	1.38
Floor, ceiling and sidewalls (by subtraction)	20.1	–	0.2177	4.26
				$\theta_{p1} = 14.99 \triangleq 15.0$

Appendix 4.2. The external wall and window surfaces are parallel to the plane element at the point P and hence subtend 'parallel' view factors (F_p). The floor, ceiling and sidewall surfaces are at right angles to the plane element at P and subtend 'orthogonal' view factors (F_o).

The difference in surface temperatures is less than 20 °C and hence Celsius temperatures can be used without serious loss of accuracy.

From Section 4.6.3, $\theta_{res,p}$ = 19.7 °C and hence ($\theta_{res,p} - \theta_{p1}$) = 4.7 °C. Also we compute θ_{ar} from equation 4.23,

$$\theta_{ar} = (0.41 \times 21.7) + (0.59 \times 15.0) = 17.7\,°C.$$

Thus based on an assumed θ_{ap} = 21.7 °C, local discomfort does not exist. This assumption makes no allowance for local cooling of the air but note that if a value of θ_{ap} = 19 °C were assumed then θ_{ar} would be about 16.6 °C.

(ii) Radiant field asymmetry check:
Inspection of the surface temperatures shows that the plane radiant temperature calculated at the point P of θ_{p1} = 15.0 °C in the direction of the cold window surface will be the minimum value for that point and that the remaining five principal plane radiant temperatures will probably lie in the range 15 °C to about 20 °C and hence it is most

unlikely that an excessive vector radiant temperature will occur at this point. In order to demonstrate the procedure however the remaining principal θ_p values have been calculated as per the method shown in Table 4.8:

$$\theta_{p1} = 15.0\,^\circ C, \quad \theta_{p2} = 19.9\,^\circ C$$
$$\text{hence } (\theta_{p1} - \theta_{p2}) = -4.9\,^\circ C$$
$$\theta_{p3} = 18.6\,^\circ C, \quad \theta_{p4} = 18.6\,^\circ C$$
$$\text{hence } (\theta_{p3} - \theta_{p4}) = 0\,^\circ C$$
$$\theta_{p5} = 16.1\,^\circ C, \quad \theta_{p6} = 19.8\,^\circ C$$
$$\text{hence } (\theta_{p5} - \theta_{p6}) = -3.7\,^\circ C$$
$$\text{and hence } \theta_v^2 = (24.0 + 13.7)$$
$$\text{and } \theta_v = 6.1\,^\circ C$$

This result confirms that no significant radiant field asymmetry exists at point P.
Using the same methods θ_{ar} and θ_v could be checked for other points of interest, (e.g. head height at x = 2.0 m y = 4.5 m and z = 1.2 m (Q) (sitting), or 1.8 m (R) (standing).

4.8.3 Proposal B (panel radiator below window)

Table 4.9 shows the revised θ_{p1} value for the point with the panel radiator placed below the window.

$$T_{p1} = 303.4 \text{ K}$$
$$\theta_{p1} = 30.4\,^\circ C$$

and assuming θ_{ap} = 21.7 °C this gives θ_{ar} = 26.7 °C, which raises the question that *warm* local discomfort may result at this location.

Table 4.9

Surface	θ_s(°C)	T_s(K)	$T_s^4 \times 10^{-8}$	F_p	F_o	$FT_s^4 \times 10^{-8}$
Radiator (under window)	70	343	138	0.1954		26.96
External wall (net)	17.4	290.4	71	0.3421		24.29
Window	5.5	278.5	60	0.2508		17.81
Floor, ceiling and sidewalls (by subtraction)	20.1	293.1	74		0.2117	15.67
						Σ = 84.73

Using the same technique as before, the remaining principal plane radiant temperatures can be found and hence θ_v can be found:

$$\theta_{p1} = 30.4\,^\circ C, \theta_{p2} = 19.9\,^\circ C \quad \therefore (\theta_{p1} - \theta_{p2}) = 10.5\,^\circ C$$

$\theta_{p3} = 33.2\,°C, \theta_{p4} = 33.2\,°C \quad \therefore (\theta_{p3} - \theta_{p4}) = 0\,°C$

$\theta_{p5} = 16.1\,°C, \theta_{p6} = 26.0\,°C \quad \therefore (\theta_{p5} - \theta_{p6}) = -10.1\,°C$

thus $\theta_v^2 = (110.3 + 0 + 102.0)$

$\therefore \quad \theta_v = 14.6\,°C$

From these results we see that the presence of the radiator has eliminated the local cooling effect of the window. Indeed the radiator may well give rise to overheating and will certainly produce a noticeable radiation vector. Again the calculations could be repeated for other points of interest such as Q and R.

4.8.4 Proposal C (panel radiator on sidewall)

Table 4.10 shows the revised θ_{p1} value for this proposal.

and thus $\quad T_{p1} = 288.1$ K

hence $\quad \theta_{p1} = 15.1\,°C$

and assuming $\theta_{ap} = 21.7\,°C$, then $\theta_{ar} = 17.8\,°C$, (or assuming $\theta_{ap} = 19\,°C, \theta_{ar} = 16.7\,°C$).

Table 4.10

Surface	θ_s(°C)	T_s(K)	$T_s^4 \times 10^{-8}$	F_p	F_o	$FT_s^4 \times 10^{-8}$
Radiator (on sidewall)	70	343	138		0.0008	0.11
External wall (net)	17.4	290.4	71	0.5375		38.16
Window	5.5	278.5	60	0.2508		15.05
Floor, ceiling and sidewalls (by subtraction)	20.1	293.1	74		0.2109	15.61
						Σ = 68.93

Repeating the procedure used previously to find θ_v

$\theta_{p1} = 15.1\,°C, \theta_{p2} = 23.8\,°C \quad \therefore (\theta_{p1} - \theta_{p2}) = -8.7\,°C$

$\theta_{p3} = 21.0\,°C, \theta_{p4} = 18.6\,°C \quad \therefore (\theta_{p3} - \theta_{p4}) = 2.4\,°C$

$\theta_{p5} = 16.1\,°C, \theta_{p6} = 20.2\,°C \quad \therefore (\theta_{p5} - \theta_{p6}) = -4.1\,°C$

hence $\quad \theta_v^2 = 75.7 + 5.8 + 16.8$

$\therefore \quad \theta_v = 9.9\,°C$

Thus we can conclude for point P that local cooling may cause discomfort and the radiator place on the sidewall increases the radiation vector which may become noticeable. Again the exercise could be repeated for points Q and R.

4.8.5 Summary

From the above worked examples we can see that, at the expense of some labour, we can determine whether a problem exists with

Proposal A and investigate alternative solutions. The chosen point P was chosen as representing the nearest practical point to the external wall at which an occupant might sit and, as can be seen, none of the proposals is really satisfactory. As a result of this analysis a number of alternative solutions could be proposed for re-valuation, such as

a revised work location,
a revised heating panel position/size/surface temperature,
a revised window size/type/position.

As stated previously the above examples do not really reveal that a serious problem exists but they do serve to show the method which can be applied readily to true longwave radiant heating systems where experience dictates that a problem may exist.

4.9 SHORTWAVE RADIANT HEATING SYSTEMS

4.9.1 Introduction

As previously discussed in Section 2.4, thermal radiation is defined as electromagnetic radiation having wavelengths of between 1×10^{-4} m and 1×10^{-7} m (see Figure 2.10). This range includes the visible wavelength range and overlaps the infra-red and ultraviolet range. Figure 2.13 indicates that as the source temperature increases the emissive power increases (i.e. area under curve) and

Figure 4.19
Distribution of radiation from a black body at different temperatures

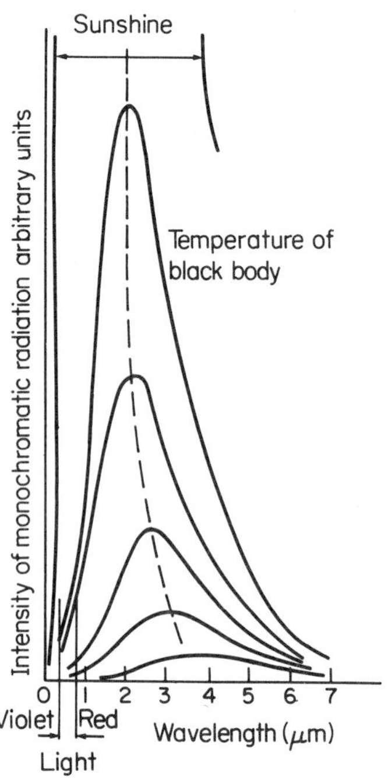

that there is also a shift in the distribution such that a greater proportion of the thermal radiation from high temperature sources is emitted at shorter wavelengths. Figure 4.19 shows that (for a black body) at low temperatures (5 °C) no shortwave radiation is emitted, while at high temperatures (> 1000 °C) there is a predominance of shortwave although longwave radiation is still present. Figure 4.20 shows that real (non-grey) emitters will tend to emit more strongly at certain characteristic wavelengths. Thus we can see that we must use the terms 'longwave' and 'shortwave' radiant heating systems with some caution.

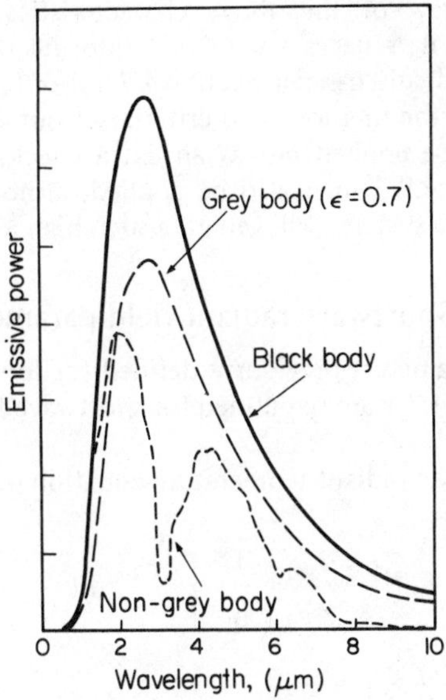

Figure 4.20 Monochromatic emissive power of a black body, a grey body and a non-grey body

The discussion about equation 2.58 in Section 2.4 highlights the fact that as the shortwave content of the heat source increases the absorptivity of the receiving body is no longer equal to its emissivity and thus the rate of heat transfer becomes in a sense independent of the temperature of the receiving body.

Although there is no clear distinction between longwave and shortwave radiant systems a general division might be:

(a) The surface temperature of a 'longwave' heating system is seldom greater than 300 °C and is usually less than 200 °C.

(b) A variety of shortwave systems exist, each characterised by a source temperature. Infra-red (electric) lamps usually have a source temperature of about 2500 °C while incandescent gas-fired heaters have a source temperature of around 1200 °C. Both of the above types also emit light and are thus known as 'bright' shortwave heat sources. However, provided surface temperatures are above about 500 °C then the shortwave component will usually be significant enough to be described as a shortwave source. Another important characteristic of true shortwave radiant heating systems is that they are generally designed to emit thermal radiation in a 'directed beam' in order to heat only the working area.

A principal design objective is to produce an acceptable level of whole body comfort without creating an unacceptable asymmetry in the radiant field. Another design objective is usually to maintain the required heat balance between the occupant(s) and the environment by creating a high mean radiant temperature with a relatively low air temperature and hence minimising infiltration/ventilation heat losses.

In view of the above characteristics of shortwave radiant systems it is necessary to re-define the radiant field parameters previously discussed in Section 4.7.1. For longwave radiant heating systems the methods and criteria set out in Sections 4.7 and 4.8 should be applied but as an extra check, the criteria set out in Section 4.9.3 may also be applied, although it is unlikely that longwave systems will generate such high levels of irradiance.

4.9.2 Shortwave radiant field parameters

The parameters previously defined for longwave radiant fields in Section 4.7.1 are re-defined for shortwave fields as follows

(i) Mean radiant temperature equation 4.16 becomes

$$T_r^4 = \frac{I_{sw}}{4\sigma} + \frac{1}{4\pi}\int_0^{4\pi} T_s^4 \, d\psi \qquad (4.25)$$

where I_{sw} = shortwave radiation (W/m^2)

σ = Stefan-Boltzmann constant
= 5.67×10^{-8} W/m^2 K^4

This can be re-written in a form similar to equation 4.18 thus

$$T_{rp}^4 = \frac{I_{sw}}{4\sigma} + (F_{ap} T_s^4) \qquad (4.26)$$

and where the longwave term can be evaluated using equation 4.19 if the shortwave source is small and the other surfaces do not differ by more than about 20 °C. The increase in mean radiant temperature (ΔT_r) caused by the irradiance from the shortwave term can be expressed as

$$\Delta T_{rp} = \frac{I_{sw}}{16\sigma T_{rp}'^3} \ (°C) \qquad (4.27)$$

where I_{sw} = irradiance at the point from the shortwave source (W/m^2)

T'_{rp} = unirradiated (longwave only) mean radiant temperature (K)

When T'_{rp} is around 293 K, (20 °C) then equation 4.27 reduces approximately to

$$\Delta T_{rp} = 0.044 I_{sw} \ (°C) \tag{4.28}$$

(ii) Plane radiant temperature
For shortwave radiant systems equation 4.20 becomes

$$T_p^4 = \left(\frac{I_{sw}}{\sigma}\right) + \Sigma\,(F_p\,T_s^4) + \Sigma\,(F_o\,T_s^4)\ (K^4) \tag{4.29}$$

and when appropriate (see equation 4.21) equation 4.29 can be re-written as

$$T_p = 64.8 I_{sw}^{0.25} + \Sigma\,(F_p\,T_s) + \Sigma\,(F_o\,T_s)\ (K) \tag{4.30}$$

(iii) Air-radiant temperature
As stated previously, no upper limits of θ_{ar} are at present recommended as a guide to overheating. Nevertheless, values of local air temperature and the plane radiant temperature evaluated as above could be used with equation 4.23 to give notional values. This may form the basis of future recommendations but perhaps with a modified weighting (i.e. different h_c and h_R values in equation 4.22).

(iv) Vector radiant temperature
Vector radiant temperature can be calculated exactly as before using equation 4.24 provided the principal plane radiant temperatures have been calculated as discussed in this Section. In practice, for shortwave radiant systems the principal longwave components are often approximately equal and, further, the radiation vector (R) can often be determined fairly accurately by inspection and in such cases,

$$\theta_v = \frac{R}{4\sigma\,(T'_{rp})^3}\ (°C) \tag{4.31}$$

where R = radiation vector (i.e. maximum difference in irradiance) (W/m^2)

where T'_{rp} = is around 293 K (i.e. 20 °C) this reduces approximately to

$$\theta_v = 0.175 R\ (°C) \tag{4.32}$$

4.9.3. Shortwave radiant heating comfort criteria

Generally, shortwave radiant heating is used for industrial applications especially for high-roofed buildings; quite often no direct air heating is employed provided the radiant heat input will provide a sufficient heat input from secondary convection from heated surfaces. The C.I.B.S. Guide criteria are framed in terms of a typical industrial application and the implication is that general radiant heating is to be provided throughout the working area as opposed to 'spot' heating (for which see Section 4.9.4). The primary aim will be to determine and provide appropriate θ_{res} criteria to ensure 'whole body' comfort (or to use Fanger's methods). The additional 'localised discomfort' criteria to be met are as follows, (these assume an air temperature of around 15 °C),

(i) the summed mean spherical irradiance at 1.8 m above floor level should not exceed 240 W/m^2; (if this value is inserted in equation 4.28 as a rough check this corresponds to a maximum ΔT_r of around 11 °C).
(ii) the total irradiance (from all heaters) at floor level should not exceed 80 W/m^2; (if this value is inserted in equation 4.32 as a rough check this corresponds to a maximum θ_v of around 14 °C),
(iii) the maximum irradiance from any one heater at floor level should not exceed 32 W/m^2 (corresponding to θ_v of around 6 °C).

From the above we can see that the irradiance values have been set as a guide to avoid local overheating ($\Delta T_r \leqslant 11$ °C) and to prevent excessive radiant field asymmetry which for industrial situations should preferably give θ_v less than 10 °C but in any case should not exceed 20 °C.

In applying the above criteria, usually fairly simple calculations for irradiances in the working zone will prove sufficient especially if the heaters are mounted at some distance from the occupants. When mounting shortwave radiant heaters at low level or if using unfamiliar heaters it is wise to seek the manufacturer's advice on emission characteristics, as the radiant intensity may vary considerably with direction from the normal. In this respect the design of radiant heating systems is very similar to the design of lighting systems.

4.9.4 'Spot' heating

In situations where one or two occupants are positioned at fixed work locations in an otherwise unheated building the use of 'spot' or beam heaters, (e.g. electrical infra-red lamps), is an economical and energy-conserving solution. The evaluation of the mean radiant temperature in such cases must take account of the directional output characteristics of the heater, the heater-to-person view factor (F_b), the absorbtivity of the occupant's clothing and a number of other considerations. The reader is referred to Fanger's book[1] in which a chapter is devoted to the solution of such problems. Fanger's solution is framed in terms of whole body comfort criteria but the localised discomfort criteria given in 4.9.3 should also be applied.

APPENDIX 4.1

A4.1 CALCULATION OF SOLID ANGLE FACTORS (F_a)*

*The authors gratefully acknowledge the assistance of Mr A. R. Veitch, BSc, Lecturer, Department of Mathematics, University of Strathclyde, Glasgow, in formulating this appendix.

Figure 4.21 (reproduced from the C.I.B.S. Guide[5]) provides a method of determining F_a but unfortunately for many cases of practical interest the method is inaccurate.

Figure 4.21
Angle factors

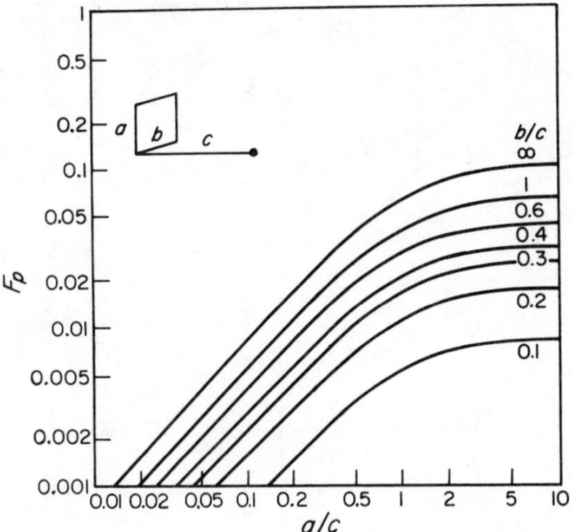

An alternative method is given in Table 4.11 which provides a basis for which most cases can be solved using simple linear interpolation and the angles altitude (α) and azimuth (β), angles which can be measured directly from scale drawings. The accuracy thus obtained should in most cases be sufficient and consistent with the technique of using measured angles, but in other cases F_a can be evaluated directly using equation 4.34 which can be derived as follows:

Referring to Figure 4.22 let O be the vortex at which the solid angles are to be calculated. ABCD lies on a plane at right angles to the line OA and subtends the angles α and β as shown. Then if θ, ϕ are the usual spherical polar angles measured from OXYZ, the solid angle Ω_c subtended at O by ABCD may be calculated by the integration

$$\Omega_c = \iint_{ABUO} \sin\theta \, d\theta \, d\phi = \int_0^{\beta}\int_{\tan^{-1}(\sec\phi\cos\alpha)} \sin\theta \, d\theta \, d\phi ,$$

$$\text{thus } \Omega_c = \sin^{-1}(\sin\alpha \sin\beta) \text{ (steradians)} \qquad (4.33)$$

Table 4.11 Solid Angle Factors (F_a)

α	β=0		10°		20°		30°		40°		50°		60°		70°		80°		90° β	α
90°	0	(139)	.0139	(139)	.0278	(139)	.0417	(139)	.0556	(138)	.0694	(139)	.0833	(139)	.0972	(139)	.1111	(139)	.1250	90°
			(02)		(05)		(07)		(11)		(14)		(20)		(31)		(57)		(139)	
80°	0	(137)	.0137	(136)	.0273	(137)	.0410	(135)	.0545	(135)	.0680	(133)	.0813	(128)	.0941	(113)	.1054	(57)	.1111	80°
			(07)		(13)		(21)		(29)		(41)		(56)		(80)		(113)		(139)	
70°	0	(130)	.0130	(130)	.0260	(129)	.0389	(127)	.0516	(123)	.0639	(118)	.0757	(104)	.0861	(80)	.0941	(31)	.0972	70°
			(10)		(21)		(33)		(46)		(62)		(82)		(104)		(128)		(139)	
60°	0	(120)	.0120	(119)	.0239	(117)	.0356	(114)	.0470	(107)	.0577	(98)	.0675	(82)	.0757	(56)	.0813	(20)	.0833	60°
			(14)		(28)		(43)		(60)		(78)		(98)		(118)		(133)		(139)	
50°	0	(106)	.0106	(105)	.0211	(102)	.0313	(97)	.0410	(89)	.0499	(78)	.0577	(62)	.0639	(41)	.0680	(14)	.0694	50°
			(17)		(35)		(53)		(71)		(89)		(107)		(123)		(135)		(138)	
40°	0	(89)	.0089	(87)	.0176	(84)	.0260	(79)	.0339	(71)	.0410	(60)	.0470	(46)	.0516	(29)	.0545	(11)	.0556	40°
			(20)		(39)		(59)		(79)		(97)		(114)		(127)		(135)		(139)	
30°	0	(69)	.0069	(68)	.0137	(64)	.0201	(59)	.0260	(53)	.0313	(43)	.0356	(33)	.0389	(21)	.0410	(07)	.0417	30°
			(22)		(44)		(64)		(86)		(102)		(117)		(129)		(137)		(139)	
20°	0	(47)	.0047	(46)	.0093	(44)	.0137	(39)	.0176	(35)	.0211	(28)	.0239	(21)	.0260	(13)	.0273	(05)	.0278	20°
			(23)		(46)		(68)		(87)		(105)		(119)		(130)		(136)		(139)	
10°	0	(24)	.0024	(23)	.0047	(22)	.0069	(20)	.0089	(17)	.0106	(16)	.0120	(10)	.0130	(07)	.0137	(02)	.0139	10°
			(24)		(47)		(69)		(89)		(106)		(120)		(130)		(137)		(139)	
0°	0		0		0		0		0		0		0		0		0		0	0°
α	β 0°		10°		20°		30°		40°		50°		60°		70°		80°		90° β	α

from which

$$F_a = \left(\frac{\Omega_c}{4\pi}\right) = \frac{1}{4\pi}\left(\sin^{-1}(\sin\alpha \sin\beta)\right) \tag{4.34}$$

Appendix 4.3 describes the actual procedure for a number of cases.

Figure 4.22

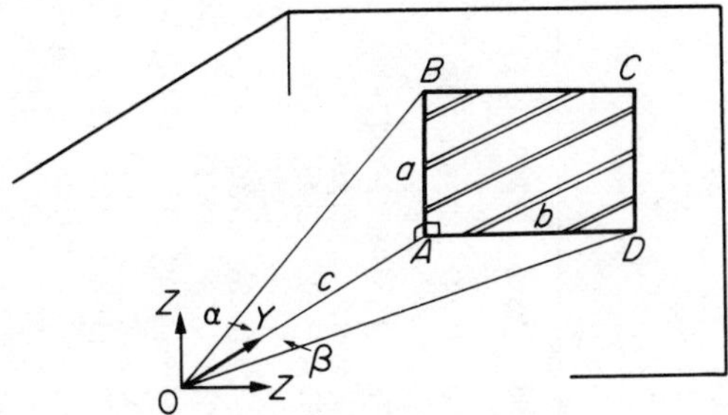

A4.2 CALCULATION OF VIEW FACTORS F_p AND F_o*

View factor for a parallel element (F_p)

Figure 4.23 (from A1.17 reproduced from the C.I.B.S. Guide[5]) provides a graphical method of determining F_p but unfortunately

Figure 4.23
Parallel form factors

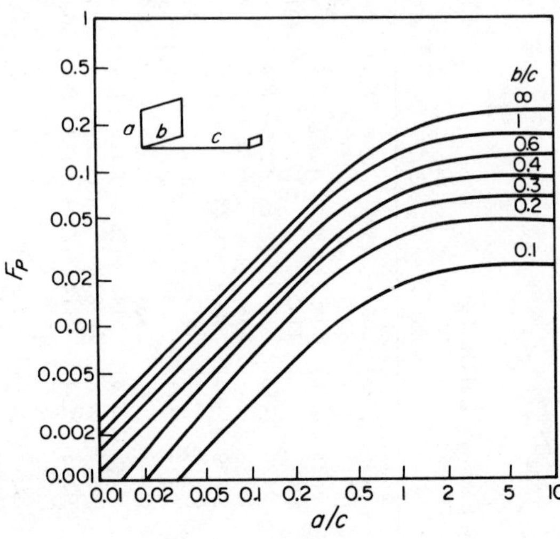

for many cases of practical interest the method is inaccurate. The graph is based on the evaluation of

$$F_p = \frac{1}{2\pi}\left(A_1 \tan^{-1} A_2 + B_1 \tan^{-1} B_2\right) \tag{4.35}$$

Table 4.12 View Factors (Parallel) (F_p)

α	β=0°		10°		20°		30°		40°		50°		60°		70°		80°		90° β	α
90°	0	(434)	.0434	(421)	.0855	(395)	.1250	(357)	.1607	(308)	.1915	(250)	.2165	(184)	.2349	(113)	.2462	(38)	.2500	90°
			(01)		(02)		(03)		(05)		(06)		(09)		(14)		(24)		(38)	
80°	0	(433)	.0433	(420)	.0853	(394)	.1247	(355)	.1602	(307)	.1909	(247)	.2156	(179)	.2335	(103)	.2438	(24)	.2462	80°
			(07)		(14)		(22)		(31)		(43)		(58)		(79)		(103)		(113)	
70°	0	(426)	.0426	(413)	.0839	(386)	.1225	(346)	.1571	(295)	.1866	(232)	.2098	(158)	.2286	(79)	.2336	(14)	.2349	70°
			(17)		(36)		(56)		(77)		(102)		(122)		(159)		(179)		(184	
60°	0	(409)	.0409	(394)	.0803	(366)	.1169	(325)	.1494	(270)	.1764	(203)	.1967	(130)	.2097	(59)	.2156	(09)	.2165	60°
			(32)		(65)		(98)		(124)		(170)		(212)		(231)		(247)		(250)	
50°	0	(377)	.0377	(361)	.0736	(333)	.1071	(299)	.1370	(234)	.1594	(120)	.1764	(102)	.1866	(43)	.1909	(06)	.1915	50°
			(44)		(96)		(145)		(201)		(234)		(270)		(295)		(307)		(308)	
40°	0	(328)	.0328	(314)	.0642	(384)	.0926	(243)	.1169	(191)	.1360	(134)	.1494	(77)	.1571	(31)	.1602	(05)	.1607	40°
			(65)		(128)		(188)		(243)		(289)		(325)		(346)		(354)		(357)	
30°	0	(263)	.0263	(251)	.0514	(224)	.0738	(188)	.0926	(144)	.1071	(99)	.1169	(56)	.1225	(23)	.1248	(02)	.1250	30°
			(78)		(155)		(224)		(284)		(333)		(366)		(386)		(395)		(395)	
20°	0	(185)	.1085	(174)	.0359	(155)	.0514	(128)	.0642	(96)	.0738	(65)	.0803	(36)	.0839	(14)	.0853	(02)	.0855	20°
			(90)		(174)		(251)		(314)		(361)		(394)		(413)		(420)		(421)	
10°	0	(95)	.0095	(90)	.0185	(78)	.0263	(65)	.0328	(49)	.0377	(32)	.0409	(17)	.0426	(07)	.0433	(01)	.0434	10°
			(95)		(185)		(263)		(328)		(377)		(409)		(426)		(433)		(434)	
0°	0		0		0		0		0		0		0		0		0		0	0°
α	β=0°		10°		20°		30°		40°		50°		60°		70°		80°		90° β	α

where

$$A_1 = \sin\alpha;\ A_2 = \cos\alpha\ \tan\beta$$

$$B_1 = \sin\beta;\ B_2 = \cos\beta\ \tan\alpha$$

To reduce the labour of computation, Table 4.12 provides a basis for interpolation; for most cases simple linear interpolation will suffice and will be consistent with the accuracy obtained from lifting angles from scale drawings. In other cases direct use of equation 4.35 can be made.

View factor for an orthogonal element (F_o)

The C.I.B.S. Guide provides no information on how to evaluate the view factors of surfaces at right angles (i.e. orthogonal) to the plane element. In this case F_o can be evaluated using

$$F_o = \frac{c}{2\pi}\left[\frac{1}{c}\tan^{-1}\left(\frac{a}{c}\right) - \left(\frac{1}{\sqrt{(b+c)}}\right)\tan^{-1}\frac{a}{\sqrt{(b+c)}}\right] \tag{4.37}$$

where a, b and c are defined in Figure 4.24.

Figure 4.24

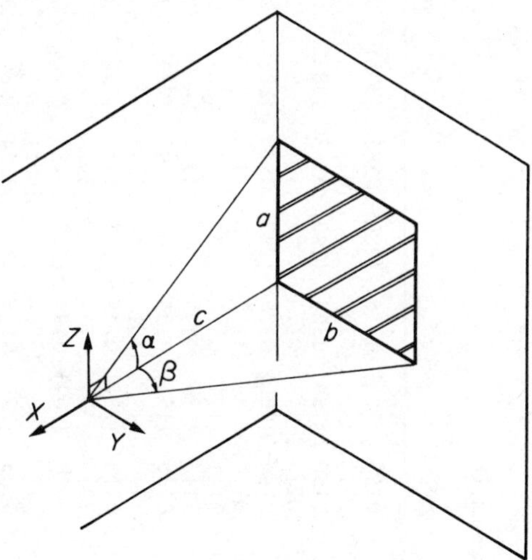

This can be transformed as to give an expression

$$F_o = \frac{1}{2\pi}\left[\alpha - \cos\beta\ \tan^{-1}\left(\cos\beta\ \tan\alpha\right)\right] \tag{4.38}$$

where angles are expressed in radians.

It is important to note that unlike equations 4.34 and 4.36, this expression is no longer symmetrical and care should be taken to ensure that α and β are correctly measured. A useful guide is

Table 4.13 View Factors (orthogonal) (F_o)

α	β=0°		10°		20°		30°		40°		50°		60°		70°		80°		90° β	α
90°	0	(39)	.0039 (0)	(145)	.0151 (01)	(184)	.0335 (01)	(250)	.0585 (02)	(308)	.0893 (04)	(357)	.1250 (08)	(395)	.1645 (19)	(421)	.2066 (59)	(434)	.2500 (278)	90°
80°	0	(39)	.0039 (02)	(145)	.0150 (02)	(184)	.0334 (06)	(249)	.0583 (13)	(306)	.0889 (24)	(353)	.1242 (47)	(384)	.1626 (92)	(381)	.2007 (186)	(215)	.2222 (278)	80°
70°	0	(37)	.0037 (01)	(111)	.0146 (07)	(180)	.0328 (16)	(242)	.0570 (31)	(295)	.0865 (57)	(330)	.1195 (96)	(339)	.1534 (158)	(287)	.1821 (235)	(123)	.1944 (278)	70°
60°	0	(36)	.0036 (03)	(105)	.0141 (11)	(171)	.0312 (27)	(227)	.0539 (52)	(269)	.0808 (88)	(291)	.1099 (138)	(277)	.1376 (198)	(210)	.1586 (254)	(80)	.1666 (277)	60°
50°	0	(33)	.0033 (04)	(97)	.0130 (17)	(155)	.0285 (40)	(202)	.0487 (72)	(233)	.0720 (115)	(241)	.0961 (166)	(217)	.1178 (219)	(154)	.1332 (261)	(57)	.1389 (278)	50°
40°	0	(29)	.0029 (06)	(84)	.0113 (23)	(132)	.0245 (51)	(170)	.0415 (89)	(190)	.0605 (135)	(190)	.0795 (185)	(164)	.0959 (232)	(112)	.1071 (265)	(40)	.1111 (278)	40°
30°	0	(23)	.0023 (07)	(67)	.0090 (27)	(104)	.0194 (59)	(132)	.0326 (102)	(144)	.0470 (150)	(140)	.0610 (198)	(117)	.0727 (239)	(79)	.0806 (268)	(27)	.0833 (277)	30°
20°	0	(16)	.0016 (08)	(47)	.0063 (31)	(72)	.0135 (66)	(89)	.0224 (110)	(96)	.0320 (158)	(92)	.0412 (204)	(76)	.0488 (243)	(50)	.0538 (269)	(18)	.0556 (278)	20°
10°	0	(08)	.0008 (08)	(24)	.0032 (32)	(37)	.0069 (69)	(45)	.0114 (114)	(48)	.0162 (162)	(46)	.0208 (208)	(37)	.0245 (245)	(24)	.0269 (269)	(91)	.0278 (278)	10°
0°	0		0		0		0		0		0		0		0		0		0	0°
α	β=0°		10°		20°		30°		40°		50°		60°		70°		80°		90° β	α

to note that here α is always measured in the same plane as the small plane element.

Table 4.13 provides a basis upon which most cases can be solved using a simple linear interpolation and gives accuracies consistent with angles measured from scale drawings. In other cases direct use of equation 4.38 can be made. Appendix 4.3 shows the actual procedure for a number of cases.

A4.3 PROCEDURE FOR DETERMINING F_a, F_p OR F_o

The same basic procedure applies to all three factors and they will be referred to henceforth simply as F.

Figure 4.25
General case

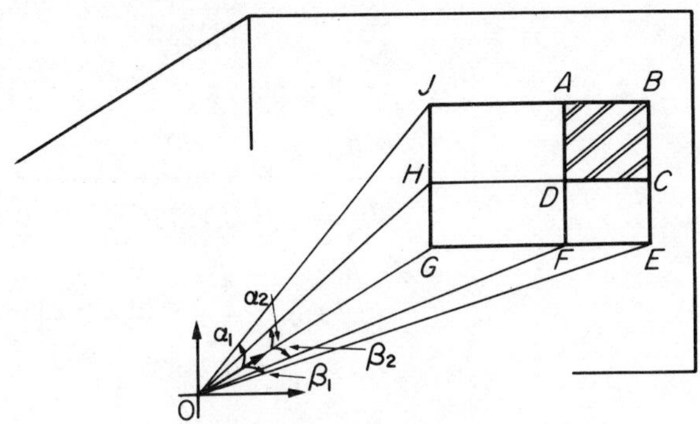

Figure 4.25 shows the diagram for the general case where the F value for the rectangle, ABCD is required. This is found by:

$$F_{\mathrm{ABCD}} = F_{\mathrm{JBGE}} - F_{\mathrm{JAFG}} - F_{\mathrm{HCBG}} + F_{\mathrm{HDFG}} \qquad (4.39)$$

(It is necessary to add the F term as a factor for this area is removed in both the F_{JAFG} and the F_{HCEG} terms.)

Each of the terms on the right hand side of equation 4.39 can be found from the graphs, tables or equations given previously.

In practice the actual procedures will often be different from those described but the above provides the basis of the method.

Figure 4.26
Gross external wall

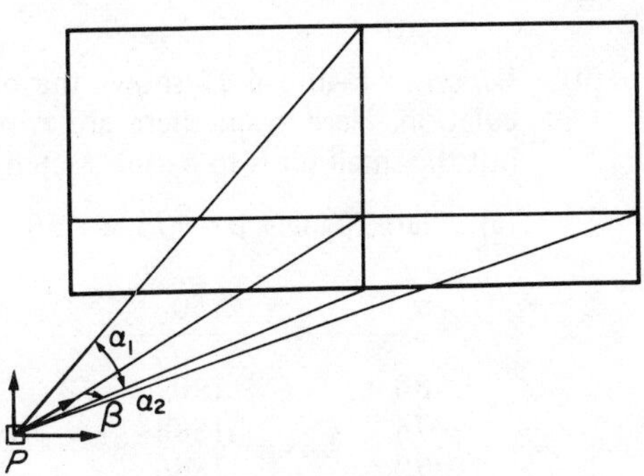

Worked example

As an example of the method, the view factors F_p given in Table 4.12 and used to evaluate θ_{p1} at point P for the fully convective case (Proposal A) were determined by measuring the required α and β angles from the scale drawing, Figure 4.12.

(i) *Gross external wall*: Figure 4.26 shows that from the point P the wall can be evaluated as four separate panels (two sets of two identical panels in this case).

(a) large panels, $\alpha = +78^\circ$ $\beta = 74^\circ$ and using Table 4.12:

α β	70	74	80
80	0.2335		0.2438
78	0.2319	0.23580	0.2417
70	0.2256		0.2335

(b) small panels, $\alpha = -40^\circ$ $\beta = 74^\circ$ and using Table 4.12:

α β	70	74	80
40	0.1571	0.15838	0.1602

$$\therefore \quad F_{P(i)} = \sum_{1}^{4} F = (2 \times 0.23580) + (2 \times 0.15838)$$
$$= 0.7883$$

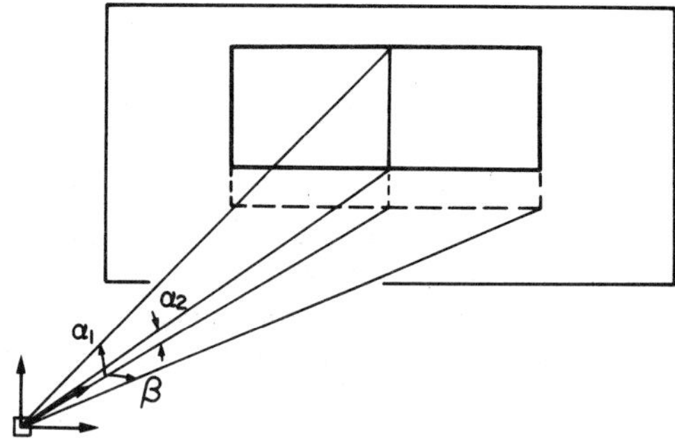

Figure 4.27
Window

(ii) *Window:* Figure 4.27 shows the basis of the window calculation. Here again there are two sets of identical panels but the small set is to be subtracted.

(a) large panels, $\alpha = +75$, $\beta = 56$

α β	50	56	60
80	.1909		.2156
75	.1888	.2031	.2127
70	.1866		.2098

(b) small panels, $\alpha = +20$, $\beta = 56$

α β	50	56	60
20	.0738	.0777	.0803

$$\therefore \quad F_{p(ii)} = \sum_{1}^{4} F = (2 \times 0.2031) - (2 \times 0.0777)$$
$$= 0.2508$$

(iii) *Net external wall:*

$$F_{p(iii)} = F_{p(i)} - F_{p(ii)} = (0.7883 - 0.2508) = 0.5375$$

REFERENCES

1. FANGER, P. O., *Thermal Comfort,* (McGraw-Hill, 1972).
2. A.S.H.R.A.E. Fundamentals, 1976.
3. BEDFORD, T., *Basic Principles of Ventilation and Heating,* (H. K. Lewis, London, 1948).
4. McINTYRE, D. A., 'A Guide to Thermal Comfort', *Applied Ergonomics,* June 1973, Vol. 4.2, pp 66–72.
5. C.I.B.S. Guide, Section A1 (Dec. 1978).
6. ANQUEZ and CROISET, M., 'Thermal comfort requirements adjacent to cold walls', *C.S.T.B.,* Jan/Feb. 1969 **96**, p 833.
7. SHILLINGLAW, J. A., 'Cold window surfaces and discomfort', *B.S.E.* July 1977, Vol. 45.
8. McINTYRE, D. A., 'The thermal radiation field', *Build. Sci.* 1974, **9**, pp 247–62.
9. CORNELL, A. A. and FORTEY, J., 'Environmental temperature and comfort', *B.S.E.,* Feb. 1976, Vol. 43.

LIST OF SYMBOLS USED IN CHAPTER 4

Temperatures

$\theta_a, \theta_{ai}, \theta_{ap}$	air dry bulb temperatures
$\theta_r, \theta_{ri}, \theta_{rp}$	mean radiant temperatures at a point
θ_{ar}	air-radiant temperature at a point
θ_b	mean body temperature
θ_{cb}	mean temperature of a clothed body
θ_{ei}	internal environmental temperature
θ_{eo}	external environmental temperature
θ_{ao}	external air dry bulb temperature
θ_s	known surface temperature
θ_r'	equivalent radiant temperature of remaining surfaces
θ_{res}	dry resultant temperature
θ_g	globe temperature
θ_{sub}	subjective temperature
$\theta_{p1}, \theta_{p2}, \theta_{p3}$ etc.	plane radiant temperature measured along a principal axis, 1,2,3 etc.

θ_v	vector radiant temperature
T_r, T_{ri}, T_{rp}	(absolute) mean radiant temperature
T_s	(absolute) surface temperatures
T'_r, T'_{ri}, T'_{rp}	(absolute) mean radiant temperature arising from longwave radiant field alone
ΔT_r, ΔT_{ri}, ΔT_{rp}	rise in mean radiant temperature due to irradiance from shortwave field

Other symbols

A_{cb}	area of clothed body
A_{DU}	Du Bois area: surface area of human body (nude)
A_s	area of room surface
clo	clothing resistance unit
F_a, F_{ai}, F_{ap}	solid angle factor subtended by a surface from a point
F_p	view factor between a surface and a small plane element *parallel* to the surface
F_o	view factor between a surface and a small plane element *orthogonal* to the surface
$\dot{H}$	internal heat production rate
h_c	convection heat transfer coefficient
h_R	radiation heat transfer coefficient
h_D	mass transfer coefficient
h_{fg}	latent energy of evaporation
I_{sw}	irradiance at a point due to shortwave radiation
M	metabolic rate
m	total number of room surfaces 'seen' by point
n	number of 'm' whose temperatures are known
p_{ss}	saturation vapour pressure
p_s	ambient vapour pressure
PMV	predicted mean vote } defined fully in text and Ref (1).
PPD	predicted percentage dissatisfied } defined fully in text and Ref (1).
LPPD	lowest possible percentage dissatisfied } defined fully in text and Ref (1).
$\dot{Q}_f$	fabric heat loss rate
Q_v	infiltration/ventilation heat loss rate
$\dot{Q}_t$	total heat loss rate
$\dot{q}_c$	rate of body heat loss by convection
$\dot{q}_R$	rate of body heat loss by radiation
$\dot{q}_E$	rate of body heat loss by evaporation
$\dot{q}_v$	ventilation allowance
v_a, v_{ai}, v_{ap}	air velocity at a point
V'	room volume
α	altitude angle
β	azimuth angle
θ, ϕ	spherical polar angles
Ω_c	solid angle
r.h.	relative humidity

5 Heating System Design

5.1 INTRODUCTION

Prior to the 1973 energy crisis it was common practice for an architect to finalise the form and fabric of a proposed building with little consideration for the subsequent thermal performance of his design. The engineer's role as the heating system designer was merely to accept this finalised design and estimate the plant size required to maintain an acceptable thermal environment. The general lack of emphasis on thermal performance prevalent at that time often gave rise to buildings which employed excessive glazing ratios, and high infiltration rates, contained numerous cold bridges and had a poor standard of thermal design and construction generally. As a result, many heating systems were deliberately designed with a large factor of safety to avoid the danger of underheating. Unfortunately many buildings thus designed will be with us for some time to come.

Post-1973, increased fuel prices and the general awareness of the need to conserve fossil fuel reserves have caused architects and engineers to cooperate at an early stage in the design to ensure improved thermal performance from the outset. The trend towards buildings which exhibit better thermal characteristics has been encouraged by various amendments to the Building Regulations and other similar legislation. Existing buildings are also being re-examined with a view to upgrading to save energy. The conventional (steady-state) heat loss calculations are now being performed with greater accuracy while in large projects the tendency is to apply the various non-steady-state calculations methods (usually computer-based) in an attempt better to understand and evaluate the various cause and effect relationships involved in intermittent plant operation and insulation strategies.

The future success of energy conserving 'building/heating system' designs rests largely on a close architect/engineer cooperations in the early stage of the design and it is therefore important for the engineer to understand the architectural design process in greater detail than was necessary hitherto.

5.2 THE DESIGN PROCESS

The design of heating systems must be seen in the context of the design of the form and fabric as well as the structure and the other

Figure 5.1

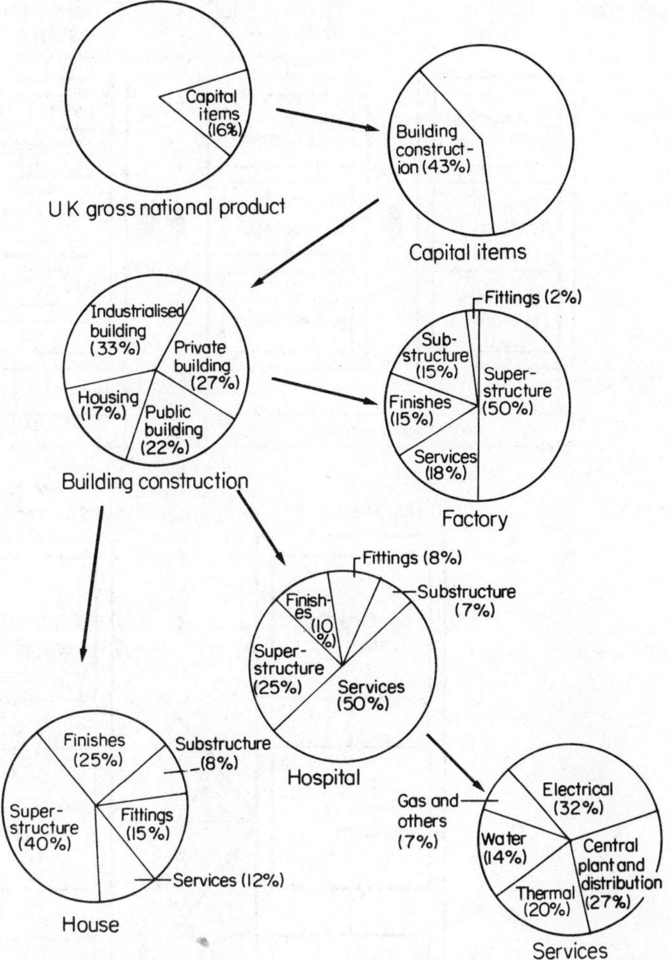

building services. Figure 5.1 (after Maver[1]) provides an overview of the importance of building services from various points of view. Figure 5.2 (after Markus[2]) provides a useful method of representing a building design problem in terms of individual sub-systems and provides a basis for cost/benefit analyses. Figure 5.3 shows the inter-relation between each of the various systems which are of primary interest to the heating system designer; there are strong links (full lines) and weak links (broken lines) between the building and environmental systems. Futher, each system must be designed taking account of its own technical constraints.

Other links to the activity and client systems must also be taken into consideration and the problem should (ideally) be resolved in a way which optimises the system as a whole. This would be done by a composite cost-benefit analysis of the system as a whole. In practice this usually proves difficult and often impossible (due to lack of data) and, as a second-best solution, optimum performance is sought for each sub-system as an entity in so far as it is possible

Figure 5.2

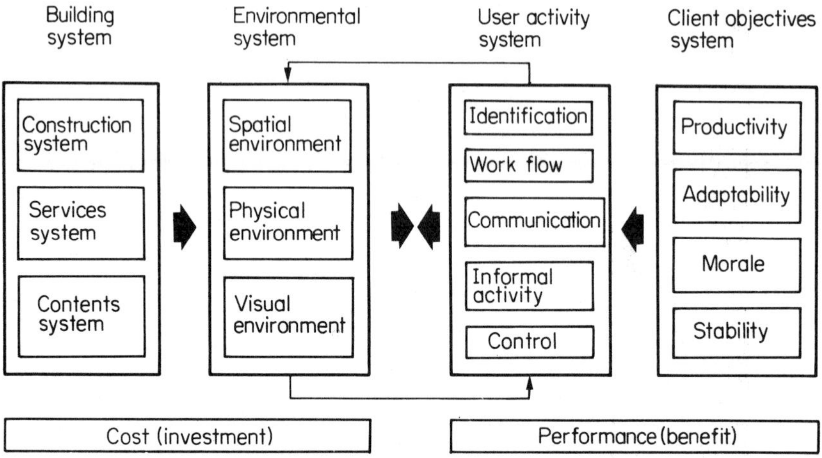

Building system
Environmental system
User activity system
Client objectives system
Construction system
Services system
Contents system
Spatial environment
Physical environment
Visual environment
Identification
Work flow
Communication
Informal activity
Control
Productivity
Adaptability
Morale
Stability
Cost (investment)
Performance (benefit)

Figure 5.3

Building system
Environmental system
Construction system
Services system
Contents system
Spatial environment
Physical environment
Visual environment

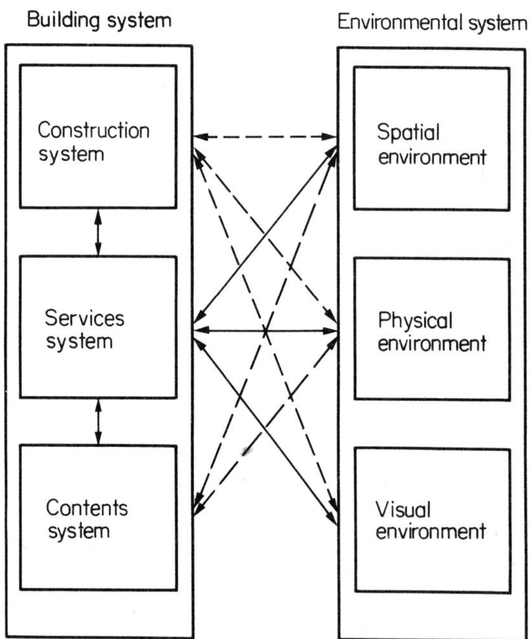

to separate all the important variables. This is discussed further later, but first the organisation of the design team must be discussed.

5.2.1 The design team

Normally a design team will comprise

(i) an architect
(ii) a structural/civil engineer
(iii) a building services engineer
(iv) a quantity surveyor.

On some projects, other consultants (e.g. for acoustics or lighting) may also be appointed. Their joint task is to resolve the overall problem but within this context each will have different roles.

(i) *Architect* Normally the architect will be appointed directly by the client. By discussion with the client he will develop the brief (i.e. set down the client's sub-system requirements). At this stage some or all of the other members of the design teams will normally be appointed to assist with development of the brief and to set out the activity sub-system requirements. Thus the architect's role is to be design team leader and client liaison officer as well as to provide an initial design hypothesis for the building and environmental systems. He will normally coordinate the design of the various specialists' sub-systems; some of these he will design himself as well as those designed by the other consultants. As normally he is responsible to the client for the overall performance of the building he will usually be the final arbiter when conflicts arise between the demands of the various sub-systems.

(ii) *Structural/Civil Engineer* responsible for the structural sub-system. This sub-system is strongly linked with the spatial layout of the building. Services routes through the building often compete with the space/location requirements of the structural sub-system. Strength requirements often constrain holes and hence routes for services.

(iii) *Building Services Engineer* responsible for the design of the various services sub-systems and increasingly involved with the architect (and client) at the preliminary design stage to assess the thermal performance of the architect's initial proposals.

(iv) *Quantity Surveyor* usually not directly involved in the technical aspects of the design as such, but must possess a wide-ranging knowledge of all the various sub-systems. He advises the design team on legal and similar matters relating to the proposals and is responsible for cost control. To do this he will develop suitable specifications and bills of quantities for many of the sub-systems, especially those designed by the architect.

5.2.2 Design team working

Figure 5.4 developed by Markus[2] shows a hierarchical design-making structure which allows the design team to proceed with a decision-making design process in order to resolve the complex design problem in a practical manner.

Figure 5.4

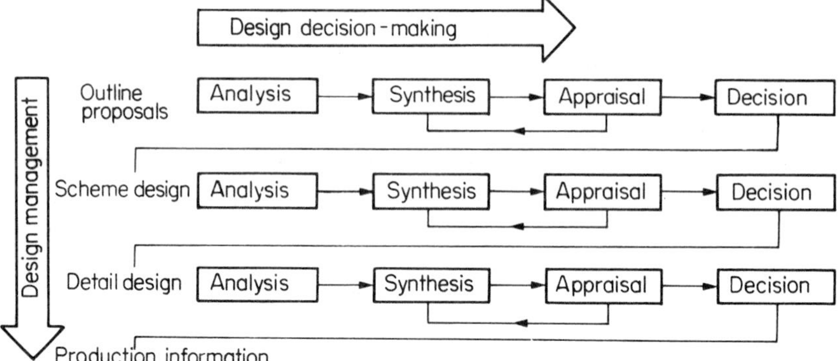

The R.I.B.A. Plan of Work management stages are set down on the left-hand side and between each stage; each system and sub-system would be subjected to a decision-making process as shown. This is a most helpful model as it allows the design team to break down the design process into manageable portions and (ideally) allows sufficient opportunity for refinement of individual sub-system design in the light of parallel developments in the designs of the other sub-systems. At the outset the architect will normally set out a schedule of design team meetings to provide sufficient opportunity for design coordination.

At each stage the design of each individual sub-system would proceed in parallel and/or in series as follows:

(i) *Analysis* of requirements including data collection and analysis from existing buildings.
(ii) *Synthesis* or generation of a design hypothesis based on the analysis of requirements and on previous successful designs.
(iii) *Appraisal* or measurement of the performance of the design hypothesis in terms of technical, economic and other criteria.
(iv) *Modification* of the design based on the appraisal.
(v) *Re-appraisal* and so on until a satisfactory solution is deemed to have been found and the design can be made to proceed to the next phase of the design.

5.2.3 The design constraints

It is evident from the above that the search for an 'optimum' solution for individual sub-systems and for their satisfactory performance is an iterative process. The fact that few building designs are in practice optimised is largely because:

(i) very few sub-systems are sufficiently well understood in themselves to allow sub-systems optimisation to be carried out,
(ii) even when (i) has been overcome there is often a lack of economic and similar crucial data required to allow a cost-benefit analysis to be carried out,

(iii) the effects of one sub-system on another are often difficult to model and evaluate,
(iv) the time required to perform numerous modification/re-appraisal loops may prove impractical.

None the less the usefulness of Figure 5.4 in presenting a logical solution procedure is invaluable and is in effect an explicit description of the intuitive design method which has long been employed in practice. Furthermore the rapid increase in the use of computer-based design methods will allow a much more rigorous application of the above procedure in the very near future. This fact has been widely recognised in building research circles and the development of such programs is now a priority.

5.2.4 Heating system design

The heating systems dealt with in this book are limited to those which employ water as the heat distribution medium (see Section 5.3.5). Steam services and air heating (and ventilating) systems are not discussed as such although many of the concepts and methods discussed here can of course be used directly or with slight modification.

Thus for our purpose we can describe the 'heating system' in terms of its three primary sub-systems:

(i) the space heating sub-system,
(ii) the domestic hot water supply sub-system,
(iii) the process heating sub-system (e.g. kitchen, laundry, swimming pool, etc.).

Generally speaking these sub-systems are strongly linked because it is usually advantageous to employ the same heat source.

Figure 5.5
Heating services sub-systems

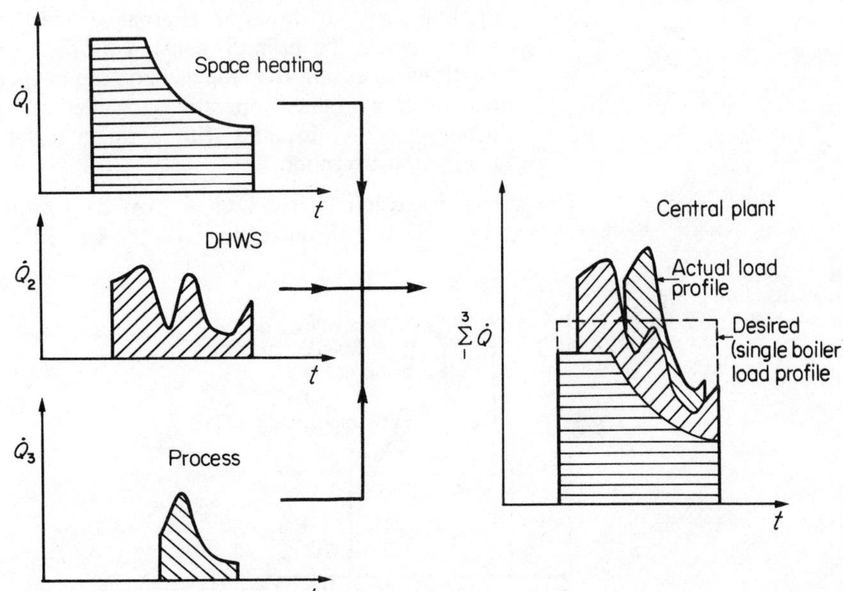

Thus we can usefully define another sub-system:

(iv) the central heat source (or central plant) sub-system. Figure 5.5 shows how the demand profile imposed on the heat source is derived from the heat demand profiles of the other sub-systems. In general the source of heat is a boiler plant and in such cases a near optimum solution *for the heat source sub-system* only will usually be obtained when

(a) the coincident heat demand is minimised (minimises capital cost), and
(b) the fluctuations about the mean are minimised (maximises boiler efficiency).

Thus the ideal central heat source sub-system demand profile for this case might be that shown by the dotted line in Figure 5.5, (achieved by manipulating the primary loads (via controls) and/or modifying the design of the primary sub-systems). In practice this will only be possible to a limited extent and in any case may result in a sub-optimal solution for one or more of the primary sub-systems.

It should also be noted that even if the central heat source sub-system is used as the primary basis of the optimising exercise, the dotted profile shown in Figure 5.5 represents only one such optimum solution (even if it were achievable in practice). This optimum solution applies really only to a single boiler plant which in many cases is undesirable for a number of reasons (e.g. lack of standby for breakdown or maintenance, variation of loads from season to season, etc.). Figure 5.6 shows an alternative 'ideal' demand profile where the primary demand profiles have been modified to ensure that the demand can be met by the equally rated boilers operating together at full load during the morning and with a single boiler at full load in the afternoon.

The above discussion is based on 'design day' profiles; the space heating profile will of course vary with the seasonal variation

Figure 5.6
Alternative load profile for two boilers

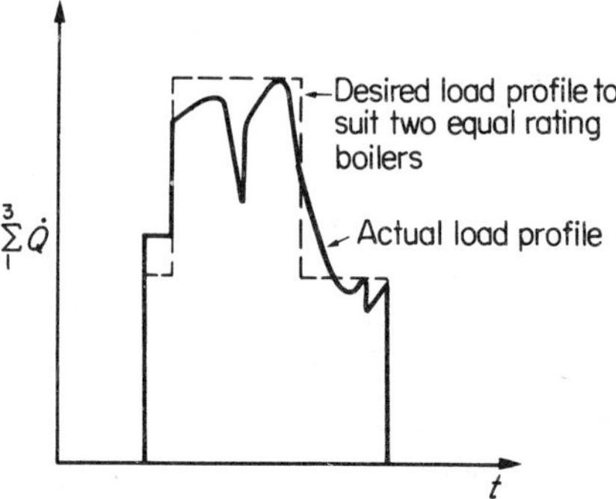

in the weather which may also affect (probably to a lesser extent) the other road profiles and we thus note that in order to evaluate fully a particular design hypothesis a great deal of data and computation is required in order to make an assessment of overall annual performance. If a manual calculation procedure were used, mean *monthly* weather data would probably be used, whereas computer based methods would employ mean *daily* or even *hourly* weather/load data. In either case it is clear that data on heat source (boiler) part-load characteristics, (i.e. efficiencies and proposed controls etc.) and fuel costs would be required in order to evaluate the running costs.

The primary sub-systems would also (ideally) be optimised in a number of series/parallel exercises in order to obtain a composite optimum solution for the heating system as a whole.

While it must be acknowledged that current design procedures do not take account of the above ideas it must be clearly recognised that the methods proposed above are presently under development and, as energy conservation measures assume more importance, will become of increasing importance in ensuring successful design solution. Maver[1] has set the scene and many computer programs are now available to assist in this and many more will become available in the next few years. One sub-set of such programs is that discussed in Section 6.2 but many others (e.g. to describe plant operation and characteristics) are required in order that all the aspects of the composite design problem be considered.

5.3 HEAT GENERATION AND DISTRIBUTION

5.3.1 Introduction

Space restrictions permit only a limited treatment of this topic and the reader is strongly recommended to read Faber and Kell[3] which provides a comprehensive introduction. References (4) and (5) are also most helpful.

5.3.2 Centralised heat sources

Heat may be produced 'directly' in the space to be heated; gas and electric heaters and coal fibres are in this category. With the exception of 'open' coal fires this method of heat generation is often quite efficient and has much to recommend in domestic and small commercial applications. In general, however, it usually makes economic and practical sense to centralise the source of heat production and to distribute the heat by means of a suitable distribution medium (e.g. water, steam or air). Such systems are known as 'indirect' heating systems and these form the basis of the discussion which follows.

The thermodynamic objections to centralising the heat source plant are:

(a) Heat losses from the distribution pipes will be incurred.
(b) Because a secondary heat emitter is used an additional thermal resistance is imposed between the heat source (the boiler) and the heat sink (the room to be heated), which results in a larger heat emitter surface area than might be required with a direct heater.

Objection (a) can be met by insulating the distribution pipes. In many cases these run through the building and to some extent heat losses from them are not entirely wasteful provide the controls can take advantage of such casual gains. In many cases the need to limit surface temperatures of 'direct' heat emitters means that objection (b) is less significant than might otherwise be the case.

The advantages of centralised heat source plants are easier to identify than to quantify but will generally include:

(i) Increased combustion efficiency.
(ii) Increased standby capacity.
(iii) Advantages of load diversity (see Section 5.3.2).
(iv) Centralised maintenance.
(v) Centralised control (see Chapter 3).
(vi) Reduced problems of fuel distribution to individual spaces.

The logical development of conventional centralised heat source plant systems is to increase the scale of operations and thus we can identify:

(a) Group heating schemes serving two or more buildings.
(b) District heating schemes (see Section 9.9 and Reference 15).

5.3.3 Boilers, fuels and combustion

The great majority of heat source plants employ *boilers* which burn gas, oil or coal. Occasionally other fuels are used such as coal tar fuel (CTF) or waste material of some kind; on rare occasions electricity is used as a fuel (electrode boilers) but generally electricity is used directly to avoid distribution losses. Other heat sources include heat pumps, waste heat boilers, solar heating systems and combined heat and power generation schemes all of which are outwith the scope of this book. Further details are given in References (3, 4, 5).

The discussion in this book is limited to those heating systems which employ water as the heat distribution medium and hence the conventional description of 'boiler' is, in a sense, inappropriate because normally it is vital to avoid boiling the water. None the less the conventional term 'boiler' is used throughout to denote the combustion chamber/heat exchanger in which heat is added to the water.

Boiler types

Hot water boilers are often classified as:

(i) Cast iron sectional.
(ii) Steel sectional.
(iii) Pressed steel welded.
(iv) Water-tube.
(v) Shell (i.e. fire-tube).

Further descriptions can be added to the above primary classifications such as 'single-pass', 'two-pass', 'three-pass', 'wet back', 'dry back', etc.

In addition, boilers can be described in terms of the fuel or fuels and/or burners for which they are suitable.

The heating system designer is thus required to comprehend and evaluate a mass of fairly detailed data pertaining to boiler types, fuels, burners and controls and it is for this reason that many boiler manufacturers market their products complete with all necessary ancillary equipment, hence the term 'packaged boilers'. Table 5.1 shows typical packaged boiler specifications. In designing a new heating system normally a packaged boiler will be chosen; for replacement, conversion or upgrading of existing plant it will often be necessary to consider the various components of the heat source system individually.

Boiler efficiency and control

A comprehensive discussion on boiler design for combustion efficiency is beyond the scope of this text. It is of value to note, however, that the combustion of the fuel in the boiler and the transfer of the heat energy from the fuel to the water being heated is a very complex process. Flame shape, size and luminosity (which controls radiant heat transfer from it) are important as well as flame stability, fuel/air ratios, gas velocities and water velocities inside the boiler. These problems are compounded when boilers have to employ two or more different fuels. There factors can be summarised as follows:

(i) Control of air/fuel ratios is important; too little air and the lack of oxygen means that fuel is discharged unburnt, too much air and heat is wasted due to excessive 'excess air'.
(ii) Control of flame shape, size and stability is important not only for good combustion and heat transfer but also to ensure that the flame is not liable to extinguish. The flame size and shape in relation to the size of the combustion chamber are important and for many boilers there exists a firing rate below which the flame stability cannot be maintained and the boiler must be shut down. Local overheating at reduced loads is also a contributory problem.

Table 5.1

Duty: Hot water

Range of f.l. outputs: 492-1148 kW [1,680,000-3,920,000 Btu/h] (9 sizes)
Max operating pressure: 400 kPa gauge [60 lb/in^{2}g]
Units for higher pressure available
Fuels: Oil, Class D fuel not exceeding 40 sec R No 1
Type of firing equipment: Pressure jet
Controls: High-Low Suitable for unattended operation
Construction: Cast iron sectional, water-cooled base; hinged furnace door; eye level control module with hours run counter
Electricity supply: 415 V, 3 ph, 50 Hz
Weight, kg: smallest dry 1854; working 2255
largest dry 3414; working 4231
Water content, l: smallest 401; largest 817
Overall dimensions, mm: smallest 2333 × 830 × 1760; largest 4027 × 830 × 1760

Duty: Hot water
Range of f.l. outputs: 100-7034 kW [340,000-24,000,000 Btu/h] (19 sizes)
Max. operating pressure: 488 kPa gauge [65 lb/in^{2}g]
Units for higher pressure available
Fuels: Oil, 28, 35, 200, 960.3500 sec R No 1
Gas, natural, towns, 1 p.g.
Type of firing equipment: Pressure jet. Dual fuel burners can be fitted
Controls: On-Off High-Low Modulating. Turn down ratio 3:1. Suitable for unattended operation
Construction: Economic wet back three pass; welded steel boiler with Ygnis flame reversal firing principle. Square covered boiler with lift off panels. Fitted with pressure, flow temperature, return temperature and flue gas temperature gauges. Very quiet unit. Luxury version of the Ay13
Electricity supply: 240 415 V, 1.3 ph, 50–60 Hz
Weight kg: smallest dry 1520; working 1920
largest dry 24 384; working N/S
Water content, l: smallest 400; largest N/S
Overall dimensions mm: smallest 2642 × 994 × 1480; largest 4572 × 2580 × 3230

Duty: Steam/hot water

Range of f.l. outputs: 2-77-22-2 kg/s [22,000-176,000 lb/h] (Any sizes)
Max. operating pressure: 11,730 kPa gauge [1700 lb/in^{2}g]
Fuels: Oil, up to 6000 sec R No 1
Gas, natural, towns, 1 p.g.
Coal
Residual oil and waste products
Type of firing equipment: Pressure jet or according to spec.
Controls: On-Off/High-Low Modulating. Turn down ratio 5:1. Suitable for unattended operation
Construction: Fully welded, self-supporting single drum, water tube, natural circulation, with pre-separation steam headers, superheater, economiser. Can be adapted for hot water. Can be combined with refuse incinerators. Supply in modules possible
Electricity supply: Any
Weight kg: smallest dry 40,000; working 46,600
largest dry 160,000; working 188,000
Water content, l: smallest 6600; largest 28,000
Overall dimensions, mm: smallest 4400 × 3000 × 4750; largest 12,200 × 4600 × 6500

(iii) In order to minimise the amount of energy loss in the flue gases the flue gas exit temperature should be as low as possible but this however is limited by the thermodynamic consideration that there must be sufficient temperature differential between the flue gas and the water being heated to maintain useful heat transfer. It is also limited by the need to avoid condensation and/or smut formation and to ensure sufficient buoyancy force in the chimney when the combustion process is not fan assisted.

While the above problems are primarily the concern of the boiler manufacturer they also have a direct bearing on the performance of the boiler in use and the heating system designer will generally ask the boiler manufacturer to supply further information to supplement that given in Table 5.1, such as:

(i) Boiler combustion efficiency at the maximum rated duty and full details of the test during which this figure was obtained.
(ii) A boiler efficiency versus load curve and details of test conditions, especially whether air/fuel ratios were controlled automatically or by hand.
(iii) In cases where it is known that a boiler will be required to fire at part-load for long periods, the manufacturer should be asked to comment on the expected life of the boiler which may be significantly reduced to localised stressing.

From the above and from the discussion in Chapter 3 regarding boiler controls it is clear that where possible boilers should be chosen to ensure that they operate at or near full load. Where they must operate under part-load conditions then due consideration should be given to the efficiency and boiler life factors given above. Not withstanding the above factors, generally speaking it is seldom economically viable to install more than two or three boilers in new plants as the cost of boilers is such that a packaged boiler rated at say 400 kW will cost considerably less than two similar boilers rated at 200 kW.

The whole problem of developing an 'optimum control strategy' for boiler plant operation has been investigated by B.S.R.U. (Building Services Research Unit, University of Glasgow)[6] and the reader is strongly recommended to study carefully this most important document pertaining to the design of central heat source sub-systems.

5.3.4 Calorifiers

As discussed in Chapter 3, the two main categories of calorifiers considered in this book are:

(a) Non-storage calorifiers.
(b) Storage calorifiers.

'Water-to-water' calorifiers only are considered here, steam heating being outside the scope of this book.

Non-storage calorifiers

For heating services, non-storage calorifiers are generally employed where it is required to distribute heat from a central heat source using pressurised hot water $> 100\,^\circ\mathrm{C}$ (as in a group or district heating scheme) for use in a conventional LPHW heating system– for reasons of economy, control, safety and convenience.

The other main application of non-storage calorifiers arises in process heating where the primary (boiler water) source may be pressurised hot water $> 100\,^\circ\mathrm{C}$ or conventional LPHW used to generate a secondary hot water supply for a special requirement. The secondary water may be consumed in the process or liable to contamination (e.g. swimming pool water) in some other way which would make it unsuitable for direct use in the boiler water circuit. Occasionally the secondary fluid will be a liquor other than water.

Non-storage calorifiers are therefore to be considered simply as heat exchangers and as such can be designed from first principles using fundamental data of the type set out in Kays and London[8]. Details of how this can be done are set out in Section 5.3.5 from which it can be seen that the initial assumption normally made for all non-storage calorifiers is that *both* primary and secondary fluids will pass over the separating heat exchange surfaces with sufficient velocity for both heat transfer coefficients to be calculated on the basis of *forced* convention.

A further assumption often made which normally incurs little error is to assume that the fluids flow in pure contra-flow. In fact this will seldom be exactly true in practice, as usually a complex contra-crossflow arrangement is employed, but for the heating system designer this assumption gives a good engineering approximation of the likely result. If necessary Reference (8) can be used or one of the many heat exchanger computer design programs can be employed. Usually however the heating designer will merely select a standard design to suit his needs from those listed in manufacturers' catalogues. Controls are discussed in Section 3.9.4.

Storage calorifiers

Generally these are used to generate the store domestic hot water from a source of pressurised hot water at $> 100\,^\circ\mathrm{C}$, or LPHW. As such they can be considered as a special case of 'process' water heating. It is interesting to note that the reason for storing hot water is to reduce the instantaneous demand on the boiler and, as discussed in Section 5.2 and Chapter 6, this is an important means of 'load profiling' and will probably be more widely used for the space heating and process heating sub-systems also in future designs.

In general the secondary fluid is deliberately introduced into the calorifier with a small entry velocity in order to promote stratification. It is perhaps surprising to note that the stratification behaviour of most calorifiers is poorly understood and exploited. Maver[1] discusses this topic in some detail and the reader is directed to References (1, 9, 10). For our present purposes we note merely that the secondary fluid motion on the outside of the separating surface will be generated by *natural* convection while in general the primary fluid velocity on the inside of the separating surface will be high enough to assume *forced* convection.

Sizing the primary coil is usually a fairly approximate business because it is difficult to estimate the value of the secondary side natural convection coefficient with precision, largely because the fluid bulk temperature will vary with operating conditions. In practice great precision is not normally required and manufacturers will usually err on the generous side in providing heat transfer surface; as temperature control is relatively easy, even at light loads (see Section 3.9.5), this over-provision generally causes no difficulty.

5.3.5 Estimating calorifier heat transfer surface

The Basic Equations. In Chapter 2, equation 2.74 was given as the expression for the heat transfer between two fluids separated by a solid surface. This is restated as:

$$\dot{Q} = U_o A_o \, \Delta\theta_m \tag{5.1}$$

$\Delta\theta_m$ is the log mean temperature difference as derived in equation 2.81 and restated as:

$$\Delta\theta_m = \frac{\Delta\theta_B - \Delta\theta_A}{\ell n(\Delta\theta_B/\Delta\theta_A)} \tag{5.2}$$

Clearly the value of $\Delta\theta_m$ is influenced by the fluid entry and leaving temperatures but these in turn are influenced by the nature of the flow (e.g. parallel, contra or crossflow) and by the presence, or otherwise, of a phase change. These matters are discussed in depth in Reference (8).

Estimation of U_o. As a consequence of Section 2.5.3, only convective effects need be considered at the inside and outside surfaces of calorifier tubes with pressurised hot water or steam within the tubes and domestic hot water in the shell. With reference to Figure 2.17, with clean tube surfaces and neglecting the tube wall resistance,

$$U_o = \frac{1}{\dfrac{1}{h_{co}} + \left(\dfrac{1}{h_{ci}} \cdot \dfrac{A_o}{A_i}\right)} \tag{5.3}$$

where A_i and A_o are the tube inside and outside surface areas and

U_o is expressed in relation to the outside surface area. For turbulent flow within pipes, equation 2.47a is appropriate in the form

$$h_{ci} = 0.0225 \frac{k}{d_i} Re^{0.8} Pr^{0.33} \tag{5.4}$$

For forced convection on the outside of a cylinder equation 2.47c is appropriate, but in relation to tube bundles the C.I.B.S. Guide suggests the use of

$$h_{co} = 0.44 \frac{k}{d_o} Re^{0.55} Pr^{0.31} \tag{5.5}$$

For natural convection on the outside of horizontal cylinders equation 2.49a is appropriate in the form

$$h_{co} = 0.53 \frac{k}{d_o} (GrPr)^{0.25} \tag{5.6}$$

The Guide gives tables of values of h_{ci} and h_{co} in relation to tube diameter, water velocity and water temperature. Additional information is presented regarding fouling resistances.

A Design Approach. A logical approach to the estimation of heat transfer surface might be as follows:

Step 1 Estimate the necessary rate of heat transfer $\dot{Q}$. In the case of a domestic hot water calorifier, for example, an analysis would require to be made of the pattern of consumption.

Step 2 Establish design primary flow and return and secondary flow and return temperatures.

Step 3 Decide on the nature of the flow within the heat exchanger (parallel, contra, forced, natural, etc.)

Step 4 Calculate the value of $\Delta\theta_m$ (equation 5.2).

Step 5 Estimate the overall heat transfer coefficient U_o for the separating solid surface (equation 5.3).

Step 6 Calculate the required heat transfer surface area from equation 5.1.

The method is illustrated by this example:

A 600 kW calorifier is to be designed with two tubes passes, using 32 mm diameter tubes, with water at a mean temperature of 100 °C flowing at a velocity of 1 m/s within the tubes. Forced convection of the water in the shell creates a velocity of flow of 0.4 m/s. The appropriate values of the

Figure 5.7

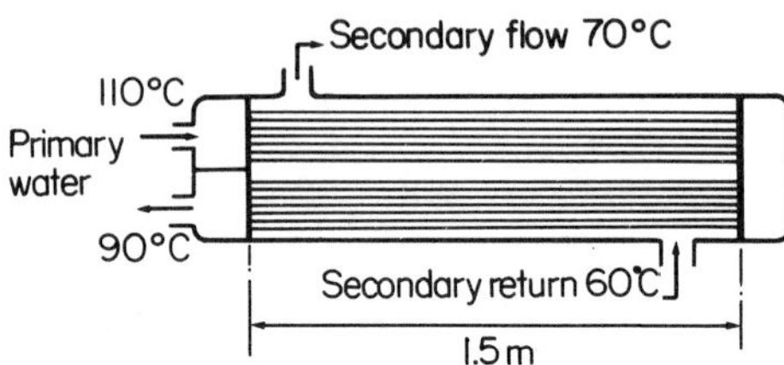

convection coefficients (from the Guide) are h_{co} = 4048 W/m^2°C and h_{ci} = 6450 W/m^2°C. For a tube length of 1.5 m determine the number of tubes required. (Ignore the thickness of the tubes.) The basic layout is shown in Figure 5.7.

Step 1 $\dot{Q} = 600$ kW

Step 2 $\theta_{pf} = 110\,°C \quad \theta_{pr} = 90\,°C$

$\theta_{sf} = 70\,°C \quad \theta_{sf} = 60\,°C$

Step 3 Basically cross flow but assume it to be contra flow

Step 4 110 → 90

A B

70 ← 60

$\Delta\theta_A = 40\,°C \quad \Delta\theta_B = 30\,°C$

$$\Delta\theta_m = \frac{40 - 30}{\ell n \dfrac{40}{30}} = 34.5\,°C$$

Step 5 Assuming clean tubes

$$U_o = \frac{1}{\dfrac{1}{h_{co}} + \left(\dfrac{1}{h_{ci}} \cdot \dfrac{A_o}{A_i}\right)}$$

$$= \frac{1}{\dfrac{1}{4048} + \dfrac{1}{6450}}$$

Thus $U_o = 2480$ W/m^2 °C

Step 6 $$A_o = \frac{\dot{Q}}{U_o \Delta\theta_m} = \frac{600 \times 1000}{2480 \times 34.5}$$

$$= 7.04 \text{ m}^2$$

Step 7 Area of one tube = $\dfrac{\pi \times 32}{1000} \times 1.5 = 0.15$ m

$\therefore$ Number of tubes = $\dfrac{7.04}{0.15} = 47$

Say 24 tubes in the top pass and 24 tubes in the bottom pass

5.3.6 Fuel costs

Figure 5.8 (after Atkinson[11]), shows an indication of the trend in fuel costs over the last few years. When all the factors have been taken into account regarding storage, ease of combustion, boiler/burner maintenance, controllability, cleanliness and boiler life,

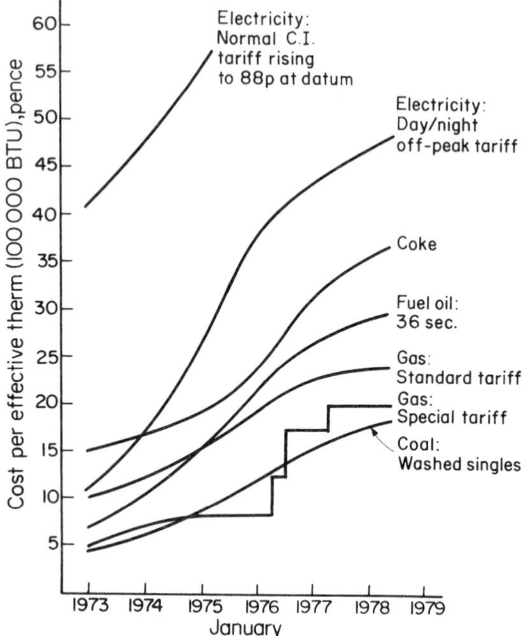

Figure 5.8 Comparative costs of fuels, Jan. 1973 to Jan. 1978

the reason why gas has increasingly become the cheapest fuel choice in almost every case is quite clear. The future cost of these main fuel sources is difficult to foretell and may well be affected by Government policy which could, for example, impose a tax on convenient fuels, (gas, oil and electricity), in order to encourage the use of less convenient fuels (such as coal) thereby conserving scarce fossile fuel resources. As most boiler plants will have a life of around twenty years, the heating system designer is faced with the task of 'crystal ball gazing' in trying to estimate future fuel costs. However, it should be noted that in terms of choosing the most economical method of firing a particular boiler plant the most important factors are:

(i) The future *relative* cost of one fuel with another which would (if known), determine to a large extent the most economic fuel choice or the viability of converting existing plants.

(ii) The future *relative* increase in fuel costs over the general inflation in costs which would (if known) determine to a large extent the value of installing additional energy saving equipment and controls.

The various methods of economic analysis and calculations for assessing the costs-in-use[7] so necessary for the numerous cost-benefit analyses discussed earlier depend implicity on such information and it is unfortunate that no clear guide lines have as yet been provided by, for example, the Department of Energy. In this context it is to some extent unnecessary that such information should be absolutely accurate provided that a *common basis* could be found for the calculations. This would provide a much more positive basis for judgement than those currently used such as:

(i) The rudimentary restraint (usually imposed by accountants and others of similar ilk) that any extra energy investment should have a simple payback period of less than *x* years (where *x* is usually any number less than 5),

(ii) the naïve assumption that all costs will remain relatively the same *vis-à-vis* each other; this seems most unlikely and traditional accounting techniques clearly do not serve for boiler plant and fuel cost analyses of this type.

At the time of writing it seems likely (in the authors' personal view), that during the next ten to twenty years fuel costs will exhibit the following trends:

(i) *Oil* fuel costs will rise steadily at not less than 10% per annum and may well rise at an overall rate of 20% or more per annum. A minimum of fuel cost increase over general inflation of around 3% per annum is suggested.

(ii) *Gas* prices will follow the trend in (i) closely at relatively the same difference in price as at present.

(iii) *Coal* prices will rise increasingly and will probably approach the cost of gas and electricity. None the less it seems likely that in a relatively short time a tax will be imposed on the convenience fuels at least in some categories of buildings (above, say, 750 kW boiler rating) which will make this fuel more or less an automatic choice.

(iv) Electricity prices will rise increasingly to increase the cost differentials between it and other forms of fuel. This trend will continue for at least fifteen years before any real relative reduction can be expected in the cost of electricity. A minimum differential rate of electricity costs over general inflation of around 4% is suggested.

Of course the above statements are controversial but they are made in order to provoke thought and increase the awareness of the need for a common basis for such predictions in order that meaningful comparisons of one design proposal with another can be made. There seems little point however in continuing with the present methods discussed above which do not take full account of the underlying trend.

5.3.7 Heat distribution

(i) *Water and other heat distribution media*. As stated previously this book describes only those systems which employ *water* as the heat distribution medium. Table 5.2 shows the characteristics of water and other heat distribution media and shows that water is a most suitable medium having a low cost, high specific heat and a small specific volume. These advantages are such

Table 5.2 Properties of liquid heating media

Liquid Heating Media	*Density* (kg/m^2)	*Specific Heat Capacity* $(kJ/kg\,°C\ at\ °C)$	*Thermal Conductivity* $(W/m\,°C\ at\ °C)$	*Boiling point at atm press.* $(°C)$	*Freezing point* $(°C)$	*Dynamic (abs.) Viscosity* $(Ns/m^2\ at\ °C)$	*Open flash point* $(°C)$
Aroclor	1441	1.16 at 25 1.365 at 200	0.115 at 30 0.115 at 200	340	6.7	1.2 at 25	193
Dowtherm A	1072 878.0	1.55 at 12.2 2.513 at 232.2	0.144 at 12.2 0.130 at 149	258	12.2	0.5 at 15 0.04 at 230	109
Ethylene Glycol (100 per cent)	1112	3.244 at 20	0.245 at 93.3	197	−11.0	0.1 at 25	108
Glycerine (99.2)	1263	2.764 average	0.281 at 30	290	13.3	1.2 at 20	—
Tetra Aryl Silicate (Typical) . .	1157 900.3	1.55 at 10 3.098 at 315.5	0.151 at 149	(used in the range 18-315°C)		0.52 at 10 0.043 at 315	—
Tetra Cresyl Silicate (Typical)	1131 982.2	1.675 at 25 2.68 at 200	0.137 at 25 0.133 at 200	440	Below 17.8	0.48 at 25 0.14 at 200	163
Water	1000	4.187 at 15	0.65 at 60	100	0	0.89×10^{-3}	—

that probably upwards of 90% of all new heating systems employ water as the heat distribution medium.

(ii) *Economic pipe sizing*. This topic is discussed more fully in Chapter 9. It is worth noting however that in practice it is difficult to determine the true economic pipe sizes in all but the simplest systems. An additional technical consideration not always considered is that the design temperature drop between flow and return should not be such that excessive thermal stress is imposed on the boiler especially under starting load conditions. The boiler manufacturer should be consulted and where necessary a separate boiler return water temperature control system should be installed, but note that this will affect the overall costing in the 'economic' pipe sizing exercise.

In many typical situations it seems unlikely that the effort of performing manually the lengthy calculations required (given the likely errors in data employed) would be cost beneficial when full account is taken of the design time involved. Computer program methods can reduce design costs and allow detailed calculations for alternative networks to be evaluated quickly and allow the methods proposed in Reference (1) to be adopted.

To some extent district heating distribution networks are a special case, (see Chapter 9 and References (3) and (5)).

(iii) *Network layouts*. As with economic pipe sizing it is often difficult to evaluate precisely the 'optimum' layout. In many cases architectural restrictions will constrain the layout used. Chapter 9 gives guidance on typical layouts in current use. Again computer based methods allow the recommendations in Reference (1) to be adopted.

(iv) *Pumps and pumping*. Chapter 9 discusses this topic in some detail, outlining pump laws, characteristics, effects of speed changes, etc.

Table 5.3 (adapted from C.I.B.S. Guide Section B16, Table B16.31) lists the principal types of pump used in heating system distribution networks; as can be seen the most important pump type is the centrifugal pump and its derivative the inline or pipeline mounting pump.

Pumping costs are closely related to pipe sizes and the values chosen for velocity and design temperature drop. As with economic pipe sizing, finding an optimum solution is often difficult. In larger schemes, particularly in district heating, variable speed pumping can be most useful in improving control performance, often giving an indirect saving in energy consumption (though this is difficult to quantify), as well as reducing pumping costs in step with load reductions.

Table 5.3

Type		Description
Centrifugal single-stage		The majority of applications involving the use of a centrifugal pump can be dealt with by a single stage pump, which has one impeller. This has backward curved vanes and is fitted in a casing usually of volute shape to provide maximum conversion of kinetic energy into pressure energy.
Direct coupled		Pump driven through flexible coupling from motor, both units mounted on steel or cast iron baseplate. The pump has its own bearings and duty is adjusted by varying the impeller diameter.
Close coupled		Motor has a specially extended shaft which enters pump casing and to which impeller is directly fitted. The pump does not have its own bearings, and the thrust developed has to be withstood by the bearings in the motor. Duty adjusted by varying the impeller diameter.
Belt drive		Pump with its own bearings driven by V-belts from motor fitted at side of pump on slide rails or mounted above pump. Pulleys can be selected to run pump at appropriate speed to suit duty required.
Inline (floor mounted)		An inline pump has its casing arranged to provide the inlet and outlet connections on the same centre line to facilitate ease of installation.
Inline (pipeline mounted)		A pipeline pump has its connections arranged in line, as above, but is of light enough construction to be fitted directly in a pipeline.
Wet rotor	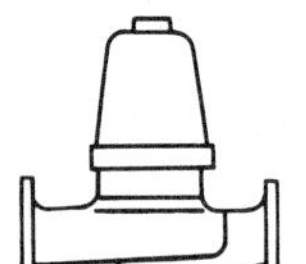	Widely used for domestic central heating. The motor is designed to allow its rotor to run in water, thus no seal is required for the motor shaft which carries the impeller where it enters the pump waterway. The stator has to be protected from water, either by a stainless steel enclosure around the rotor, known as a can, or by embedding the stator windings in the heat resistant moulding. The small gap between the rotor and stator is thereby reduced with either method, making the unit susceptable to seizure if particles of matter such as rust or scale should become lodged in this gap. These pumps usually incorporate the means of adjusting output, either in the form of a restriction at the pump discharge or by measuring leakage loss internally form discharge to suction inlet.
Centrifugal multi-stage	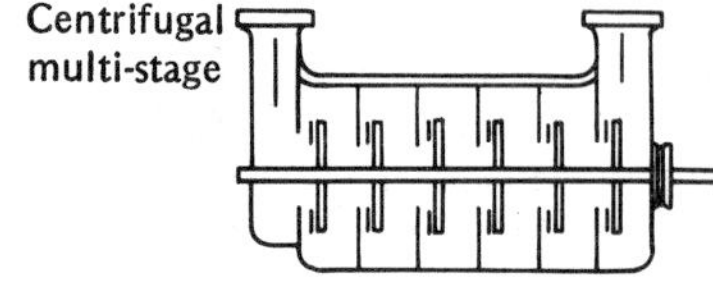	When higher pressures are required than can be reasonably obtained by the use of a single impeller, several impellers are fitted to a common shaft to form a multi stage pump, see page B16.27. It is usual for multi stage pumps to incorporate a hydraulic balancing device to relieve the pump bearings from the thrust produced by the high pressure generated at the pump outlet.

REFERENCES

1. MAVER, T.W., *Building Services Design*, (R.I.B.A. Publications, London, 1971).
2. MARKUS, T. A., 'The role of building performance, measurement and appraisal in design method', *Archit. J.* 1967, **146**, pp. 1567–73.
3. FABER, O. and KELL, J. R. (rev. KELL, J. R. and MARTIN, P. L.) *Heating and Air Conditioning of Buildings*, 6th Edn. (Architectural Press, London, 1979).
4. 'Efficient use of Energy', H.M.S.O., 1978.
5. DIAMANT, R. M. E. and McGARRY, J., 'Space and District Heating' (1978).
6. ROBERTSON, P., McKENZIE, E. and RAVENSCROFT, R. P., 'The choice and operation of space heating boiler plant for overall economy', *B.S.R.U. Report*, 1969.
7. C.I.B.S. Guide (1970), Sections B1, B2, B13 and B16.
8. KAYS, W. M. and LONDON, A. L., *Compact Heat Exchangers* (McGraw-Hill, New York, 1964).
9. MAVER, T. W., 'Some techniques of operational research illustated by their application to the problem of hot and cold water plant sizing', *J.I.H.V.E.*, 1966, **33**, pp. 301–13.
10. MAVER, T. W. and GRATTON, B., 'Study of the water and energy services in the hospital as a whole', *B.S.R.U. Report*, 1964.
11. ATKINSON, J., 'Managing an authority's needs', *Build. Serv. and Environ. Eng*, June 1979, pp. 10–12.

List of Symbols used in Chapter 5

A_o	heat transfer surface area referred to outside of tube
d_i	inside diameter of tube
d_o	outside diameter of tube
h_{ci}, h_{co}	inside and outside surface convection heat transfer coefficients
k	thermal conductivity (of fluid)
$\dot{Q}$	plant load
t	time
U_o	overall heat transfer coefficient referred to A_o
$\Delta\theta_A, \Delta\theta_B$	temperature differences between fluids
$\Delta\theta_m$	logarithmic mean temperature difference between fluids
θ_{pf}, θ_{pr}	primary fluid flow and return temperatures
θ_{sf}, θ_{sr}	secondary fluid flow and return temperatures

6 Estimating the Heating System Load and Energy Consumption

6.1 INTRODUCTION

The fundamental problem in heating system design is that of estimating the loads on the various sub-systems (see Section 5.2.4). The more certain that the designer can be about the accuracy of these basic load estimates the more meaningful will be his decisions about load profiling and his estimates of boiler efficiency during the various load conditions which will arise throughout the heating season. It must be noted that all subsequent detail design (sizing/selection of emitters, pumps, piping, etc.) and the control quality and overall plant efficiency throughout the rest of the working life of the heating system depend largely on the accuracy and comprehension of this primary data. Indeed the study of these basic load profiles against a background knowledge of the characteristics, limitations and expected performance of system hardware is probably the single most important design contribution that the engineer can make to the design of a heating system.

In most schemes, space heating requirements constitute the most important part of the load; furthermore they are often difficult to assess accurately. Chapter 2 discusses the fundamental laws governing the various modes of heat transfer from which it can be seen that a large number of parameters are involved even in the simplest cases. It is to be expected therefore that some simplification will be required, especially when manual calculation methods are employed. During the last decade or so there has been rapid development in heating and cooling load calculation methods, all aimed at providing greater accuracy albeit at the expense of additional computation. Some of these methods are inherently computer-orientated while others, although designed originally for manual calculation, have been computerised in order to free the designer from the increased computation involved.

Many of the methods were developed originally to provide

cooling load estimates but can be used equally well for heating load calculation, particularly when intermittent operation is to be employed. Further, the current trend in fabric design is to employ much higher insulation standards and in this context is becomes increasingly important that thermal performance during summer is also considered in order that summer overheating problems do not result as a consequence of designing to reduce winter heat losses. Such overheating problems if not detected at the design stage are often difficult to solve without some measure of mechanical ventilation or even cooling which would of course result in a more costly and energy intensive overall solution.

Before the advent of suitable computer programs it was not possible to take proper account of the many variables involved in computing heating load profiles. This is no longer the case but students and many practising engineers are largely unaware of the basis and assumptions inherent in such programs. A large part of this Chapter is therefore devoted to describing the problem variables and the program bases in what is (hopefully) a clear and readily understandable way. However, manual methods are not neglected and the C.I.B.S. approach is also discussed in some detail.

The final sections of the Chapter deal briefly with load profiling and estimating annual energy consumption.

6.2 SPACE HEATING SUB-SYSTEM LOADS

6.2.1 Preliminary considerations

The thermal performance of real buildings is a complex problem. Chapter 2 sets out the fundamental laws which govern the various heat flow rates to, from and within a given building, from which it can be seen that all three modes of heat transfer are involved.

Conduction. Although in reality the majority of building conduction problems will approximate closely to the three- or two-dimensional transient cases discussed in Chapter 2, the only practical solution procedures which have so far been developed employ the one-dimensional transient equation (see Section 2.2.2, equation 2.11). Exact analytical solutions of this equation are possible for certain convenient geometric shapes but the majority of cases are mathematically inconvenient and in almost every case some form of approximation is necessary, particularly in multi-layer construction. It is the difference in the way in which equation 2.11 is solved which gives rise to the many different procedures which have been developed, and the most important ones are outlined below in Sections 6.2.4 to 6.2.7.

Outside surface convection heat transfer coefficient (h_{co}). The basis for calculating values of surface heat transfer coefficients is given in Sections 2.3.3 and 2.3.4. The various recommended values for convection heat transfer coefficients (h_{co}) are discussed and compared in Section 7.2.2 as a function of wind speed and characteristic length.

Solar Radiation on external surfaces. In steady-state heat loss calculations it is conventional to ignore the effects of solar radiation incident on external surfaces, i.e. $\theta_{eo} = \theta_{ao}$ is assumed. More sophisticated methods require that some account be taken of solar radiation either directly (see Section 6.2.2) or by means of sol-air temperature which is defined:

$$\theta_{eo} = \theta_{ao} + \frac{1}{h_{co}} (\alpha I_t - \epsilon I_e) \tag{6.1}$$

where θ_{eo} = sol-air temperature (°C)
θ_{ao} = outside air temperature (°C)
h_{co} = outside surface convection heat transfer coefficient (W/m^2 °C)
α = absorption coefficient of outside surface
ϵ = emissivity of outer surface to long wave radiation
I_t = intensity of direct plus diffuse radiation on the outer surface (W/m^2)
I_e = longwave radiation from a black surface at surface temperature (normally assume $\simeq$ air temperature) (W/m^2)

Section A6 of the C.I.B.S. Guide gives further details of sol-air temperatures and tabulates values for use during months March to September; Table A6.8 gives details of I_t for all other months from which θ_{eo} can be computed.

It should be noted that the consequence of including solar radiation effects in the steady-state procedure will have only a very small effect on the calculated actual peak load. However in spring and autumn the contribution of solar radiation via large glazed areas can be quite significant in reducing loads thereby making an important saving in annual energy consumption (see Section 6.5).

Night-time radiation to the sky vault. Equation 6.1 includes a net longwave radiation term and can therefore be used (with caution) to evaluate an equivalent night-time external environmental temperature (θ_{eo}), but only for roofs and the high parts of exposed walls which 'see' the sky vault. In general, vertical walls do not 'see' the sky vault and therefore during night-time they exchange radiation with the ground and other walls, etc. which are at about the same temperature and thus the net radiation exchange will be virtually zero and $\theta_{eo} = \theta_{ao}$ will be a good approximation.

C.I.B.S. recommends that $\theta_{eo} = \theta_{ao}$ be assumed for winter steady-state heat loss calculations (see Chapter 7); this is a valid assumption during daytime when incoming solar radiation will probably balance outgoing longwave radiation and hence the second term in equation 6.1 will be zero or small. In winter this applies on both clear and overcast days because the presence of cloud cover affects both the incoming solar and the outgoing longwave radiation alike. During overcast nights cloud cover emits longwave radiation to the building which more-or-less balances the outgoing radiation and again $\theta_{eo} = \theta_{ao}$ is a reasonable assumption.

On *clear* nights however the incoming sky vault radiation is much reduced to the extent that the net outgoing longwave radiation from the surface is large enough to require consideration. Tables 6.1 and 6.2 give data abstracted from Kew meteorological

Table 6.1 Typical summer night θ_{eo} values (17/7/67)

Time	θ_{ao}(°C)	*net* I_e(W/m²)	*roof* θ_{eo}(°C)	$(\theta_{ao} - \theta_{eo})$ (°C)
0001	16.3	−36	15.0	1.3
0002	16.2	−29	14.9	1.1
0003	15.2	−26	14.3	0.9
0004	15.9	−27	14.9	1.0

Table 6.2 Typical winter night θ_{eo} values (17/1/67)

Time	θ_{ao}(°C)	*net* I_e(W/m²)	*roof* θ_{eo}(°C)	$(\theta_{ao} - \theta_{eo})$ (°C)
0001	−2.1	−58	−4.2	2.1
0002	−2.6	−75	−5.3	2.7
0003	−2.6	−74	−5.3	2.7
0004	−2.7	−72	−5.3	2.6

records (with derived values computed from equation 6.1 with standard h_{co} and ϵ values) and show *typical* (not severe) effects of night-time radiation to the sky values. The difference in net outgoing radiation in summer is due to the combined effect of higher temperatures and higher moisture content in the upper atmosphere.

In well-insulated heavyweight structures this effect is probably small but for lightweight buildings with poor roof insulation and extensive roof glazing the effect will be marked especially if the building is to be occupied during the night. In these circumstances it would be advisable to use a design $\theta_{eo} = -7\,°C$ (say) for the roof fabric heat loss calculations and a normal value of −1 °C (or even −3 °C) for other fabric losses and infiltration/ventilation losses. Most sophisticated computer programs employing 'real' weather tapes take account of this automatically.

Inside surface convection heat transfer coefficient (h_{ci}). See Sections 2.3.3, 2.3.4 and 7.2.2 for the basis of calculating h_{ci}.

Radiation exchange at inside surfaces. Solar radiation (entering via windows) and other high temperature sources give rise to 'shortwave' radiation which can have a marked effect on the thermal comfort of occupants (see Sections 4.7 and 4.9). In general, in most rooms solar radiation strikes floors and walls causing them to heat up and subsequently to convect heat to the room air and re-radiate at low temperatures (longwave radiation) to other room surfaces including 'exposed' surfaces (such as inside surfaces of external walls). Furthermore, as most 'radiant' heating systems (both high- and low-temperature types) are arranged to give downward radiation, the overall effect is similar to that of solar radiation. For the cases outlined above, the (manual) steady-state heat loss calculation procedure uses the environmental temperature concept (see Section 7.4).

In the case of high-temperature radiant heating systems which directly irradiate inside surfaces of exposed elements causing a significant increase in surface temperatures then heat flow into such elements will naturally be greater than those indicated by simplified assumptions such as those in Section 7.4. A similar comment applies to some wall, floor and ceiling longwave heat source systems. The more sophisticated computer programs inherently take account of such radiation exchange using view factors, surface emissivities, etc.; for manual calculation no simple guidance can be given except to say that probably the effect is small in most cases. A first-order estimate can be made by returning to basic view factor radiation exchange calculations (see Chapter 2 and Reference (1)), provided steady-state conditions are assumed.

Wind and infiltration. Without question the problem of modelling the correlation between wind speed and infiltration rate is the least well understood and most intractable of the problems facing the designer as he attempts to estimate space heating subsystem loads. Chapter 7 discusses the two basic methods:

(i) Nominal air-change rate.
(ii) Crack length method.

As stated in Chapter 7 there are quite major difficulties with both methods and many complicating factors, e.g. stack effects and/or building aerodynamics which give rise to quite large pressure differences; occupants use of doors and windows also has a major influence but this is difficult to predict at the design stage. The position can be summarised as follows:

(i) Infiltration heat load estimates have a low accuracy, e.g. room under leeward conditions may suffer exfiltration while the same room under windward conditions may exhibit infiltration rates of 3, 5 or even 10 or more times the estimated design value.
(ii) As insulation standards increase (even given better weather stripping) the infiltration heat loads will become a larger and even more volatile *percentage* component of the heat load and hence better controls are necessary to ensure full energy savings.

(iii) Paradoxically the basic inaccuracy of the infiltration heat load calculation increases the need for accuracy in modelling the fabric heat load even though the accuracy of the combined (air and fabric) heat load is reduced as insulation is increased; this is especially so where heating systems operate intermittently. Thus computer based techniques should prove even more useful in future designs.

(iv) There is quite an obvious need for reduced infiltration rates in real buildings and for better modelling techniques for evaluating proposed designs.

Rain and moisture. The effects of moisture content on the conductivity of building materials is discussed in Chapter 2 while Chapter 7 deals with condensation. Rain, hail, snow and fog effects on the conductivity of exposed elements are difficult to model precisely and to date have only been modelled in a simple way by assuming a (fixed) higher moisture content and corresponding conductivity for all exposed porous surfaces. The drying effect (latent heat flow) from surfaces is not at present considered. A few computer programs take account of the insulating effect of a nominal snow blanket during certain winter months for some locations.

At the moment it is believed that rain and moisture effects are relatively small although this remains to be proved. For the present however, no recommendations can be given other than to follow the current practice described above.

6.2.2 Climate

Table 6.3 lists the climate parameters currently recorded at the larger British weather stations. Even sophisticated computer programs take account of only those which are believed to be the

Table 6.3 Typical Values Recorded by the Larger Meteorological Stations

	Parameter	*Unit*	*Type of measurement*
1	Global solar radiation	$W.m^{-2}$	Total
2	Diffuse solar radiation	$W.m^{-2}$	Total
3	Sunshine duration	0.1 h	Total
4	Dry bulb temperature	0.1 °C	Mean
5	Wet bulb temperature	0.1 °C	Mean
6	Specific enthalpy	$0.1\ kJ.kg^{-1}$	Mean
7	Atmospheric pressure	0.1 mb	Mean
8	Wind speed	$0.1\ m.s^{-1}$	Mean
9	Wind direction	Degrees	Mean
10	Rainfall amount	0.1 mm	Total
11	Rainfall duration	0.1 h	Total
12	Dew point	0.1 °C	Mean
13	Vapour pressure	0.1 mb	Mean
14	Relative humidity	per cent	Mean
15	Direct solar radiation	$W.m^{-2}$	Total
16	Net radiation	$W.m^{-2}$	Total
17	Global illumination	0.1 kilolux	Total
18	Diffuse illumination	0.1 kilolux	Total

most important parameters and the basis for using 'simplified' weather data is now discussed briefly.

In December and January the various radiation terms are usually quite small and thus steady-state and other simplified methods use only air dry bulb temperature in peak load calculations. They also ignore the effects of wind speed and direction on h_{co} and infiltration rate. (As stated in Section 6.2.1, account should be taken of night-time radiation from roofs in certain cases.)

As stated elsewhere the main disadvantage of the steady-state approach is that it does not allow load 'profiles' to be computed. This is especially important for light loads in September, October and March, April and May.

Normally, radiation values are measured with reference to a flat horizontal surface from which (using astronomical data about the sun's position at the time of measurement) corresponding values of irradiance on vertical or inclined surfaces of different orientation can be computed using appropriate formulae.

Sections A2 and A6 of the C.I.B.S. Guide give comprehensive *design weather* data for winter and summer which are useful for most steady-state and some other methods. For more sophisticated calculation methods, hourly *real weather* data are necessary and weather tapes produced by the Meteorological Office and others are now available which provide yearly records of averaged hourly values for each of the above variables on magnetic tape for direct computer use.

One fundamental advantage in using such data is that, as yet, the intercorrelation of the above weather variables with each other is not well understood and work is at present in progress to produce 'severe', 'average' and 'mild' test reference years (TRY) for design purposes.

Weather patterns. Although there is a fundamental cyclic pattern in weather variations based on the solar day, in Britain the rapid growth and decay of warm and cold fronts and associated cloud cover has a highly disruptive effect.

British weather is thus characterised mainly by its inconsistency; 'warm' winter days and 'cold' summer days are common. Prolonged spells ($\geqslant$ 5 days) of extreme hot or cold weather are uncommon. The response of most of buildings normally takes more than five days to come into harmony with imposed 'steady periodic' weather patterns of this type and the simplifying assumptions which are suitable for other countries are not ideally suited to British practice, although such an assumption is required in the 'harmonic methods' discussed in Section 6.2.6.

6.2.3 Problem definition

Perhaps the foregoing sections of this Chapter will have persuaded the reader that the number of difficulties involved in attempting to model the thermal performance of buildings is such that almost every attempt so to do will be doomed to failure. Although it is probably true that very high precision is not really possible, Clarke[2] and others have shown that by using sophisticated computer programs it is possible to predict thermal performance with satisfactory accuracy and while increased rigour is desirable, and is of course sought after in all such research, in most practical design situations what is required of such computer programs is that they should allow a *fair comparison* of one proposed design solution with another and be sophisticated and *flexible* enough to handle *detailed differences* in form and fabric, plant operating strategy, casual gains, etc.

Burberry[3] provides an excellent introduction to the problem as a whole and the reader is strongly recommended to this source which gives a clear and concise overview of the problem; as this source is readily available little further need be done than to súmmarise the major factors which have now been identified:

(i) The full implications of all the above complexities of thermal modelling are not yet fully understood and hence even the most sophisticated and accurate of the present methods incorporate numerous approximations (in particular see (iv) below).

(ii) Steady-state (Section 6.2.8) and other simplified manual methods (Section 6.2.4) although providing an acceptable estimate of peak loads, do not allow load profiles to be assessed accurately nor do they allow accurate prediction of energy consumption; however, perhaps their most serious criticism is that they do not provide a satisfactory basis for making detailed comparisons of a number of alternative design proposals.

(iii) Sophisticated computer programs are now available (especially those based on the techniques described in Sections 6.2.4 and 6.2.5) which can provide results of moderately high accuracy on all aspects of building thermal performance; in future such programs will almost certainly replace manual calculation methods in all but the simplest cases.

(iv) The final accuracy of all present calculation methods is dependent largely on the accuracy in modelling wind/infiltration effects; unfortunately present modelling techniques give poor accuracy.

6.2.4 Harmonic methods (C.I.B.S. procedures)

There are a number of different harmonic methods in use; space restrictions prohibit a detailed review and the reader is referred to Gupta *et al*[4] for a comprehensive introduction.

The discussion here is intended to provide a basic understanding of the assumptions and limitations inherent in the 'harmonic method' as currently recommended in the C.I.B.S. Guide Sections A8 and A9 (1970 edition).

Although it may not be readily apparent, the C.I.B.S. method really consists of two separate calculation routines:

(i) Calculation of heat flows through external surfaces using a simple harmonic method based on that proposed originally by Mackey and Wright[5] and others. This part is similar to that proposed in the 1965 Guide and although intended primarily for estimating summer cooling loads it can also be used for other seasons of the year.

(ii) Calculation of the heat flows inside the room between the various internal surfaces using the 'environmental temperature concept' and the 'admittance procedure'.

Heat flow through external surfaces. Basically the Guide method makes the following assumptions and approximations regarding heat flow through external walls and roofs:

(a) Although not explicitly stated, the effects of wind on h_{co} are ignored; rain and moisture effects are ignored and weather is considered only in terms of solar radiation and external air dry bulb temperature (to give θ_{ao}); night-time radiation to the sky vault is not considered.

(b) For 'opaque' fabric, solar radiation and θ_{ao} are combined to give sol-air temperature θ_{eo} for which 'design' values are tabulated in the Guide (Section A6); alternatively values of θ_{eo} can be computed from known weather data using equation 6.1.

(c) Heat flow through glazing is considered to have a solar radiation component (discussed in fuller detail below) and a normal transmission heat load calculated using $(\theta_{ao} - \theta_{ei})$ as the driving potential (cf. (d)).

(d) In order to simplify calculations (see separate discussion below on heat flows at internal surfaces) the heat is considered to flow through the fabric to the 'environmental point', at which the internal environmental temperature (θ_{ei}) is used as a measure of the internal conditions (see Section 7.4). Thus for heat flow through the external opaque fabric the approximation used is that $(\theta_{eo} - \theta_{ei})$ is the driving potential (cf. (c)).

(e) What characterises the harmonic method (and gives it its name) are the further assumptions about the weather (i.e. the behaviour of θ_{eo}). As has been stated previously, the central problem is to solve equation 2.11. In order to obtain an analytical solution for this the C.I.B.S. method assumes that the daily variation in θ_{eo} can be expressed by

$$\theta_{eo,t} = \bar{\theta}_{eo} + \Delta\theta_{eo} \sin(\omega t) \qquad (6.2)$$

where $\theta_{eo,t}$ = sol-air temperature at time t (°C)

$\bar{\theta}_{eo}$ = daily mean sol-air temperature (°C)

$\Delta\theta_{eo}$ = amplitude of dialy sol-air temperature variation (°C)

ω = fundamental steady cycle frequency (see discussion below) (rad/h)

t = time (h)

C.I.B.S assume that weather varies 'smoothly' with a 24-hour cycle period; this is known as the 'fundamental' period, which probably gives an accuracy appropriate to the Guide's (manual) method. A more accurate representation of real weather can be achieved by adding one or more 'harmonic' frequency terms (having periods of 12 h, 6 h, 3 h and so on). In many cases a very good accuracy (appropriate to computer use) can be obtained by adding the first (12 h) and second (6 h) harmonics only.

(f) A *crucial* supplementary assumption which is required to solve equation 2.11 is that the weather during the *five days* previous to the day for which the analysis is performed has followed the *same weather pattern* (i.e. obeyed equation 6.2). This is required in order that the fabric will be in 'harmony' with the weather. As British weather seldom conforms to the assumption (see Section 6.2.2) this is an obvious and crucial weakness inherent in the method.

(g) Assumption (f) also means that the method should only be used with average hourly sol-air data (such as that in Guide Section A6) and should be used to estimate typical loads or conditions. Unlike the numerical and response factor methods, it is inherently *unsuitable* for use with real weather data for evaluation annual energy consumptions. This is because real weather will not allow the structure to respond in 'harmony' as is required for the validity of the solution discussed below.

Solving the heat flow problem. For a simple homogenous slab, analytical solutions of equation 2.11 have been obtained usi^ the above assumptions; solutions for multi-layer slabs involve . number of assumptions and approximation and are expressed in terms of an 'equivalent' homogenous slab (see Appendix 6.1). Omitting a discussion of the mathematics involved we simply note that these solutions have the same form as equation 6.2 but that the temperature wave leaving the inside surface of the exposed wall or roof has a reduced amplitude and arrives at some time after the occurrence of the corresponding variation of θ_{eo} at the outside surface. This is shown graphically in Figure 6.1 and gives rise to the terms decrement factor (f) and phase lag (ϕ). Superimposed (dotted) on Figure 6.1 is a plot of the actual variation which might occur at the external surface of a typical wall or roof and the corresponding (approximate) response which would follow using f and ϕ, and hence assuming that this weather could be represented by the 'fundamental steady cycle' on which f and ϕ are based.

In view of the above approximation and assumptions and to avoid the mathematics involved, the Guide provides simple graphs in Sections A8 and A9 from which approximate values of f and ϕ

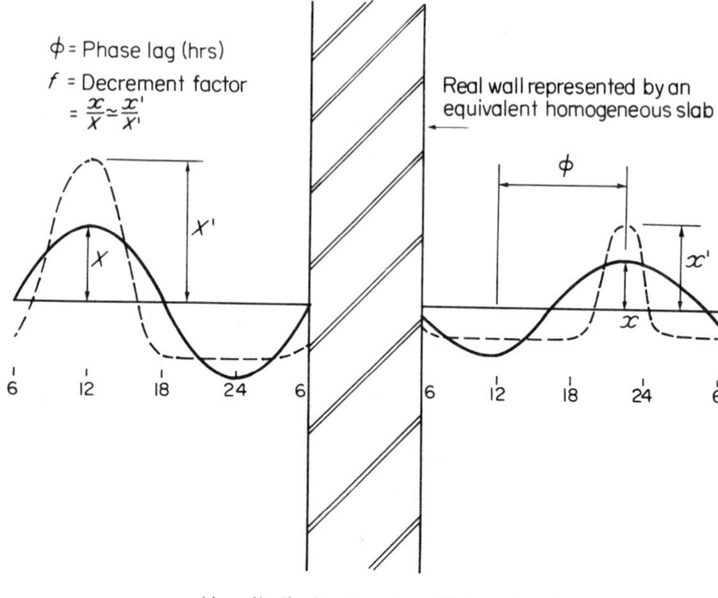

Figure 6.1 Harmonic response method

can be obtained directly. Assuming f and ϕ are known, the flow of heat to the environmental point at any time of interest (e.g. 1600 hours) can be found provided that the value of θ_{eo} is known at time $(1600 - \phi)$ hours.

Problems arise when θ_{ei} varies with time. Indeed this is the common case arising from

(i) The intermittent operation of cooling and heating plant.
(ii) The case of a non-air conditioned or free-running building.

Both of these cases are discussed further later but for the present the method can be demonstrated by assuming θ_{ei} = constant (say 22 °C) due to continuous plant operation. Thus we can make a simple estimate of heat flow through exposed fabric at 1600 h as:

$$\frac{\dot{Q}}{A} = Uf\,(\theta_{eo(1600-\phi)} - \theta_{ei}) \qquad (6.3)$$

Note that the direction of heat flow may well be *from* the room depending on the value of $\theta_{eo(1600-\phi)}$.

Simplified heat flow at internal surfaces (the admittance method)

(a) *Heat flow through exposed fabric*. The above discussion on harmonic methods is focussed on heat flow *through* opaque fabric elements exposed to climate (i.e. walls and roofs).

These heat flows give rise to heat gains or losses to or from the room at each hour during the calculation period. To account for convection and radiation exchange between a surface, room air and all other surfaces according to the fundamental methods set out in Chapter 2 requires that view factors be accounted for and that a heat balance be struck. This is difficult and tedious and best suited to computer solution; C.I.B.S. propose a simplified approach based on the 'environmental temperature concept' and 'the admittance procedure'.

(b) *Solar insulation via glazing.* Not all of the solar radiation arriving at the outside of a window will gain access to the inside of the room. The amount transmitted depends on the angle of incidence, type of glass, blinds shading devices, etc.; the C.I.B.S. Guide provides simplified average values known as solar gain factors (S) for a variety of cases; these are listed in Sections A8 and A9.

Transmitted solar radiation strikes occupants, affecting their comfort (see Chapter 4), and furniture, floors and walls, causing their surface temperatures to rise. As a result a fraction of the radiant energy will be released to the environmental point within an hour or so of entering the room; the actual fraction can be estimated using the 'surface factors' (F) listed in Guide Sections A8 and A9 (e.g. $F = 0.4$ for a bare concrete floor and $F = 0.85$ for a carpeted timber floor; furniture will tend to have high surface factors of the order of 0.85). The remaining fraction of the radiant energy is 'admitted' into the fabric and is stored until later (see 'Admittance Method' below).

This procedure is semi-empirical; for ease of manual computation the Guide Sections A8 and A9 give modified solar gain factors known as 'alternating solar gain factors' (S_a) which incorporate an average surface factor for 'lightweight' structures (high F values) and 'heavyweight' structures (low F values). Because the time-lags associated with the F (and hence S_a) values are small they are conveniently ignored in manual calculations.

(c) *Internal heat gains and losses.* Normally heat gains occur from people, lights and machinery. Although most will be seen as inputs to the room air it is convenient to handle these as direct heat inputs to the environmental point. (Certain types of scientific equipment can act as heat sinks but this is unusual).

(d) *Heating (and cooling) plant.* Where necessary these can be handled as inputs at the air point but normally they are considered to flow directly to the environmental point for convenience.

The resultant effect of all these heat flows to the environmental point is discussed below under the 'Admittance Procedure' but before this is done the concept of admittance will be introduced.

Surface admittance (Y): The original 'admittance' concept was defined in a semi-empirical fashion to deal with electric underfloor heating which emitted heat continuously to a room in a roughly sinusoidal manner over 24 h, including a charging period of about 12 h from which the original room admittance concept was defined simply as:

$$\text{room admittance (W/°C)} = \frac{\text{range of heat outputs (from floor) (W)}}{\text{swing in room temperature (°C)}}$$

Modern practice is to define admittance by analogy with electrical *RC* networks. Such electrical networks when subjected to a (sinusoidally) alternating current do not obey Ohm's law but exhibit a phase *lead* of current in advance of voltage. The electrical term used to characterise this response is known as *impedance* (Z_e). The reciprocal property of electrical impedance is called *admittance* ($Y_e = 1/Z_e$).

Consider a simple homogeneous concrete floor slab; the thermal admittance (Y) of the slab can be calculated (to a first order approximation) by considering it as a simple equivalent thermal *RC* network having a thermal resistance R and a thermal capacitance C. A sinusoidal heat flow with a periodic time of 24 h is then assumed from which a value of the thermal admittance and its associated time lead can be calculated. The mathematical derivation of these values is not straightforward however and the reader is referred to Reference (6) for further guidance.

Admittance values for multi-layer slabs can also be calculated; this is facilitated by matrix multiplication and is more suited to evaluation by computer or programmable calculator (C.I.B.S. now offer a suitable program for the latter). Alternatively, tables of computed Y values are listed in Guide Sections A5, A8 and A9.

From the above we see that the approach is similar to that of the harmonic response method used for heat flow *through* exposed walls and roofs; in fact it is exactly the same problem as that expressed by equation 2.11 but in this case we consider the problem essentially as a *surface effect* of heat flowing *into* the slab as opposed to *through* it as previously.

To summarise, three general comments are required at this point:

(i) In all cases the admittance of a fabric (or furniture) surface is determined very largely by the surface layer; thus a 'heavyweight' bare concrete floor ($Y \simeq 6.0$ W/m °C), will become much 'lighter' ($Y \simeq 3.0$ W/m °C) when carpetted.

(ii) The underlying assumption in all calculations is that the heat flows are 'periodic' if not actually sinusoidal, (the method will prove inaccurate if heat flows are highly 'peaky').

(iii) We note that the use of surface admittance (Y) complements the use of surface response factor (F) and that, as before, the time 'lead' of Y is the time lag of F.

Although such 'leads' are not always small they are usually conveniently ignored in manual calculations.

The admittance procedure

The simple concept introduced by equation 6.3 has now been developed to provide a simplified method of assessing the dynamic performance of buildings which can be used at various levels of sophistication and corresponding accuracy. The main applications are now discussed briefly in order to convey the details of how the admittance procedure can be used in practice.

(i) *Summertime temperatures in buildings*. In view of the increasingly high insulation levels being proposed for many new buildings, and bearing in mind that windows opening to obtain relief via increased ventilation rates may give rise to ingress of noise and dirt, clearly some method of checking possible summer overheating at the design stage is desirable to allow a rough check to be made before finalising the design of the heating system.

Section A8 of the Guide (1970) sets out a simplified manual procedure based on proposals by Louden[6]. Here the basic idea is to simplify the climate, external wall and roof response and the heat flows at internal surfaces as discussed above using harmonic and periodic (i.e. admittance) methods. The Guide procedure provides a means of making a first order estimate of the likely internal conditions in a building which has no cooling plant. Briefly the main steps are:

(a) For the month during which peak gains are expected, average solar radiation data are used to estimate the daily mean solar heat gain ($\bar{\dot{Q}}_s$) to the room via glazing.

(b) The daily mean casual heat gain ($\bar{\dot{Q}}_c$) is estimated arising from lights, machinery, people, etc.

(c) Summing these gives the total daily mean gain ($\bar{\dot{Q}}_T$). As a result of this total heat gain the interior of the room is maintained at a daily mean internal environmental temperature ($\bar{\theta}_{ei}$) which is above that of the daily mean value of the external sol-air temperature ($\bar{\theta}_{eo}$), which in turn governs fabric transmission heat loss, and daily mean external air dry bulb ($\bar{\theta}_{ao}$), which governs infiltration/ventilation heat loss and glazing transmission heat loss.

(d) $\bar{\theta}_{eo}$ is a function of the (simplified) design weather data and is therefore known, hence $\bar{\theta}_{ei}$ can be computed using equation 6.4 (after Guide equation A8.4) assuming an average value of C_v.

$$\bar{\theta}_{ei} = \bar{\theta}_{eo} + \frac{\bar{\dot{Q}}_T - [\Sigma(A_G U_G) + Cv] \; (\bar{\theta}_{ei} - \bar{\theta}_{ao})}{\Sigma (A_f U_f)} \qquad (6.4)$$

where $\Sigma(AU)$ = sum of products of areas and corresponding U values (W/°C)
C_v = ventilation conductance (W/°C) (see Chapter 7)

and suffixes are f = fabric, $\dot{G}$ = glazing.

(e) The internal environmental temperature will *swing* about this mean value, the swing normally being greatest at times of maximum solar gain; this provides a guide to choosing particular hours of the day for which to calculate the various 'mean to peak' swings.

(f) For the most likely times 'mean to peak' swings are calculated for each component of the load and summed to give $\Delta\dot{Q}_{T,t}$ the total swing in heat gain at time t.

(g) A simple estimate of the swing in internal environmental can now be found from equation 6.5 (after Guide equation A8.12)

$$\Delta\theta_{ei,t} = \frac{\Delta\dot{Q}_{T,t}}{(\Sigma AY + Cv)} \qquad (6.5)$$

(h) To a first approximation the value of the actual environmental temperature at each time t is thus given by adding the results of equation 6.4 and equation 6.5, thus

$$\theta_{ei,t} = \bar{\theta}_{ei} + \Delta\theta_{ei,t} \qquad (6.6)$$

Examples of the application of the procedure are given in the Guide, Section A8.

The above provides a simplified manual method which can be used to check whether summer overheating is likely to occur (although the Guide suggests a rather high comfort criterion for this as $\hat{\theta}_{ei} > 27$ °C and perhaps a value of $\hat{\theta}_{ei} > 25$ °C would be more realistic).

(ii) *Application to heating and cooling load calculations.* Section A9 outlines separate procedures for calculating

(a) 'design' heating loads
(b) 'design' cooling loads

In this context 'design' means the 'expected peak' load and although these peak values are of importance it is really 'load profiles' that are required (see Section 5.2.4) in order to develop an optimum control strategy for central plant operation. In practice the straightforward manual method is inflexible and cannot allow a detailed comparison of different plant operating and sizing strategies. The results obtained from these procedures should be considered as the absolute minimum acceptable to a serious designer and do not give a fraction of the data provided by the computer programs based on the methods discussed in Sections 6.2.5 and 6.2.6. They are however, very useful in providing a rough manual check and are always worthwhile as preliminary and/or final checking procedures to complement computer results.

(a) *Space heating sub-system loads*. Chapter 7 outlines the steady-state environmental temperature method set out in Guide Section A9. This is discussed fully in section 6.2.7; approximations can be used to estimate 'peak' loads to be used as the basis for intermittent plant operation discussed in Section 6.3.
(b) *Air conditioning (cooling)sub-system loads*. It is instructive to discuss this briefly as it helps to explain the basis of further applications of the Guide 'harmonic method' to winter heating to be discussed later in this Section.

Section A9 is similar to Section A8, but here the internal environmental temperature (θ_{ei}) is assumed to be *controlled* at a constant value. Heat flows to the environmental point can then be evalulated on an hourly basis for constant θ_{ei} in order to determine the 'peak' cooling load. Note however, that unlike the winter heating procedure this inherently provides a method of 'load profiling' and when this is done room by room then a composite load profile taking account of diversity can be found and an optimum cooling plant control strategy can be devised.

Notice that as θ_{ei} is constant there is no 'mean to peak' swing in θ_{ei} and hence surface admittances do not feature in the calculation procedure; strictly this requires 24 h plant operation (with wasteful heating at night in most cases). However, the main components of the cooling load (solar and casual) are present only during the normal 12 h occupancy period and the Guide method assumes 12 h plant operation; in practice this cooling load is little different from the wasteful 24 h operation load (for which correction factors are given).

It is important to note that the procedure just described for approximate cooling load profiling can also be used for heating load profiling but with reduced accuracy. This is not carried out in practice because solar radiation and casual gains are generally considered fortuitous (i.e. contributing to a reduction in peak load) and are hence to be considered as a safety factor. While this may well be true in December/January for many cases, 'load profiling' during the rest of the year is also of interest and can be calculated using the same basic procedure with appropriate winter weather factors and incorporating the refinements outlined below.

(iii) *Refinements to the basic method.* The above procedures for heating and cooling load estimation incorporate simplifying assumptions to suit manual calculations. These restrictive assumptions can be removed when the method is computerised.

The two basic refinements which the computer-based harmonic response/admittance procedures provide are

(a) variable ventilation (Cv) rates throughout the 24 h period
(b) variable θ_{ei} values throughout the 24 h period to suit different plant operating strategies.

(Papers by Harrington-Lynn[7,8] give details)
In fact both of these refinements can be incorported using manual methods, but this is tedious. The computer-based methods are very simple and computing time is minimal. The results however must be considered as approximate only and at best can give only expected 'monthly average' performance. Accurate energy consumption estimates cannot be obtained from this method (cf. the methods in Sections 6.2.4 and 6.2.6).

A further refinement shortly to be included in the revised (1979) Guide Sections A5 and A9 is that heating and cooling load calculations are to be conducted directly in terms of θ_{res} (as measured at the index point). While this adds a further complication to the actual method of framing the various heat flow equations it does *not* alter in any way the fundamental concepts discussed above. While this refinement is advantageous in some ways, it does tend to conceal the full implications of assessing comfort as set down in Guide Section A1 and Chapter 4 of this book.

6.2.5 Response factor methods (A.S.H.R.A.E. procedures)

Space restrictions allow only a very brief review of the A.S.H.R.A.E. method. Indeed, Kimura[9] in his excellent book outlines the fundamental mathematical formulation and solution basis of the method. The actual procedures have been set out clearly by A.S.H.R.A.E.[16].

Simplified weather. Weather is represented using sol-air temperature. Although in fact only specific (hourly) values are used, the underlying concept is to represent the 'continuous real weather curve' by means of a series of equivalent triangular pulses as shown in Figure 6.2. This forms the 'input' to the wall and the record of hourly θ_{eo} values is known as a 'time series'.

Figure 6.2
Response factor method

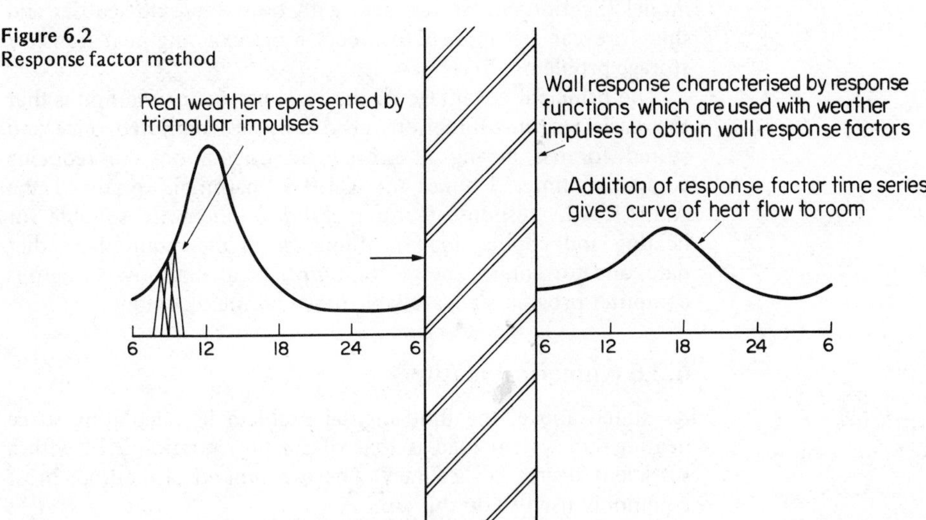

Simplified response of exposed fabric. The heat flow through the exposed walls and roofs is governed by both the structure and internal conditions which the inside surface of the wall experiences (i.e. internal radiative exchange with other surfaces and convective exchange with room air). The mathematical solution of this problem is complex (Kimura[9]).

The resulting solution gives a set of response *functions* which can then be used to obtain the response *factors* for the structure. These response factors are in fact a time series which when added represent heat flow *from* the inside surface to the room. The effect of the structure is to damp the external pulses, reducing the magnitude of their internal response and to cause a time lag in the arrival of the pulses (cf. decrement factor and phase lag in Section 6.2.4).

Simplified heat flow at internal surfaces. Response functions can also be derived for internal surfaces and from these, response factors can be obtained. Triangular pulses are used to model solar and other shortwave radiation inputs to the surfaces affected, while convective casual gains are input directly to the room air; a set of heat balance equations is solved simultaneously to account for convective and radiative heat flows between surfaces and room air.

Summary

The method can be simplified for manual solution but for rigorous application it requires a computer. The principal disadvantage, although not crucial in conventional building design, is that it can be applied only to a system which can be represented by equations which are linear and invariable whereas the 'heat balance method' model (Section 6.2.6) can deal with both these difficulties and therefore can be applied to model more exacting heat transfer/storage problems.

The principal advantage of the response factor method is that the basic response functions need only be computed once and stored for use during all subsequent calculations, so reducing computer time. Unlike the C.I.B.S. harmonic method, the A.S.H.R.A.E. response factor method is inherently suitable for heating and cooling load profiling, given the required weather data, and for annual energy consumption calculations. Numerous computer programs are available based on the technique.

6.2.6 Numerical methods

As stated above, the fundamental problem in calculating space heating sub-system load is that of solving equation 2.11 with a sufficient degree of accuracy. The two numerical methods most commonly used to do this are

(i) the finite difference method
(ii) the heat balance method.

These are essentially computer-based methods; both methods model the 'real distributed system' (in which thermal resistance and thermal capacitance are distributed throughout the fabric), by means of a 'lumped system', where the real distributed system is considered as a series of separate lumps each having a pure thermal capacitance separated by pure thermal resistance. This gives rise to 'lumping errors'; these can be reduced by increasing the number of 'lumps' (i.e. a large number of small lumps) but this gives rise to large numbers of equations; the time step chosen (normally, but not necessarily, one hour) also introduces 'lumping' errors. High computation speed (i.e. low user cost) requires that the number of 'lumps' be small and the time step be large and their choice involves an accuracy/cost/speed compromise.

Unlike analytical methods, numerical methods can only provide information about the temperature variations and heat flows at the chosen points of interest (i.e. 'nodes') within the structure. Usually this is sufficient for most purposes but when data about other nodes is required (e.g. in condensation problems) this can be obtained usually by careful positioning of nodes.

(i) *Finite difference methods.* The complexity of boundary conditions (see Section 6.2.1) and the multi-layer construction of most real cases is such that the 'finite difference approximation method' can provide a useful basis for setting out the numerical approximations used for the conduction equations for each fabric layer and for the boundary conditions. To convey the basic idea of the technique consider one layer (of a wall for example) for which equation 2.11 is considered to provide a suitable mathematical representation.

$$\frac{\partial \theta}{\partial t} = \alpha \, \frac{\partial^2 \theta}{\partial x^2} \qquad (2.11)$$

It is possible to formulate both 'explicit' and 'implicit' finite difference approximation equations but explicit equations are normally easier to solve although the time step required to ensure a satisfactory solution will normally be prohibitively small (i.e. expensive in calculation time) and hence implicit approximations are more convenient.

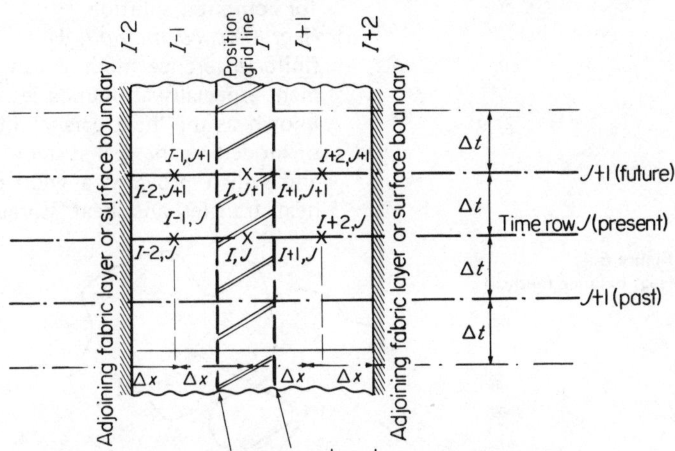

Figure 6.3
Formulating the finite difference approximation equation

Figure 6.3 shows a representation of a fabric layer for which equation 2.11 holds true; Reference (10) shows how to obtain the following finite difference approximation equation for this case:

$$(\theta_{I,J-1} - \theta_{I,J}) = \beta \left(\frac{\alpha \Delta t}{\Delta x^2}\right) (\theta_{I+1,J+1} - 2\theta_{I,J+1} + \theta_{I-1,J+1}) + (\beta - 1) \left(\frac{\alpha \Delta t}{\Delta x^2}\right) (\theta_{I+1,J} - 2\theta_{I,J} + \theta_{I-1,J}) \qquad (6.6(a))$$

If $\beta = 0$ is chosen equation 6.2 becomes an explicit equation; for $\beta = 1.0$ equation 6.6a is fully implicit. Commonly $\beta = 0.5$ is chosen which gives an implicit equation known as the Crank-Nicholson approximation and which provides a stable convergent solution. Assuming α (thermal diffusivity) to be constant and choosing convenient values for Δx and Δt (normally 1 h), equation 6.6a can be re-arranged to give equation 6.6b (where K_1, K_2 and K_3 are suitably chosen constants) and which the reader can formulate himself from equation 6.6a.

$$-K_1\theta_{I-1,J+1} + K_2\theta_{I,J+1} - K_1\theta_{I+1,J+1} = K_1\theta_{I-1,J} + K_3\theta_{I,J} + K_1\theta_{I+1,J} \tag{6.6b}$$

Note that the left-hand side of equation 6.6b contains all the unknown temperatures of interest expressed in terms of all the known temperatures of interest on the right-hand side. The above procedure can then by repeated at all other points ('nodes') of interest and a large set of simultaneous equations obtained suitable for computer solution.

(ii) *Heat balance method.* Unfortunately in practice the finite difference method can prove difficult to implement especially at boundaries. Another related method, known as the 'heat balance method' provides a means of modelling the real system in a way which is realistic and which provides a good physical insight into the heat transfer and heat storage mechanisms involved.

Figure 6.4
Heat balance method

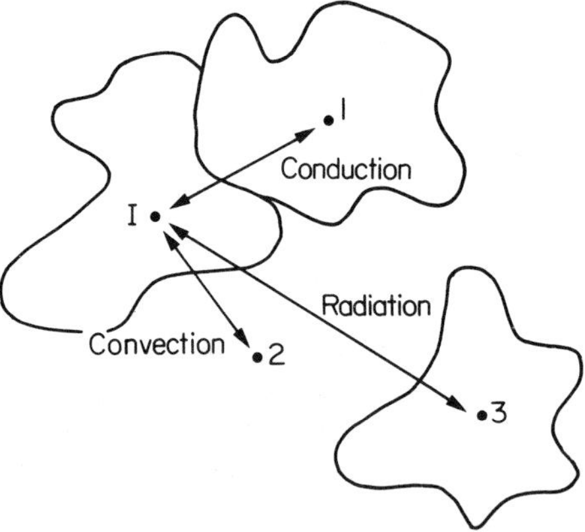

Consider Figure 6.4 which depicts the general case of a node, I, which is in thermal contact with its surroundings in which conduction, convection and radiation exchange take place (this is the general case of a surface node; embedded nodes will be in contact with both adjacent nodes by conduction only. For the general case heat flow to node I can be defined as follows:

(*a*) *Conduction*

$$\dot{q}_{1,I} = K_{1,I}(\theta_1 - \theta_I) \text{ (W/m}^2\text{)} \tag{6.7a}$$

where $K_{1,\mathrm{I}} = \left(\dfrac{k}{\Delta x}\right)$ W/m² °C

k = thermal conductivity (W/m °C)

Δx = finite difference as in Figure 6.1 (m)

(*b*) *Convection*

$$\dot{q}_{2,\mathrm{I}} = K_{2,\mathrm{I}}\,(\theta_2 - \theta_\mathrm{I})\ (\mathrm{W/m^2}) \qquad (6.7b)$$

where $K_{2,\mathrm{I}} = h_{\mathrm{c,I}}$ = convection coefficient (W/m² °C)

(*c*) *Radiation* (longwave)

$$\dot{q}_{3,\mathrm{I}} = K_{3,\mathrm{I}}\,(\theta_3 - \theta_\mathrm{I})\ (\mathrm{W/m^2}) \qquad (6.7c)$$

where $K_{3,\mathrm{I}} = h_{\mathrm{R,I}}$ = linearised radiation coefficient (W/m² °C) (Section 2.4.3)

(d) Solar and other shorwave radiation heat transfer will give rise to a direct heat input to the surface nodes; similarly, embedded electrical heat inputs will give rise to heat generation at internal nodes; if these heat flows are considered to be independent of the node temperature then this can be accounted for by

$$\text{heat generation rate} = \ddot{q}_{\mathrm{I,t}}\ (\mathrm{W/m^2}) \qquad (6.7d)$$

(e) To account for the rate at which heat is stored in the 'lump', which has unit area normal to heat flow and is Δx thick for internal nodes and $(\Delta x/2)$ for surface nodes, as in this case, during the chosen time step period (Δt), a further term is added:

$$\text{heat storage rate} = \dot{q}_{\mathrm{s,I}} = K_\mathrm{s}\,(\theta_{\mathrm{I,t+\Delta t}} - \theta_{\mathrm{I,t}})\ (\mathrm{W/m^2})$$

$$\text{where } K_\mathrm{s} = \frac{\rho c\,(\Delta x')}{t}\ \mathrm{W/m^2\,°C} \qquad (6.7e)$$

where ρ = density (kg/m³)

c = specific heat (J/kg °C)

$\Delta x' = \Delta x$ for internal nodes and $\dfrac{\Delta x}{2}$ for surface nodes

Δx = finite difference (m) (as per Figure 6.3)

Δt = time step expressed in seconds.

The above terms can now be expressed as a 'heat balance' at the general node I:

$$K_\mathrm{s}(\theta_{\mathrm{I,t+\Delta t}} - \theta_{\mathrm{I,t}}) = \sum_{x=1}^{n} K_{x,\mathrm{I}}\,(\theta_x - \theta_\mathrm{I})_\mathrm{t} + \ddot{q}_{\mathrm{I,t}} \qquad (6.8)$$

where n = the number of thermal contact paths between nodes.

The above outline illustrates the simple elegance and physical insight which the 'heat balance' model provides; two principal problems arise in its implementation:

(i) Insolation will vary and may only strike part of a surface.
(ii) The calculation of a linearised value of $h_{R,I}$ must take account of surface temperatures, emissivities and view factors between all surfaces.

Both of these can be dealt with fairly readily using computer-based methods.

In order to formulate a computer-based simulation model of the building, nodes are first placed at all necessary points and then equations similar to equation 6.8 are written and formed into a large matrix and solved to yield data on internal air and surface temperatures, used to assess thermal comfort, in addition to heating and cooling loads, (and temperatures at all node points if required).

For those unfamiliar with, and perhaps wary of, computer-based techniques it is perhaps worth stressing the ease with which the basic 'heat balance equation' was formulated and compare this both with the inherent complexity of the problem which it solves (as outlined in Sections 6.2.1, 6.2.2 and 6.2.3) and the difficulties and qualifications attached to the other so-called 'simplified' methods outlined in Section 6.2.4.

Energy Simulation Package (ESP)

ABACUS[2] have developed a sophisticated computer program called ESP based on the work of Clarke[15] and others. ESP is based on the 'heat balance method' and has many desirable and sophisticated features. It uses real weather data directly (see Section 6.2) and can accommodate a variety of plant operating strategies. Perhaps its most significant feature (in the light of Section 5.1) is that it is specifically designed for interactive use (via a visual display unit terminal) allowing rapid appraisal and easy modification by the design team.

6.2.7 Steady-state methods

These methods are fully discussed in Chapter 7 and only a few brief points will be made here.

(i) *Air temperature method*. This is the original, conventional method; in modern practice it can be recommended as a method for obtaining only a very aprroximate assessment of heat load. In some ways perhaps it is unfortunate that this is the method used as the basis for the current R.I.B.A./C.I.B.S. programmable calculator programs; it is therefore vital that designers be aware that such a basic program can do no more than give very simple and not necessarily accurate results, given the many assumptions and simplifications which it incorporates. Such a program does have its place however (see 'Summary' below) at the preliminary stage in design but should be considered as no more than a 'blunt instrument' for the serious designer.

(ii) *Environmental temperature method.* From the discussion on the 'admittance procedure' in Section 6.2.4 it can be seen that in fact the 'environmental temperature method' as recommended in the Guide Sections A5 and A9 for estimating winter heat losses is in fact exactly the same as the admittance procedure used for summer cooling load calculations but with the following four principal simplifying assumptions designed to reduce the calculation time involved:

(a) External conditions are represented by θ_{eo} only which is assigned by a constant (i.e. steady-state) design value.

(b) A further simplifying assumption is that a 'design value' of $\theta_{eo} = \theta_{ao}$ (often taken as $-1\,^{\circ}C$ but see Sections 6.2.1 and 6.3) is to be assumed; this means (see equation 6.1) that for design purposes, net radiation exchange between external surfaces and surrounds is assumed to be zero for design purposes. In the light of the discussion in Section 6.2.1 this will be clearly a poor assumption in certain cases.

(c) Internal conditions are represented by θ_{ei} which does improve the modelling of convective and radiation heat exchanges between inside surfaces (as discussed in Chapter 7). From the previous discussion in Sections 6.2.3 *et seq.* there should be little doubt however that the use of θ_{ei} is inferior to the methods used by the more sophisticated computer programs.

(d) Although the Guide does recommend that solar and casual gains should be accounted for in steady-state calculations, designers are rightly conservative in considering them as fortuitous in many cases; there intermittancy and the lack of precision involved in estimating such gains makes accounting for them with a satisfactory degree of accuracy something of a hazard in view of the many other assumptions involved.

6.2.8 Summary

In view of the many complexities involved in modelling building thermal performance (see Sections 6.2.1, 6.2.2 and 6.2.3) it is clear that future designs will be based increasingly on the more exact methods now available using sophisticated computer programs (based probably on those outlined in Sections 6.2.4 and 6.2.6 rather than the current C.I.B.S. procedures in Section 6.2.5).

None the less, conventional steady-state heat losses are still widely used and provide a useful reference point at the initial stages of design from which the designer can proceed from a well established base value of building performance to establish much more accurate measures of building thermal performance using computer programs.

Conventional steady-state heat loss methods also form the basis for the more traditional, often semi-empirical methods of sizing plant and estimating annual energy consumption which are discussed in the remainder of this Chapter. In this context it should be noted that these methods form suitable bases for many fairly simple programs for use with in-house minicomputers or programmable calculators.

6.3 INTERMITTENT OPERATION OF SPACE HEATING SUB-SYSTEMS

6.3.1 Introduction

Following the discussion in the foregoing Sections of this Chapter it will be apparent that only by means of detailed computer programs can accurate modelling of intermittent plant operating strategies be carried out. None the less it is useful to discuss the conventional manual method recommended by C.I.B.S. as this provides a useful introduction to the subject; the method is based on H.V.R.A. Lab. Report No. 26[11]; this was published prior to the introduction of the environmental temperature concept and the admittance procedure but the underlying assumptions hold good. The recommendations are *based* on simplified finite difference calculations.

The Guide lists the following factors as being of primary interest in intermittent heating:

(i) thermal response of the plant
(ii) thermal response of the building
(iii) duration of heating, of cooling and of pre-heating
(iv) relative capital and running costs.

Plant and building response are considered simply as 'short' or 'long' and 'lightweight' or 'heavyweight'; this is consistent with the accuracy expected of a manual method. The Guide method is based on a standard schedule of operation of 8 h heating each day for a seven-day week; in a sense this is unfortunate because, typical, buildings operate on a five-day week cycle. Short pre-heating times require large plant size ratios (i.e. large boilers and emitters and hence high capital costs) in order to reduce fuel consumption (and hence give low running costs). The combined effects of these costs will therefore have a strong influence in designing for optimum performance of the space heating and associated sub-systems. As outlined previously in Section 5.2., this is a difficult problem to resolve and requires the designer to make predictions about future fuel costs in order to obtain realistic results. Clearly the future trend will be to use rapid response plant to heat 'lightweight' buildings and to use shorter pre-heating times with larger plant size ratios, (but see also Section 6.5).

Figure 6.5

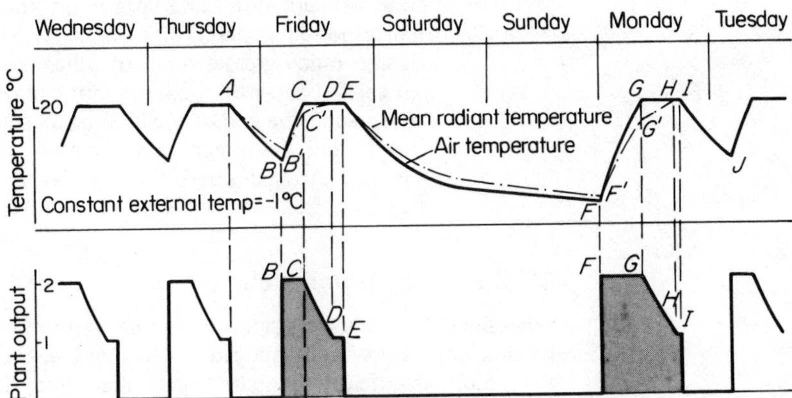

Figure 6.5 shows the hypothetical curves for a simplified building and heating system with a typical plant operating schedule where the external temperature is assumed to be constant at the design value:

A – B Following plant shut-down the air temperature falls as a result of 'natural' cooling processes (see further discussion below).

A – B′ The mean radiant temperature response as heat flows to the cooler room air from the structure.

B – C Pre-heating period during which room air temperature rises to the controlled value of 20°C at the start of the occupancy period; heating plant output is at maximum (here twice the design steady-state heat loss estimate).

B′ – C′ When room air temperature rises above mean radiant temperature, heat flow is reversed and structure is now being heated by room air.

C – D – E Plant output is regulated to maintain constant room air temperature, (chosen as the controlled variable for the purposes of this discussion). From C – D, gradual reduction from maximum plant output to design steady-state heat loss value which is maintained during D – E.

C – D Because the plant output is controlled on air temperature, the rate of rise of mean radiant temperature is reduced during C – D until the structure is warmed to the design internal mean radiant temperature, (for convenience assumed here to be equal to the air temperature).

The plant output during the total 'on' period B – C – D – E is shown shaded; this represents a typical design weekday.

The response curves (E – F – G – H – I – J) for the weekend shut-down and Monday morning start-up period shows that a longer pre-heating period is necessary and that the building fabric is not fully warmed up until late afternoon (i.e. at I); the corresponding energy input under (F – G – H – I) is also much greater than for other days. Note that in certain cases the building fabric may not be fully warmed at the end of the Monday occupancy period (at J) and the effects of weekend cooling may not be fully ironed out until Tuesday or Wednesday.

6.3.2 Cooling response curves

The concept of characterising the thermal response of a mass by its time-constant was introduced in Chapter 3 where it was noted that multi-stage systems have more than one time-constant. Following the discussion of the complexities of modelling building thermal response outlined in Sections 6.1 and 6.2, it is not surprising that experiments indicate that during the cooling phase following plant shut-down most buildings exhibit two or three time constants (see Figure 6.6).

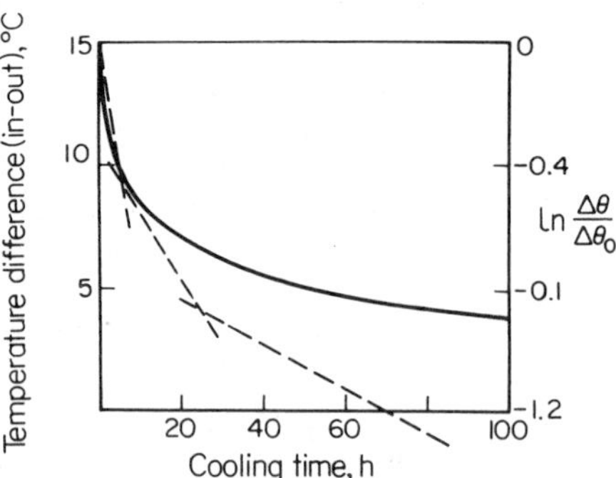

Figure 6.6 Typical building cooling curves. Solid line, temperature; broken line, log plot

The initial rapid drop characterises the room air and heating plant response; this is of primary interest in automatic control during the hours of occupation. The intermediate time-constant is often used to characterise overnight cooling while the final one characterises cooling during weekend shut-down; both these latter values are of importance in optimum start control. The actual

Table 6.4 Approximate Cooling Time-constants

Factors Influencing Time-constant	*Lightweight*	*Heavyweight*
Cooling of air and heating plant	30 min – 2 h	up to 10 h
Cooling of contents (e.g. furniture)	2 – 8 h	40 h
Cooling of structure	8 – 12 h	120 h

values vary depending on the internal surface properties; Table 6.4 gives approximate orders of magnitude.

The important point about cooling time-constants is that they occur as a result of a *natural* process and are governed by the *rates* of:

(i) infiltration and external air temperature
(ii) heat transfer from inside surfaces to infiltration air
(iii) heat loss throuch glazing (especially during initial stages)
(iv) heat loss from external surfaces (affected by θ_{eo}).

6.3.3 Pre-heating response curve

Figures 6.7 and 6.8 show hypothetical pre-heating curves for two buildings and demonstrate that short pre-heating time (T_p) requires

Figure 6.7
P, T_p for lightweight building

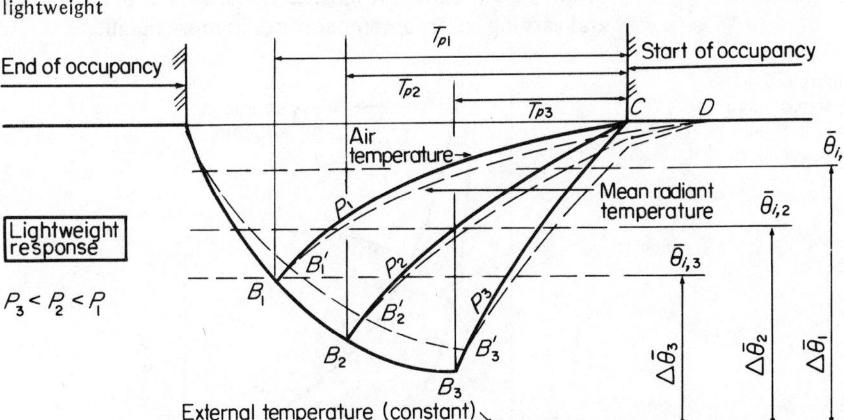

a large plant size ratio (P) and vice versa. Furthermore each combination of T_p and P gives rise to a particular average value of internal air (or environmental) temperature $\overline{\theta}_i$; the corresponding average temperature, difference $\Delta\theta_i$ in each case can be used as a rough indicator of the heat loss from the building during the unoccupied period. Large values of P give small T_p and $\Delta\theta_i$ values and hence tend to give lower fuel consumption and fuel costs. A more accurate assessment of the same trend would be obtained by calculating the weekly energy input for each case from the weekly cycle curves as shown in Figure 6.5.

Figure 6.8
P, T_p for heavyweight building

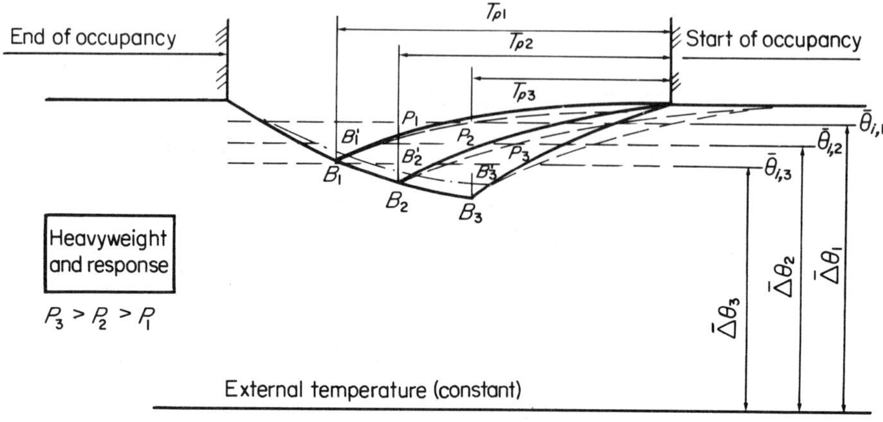

6.3.4 Choosing T_p and P

The C.I.B.S. Guide (1970) Table A9.3 is reproduced here as Table 6.5. This provides a simplified method of estimating the various parameters involved. Figures 6.9, 6.10 and 6.11 present the same information in graphical form and allow easier interpretation of the data and include the effects of plant time-constants and varying external temperatures in more detail.

Figure 6.9
Design diagram for buildings with plant time constant $1\frac{1}{2}$ h

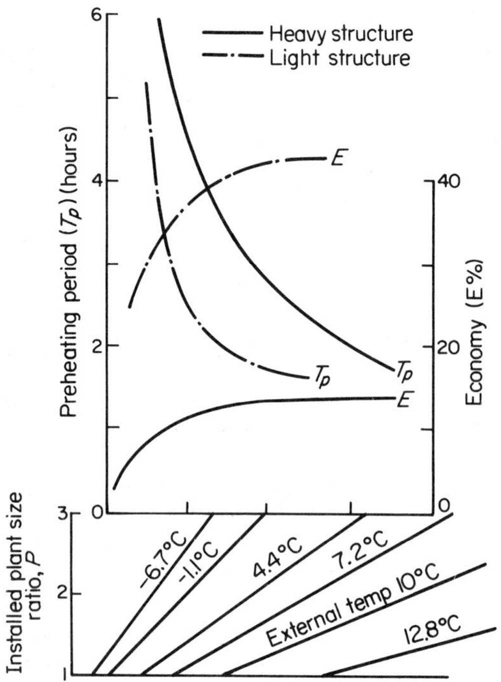

Table 6.5 Allowances for intermittent heating

Plant characteristics		*Heavyweight building*				*Lightweight building*			
		−1 °C outside		*5 °C outside*		*−1 °C outside*		*5 °C outside*	
Response	*Size ratio*	*Preheating time* (h)	*Fuel consumption* (%)	*Preheating time* (h)	*Fuel consumption* (%)	*Preheating time* (h)	*Fuel consumption* (%)	*Preheating time* (h)	*Fuel consumption* (%)
	1.2	v. long	96	5.0	89	7.0	75	1.5	56
	1.5	6.5	91	2.8	84	3.0	60	0.8	53
Short	2.0	3.3	85	1.4	81	0.9	54	0.5	52
	2.5	1.8	82	0.7	78	0.6	53	0.4	51
	3.0	1.1	80	0.3	76	0.4	52	0.3	50
	1.2	v. long	96	5.9	90	long	80	3.5	66
	1.5	6.5	91	4.2	88	4.0	69	2.3	62
Long	2.0	4.4	89	3.0	87	2.4	62	1.8	58
	2.5	3.4	87	2.3	86	1.9	59	1.6	57
	3.0	2.8	86	1.9	85	1.7	58	1.5	56

Notes:

1. The Table is based on an indoor-outdoor temperature differential of 20 °C and an occupied period of 8 h, for a seven day week.
2. Plant size ratio = $\dfrac{\text{Normal maximum plant output}}{\text{Design load for 20 °C rise}}$
3. Examples of short response plant—direct warm air heating, forced convection; gas or electric radiant panels. Examples of long-response plant—hot water system with radiators, convectors or radiant panels. Embedded panels have time-constants of several hours, and intermittent operation leads to very little economy in fuel.
4. Heavy structure: curtain walling, masonry or concrete (especially multi-storey), subdivided with heavy partitions or floors (a high value of ΣAY. See Table A9.4).
5. Light structure: single-storey, factory-type construction; little or no solid partition; structures lined with insulating materials (a low value of ΣAY. See Table A9.4).
6. Fuel consumption is expressed as a percentage of fuel needed for continuous operation.

Figure 6.10
Design diagram for buildings with plant time constant $\frac{3}{4}$ h

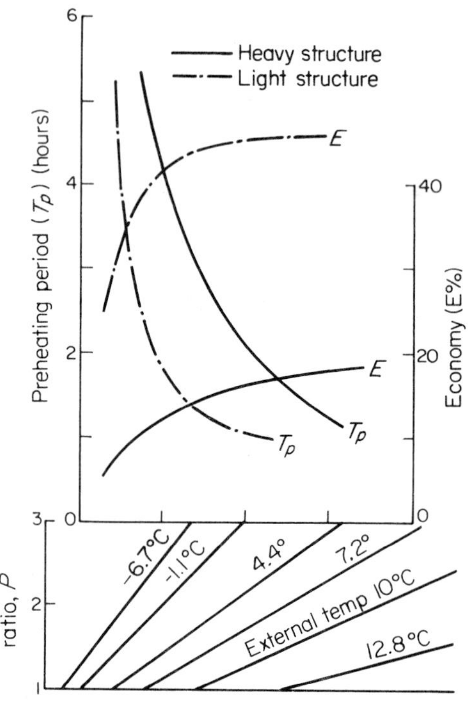

Figure 6.11
Design diagram for buildings with plant time constant 0 h

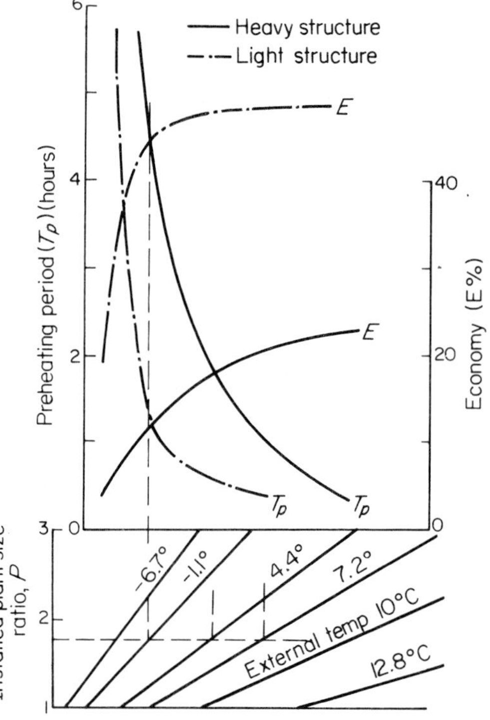

There are many assumptions and approximations inherent in the method and it should be used only to obtain a first estimate of the data. In particular it should be noted that predictions on fuel savings are far from exact and a more meaningful assessment of annual energy consumption compared with continuous operation can be obtained using the methods outlined in Section 6.5. Note: In certain cases the C.I.B.S. definition of plant size ratio (P) given in Table 6.5 may be misleading and a more helpful definition might be

$$P = \frac{\text{installed or available power input (kW)}}{\text{steady-state design heat loss rate (kW)}}$$

6.3.5 Optimum start control and other plant operating strategies

When a plant size has been chosen for a particular building the plant size ratio (P) is then fixed. As can be seen from Table 6.5 and Figures 6.9, 6.10 and 6.11, the required pre-heating time for a particular day varies with the prevailing mean external temperature. As discussed in Section 6.3.2, the cooling response curve will vary with external conditions and in particular wind speed and air temperature will have an important effect. In addition cooling response will vary from room to room. None the less, optimum start control (OSC) seeks to cater for variations in external and internal conditions by monitoring both - internal conditions usually being represented by the temperature monitored in a 'typical' room.

From a knowledge of P (fixed by the design) and an estimate of the building characteristics (i.e. light, medium or heavy) the OSC can be set to start the heating plant at the required 'optimum' start time in order that the end of the pre-heating time coincides with the start of the occupancy period. In practice some trial-and-error adjustment of the controller is normally required during commissioning of the plant and yearly checks should be made to

Figure 6.12
Effect of θ_o on T_p

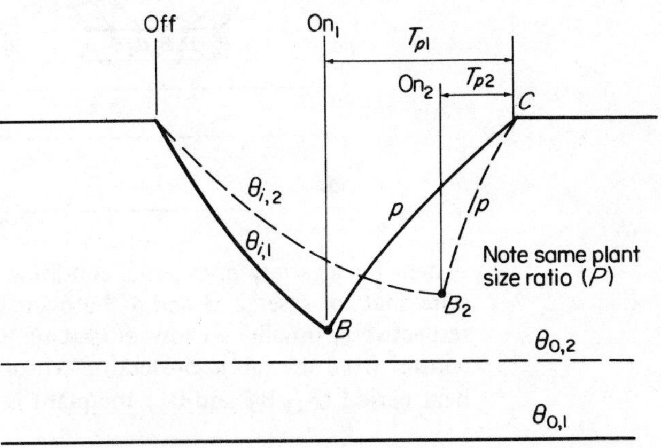

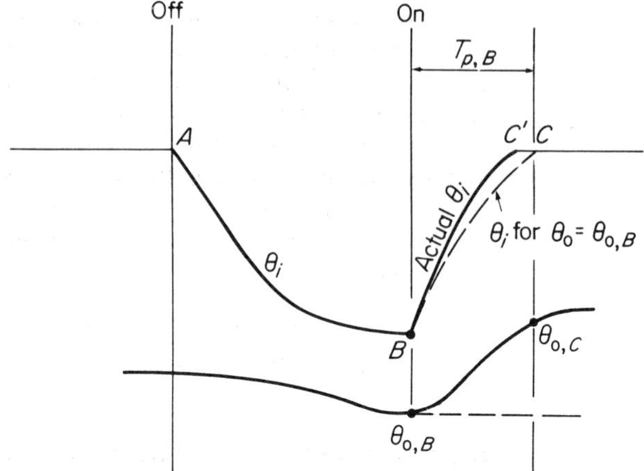

Figure 6.13
Effect on varying θ_o on OSC

ensure that satisfactory operation is being maintained.

Figure 6.12 oultines the above points graphically on the basis of idealised constant values of external temperature (θ_o). Figure 6.13 shows the probable likely response with diurnal variation in θ_o from which it can be seen that both the cooling and heating response curves are affected. OSC controllers are usually built to start the plant at B on the assumption that $\theta_o = \theta_{o,B}$ will be constant for the remainder of the pre-heating period. In general this is a conservative estimate and leads to early attainment of the desired internal conditions at C′ rather than C and hence effectively increases fuel consumption. Late arrival at the desired internal conditions can of course also occur when external temperature falls and/or wind speed increases after plant start at B.

A common variation of basic OSC imposes a lower permissable limit on internal temperature for fabic protection purposes (see Section 3.9.3). Figure 6.14 shows the idealised response for this

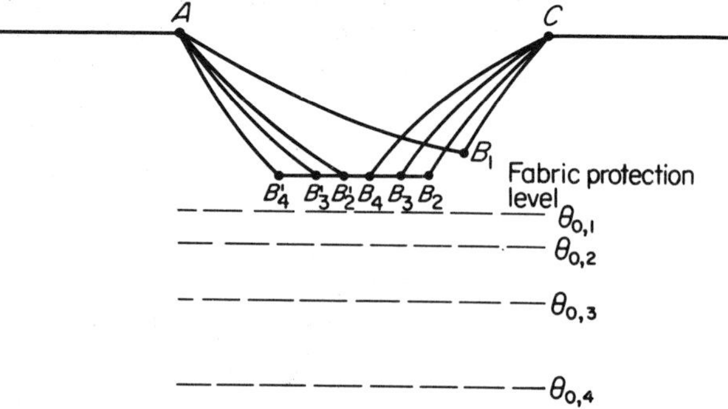

Figure 6.14
OSC with fabric protection thermostat

system for a variety of external conditions. From this diagram we note that for cases 2, 3 and 4 the plant starts at B'_2, B'_3 and B'_4 respectively, usually on low output or under reset thermostatic control from the fabric protection sensor; at the start of the pre-heat period (B_2, B_3 and B_4) the plant is switched to full output under normal thermostatic control. In general this operational strategy consumes more energy than the basic OSC (particularly in 'lightweight' structures).

6.3.6 Emitter loadings

The topic of emitter sizing is discussed in detail in Chapter 8 but some comments regarding emitter loads to be used as the basis of the sizing exercise are required here.

The term 'plant size ratio' refers not only to the central heat source sub-system but also to the rate of heat output at the emitters. This means that during the pre-heating period it is necessary to ensure that emitter heat output is increased pro rata. However this does not necessarily mean that the installed emitter size for intermittent operation need necessarily be equal to P times the size for steady-state design heat loss because:

(i) During pre-heating the emitter output will be increased above the steady-state design value because the room internal temperature is low, thus tending to increase the difference between the emitter mean water or surface temperature and the room temperature.

(ii) As a consequence of (i), lower return water temperatures occur at the central heat source but to counter this, additional heat input is available and hence it is often possible to elevate the boiler flow temperature from the normal working temperature, say 80°C, to a pre-heating 'boost' temperature of say 90–95°C thus providing extra potential for heat transfer at the emitters.

An estimate of the above influences can be made using the principles outlined in Chapter 8 and can often lead to useful reductions in emitter sizes and consequent savings in capital costs.

6.3.7 Highly intermittent systems

The discussion above refers to buildings which are used regularly. Buildings such as churches which are used only once a week or so require special consideration.

There are two methods of tackling this problem:

(a) Use a radiant heating system to create a high mean radiant temperature to compensate for low air temperatures. Short-wave radiant heating systems are ideal for this application but require special care to ensure comfort (see Section 4.9. and Fanger[12]).

(b) The alternative solution is to install a convective heating plant with a large plant size ratio to give a high air temperature to compensate for low mean radiant temperatures.

In such cases the conventional heat loss calculations outlined in Chapter 7 are inappropriate because for most of the period of plant operation heat flow is into rather than through the structure.

In order to arrive at a first order estimate of the required plant size C.I.B.S. recommend that the heat load in such cases should be calculated using equation 6.9:

$$\dot{Q} = C(\Sigma(AY) + Cv)\,(\theta_{ei} - \theta_{eo}) \tag{6.9}$$

Some comments on the use and application of this equation may prove helpful:

(i) As will be seen from Section 6.2.4 the use of Y values (which are based on steady period (24 h cycle) heat flows), will give only a very approximate result in this case and hence C.I.B.S. recommend the use of the correction factor (C) computed from

$$C = \left(\frac{24}{T_t}\right)^{0.5} \tag{6.10}$$

where T_t = total plant operating time (i.e. pre-heating plus occupancy periods).

As can be seen, T_t has a marked effect on C:

T_t	24	20	16	12	8	4	2	1
C	1.00	1.10	1.22	1.41	1.73	2.45	3.46	4.90

(ii) The use of Y in place of U has two effects; firstly, if we compare U for a typical cavity wall with its corresponding Y values we see:

$$U \simeq 1.5\ \text{W/m}^2\ {}^\circ\text{C}$$
$$Y \simeq 4.0\ \text{W/m}^2\ {}^\circ\text{C}$$

and note that $(Y/U) \simeq 2.67$; secondly, $\Sigma\ (AY)$ is to take account of all room surfaces and not just external surfaces.

Designing for comfort

Unfortunately C.I.B.S. give no guidance on the design value of θ_{ei} to be used in this case and as the object will of course be to ensure comfort during the occupied period, the designer will be faced with some difficulty in performing the comfort checks recommended in Chapter 4. During the pre-heating period surface temperatures will be continually rising and hence θ_{ei} will alter; further, the conventional steady-state methods of estimating θ_{ai} and θ_{ri} as outlined in Chapter 7 will be inappropriate.

To make an approximate (manual) estimate of comfort conditions at the start of the occupied period the following iterative procedure is therefore recommended:

(a) Assume an initial value of $\theta_{ei,B}$ = 20°C (say) at start of pre-heat period (B) and assume all surfaces to be equal to θ_{eo}

(and hence $\theta_{ri,B} = \theta_{eo}$ can be assumed), from which a first estimate of $\theta_{ai,B}$ can be obtained from:

$$\theta_{ai,B} = 3\theta_{ei,B} - 2\theta_{ri,B} \; (= \theta_{eo})$$

Use $\theta_{ai,B}$ to make a first estimate of ventilation heat loss, $\dot{Q}_v = \rho N V' C_p \, (\theta_{ai,B} - \theta_{ao})$

(b) Now estimate heat loss through glazing during pre-heating from

$$\dot{Q}_G = \Sigma (A_G U_G) \, (\theta_{ai,B} - \theta_{ao})$$

(c) Use equation 6.9 to estimate plant size $\dot{Q}$ and hence estimate heat flow rate *to* fabric from

$$\dot{Q}_{f,B} = (\dot{Q} - \dot{Q}_v - \dot{Q}_G)$$

The heat flow rate *to* the fabric during the pre-heating period will tend to decrease to a lower value of $\dot{Q}_{f,e}$ at the start of occupancy but this in turn will tend to increase the value of θ_{ai}; the relationship is not easy to evaluate and will depend in part on plant characteristics. We recommend therefore that $\bar{\dot{Q}}_f = \dot{Q}_{f,B}$ be assumed and that air temperature be taken as constant throughout at the value assumed at the start of the pre-heat period.

(d) Use equation 6.11 (for details see Appendix 6.1) to estimate the rise in average fabric surface temperature during the pre-heat period and hence obtain a first estimate of opaque fabric surface temperature at the end of the pre-heat period (C).

$$\Delta\bar{\theta}_{si} = \frac{\bar{\dot{Q}}_f}{\Sigma A_f} \sqrt{\left(\frac{5T_p}{4(k\rho c)'_e}\right)} \qquad (6.11)$$

where $\Delta\bar{\theta}_{si}$ = average increase in opaque fabric inside surface temperature during T_p (°C)

$\bar{\dot{Q}}_f$ = average heat flow rate to opaque fabric during T_p (W)

ΣA_f = surface area of opaque fabric (m^2)

$(k\rho c)'_e$ = equivalent average thermal properties of opaque surfaces as defined in Appendix 6.1

(e) Assuming $\theta_{ai,C} = \theta_{ai,B}$ estimate a value of inside surface temperature for glazing; this information together with that from step (d) can be used to estimate $\theta_{ri,C}$, and hence give a rough check on comfort at the start of occupancy (C) using $\theta_{res,C} = 0.5\theta_{ai,C} + 0.5\theta_{ri,C}$.

(f) If steps (a) to (e) give too high or too low a value of $\theta_{res,C}$ either alter $\theta_{ei,B}$ chosen in step (a) or T_p chosen in step (d), and iterate until a satisfactory value of plant size and/or T_p is found.

Obviously the above procedure can only be regarded as very approximate and a detailed computer program is recommended. It has been presented mainly to illustrate the *difficulties inherent* in using *simplified* approximate methods!

6.4 PROFILING CENTRAL PLANT LOADS

The method of computing load profiles was discussed in Section 5.2.4 and the necessity of evaluating the performance of the central heat source (boiler plant) sub-system over the whole heating season was also stressed in Sections 5.2.4 and 5.3, and hence only a summary of the position with regard to each of the sub-systems is given here to outline the designer's main tasks:

(i) Sections 6.2 and 6.3 are aimed at explaining the techniques (together with their limitations), which can be used to obtain design and part-load estimates of the heat load profiles for the space heating sub-system. From what has been written it is obvious that accurate assessments required computer-based methods.

(ii) Chapter 10 outlines the basic method of computing DHWS storage and power input requirements. Maver(13) also discusses the methods of optimising this sub-system in some detail. What is not always clearly recognised by designers is that in most cases, (excepting perhaps hospitals, hotels and the like), DHWS can and should be regarded as a low-priority demand on the central heat source sub-system. This would be a great aid to load profiling because the designer could, by careful choice of DHWS calorifier size and primary heating surface, arrange to supply heat to it at times which would suit the central heat source load profile. This could be done simply by either a time clock or a low-temperature detector on the space heating sub-system return arranged to switch off DHWS primary pumps and perhaps also to close a solenoid valve in the primary circuit. There are benefits to be obtained both in capital and running costs from doing this and should the DHWS supply 'fail' on occasions this would be normally of little consequence.

(iii) Profiling the process heating sub-system load will be a matter of separate consideration in each individual case but it may well be possible to use the same techniques as in (ii) by employing a storage buffer against times of peak demand.

(iv) It follows from (ii) that heat storage in calorifiers can also be used to profile the space-heating sub-system load. In the past this has normally be restricted to systems employing off-peak electricity via electrode boilers but clearly this is a technique for greater consideration in future designs, for example those employing coal-fired plant which give better efficiency when continuously fired.

(v) Again it must be stressed that there is little point in profiling the 'design' heat load alone; the very least that would be of value here would be typical weekly load profiles for each month of the year (again see Section 5.3).

6.5 ESTIMATING ANNUAL ENERGY CONSUMPTION

Even the very best computer-based estimates taking account of all the many variables are unlikely to predict annual energy consumption with an accuracy greater than about ±15 to ±20%. The reasons why this is so are partly technical and partly sociological but, briefly, major causes are believed to be:

(i) Inadequate modelling of infiltration/ventilation losses which is probably due in turn to deliberate or accidental window and door opening as well as poor modelling of the wind speed/infiltration rate in most programs.

(ii) Individual preferences regarding comfort affecting (i) and causing unnecessary and/or undesirable adjustment of controls.

(iii) Variations in plant maintenance and supervision, particularly with respect to regular checks (3, 6 or 12 monthly) on control settings and performance.

Energy targets and energy audits

Research has shown that similar buildings on the same site with similar heating systems, controls, occupancy, etc. can have quite large variations in energy consumption. Further investigation has usually found some good reason for the discrepancy which when corrected has led to energy savings. Unfortunately most buildings tend to be unique in one or more of the major factors influencing thermal performance and hence it is usually difficult to find another similar building with which to make comparisons.

It is therefore beneficial that an 'energy audit' be carried out to determine (or estimate) the amounts and costs for the various types of fuel which have been (or will be) used during a certain period. Generally this is done after the building has been in use for some time but an estimated energy audit can be prepared at the design stage. As a result of such an audit and by comparison with similar buildings, the designer can then outline hypothetical 'energy targets' for the guidance of the maintenance personnel. Such targets may be daily, weekly or monthly and should be capable of adjustment to account for weather factors.

Degree-day estimates

The basis for estimating annual energy consumption is clearly set out in C.I.B.S. Guide Section B.18 which provides all necessary tables and graphs and gives worked examples. This provides a simple manual method appropriate for use with the conventional heat loss calculations described'in Chapter 7. The Guide method is framed in terms of annual estimates but calculations can be performed readily in terms of monthly degree day values which are published regularly in the C.I.B.S. Journal and elsewhere although clearly the method lacks the precision and sensitivity to provide 'daily' or 'weekly' estimates.

Computer estimates

Although computer results will in general be more accurate, their real value is in providing the load profiles for the plant because real weather is used and proper account is taken of ups and downs in external conditions. It is important to note however that at the present time there is little real advantage in performing a full hour by hour simulation for the complete heating season. Such a result would be best only indicate the requirement for the particular weather tape used and in any case the eventual overall accuracy would still be fairly approximate (± 15% to ± 20% as stated previously). Obviously we hope for improvements in accuracy in future computer models but for the present it is advisable only to obtain typical days or weeks and then to multiply these in the ratio of the degree days for the month considered divided by the degree days during the period selected for computer evaluation.

Lastly, most computer programs gives results of energy consumption in terms of energy inputs to the room without modification to account for distribution losses and combustion efficiencies. It is important therefore for the designer to check this and account for these factors either simply using average values, or perhaps in more detail using the typical boiler load profiles for each month or week - which information is the real benefit of using 'accurate' computer models.

APPENDIX 6.1 DUFTON'S EQUATION

Equation 6.11 is simply a re-expression of the equation proposed originally by Dufton[14]. It is given here in its more conventional form for reference:

$$\theta = (p\dot{q})\sqrt{\frac{5t}{4(k\rho c)}} \tag{A6.1}$$

where θ = temperature difference between inside and outside wall surface temperatures (°C)

$(p\dot{q})$ = heat input rate to wall (W)

t = duration of heat input (s)

k = conductivity (W/m^2 °C)

ρ = density (kg/m^3)

c = specific heat (J/kg)

Equation A6.1 was obtained by Dufton by solving equation 2.11 with a number of restrictions (i.e. approximations) for a simple homogeneous wall (having thermal properties k, ρ and c). Equation A6.1 is thus an approximate equation describing the rise in temperature to be expected after a given time t as a result of a steady heat flow ($p\dot{q}$) into the simple homogeneous wall.

In order to use equation A6.1 for multi-layer construction it is necessary to define an average equivalent homogeneous wall denoted by $(k\rho c)'_e$ in equation 6.11. This can be done using equations A6.2, A6.3 and A6.4; equation A6.3 is an empirical expression proposed by Mackey and Wright[5] from a consideration of the results of a number of analytical solutions.

For an n-layer construction,

$$R_e = \sum_{i=1}^{n} R_i \tag{A6.2}$$

where R_e = equivalent thermal resistance of n-layers (m^2 °C/W)
R_i = thermal resistance of ith layer (m^2 °C/W)

For layers numbered 1 = inside to n = outside,

$$(k\rho c)_e = \frac{1.1R_1(k\rho c)_1 + 1.1R_2(k\rho c)_2 + 1.1R_3(k\rho c)_3 \ldots}{R_e}$$

$$\ldots + \frac{(k\rho c)_n}{R_e}(R_n - 0.1R_1 - 0.1R_2 - 0.1R_3 \ldots) \tag{A6.3}$$

In cases where R_n is small compared to that of the other layers the second term of equation A6.3 becomes negative, in which case it should be neglected; for cavities $(k\rho c)$ should be taken as zero.

This process must be repeated for each surface forming the enclosure; difficulties arise with partitions heated from both sides and these can be treated approximately by evaluating $(k\rho c)_e$ values by the above methods by considering the semi-thickness of each panel separately. When $(k\rho c)_e$ values are known for each opaque surface the 'average' value can be found from

$$(k\rho c)'_e = \frac{\sum_{i=1}^{m} (A(k\rho c)_e)_i}{\sum_{i=1}^{m} (A_i)} \tag{A6.4}$$

The final steps in obtaining equation 6.11 from equation A6.1 involve simply changing t to T_p the pre-heating time, noting that this must be expressed in seconds, not hours as is normally used, and estimating $(p\dot{q})$ (assumed equal to $\bar{Q}_f$ in equation 6.11). Finally, we can assume that the outside surface temperature remains constant throughout the pre-heating time T_p and hence θ can be re-interpreted as the rise in inside average surface temperature $(\Delta\bar{\theta}_{si})$ during T_p.

REFERENCES

1. WRANGHAM, D. A. *The Elements of Heat Flow*, (Chatto & Windus, London, 1961).
2. CLARKE, J. A. 'Validation of the ESP Thermal Simulation Program', *ABACUS Occasional Paper No. 61*, University of Strathclyde.
3. BURBERRY, P. *Building for Energy Conservation*, (Architectural Press, London, 1978).
4. GUPTA, C. L., SPENCER, J. W. and MUNCEY, R. W. R. 'A Conceptual Survey of Computer-Oriented Thermal Calculation Methods', *Proc. Second Int. Symp. on Use of Computers for Environmental Enging Related to Buildings, Paris*. Building Science series 39, 4, 1970.
5. MACKEY, C. O. and WRIGHT, L. T. 'Periodic Heat Flow – Composite Walls or Roofs', *Heating, Piping and Air Conditioning*, Vol. 18, Part 6, 1946, pp 107–10.
6. LOUDEN, A. G. 'Summertime temperatures in buildings without air conditioning', *B.R.S. Current Paper* No. 46/68.
7. HARRINGTON-LYNN, J. 'The Admittance Procedure: Variable Ventilation', *J.I.H.V.E.*, Vol. 42, November 1974, pp 199–200
8. HARRINGTON-LYNN, J. 'The Admittance Procedure: Intermittent Plant Operation', *J.I.H.V.E.*, Vol. 42, December 1974, pp 219–21.
9. KIMURA, K.-I. *Scientific Basis of Air Conditioning*, (Applied Science Publishers, 1978).
10. CARSLAW, H. S. and JAEGER, J. C. *Conduction of Heating Solids*, (Oxford University Press, New York, 1947).
11. BILLINGTON, N. S., COLTHORPE, K. J. and SHORTER, D. N. 'Intermittent Heating', *H.V.R.A. Lab. Report No. 26*, 1964.
12. FANGER, P. O. *Thermal Comfort*, (McGraw-Hill, New York, 1972).
13. MAVER, T. M. *Building Services Design*, (R.I.B.A. Publications, London, 1971).
14. DUFTON, A. F. 'The Warming of Walls', *J.I.H.V.E.*, 2(21), 1934, pp 416–7.
15. CLARKE, J. A. 'Environmental Systems Performance,' Ph.D. Thesis, University of Strathclyde, 1977.
16. A.S.H.R.A.E. Fundamentals, 1976.

LIST OF SYMBOLS USED IN CHAPTER 6

Temperature

θ	temperature
θ_i	internal temperature
θ_o	external temperature
$\theta_{ai}, \theta_{ao}, \overline{\theta}_{ao}$	air temperatures
$\theta_{eo}, \overline{\theta}_{eo}$	sol-air or 'equivalent' external temperatures
$\theta_{ei}, \hat{\theta}_{ei}, \overline{\theta}_{ei}$	internal environmental temperatures measured at the room centre (environmental) point
θ_{ri}	internal mean radiant temperature measured at the room centre (environmental) point
$\Delta\theta_{ei}$	'swing' in θ_{ei}
$\Delta\theta_{eo}$	'swing' in θ_{eo}
$\Delta\overline{\theta}_{si}$	rise in average internal fabric surface temperature
θ_{res}	dry resultant temperature

Other Symbols

A	area
C	a correction factor
c	specific heat of fabric
C_p	specific heat of moist air
f	decrement factor; subscript denoting 'fabric'
F	surface factor
$h_{c,I}$	convection heat transfer coefficient to node I
h_{ci}	inside surface convection heat transfer coefficient
h_{co}	outside surface convection heat transfer coefficient
$h_{R,I}$	radiation heat transfer coefficient to node I
$I, I-1, I+1$	nodes
I_t	total solar radiation striking an outside surface
I_e	net longwave radiation at an outside surface
$J-1, J, J+1$	past, present and future time rows
$K_{1,I}; K_{2,I}; K_{3,I}$	conduction, convection and radiation heat transfer coefficients referred to node I
$K_1; K_2; K_3$	derived constants in heat balance equation
k	thermal conductivity of fabric
N	air change rate
n	hourly air change rate; number of fabric layers
m	number of opaque fabric surfaces
$\underline{P}, P_1, P_2$, etc.	plant size rates
$\underline{\dot{Q}}_c$	mean casual heat gains
$\underline{\dot{Q}}_f$	mean fabric heat gain (or loss)
$\underline{\dot{Q}}_s$	mean solar heat gains
$\underline{\dot{Q}}_r$	mean total heat gain
$\dot{Q}_f$	fabric heat gain (or loss)
$\dot{Q}_G$	heat loss through glazing
$\dot{Q}_v$	infiltration/ventilation heat loss
$\Delta\dot{Q}_T$	'swing' in total heat gain
$\dot{q}_{1,I}; \dot{q}_{2,I}; \dot{q}_{3,I}$	conductive, convective and radiative heat flow rates at node I
$\ddot{q}_I$	heat generation at node I
$\dot{q}_{s,I}$	heat storage rate at node I
$(p\dot{q})$	rate of heat flow into wall (and other fabric)
R_1, R_2, R_3, etc.	thermal resistance of each layer
R_e	equivalent thermal resistance of multi-layer construction
S	solar gain factor for glazing system/building weight classification
S_a	alternating solar gain factor for glazing system/building weight classification
T_p	pre-heating time

T_t	total plant operating time (pre-heat plus occupancy)
t	at time t; a period of time
Δt	time step in computer program
U	overall (steady-state) heat transfer coefficient
V	room volume
x	a number
Δx, $\Delta x'$	finite differences
Y	surface admittance
Z	surface impedance
α	thermal diffusivity
ϵ	surface emissivity
ϕ	phase lag
ρ	density
ω	fundamental frequency

7 Steady-state Heat Losses

7.1 INTRODUCTION

The generally accepted basis for the design of a space heating subsystem is the evaluation of the building heat losses. The object of this Chapter is to examine in some detail the conventional procedures of heat loss computation. The major simplifying assumption underlying these procedures is that of a steady-state condition, i.e. it is assumed in the calculations that the internal and external temperatures are invariant with respect to time. In reality, of course, steady-state conditions seldom exist and the results obtained from the procedures outlined in this Chapter may require to be modified, as discussed in Chapter 6, in order to obtain the required heat emitter loads and the central plant size. Nevertheless, although steady-state transmission rarely occurs in practice, the assumption of such a condition does provide a convenient basis for an initial assessment of heating requirements.

In the heat loss procedure, each individual space in the building is considered in turn and a calculation is made of the rate of heat supply necessary to maintain the selected internal design temperature assuming that the outdoor temperature is at the appropriate winter design value. This heat requirement has two components:

(i) The heat required to balance the heat transmission through the exposed parts of the building fabric (the fabric heat loss) and,

(ii) the heat required to raise the temperature of the outside air infiltrating into and through the space up to the internal air temperature (the ventilation heat loss).

Also included in this present Chapter is a section which deals with condensation problems in buildings.

7.2 FABRIC HEAT LOSS: AIR TEMPERATURE METHOD

7.2.1 The basic equations

Consider the typical room of a building shown in Figure 7.1a. This particular space will experience a transmission heat loss to the outside environment through its roof and external wall. Let

Figure 7.1

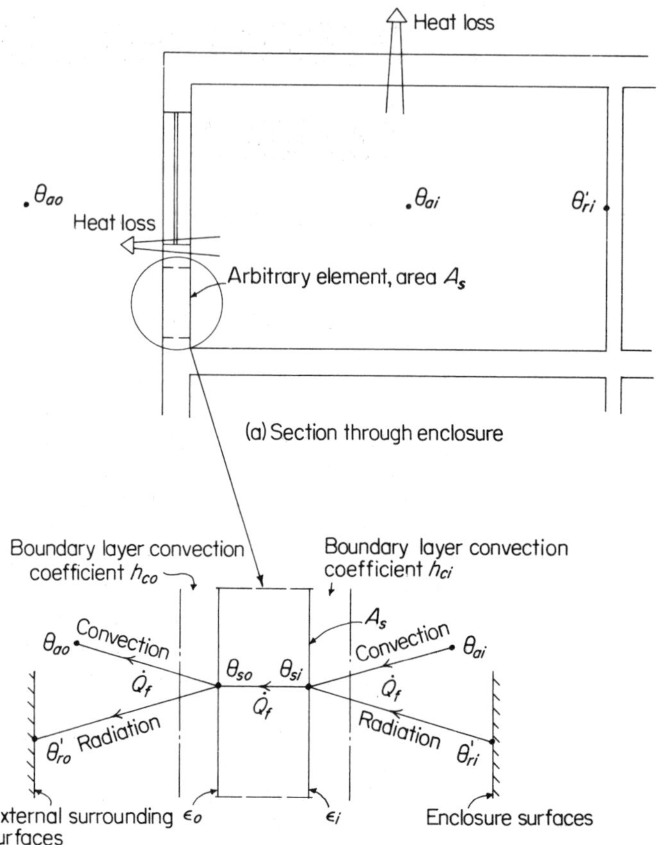

us examine the computation of this loss. The air temperature inside the room is θ_{ai}. Outside, the air temperature is θ_{ao} and the mean temperature of the surrounding surfaces (nearby buildings, sky, etc.) is taken as θ'_{ro}. In the winter design situation, sunless conditions are assumed and solar radiation effects on the outside surface of the building are ignored.

Consider firstly the heat loss associated with an arbitrary element of the external structure of area A_s (Figure 7.1b). The inside surface temperature of this element is θ_{si} and the mean temperature of the remaining enclosure surfaces as seen by it is θ'_{ri}. Let the rate of transmission through the element be $\dot{Q}_f$. By referring back to Chapter 2, this problem can be identified immediately as a particular example of the general heat transfer process between separated fluids discussed in Section 2.5.2. Comparing Figure 7.1b with Figure 2.15, we see that the inside air corresponds to Fluid 1, the outside air corresponds to Fluid 2 and that the notation of the quantities involved has been modified appropriately.

If we assume $\theta_{ai} = \theta'_{ri}$ and $\theta_{ao} = \theta'_{ro}$, then the present problem becomes identical to that considered in Section 2.5.2 and the equations derived there may be applied to the evaluation of $\dot{Q}_f$. The rate of heat loss through A_s is therefore given by

$$\dot{Q}_f = UA_s\,(\theta_{ai} - \theta_{ao}) \tag{7.1}$$

where
$$U = \frac{1}{R_{so} + R_t + R_{si}} \tag{7.2}$$

R_{so} and R_{si} are referred to respectively as the outside and inside surface resistances. The total fabric heat loss from a room is obtained by summating equation 7.1 over all the external elements of the structure (in this particular example the roof, glazing and external wall), thus

$$\Sigma\dot{Q}_f = \Sigma\,UA_s\,(\theta_{ai} - \theta_{ao}) \tag{7.3}$$

Equation 7.3 forms the basis of the traditional heat loss procedure in which air temperature is used as the sole parameter for computation purposes. Certain comments require to be made here. Firstly, it is implicit in equation 7.3 that the fabric heat loss from a space is associated wholly with the external parts of the structure. However, this will be the case only if the adjacent internal spaces of the building are maintained at the same temperature. Where an adjacent space is at a lower temperature, there will be a transmission of heat from the warmer space through the internal parts of the structure to this cooler one. Such internal losses, where they occur, should be calculated and incorporated into $\Sigma\dot{Q}_f$. The overall heat transfer coefficients for use in this respect are evaluated using two appropriate internal surface resistances. Secondly, if certain surfaces of a space, such as the ceiling or the floor, are used for heating purposes, then these areas can be omitted from the heat loss calculations.

7.2.2 Overall heat transfer coefficients

The evaluation of the thermal resistance R_t in equation 7.2 has been well covered in Chapter 2. Let us consider now the evaluation of the outside and inside surface resistances. From Section 2.5.2 we see that R_{so} and R_{si} are defined by the expressions

$$R_{so} = \frac{1}{h_{co} + \epsilon_o h_{Ro}} \tag{7.4a}$$

$$R_{si} = \frac{1}{h_{ci} + \epsilon_i h_{Ri}} \tag{7.4b}$$

The values of h_{Ro} and h_{Ri} recommended for heat loss computation purposes have already been quoted in Section 2.4.3 (4.6 and 5.7W/m^2 °C respectively) and it has been demonstrated how such values may be arrived at from fundamental radiation theory. Values of emissivity for various surfaces have also been quoted and it is seen that for most ordinary building materials a representative value of 0.9 is appropriate. The outside and inside convection heat transfer coefficients will now be examined.

Outside coefficient, h_{co}

At the outside surface of a building the convective heat transfer process is primarily one of forced convection and the main factor influencing h_{co} is the wind speed. Reference (1) proposes the use of the following relation

$$h_{co} = 5.8 + 4.1\ V \tag{7.5}$$

where V is the wind speed in m/s. If values of h_{co} computed for various wind speeds are substituted into equation 7.4a, the design values of R_{so} given in Table 7.1 are obtained. It is unfortunate

Table 7.1 Outside Surface Resistance R_{so}, m² °C/W

Surface	*Exposure**		
	Sheltered	*Normal*	*Severe*
Roof (ϵ_0 = 0.9)	0.071	0.C45	0.021
(ϵ_0 = 0.05)	0.099	0.055	0.023
Wall (ϵ_0 = 0.9)	0.079	0.055	0.029
(ϵ_0 = 0.05)	0.114	0.07	0.033

*'Sheltered', 'normal' and 'severe' exposures are the classifications used by the C.I.B.S. and correspond to wind speeds of 1.0, 3.0 and 9.0 m/s respectively in the case of a roof and 0.66, 2.0 and 6.0 m/s respectively in the case of a wall. Information on how these classifications are applied is given in the Guide (A3).

that equation 7.5 is unreferenced in Reference (1). However, some interesting information concerning the nature of the equation may be deduced as follows.

The fundamental theoretical relation for turbulent flow over a plane surface quoted by Kreith[(2)] and many other authorities is

$$Nu = 0.036\ Pr^{0.33}\ Re^{0.8} \tag{7.6}$$

This equation is in fact based on the Reynolds analogy between the transfer of heat and momentum. Taking 0 °C as a representative outdoor winter temperature, the Prandtl number evaluates to 0.715 and equation 7.6 simplifies to

$$h_c = \frac{6.4\ V^{0.8}}{\ell^{0.2}} \tag{7.7}$$

where ℓ is the characteristic dimension (i.e. the length) of the plane surface being considered. In order to arrive at an equation which contains only the velocity as an independent variable, a value must be attributed to the length ℓ. Putting ℓ = 0.5 m

$$h_c = 7.36\ V^{0.8} \tag{7.8}$$

Equations 7.5 and 7.8 are shown plotted in Figure 7.2. For the

Figure 7.2

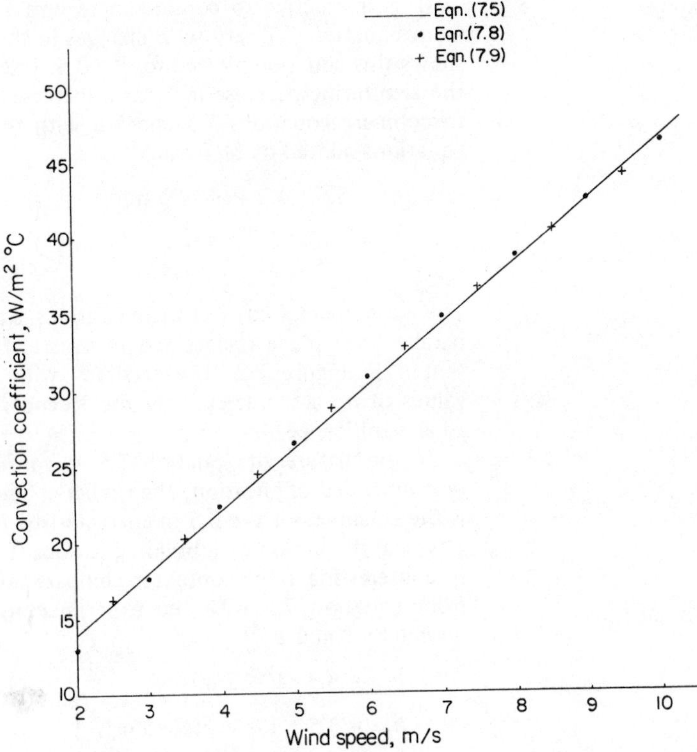

Figure 7.3

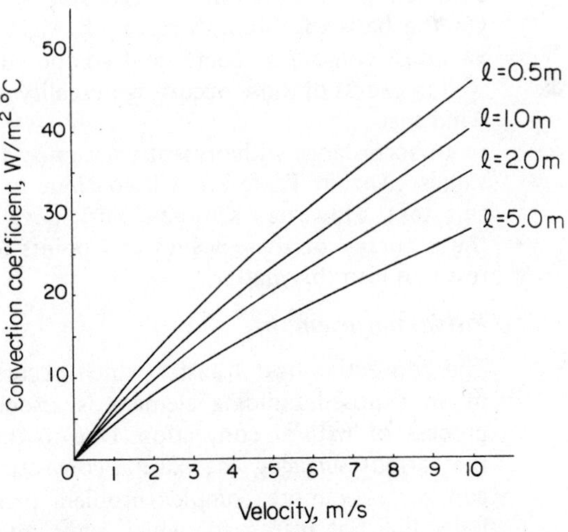

velocities considered the close correspondence between the equations is readily apparent. It can therefore be concluded that within this velocity range the form of equation 7.5 is essentially that of the basic Reynolds analogy relation for a characteristic dimension of approximately 0.5 m.

It is instructive to consider here how values of h_c obtained from equation 7.7 vary with changes in the length ℓ. Figure 7.3 shows this equation plotted for $\ell = 0.5$, 1, 2 and 5 m and we note the significant decrease in h_c as ℓ increases. It is also instructive to compare equations 7.5 and 7.8 with the following empirical equations quoted by McAdam[3]

$$\left.\begin{aligned} h_c &= 5.6 + 4.2\,V;\ V < 5 \text{ m/s} \\ h_c &= 7.63\,V^{0.78};\ 5 \leqslant V < 30 \text{ m/s} \end{aligned}\right\} \tag{7.9}$$

The experimental basis of these relations is the flow of air at 0 °C parallel to a plane surface 0.5 m square. Equation 7.9 is represented in Figure 7.2. The accuracy with which the empirical values of h_c are predicted by the Reynolds analogy is of particular significance.

If the nature of equation 7.5 is considered relative to its recommended application, then what is immediately noticeable is the smallness of the 0.5 m characteristic length compared with a typical dimension of a building surface (Figure 7.3). It is therefore interesting in this context to compare values of h_{co} computed from equation 7.5 with the experimentally determined values quoted by Kimura [4]:

$$\left.\begin{aligned} h_{co} &= 6.3\,;\ V \leqslant 2 \text{ m/s} \\ h_{co} &= 3.5 + 1.4\,V\,;\ V > 2 \text{ m/s} \end{aligned}\right\} \tag{7.10}$$

For wind speeds of 1, 3 and 9 m/s, equation 7.10 gives values of 6.3, 7.7 and 16.1 W/m^2 °C respectively. The corresponding values obtained from equation 7.5 are 9.9, 18.1 and 42.7 W/m^2 °C. On the basis of Kimura's results it would thus appear that the values of convection coefficient computed from equation 7.5 are well in excess of those occurring in reality, especially at the higher wind speeds.

In accordance with presently accepted design practice, the R_{so} values given in Table 7.1 will continue to be used in the rest of this text. However, Kimura's work does cast some doubt over the accuracy of these values and points to the need for further research into this matter.

Inside coefficient, h_{ci}

The convective heat transfer which occurs at the inside surface of an exposed building element is associated primarily with a process of natural convection. Due to the interaction between the various surfaces, the natural convection within an enclosure constitutes a more complex problem than the heat transfer at individual flat plates, for which some information was given in Chapter 2. However, the recommended design values of h_{ci} quoted in Reference (1) do reflect two general results which relate to the natural convection processes at such surfaces. The first one is simply that the heat transfer coefficient depends on the orientation of the surface and the direction of the heat flow.

A clear illustration of this is given by the empirical formulae listed in Section 2.3.4. The second result is one suggested initially by Fishenden and Saunders [5] and is that, in the turbulent range, natural convection from vertical surfaces is not far from the mean for horizontal surfaces facing upwards and downwards.

The values of h_{ci} given in the Guide are:

Walls	;	3.0 W/m² °C
Ceiling (upward heat flow)	;	4.3 W/m² °C
Floor (downward heat flow)	;	1.5 W/m² °C

There is some experimental support for these convection coefficients[6], which compare closely with those suggested by A.S.H.R.A.E.[7] Substituting the above values into equation 7.4b we obtain the internal surface resistances given in Table 7.2.

Table 7.2 Inside Surface Resistance R_{si}, m² °C/W

Surface orientation	*Direction of heat flow*	*Surface emissivity* $\epsilon_i = 0.9$	$\epsilon_i = 0.05$
Vertical	Horizontal	0.123	0.304
Horizontal	upward	0.106	0.218
Horizontal	downward	0.15	0.56
*Sloping 45°	upward	0.11	0.24
*Sloping 45°	downward	0.133	0.39

*Sloping surface values have been adapted from A.S.H.R.A.E.

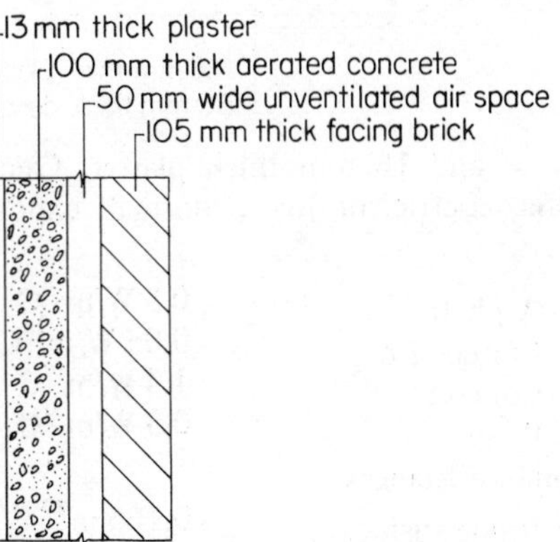

Figure 7.4
Wall for Example 7.1

Example 7.1 Figure 7.4 shows a wall consisting of 13 mm thick plaster, 100 mm thick aerated concrete, 50 mm wide unventilated air space and 105 mm thick external facing brick. Calculate the overall heat transfer coefficient for a wind speed of 2.0 m/s.

Thermal conductivities:

Plaster	0.5 W/m °C
Aerated concrete	0.14 W/m °C
Brick	0.84 W/m°C

Thermal resistances:

Inside surface	0.123 m^2 °C/W
Air space	0.18 m^2 °C/W
Outside surface	0.055 m^2 °C/W

$$U = \frac{1}{R_{si} + R_a + R_{so} + \Sigma\frac{L}{k}}$$

$$= \frac{1}{0.123 + 0.18 + 0.055 + (0.013/0.5) + (0.1/0.14) + (0.105/0.84)}$$

$$= \frac{1}{1.223}$$

$$= 0.82 \text{ W/m}^2\,°\text{C}$$

Example 7.2 A flat roof (Figure 7.5) consists of 20 mm thick asphalt, 75 mm thick fibre insulating board, 150 mm thick dense concrete and 16 mm thick plaster. Calculate the overall heat transfer coefficient for a 'normal' exposure. Thermal conductivities:

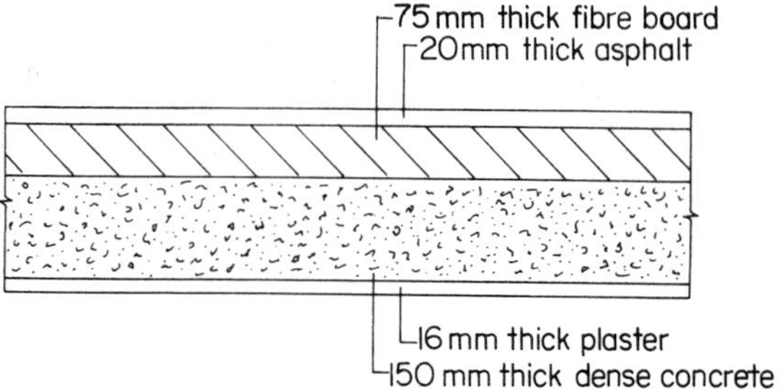

Figure 7.5
Roof for Example 7.2

Asphalt	0.5 W/m °C
Fibreboard	0.05 W/m °C
Concrete	1.4 W/m °C
Plaster	0.5 W/m °C

Thermal resistances:

Inside surface	0.106 m^2 °C/W
Outside surface	0.045 m^2 °C/W

$$U = \frac{1}{R_{si} + R_{so} + \Sigma\frac{L}{k}}$$

$$= \frac{1}{0.106 + 0.045 + (0.02/0.5) + (0.075/0.05) + (0.15/1.4) + (0.016/0.5)}$$

$$= \frac{1}{1.83}$$

$$= 0.546 \text{ W/m}^2\,°\text{C}$$

An extensive tabulation of computed U values is given in References (1,7).

7.2.3 Some special considerations

The evaluation of the overall heat transfer coefficients for most building elements may be carried out in the manner illustrated by the foregoing worked examples. The designer will however inevitably encounter certain structural arrangements which require special consideration. Some of these arrangements are examined here.

(i) *Pitched roof with ceiling*

The calculation of the U value for a pitched roof follows the same procedure as that for a flat roof (Example 7.2). However, in the case of a pitched roof combined with a ceiling, some additional refinement is necessary to take account of the different areas involved (this problem is basically one of series heat flow through unequal areas).

Figure 7.6
Pitched roof with ceiling

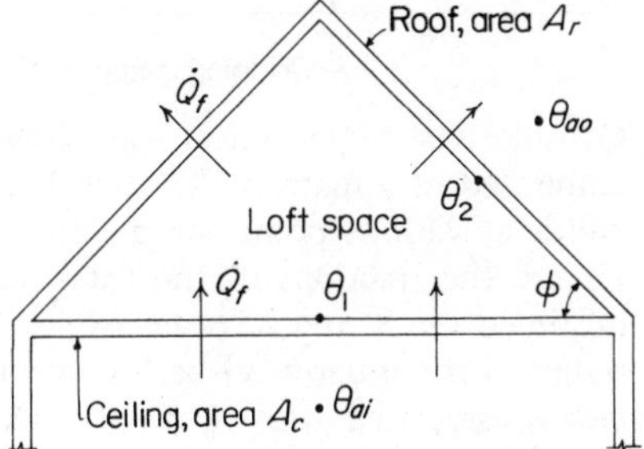

Consider the arrangement shown in Figure 7.6. A_c and A_r are the areas of the ceiling and roof respectively. Let the total rate of heat flow through the ceiling be $\dot{Q}_f$. For steady-state conditions the same amount will also pass through the roof. We can therefore write

$$\dot{Q}_f = \frac{A_c(\theta_{ai} - \theta_1)}{R_{si} + R_{t(c)}} = \frac{A_c(\theta_1 - \theta_2)}{R_a} = \frac{A_r(\theta_2 - \theta_{ao})}{R_{so} + R_{t(r)}}$$

where $R_{t(c)}$ and $R_{t(r)}$ are the thermal resistances of the ceiling and roof respectively and R_a in this case is the thermal resistance of the loft space. Now $A_r = A_c/\cos\phi$ and the above equation becomes

$$\dot{Q}_f = \frac{A_c(\theta_{ai} - \theta_1)}{R_{si} + R_{t(c)}} = \frac{A_c(\theta_{\mathrm{i}} - \theta_2)}{R_a} = \frac{A_c(\theta_2 - \theta_{ao})}{\cos\phi\,(R_{so} + R_{t(r)})}$$

Eliminating the surface temperatures θ_1 and θ_2

$$\dot{Q}_f = \frac{A_c(\theta_{ai} - \theta_{ao})}{R_{si} + R_{t(c)} + R_a + (R_{so} + R_{t(r)})\cos\phi} \qquad (7.11)$$

The overall heat transfer coefficient based on the ceiling area is therefore given by

$$U = \frac{1}{R_{si} + R_{t(c)} + R_a + (R_{so} + R_{t(r)}) \cos \phi} \quad (7.12)$$

The thermal resistances of various types of loft space are given in Reference (1).

(ii) *Heat bridges*

In certain structures some of the elements are not continuous but are bridged (either wholly or partly) by material of a different thermal conductivity. Framed and hollow block constructions are examples of this type of problem. In a large number of cases a simple area-weighting method may be used to determine the overall thermal transmittance.

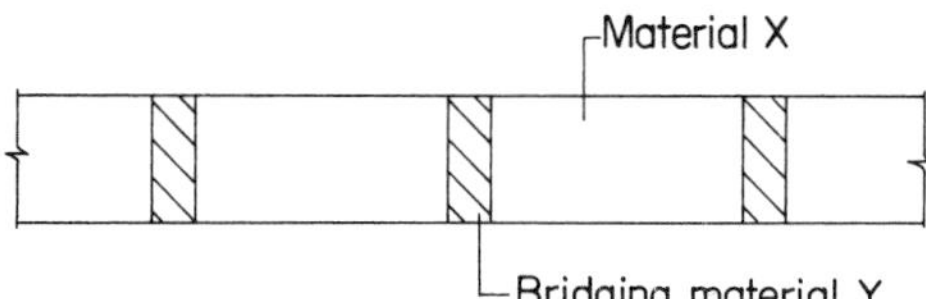

Figure 7.7
Hypothetical bridged wall

Consider the hypothetical wall shown in Figure 7.7. It is composed of a material X which is bridged across its entire width at various points by a different material Y. F_x and F_y are the fractions of the total surface area of the wall occupied by X and Y respectively. Let U_x and U_y (computed in the usual way) be the overall heat transfer coefficients associated with X and Y respectively. The resultant overall heat transfer coefficient for the wall is given by

$$U = F_x U_x + F_y U_y \quad (7.13)$$

The assumption implicit in this equation is that the heat flow is everywhere normal to the surfaces of the wall (i.e. fully one-dimensional) and that there is no lateral flow of heat across any material interface. This assumption is certainly valid for the case where the conductivities of X and Y do not differ greatly. However, the method may also be applied with reasonable accuracy to various other situations, such as hollow block constructions, partly-bridged structures and even to some cases where building elements are traversed by highly conducting materials (metals). For guidance the reader should consult References (1,7). Reference (7) is particularly recommended as it gives some excellent worked examples.

(iii) *Solid floors*

The typical heat flow pattern from inside a building to the outside through a solid floor in contact with the ground is shown in Figure 7.8. The determination of the heat loss in such a case is not a simple matter because, as we see, the

Figure 7.8
Heat flow through solid ground floor

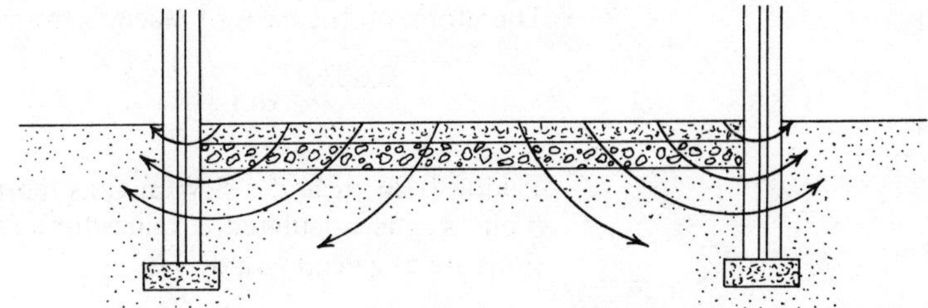

lines of heat flow are not straight and parallel but form a series of curves. The heat loss is greatest near the edges of the floor, since here the heat flow path is shortest, and decreases as the distance from the outside wall increases. The total heat loss through a solid floor therefore depends on its dimensions and edge conditions and the associated U value for the floor must necessarily take account of these factors.

An interesting theoretical study of this problem was made by Macey[8], who derived a formula for the rate of heat loss through a rectangular concrete floor with all four edges terminating at an outside wall. Macey's formula, in an adapted form, is

$$\dot{Q}_f = 1.26kBL_1\,(\theta_{ai} - \theta_{ao})\,\tanh^{-1}\left(\frac{L_2}{L_2 + b}\right) \quad (7.14)$$

where k = thermal conductivity of ground (W/m °C)
L_1 = inside longer dimension of floor (m)
b = thickness of outside wall (m)
L_2 = inside shorter dimension of floor (m)
B = a shape factor as given below

L_1/L_2	1	2	3	4	5	6	10	∞
B	1.6	1.3	1.2	1.15	1.12	1.1	1.06	1.00

The conductivity of earth is dependent to a large extent on its water content. For ground of normal dampness a typical value of 1.5 W/m °C may be assumed.
The heat loss through a floor may be written as

$$\dot{Q}_f = U(L_1 L_2)\,(\theta_{ai} - \theta_{ao})$$

Comparing this equation with equation 7.14 gives the following expression for the overall heat transfer coefficient

$$U = \frac{1.26kB}{L_2}\,\tanh^{-1}\left(\frac{L_2}{L_2 + b}\right)$$

Now

$$\tanh^{-1}\left(\frac{L_2}{L_2 + b}\right) = \frac{1}{2}\,\ell n\,\frac{1 + [L_2/(L_2 + b)]}{1 - [L_2/(L_2 + b)]}$$

$$= \frac{1}{2}\,\ell n\,\frac{2L_2 + b}{b}$$

Therefore, on the basis of Macey's formula

$$U = \frac{0.63kB}{L_2} \ln\left(\frac{2L_2 + b}{b}\right) \tag{7.15}$$

Taking 0.26 m and 1.5 W/m°C as representative values of b and k, the calculated U values for a range of floor dimensions are as given in Table 7.3.

Table 7.3 Calculated U Values for Solid Floors having Four Exposed Edges (Macey)

Dimensions of floor (m)	U (W/m^2 °C)
150 × 60	0.12
150 × 30	0.19
100 × 100	0.10
100 × 50	0.15
100 × 25	0.23
60 × 60	0.16
60 × 30	0.22
60 × 15	0.35
30 × 30	0.27
30 × 15	0.39
30 × 7.5	0.59
15 × 15	0.48
15 × 7.5	0.67
10 × 10	0.66
7.5 × 7.5	0.82
3 × 3	1.60

A detailed experimental study of the heat loss through solid ground floors was carried out by Billington[9] and it is the results of this work which form the basis of the design recommendations quoted in Reference (1). The basic overall heat transfer coefficients for various floor sizes having two or four exposed edges are listed in Table 7.4. We note that Billington's experimentally determined coefficients for

Table 7.4 U Values for Solid Floors (Billington)

Dimensions of Floor (m)	U (W/m^2 °C)	
	Four exposed edges	*Two exposed edges at right angles*
150 × 60	0.11	0.06
150 × 30	0.18	0.10
60 × 60	0.15	0.08
60 × 30	0.21	0.12
60 × 15	0.32	0.18
30 × 30	0.26	0.15
30 × 15	0.36	0.21
30 × 7.5	0.55	0.32
15 × 15	0.45	0.26
15 × 7.5	0.62	0.36
7.5 × 7.5	0.76	0.45
3 × 3	1.47	1.07

floors with four exposed edges are closely predicted by equation 7.15. Two comments concerning the U values given in Table 7.4 are appropriate at this point.

(a) The values apply to floors of dense concrete and as the thermal conductivities of concrete and earth are similar (this incidentally was an assumption made by Macey in his analysis), they are independent of the floor slab thickness.

(b) The values (like those in Table 7.3) are based on the whole floor area and the full temperature difference between the inside and outside air. If an overall layer of insulation is incorporated into a floor (Figure 7.9)

Figure 7.9
Overall insulation of floor

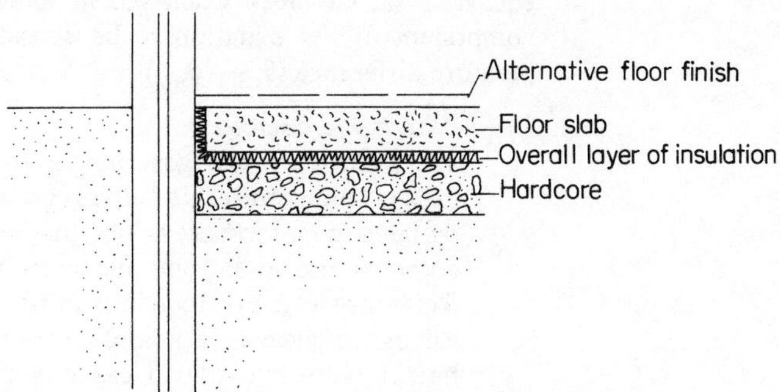

(or if a floor finish affording a useful degree of thermal resistance is applied), the U value for the floor may be calculated from the basic U value (Table 7.4):

$$U = \frac{1}{1/U_{\text{basic}} + L/k} \tag{7.16}$$

where L and k are the thickness and thermal conductivity of the insulating layer respectively. As the greatest part of the heat loss occurs at the edges of a floor, the use of an overall layer of insulation will rarely be justified.

Figure 7.10
Edge insulation of floor

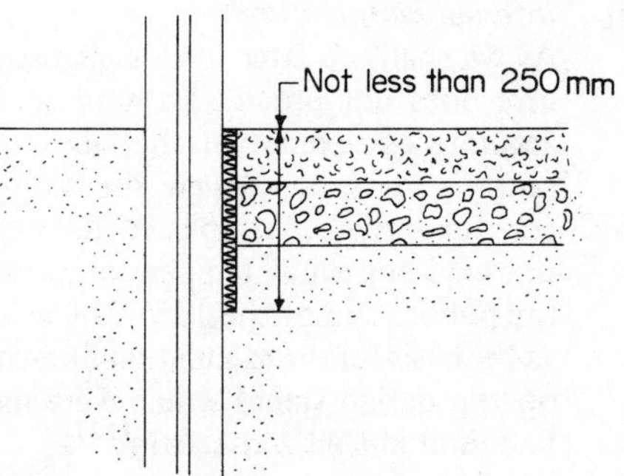

An alternative approach which gives almost as good results is the provision of a vertical layer of edge insulation (Figure 7.10). This should extend from finished floor level to a depth of at least 250 mm. Corrections to be applied to the U values contained in Table 7.4 for various depths of edge insulation are given in Reference (1). These corrections correspond to an insulating layer having a minimum thickness of 25 mm.

7.2.4 Design temperatures

The evaluation of the overall thermal transmittance for use in equation 7.3 has been examined in some detail. The remaining component of the equation to be considered is the design temperature difference $(\theta_{ai} - \theta_{ao})$.

(i) *External temperature*
The principal factor governing the selection of the external design temperature is of course geographical location. Comprehensive information on the winter design values in common use in Europe and elsewhere overseas is given in Reference (1). For locations in the UK, the recommended values are given as functions of the type of building and the heating system overload capacity (Table 7.5). The specification of these temperatures in this way follows basically from the approach laid down in References (10,11) and is considered in Chapter 6.

Table 7.5 External Design Temperatures

Type of Building	*Heating System Overload Capacity* (%)	*External Design Temperature* (°C)
Multi-storey buildings with solid intermediate floors and partitions	20	−1
	0	−4
Single-storey buildings	20	−3
	0	−5

(ii) *Internal temperature*
As we shall see later in this chapter, the inside air temperature does not provide us with an index which permits an accurate assessment of the fabric heat loss over the whole range of possible design situations. This temperature has consequently been replaced for computation purposes by the environmental temperature (7.4.1.). However, prior to the publication of the 1970 Guide, it was generally accepted as the basis for the calculation of heat losses and information on the design values which were used in this context is to be found in Kell and Martin[12].

7.3 VENTILATION HEAT LOSS

The infiltration of outside air into a building through doors, windows, interstices and cracks, generally accounts for a significant part of the building's heating requirement. This heating load is, in fact, more difficult to assess accurately than that associated with the building fabric loss.

The basic relation defining the ventilation heat loss for a given room of a building is

$$\dot{Q}_v = \rho C_p \dot{G} (\theta_{ai} - \theta_{ao}) \qquad (7.17)$$

where $\dot{Q}_v$ = heat required to raise the temperature of the infiltration air from θ_{ao} to θ_{ai} (W)
ρ = density of air (kg/m^3)
C_p = specific heat of air (J/kg°C)
$\dot{G}$ = rate of air flow into the room (m^3/s)

This ventilation loss, together with the fabric heat loss evaluated from equation 7.3, gives the total steady-state heat loss for the room

$$\dot{Q}_t = \Sigma \dot{Q}_f + \dot{Q}_v \qquad (7.18)$$

It is common practice to express the infiltration rate in terms of the room volume V'(m^3) and the air change rate per hour n as

$$\dot{G} = \frac{n V'}{3600}$$

Substituting this expression into equation 7.17 and taking 1.2 kg/m^3 and 1000 J/kg°C as representative values of ρ and C_p respectively, the following modified equation is obtained

$$\dot{Q}_v = 0.33 n V' (\theta_{ai} - \theta_{ao}) \qquad (7.19)$$

The group $0.33n$ is known as the *ventilation allowance*, $\dot{q}_v$, thus

$$\dot{Q}_v = \dot{q}_v V' (\theta_{ai} - \theta_{ao}) \qquad (7.20)$$

There are two methods of assessing the air infiltration rate for substitution in the foregoing equations. The first method is empirical and consists simply of assuming a certain number of air changes per hour for each room, the number of changes selected being dependent on the characteristics of the room. Some generally-accepted air change rates are listed in Table 7.6 and additional empirical values are given in References (13,7). It should be obvious that although this air change method is simple to use, satisfactory estimates of ventilation heat loss will be obtained only if care and judgement are exercised by the designer. Another point to note is that the air change rates given in Table 7.6 and in the quoted references correspond principally to the ventilation allowances (required for emitter sizing) for the individual spaces of a building. In assessing the total infiltration

Table 7.6 Some Empirical Values of Air Infiltration

Type of Space	*Air Change Rate* (h^{-1})
Residences	
Living rooms	1
Bedrooms	½
Entrance halls; staircases	2
Kitchens; bathrooms	2
Lavatories; cloakrooms	1½
Hotels	
Bedrooms	1
Public rooms	1
Lobbies	3
Corridors	2
Offices	1
Schools	
Classrooms	2
Assembly Halls	1½
Law Courts	1

allowance for the building as a whole (as part of the central plant sizing procedure), some modification of these values will generally be appropriate. This is because the infiltration air which enters a building on the windward side normally leaves on the leeward side. A heating load due to infiltration will therefore seldom exist simultaneously in all the rooms of a building.

The second method of assessment involves the actual calculation of the infiltration rate on the basis of the leakage characteristics of the building. A detailed procedure for this calculation is clearly described by the C.I.B.S.[(13)]. This particular procedure is restricted to windows, which usually form the major source of infiltration into buildings. For information on the leakage due to doors and other structural components, the reader is referred to Reference (7).

The calculation method is, of course, more precise than the empirical air change method and its application is recommended. However, it should be noted that its usefulness depends to a great extent on the accuracy with which the leakage characteristics of the windows can be specified at the design stage.

7.4 ENVIRONMENTAL TEMPERATURE THEORY

7.4.1 Introduction

Consider again the arbitrary exposed structural element of area A_s shown in Figure 7.1. From 2.5.2, it is seen that the rate of heat flow at the inside surface of this element is given by

$$\dot{Q}_f = A_s h_{ci} (\theta_{ai} - \theta_{si}) + A_s \epsilon_i h_{Ri} (\theta'_{ri} - \theta_{si}) \tag{7.21}$$

As we have already observed, one of the fundamental assumptions on which the air temperature method of heat loss computation is based is that $\theta'_{ri} = \theta_{ai}$. This assumption leads to the simplification

$$\dot{Q}_f = A_s(h_{ci} + \epsilon_i h_{Ri})(\theta_{ai} - \theta_{si}) \tag{7.22}$$

or
$$\dot{Q}_f = \frac{A_s(\theta_{ai} - \theta_{si})}{R_{si}} \tag{7.23}$$

where R_{si} is as specified in equation 7.4b

$$R_{si} = \frac{1}{h_{ci} + \epsilon_i h_{Ri}}$$

Although the simplification from equation 7.21 to equation 7.22 is a convenient one to make, it does obviously introduce errors into the calculation of the fabric heat loss in situations where θ_{ai} and θ'_{ri} are not equal. Overestimation of the fabric loss will occur if $\theta_{ai} > \theta'_{ri}$, as in the case of convective heating, while underestimation will occur if $\theta_{ai} < \theta'_{ri}$, as in the case of radiant heating. The magnitude of the error involved will of course increase as the difference between the temperatures increases.

The assumption that $\theta'_{ri} = \theta_{ai}$ is a reasonable approximation under certain conditions, for example, in the convective heating case where the standard of thermal insulation is high or the number of exposed structural elements is small. However, many other conditions are encountered for which this assumption is far from valid. Taking again the convective heating case as an example, the difference between θ_{ai} and θ'_{ri} will be of significance in spaces which are poorly insulated or have a large number of exposed surfaces. There are consequently a large number of situations in which significant errors (15-20%) will arise in the calculation of the fabric heat loss using the air temperature as the internal index. In extreme cases the errors involved may be of the order of 40%.

7.4.2 A new index

In the light of the foregoing remarks, it is obvious that there is a need for an alternative index to be used in heat loss calculations in place of the inside air temperature. The basic requirement is for a parameter which will permit the accurate prediction of the steady-state fabric loss over a wide range of possible design situations. Let us consider now how such a parameter may be arrived at.

In order that the form of the heat loss procedure for the new method remains similar to that for the traditional air temperature method, an equation is desirable of the type

$$\dot{Q}_f = A_s h_i (\theta_x - \theta_{si}) \tag{7.24}$$

Additionally, in order that the new procedure be not unduly complex, h_i will require to be a coefficient which does not depend on the distribution of surface temperature within the enclosure and θ_x will require to be an index temperature which is independent of the particular exposed structural element being considered.

As a first step in determining such an index temperature, the temperature θ'_{ri} in equation 7.21 is replaced in terms of the mean radiant temperature of the enclosure, θ_{ri}. This is done using an area weighting method. On this basis, taking A_s as a component plane surface of the enclosure, we may write

$$\Sigma A\,\theta_{ri} = A_s\,\theta_{si} + (\Sigma A - A_s)\,\theta'_{ri}$$

where ΣA is the total internal surface area of the enclosure; thus

$$\theta'_{ri} = \frac{(\Sigma A\,\theta_{ri} - A_s\,\theta_{si})}{(\Sigma A - A_s)}$$

and substituting this into equation 7.21 gives

$$\frac{\dot{Q}_f}{A_s} = h_{ci}\,(\theta_{ai} - \theta_{si}) + \epsilon_i\,h_{Ri}\left[\frac{(\Sigma A\theta_{ri} - A_s\theta_{si})}{(\Sigma A - A_s)} - \theta_{si}\right]$$

This equation simplifies to

$$\frac{\dot{Q}_f}{A_s} = h_{ci}\,(\theta_{ai} - \theta_{si}) + h'_{Ri}\,(\theta_{ri} - \theta_{si}) \qquad (7.25)$$

where

$$h'_{Ri} = \frac{\epsilon_i\,h_{Ri}\,\Sigma A}{(\Sigma A - A_s)} \qquad (7.26)$$

The second step is the replacement of θ_{ai} and θ_{ri} in equation 7.25 by the index temperature θ_x, to give

$$\dot{Q}_f = A_s\,(h_{ci} + h'_{Ri})\,(\theta_x - \theta_{si}) \qquad (7.27)$$

Equating equations 7.25 and 7.27 it is found that

$$\theta_x = \frac{h_{ci}\,\theta_{ai} + h'_{Ri}\,\theta_{ri}}{h_{ci} + h'_{Ri}} \qquad (7.28)$$

We see from the above equation that θ_x is a function of h'_{Ri} and h_{ci}, the values of which depend clearly on the surface to which the heat loss procedure is being applied. In order that θ_x meets the requirement of being independent of the surface under consideration, some representative values must be attributed to the heat transfer coefficients in this equation.

Considering some typical room shapes, the value of the ratio $\Sigma A/(\Sigma A - A_s)$ may be computed for the following configurations:

	Ratio $\Sigma A/(\Sigma A - A_s)$			
Dimensions	Long wall	Short wall	Ceiling	Simple average
1 x 1 x 1 high	1.20	1.20	1.20	1.20
3 x 2 x 1 high	1.16	1.10	1.38	1.21
10 x 2 x 1 high	1.18	1.03	1.46	1.22

Accepting 1.2 as a representative value of $\Sigma A/(\Sigma A - A_s)$ and recalling that h_{Ri} has been evaluated as 5.7 W/m² °C (2.4.3), the following value of h'_{Ri} is established for high emissivity surfaces

$$h'_{Ri} = 1.2\,\epsilon_i\,h_{Ri} = 1.2 \times 0.9 \times 5.7 = 6.2 \text{ W/m}^2\,^\circ\text{C}$$

The generally accepted values of h_{ci} have been quoted previously in 7.2.2 as 3.0, 4.3 and 1.5 W/m² °C. Taking 3.0 W/m² °C as an approximate average value of h_{ci} and substituting this and the calculated value of h'_{Ri} into equation 7.28, we obtain

$$\theta_x = \frac{3\theta_{ai} + 6.2\theta_{ri}}{3 + 6.2}$$

or

$$\theta_x = 0.326\theta_{ai} + 0.674\theta_{ri}$$

The environmental temperature, θ_{ei}, may now be introduced and defined as follows

$$\theta_{ei} = 1/3\theta_{ai} + 2/3\theta_{ri} \qquad (7.29)$$

If the environmental temperature is used as the parameter in heat loss calculations, instead of the air temperature, then a more accurate prediction of the fabric heat loss is obtained over the whole range of design situations, with errors being limited to approximately 5%. Some typical recommended design values of internal environmental temperature are shown in Table 7.7. A comprehensive list is to be found in Reference (16).

Table 7.7 Some Recommended Design Values of Internal Environmental Temperature

Type of Space	θ_{ei} (°C)
Residences	
Living rooms	21
Bedrooms	18
Entrance halls; staircases	16
Bathrooms	22
Lavatories; cloakrooms	18
Hotels	
Bedrooms	22–24
Public rooms	21
Lobbies; corridors	18
Offices	20
Schools	
Classrooms	18
Assembly Halls	18
Law Courts	20

Using the environmental temperature as the design parameter, equation 7.27 becomes

$$\dot{Q}_f = A_s\,(h_{ci} + h'_{Ri})\,(\theta_{ei} - \theta_{si}) \qquad (7.30)$$

where h_{ci} is the convection coefficient appropriate to the particular surface being considered. An alternative form of this equation is

$$\dot{Q}_f = \frac{A_s(\theta_{ei} - \theta_{si})}{R'_{si}} \tag{7.31}$$

where R'_{si} is the inside surface resistance based on environmental temperature

$$R'_{si} = \frac{1}{h_{ci} + h'_{Ri}} \tag{7.32}$$

Recalling that $h'_{Ri} = 1.2\ \epsilon_i\ h_{Ri}$, then strictly this surface resistance is given by

$$R'_{si} = \frac{1}{h_{ci} + 1.2\ \epsilon_i\ h_{Ri}} \tag{7.33}$$

However, the omission of the factor 1.2 leads to very small errors in the evaluation of the overall thermal transmittance. For example, in the case of a cavity brick wall, the error involved is of the order of 2%. It is therefore generally accepted that the introduction of this factor is unnecessary and in Reference (1) values of inside surface resistance have been computed from

$$R'_{si} = \frac{1}{h_{ci} + \epsilon_i\ h_{Ri}}$$

which as we have seen is the definition of the inside surface resistance based on air temperature, i.e. it is assumed that $R'_{si} = R_{si}$.

Consider now the specification of the heat flow $\dot{Q}_f$ in terms of the internal and external temperatures. The appropriate equation is

$$\dot{Q}_f = A_s\, U(\theta_{ei} - \theta_{eo}) \tag{7.34}$$

where θ_{eo} = outside environmental temperature (°C),
and U = overall thermal transmittance based on the difference in environmental temperatures, (W/m^2 °C).

In the case of winter heat loss calculations it is acceptable to take θ_{eo} as being equal to the outside air temperature, θ_{ao}. Thus equation 7.34 becomes

$$\dot{Q}_f = A_s\, U(\theta_{ei} - \theta_{ao}) \tag{7.35}$$

This equation may be used to compute the heat loss for each element of the external structure. As before, the total fabric heat loss is given by

$$\Sigma\dot{Q}_f = \Sigma A_s\, U(\theta_{ei} - \theta_{ao}) \tag{7.36}$$

It is instructive at this point to compare the environmental temperature approach with the traditional air temperature approach discussed in 7.2.1. In the air temperature method the equation corresponding to equation 7.36 is equation 7.3

$$\Sigma\dot{Q}_f = \Sigma A_s\, U(\theta_{ai} - \theta_{ao})$$

where in this case the overall thermal transmittance U is based on the air-to-air temperature difference. However, because of the assumption that $\theta_{eo} = \theta_{ao}$ and because values of internal surface resistance may be computed for either approach from equation 7.4b, the value of the overall thermal transmittance is the same in both equations. Values of U determined as shown in 7.2 may therefore be used with either method, as may the calculated U values given in the quoted references.

7.4.3 Heat flow paths

Although equation 7.29 defines the environmental temperature in terms of the air and mean radiant temperatures in an enclosure, it does not in itself permit the evaluation of unique values of these parameters for a given θ_{ei}. The relationship between θ_{ei}, θ_{ai} and θ_{ri} is dependent on the nature of the heat input to the space. In the case of convective heating it is assumed that the heat input is at the air temperature while in the case of radiant heating it is assumed to be at the environmental temperature.

In order to establish relations between the above temperatures it is firstly necessary to consider again the heat transfer processes occurring at the inside surface of an exposed structural element (Figure 7.1). These actual processes may now be replaced by the equivalent hypothetical process defined in terms of the environmental temperature by equation 7.31. This equivalent process between θ_{ei} and θ_{si} (Figure 7.11) applies whatever the nature of the heat input.

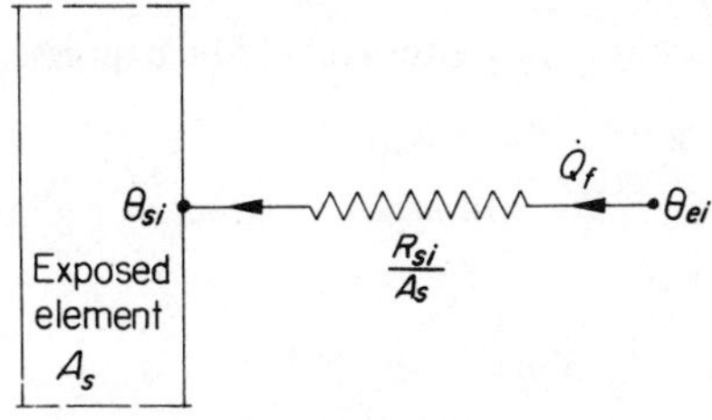

Figure 7.11
Diagrammatic representation of equivalent heat transfer to the fabric

A complication immediately becomes apparent in the convective heating situation. In this case the heat input is at the air temperature and this fact obviously requires the introduction of a hypothetical resistance R_e between θ_{ai} and θ_{ei} (based on the total area of the surfaces bounding the enclosure), through which the total fabric loss $\Sigma\dot{Q}_f$ may be transmitted (Figure 7.12). The requirement for such a resistance can also be established by considering the radiant heating situation. The ventilation loss $\dot{Q}_v$ is

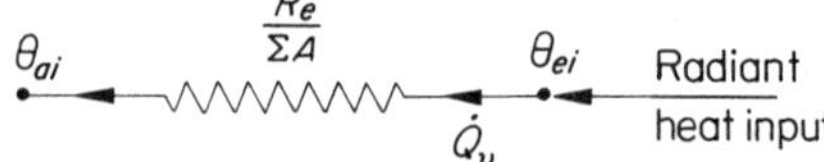

Figure 7.12 Diagrammatic representation of hypothetical resistance for convective case

defined by equation 7.19 in terms of the air temperature difference. As the heat input in a radiant system is at the environmental temperature there again arises a need for the introduction of the hypothetical resistance R_e through which the ventilation loss may be transmitted (Figure 7.13).

Figure 7.13 Diagrammatic representation of hypothetical resistance for radiant case

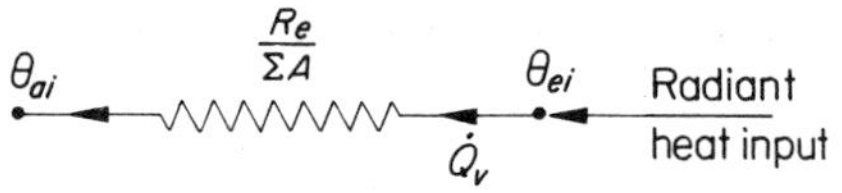

An equation for the ventilation heat loss in terms of the internal environmental temperature may be obtained by examining the heat flow $\dot{Q}_v$ in the case of the radiant system (Figure 7.14).

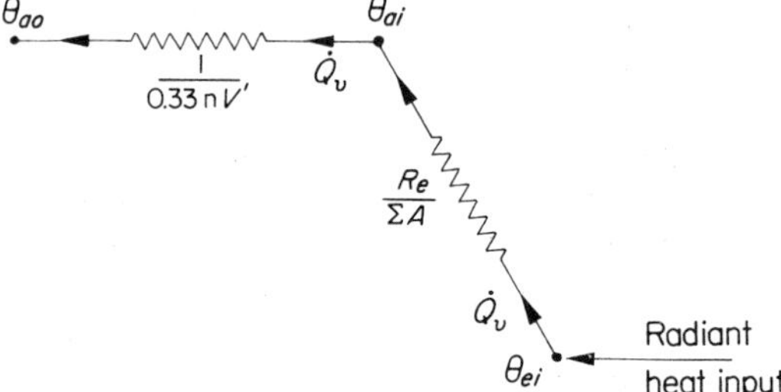

Figure 7.14 Diagrammatic representation of ventilation heat loss $\dot{Q}_v$ for a radiant system

From the figure we see that

$$\dot{Q}_v = \frac{(\theta_{ei} - \theta_{ao})}{\dfrac{1}{0.33nV'} + \dfrac{1}{h_e \Sigma A}}$$

where $h_e = 1/R_e$.

This equation may alternatively be expressed as

$$\dot{Q}_v = C_v(\theta_{ei} - \theta_{ao}) \tag{7.37}$$

where C_v is the *ventilation conductance* given by

$$\frac{1}{C_v} = \frac{1}{0.33nV'} + \frac{1}{h_e \Sigma A} \tag{7.38}$$

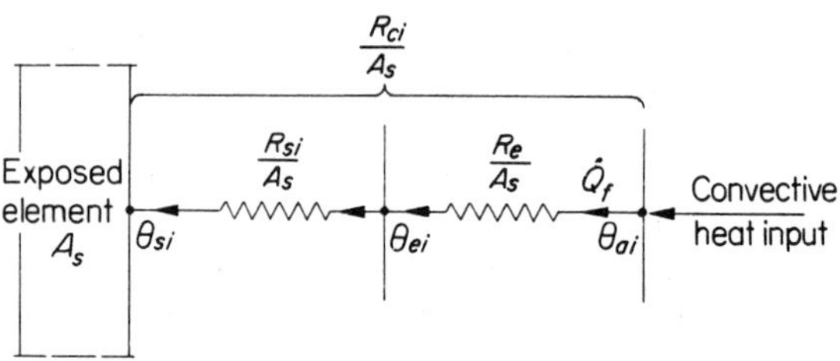

Figure 7.15 Diagrammatic representation of the relationship between R_e, R_{si} and R_{ci}

The relationship between the hypothetical resistance R_e, the convection resistance $R_{ci} = 1/h_{ci}$ and the inside surface resistance R_{si} may be established with reference to Figure 7.15 which represents the flow of the fabric loss for an exposed element, $\dot{Q}_f$, from the inside air temperature to the inside surface temperature for the convective heating situation. It is seen that

$$R_e = R_{ci} - R_{si}$$

or $$1/h_e = 1/h_{ci} - R_{si},$$

where h_e is the hypothetical conductance for this particular exposed element.

A mean value for h_e must now be established. In 7.4.2 the mean value of h_{ci} for all room surfaces was taken to be 3.0 W/m^2 °C. The value of R_{si} for vertical surfaces (i.e. 0.123 m^2 °C/W) is about the mean of the floor and ceiling values (Table 7.2) and hence may be taken as a representative value for the enclosure. Thus

$$\frac{1}{h_e} = \frac{1}{3} - 0.123$$

from which we find that h_e = 4.8 W/m^2 °C.

Having obtained this mean value for h_e, the heat flow paths between the air, environmental and outside temperatures may now be considered for both convective and radiant heating systems.

(i) *Convective Systems*

The heat input in such systems is taken to be at the internal air temperature, from which the ventilation loss $\dot{Q}_v$ flows through the resistance $1/0.33nV'$ to the outside air temperature (equation 7.19). The total fabric loss $\Sigma\dot{Q}_f$ flows through the hypothetical conductance h_e to the environmental temperature and thence to the outside air temperature through the summated overall thermal transmittance of the exposed structure $\Sigma A_s U$ (equation 7.36). The heat flow paths and resistances for the convective situation may thus be represented as shown in Figure 7.16.

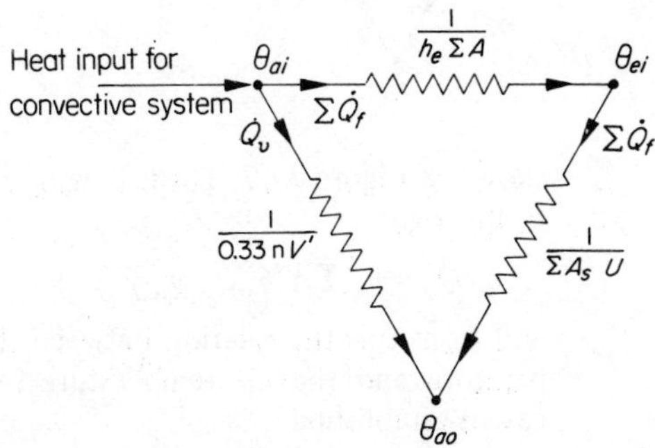

Figure 7.16
Heat flow paths and resistance for a convective heat input

The following equation is appropriate to this figure

$$\Sigma \dot{Q}_{\mathrm{f}} = h_{\mathrm{e}}\, \Sigma A\, (\theta_{\mathrm{ai}} - \theta_{\mathrm{ei}}) \tag{7.39}$$

Writing this equation in another form establishes the relation between the environmental temperature and the air temperature for the convective heating case

$$\theta_{\mathrm{ai}} - \theta_{\mathrm{ei}} = \frac{\Sigma \dot{Q}_{\mathrm{f}}}{4.8\, \Sigma A} \tag{7.40}$$

We note that this relation is dependent on the total fabric heat loss and is not influenced by the infiltration rate.

By making the appropriate substitution from equation 7.29, the relation between the environmental temperature and the mean radiant temperature is obtained:

$$\Sigma \dot{Q}_{\mathrm{f}} = 2h_{\mathrm{e}}\, \Sigma A\, (\theta_{\mathrm{ei}} - \theta_{\mathrm{ri}}) \tag{7.41}$$

or

$$\theta_{\mathrm{ei}} - \theta_{\mathrm{ri}} = \frac{\Sigma \dot{Q}_{\mathrm{f}}}{9.6\, \Sigma A} \tag{7.42}$$

(ii) *Radiant Systems*

The heat input in a radiant system is taken to be at the environmental temperature, from which the ventilation loss $\dot{Q}_v$ flows through the hypothetical conductance h_{e} to the internal air temperature and thence through the resistance $1/0.33nV'$ to the outside air temperature. The total fabric loss $\Sigma \dot{Q}_{\mathrm{f}}$ flows from the environmental temperature to the outside air temperature through the summated overall thermal transmittance $\Sigma A_{\mathrm{s}} U$. The heat flow paths and resistances for the radiant situation may thus be represented as shown in Figure 7.17. The following equation is appropriate to this figure

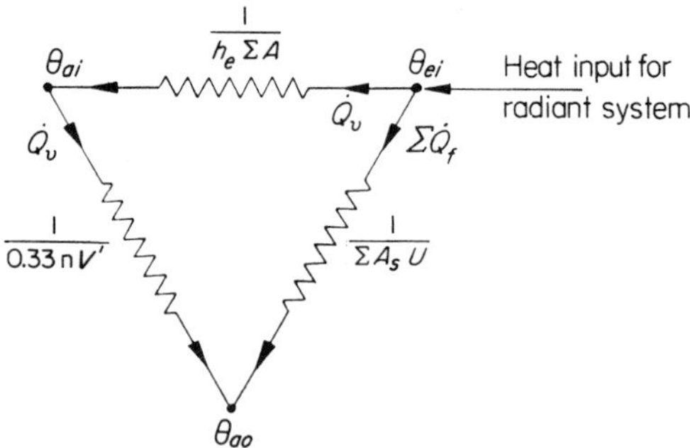

Figure 7.17 Heat flow paths and resistances for a radiant heat input

$$\dot{Q}_v = h_{\mathrm{e}}\, \Sigma A\, (\theta_{\mathrm{ei}} - \theta_{\mathrm{ai}}) \tag{7.43}$$

and from this the relation between the environmental temperature and the air temperature for the radiant heating case is established

$$\theta_{ei} - \theta_{ai} = \frac{\dot{Q}_v}{4.8\,\Sigma A} \tag{7.44}$$

We see that this relation depends on the infiltration rate and is not influenced by the fabric loss. The relation between the environmental temperature and the mean radiant temperature is obtained as before:

$$\dot{Q}_v = 2h_e\,\Sigma A\,(\theta_{ri} - \theta_{ei}) \tag{7.45}$$

or

$$\theta_{ri} - \theta_{ei} = \frac{\dot{Q}_v}{9.6\Sigma A} \tag{7.46}$$

A further point of interest in connection with radiant systems relates to the situation where part of the enclosure surface is unheated and at an average temperature $(\theta_{ri})_u$ while the remainder is heated and at an average temperature $(\theta_{ri})_h$. The mean radiant temperature within the enclosure, θ_{ri}, may be taken as an area weighted mean of the average temperature of heated surfaces (ΣA_h) and unheated surfaces (ΣA_u). Thus

$$\theta_{ri} = \frac{(\theta_{ri})_u\,\Sigma A_u + (\theta_{ri})_h\,\Sigma A_h}{\Sigma A_u + \Sigma A_h} \tag{7.47}$$

To enable the heated surface temperature $(\theta_{ri})_h$ to be calculated, a value for $(\theta_{ri})_u$ must be established. This can be done by examining the fabric heat flow $\Sigma\dot{Q}_f$ from the environmental temperature to the enclosure surfaces.

Consider firstly the convective system. The appropriate relation in this case would be equation 7.41, which specifies the total fabric heat flow $\Sigma\dot{Q}_f$ from the environmental temperature θ_{ei} to the mean fabric surface temperature θ_{ri} (Figure 7.18). When considering the radiant system it

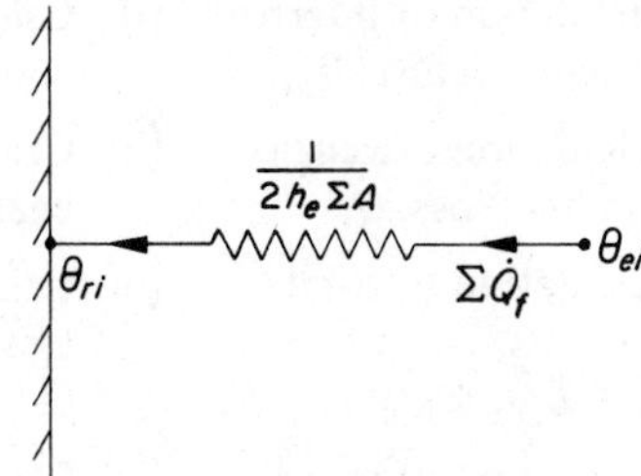

Figure 7.18
Diagrammatic representation of total fabric heat flow for convective heating

would seem reasonable to suggest that this heat flow will be associated with the unheated areas of the fabric. The appropriate equation in this case is therefore

$$\Sigma\dot{Q}_f = 2h_e\,\Sigma A_u\,(\theta_{ei} - (\theta_{ri})_u) \tag{7.48}$$

which yields the relation between θ_{ei} and $(\theta_{ri})_u$

$$\theta_{ei} - (\theta_{ri})_u = \frac{\Sigma \dot{Q}_f}{9.6\ \Sigma A_u} \qquad (7.49)$$

Thus, once θ_{ri} has been established for a radiant system from equation 7.46, $(\theta_{ri})_u$ may be calculated from equation 7.49 and finally equation 7.47 may be used to determine the required temperature of the heated surface $(\theta_{ri})_h$. This aspect of the design process for radiant systems will be considered further in Chapter 8.

7.4.4 Heat loss computation procedure

The computation of heat losses on the basis of environmental temperature provides the designer with two advantages over the traditional air temperature approach. Firstly, as has already been made clear, a more accurate assessment of the steady-state heat loss is achieved. Secondly, as is apparent from the foregoing analysis of the heat flow paths, there is provision for the determination of both the air and mean radiant temperatures in an enclosure for a given design environmental temperature. As we have seen in Chapter 4, this places the designer in a more favourable position with respect to the assessment of thermal comfort in a space. Let us now consider the heat loss computation procedure as proposed in Reference (17):

1. Categorisation of the heating appliances according to their heat emission characteristics.
2. Selection of design outdoor temperature and design internal environmental temperature.
3. Determination of U values, total surface area of enclosure ΣA, and total fabric heat loss $\Sigma \dot{Q}_f$.
4. Determination of air infiltration rate n (air changes/h) and the ventilation allowance $\dot{q}_v = 0.33nV'$
5.

Convective systems	*Radiant systems*
(i) Calculation of internal air temperature, θ_{ai}	(i) Calculation of ventilation conductance, C_v
(ii) Calculation of ventilation heat loss, $\dot{Q}_v$	(ii) Calculation of ventilation heat loss, $\dot{Q}_v$
(iii) Calculation of total heat loss, $\dot{Q}_t = \Sigma \dot{Q}_f + \dot{Q}_v$	(iii) Calculation of total heat loss $\dot{Q}_t = \Sigma \dot{Q}_f + \dot{Q}_v$
(iv) Calculation of mean radiant temperature, θ_{ri}	(iv) Calculation of internal air temperature, θ_{ai}
	(v) Calculation of mean radiant temperature, θ_{ri}

6. Assessment of thermal comfort (this part of the procedure has been covered in depth in Chapter 4.)

In accordance with the above procedure, heat emitters require to be categorised as either convective or radiant. Although emitters in general do not fall rigidly into these classifications, it is accepted that, for the purposes of the procedure, emitters such as natural convectors, radiators, forced convectors, unit heaters and low temperature panels should be considered as being convective, while heated floors, heated ceilings and high temperature panels should be considered as radiant.

The application of the heat loss procedure to convective and radiant systems is illustrated by the following worked examples.

Example 7.3: Convective system. Private office in a single-storey structure in a suburban location. Details of the office construction are shown in Figure 7.19.

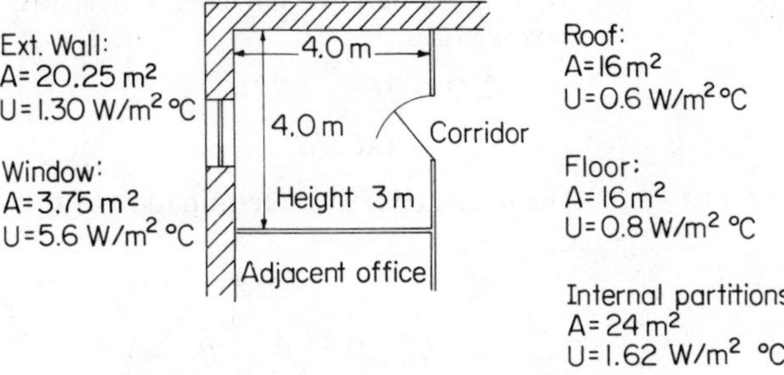

Figure 7.19 Details of office construction for Example 7.3

Design outdoor temperature, $\theta_{ao} = -3\,^\circ C$
Design internal environmental temperature
θ_{ei}, office $= 20\,^\circ C$
θ_{ei}, corridor $= 16\,^\circ C$

Item	*Area* (m^2)	*U Value* $(W/m^2\,^\circ C)$	*Temp. Diff.* $(^\circ C)$	*Heat Loss* (W)
External wall	20.25	1.30	23	605
Window	3.75	5.60	23	484
Int. partition	12.00	1.62	4	78
Floor	16.00	0.80	23	295
Roof	16.00	0.60	23	221
Other partitions	12.00	—	—	—
	$\Sigma A = 80$			$\Sigma \dot{Q}_f = 1683$

$$\theta_{ai} - \theta_{ei} = \frac{1683}{4.8 \times 80} = 4.4\,^\circ C$$

$$\therefore \quad \theta_{ai} = 24.4\,^\circ C$$

Air infiltration rate, $n = 1\ h^{-1}$

Ventilation allowance, $\dot{q}_v = 0.33\ W/m^3\,^\circ C$

$$\dot{Q}_v = 0.33 \times 48 \times [(24.4 - (-3))]$$

$$= 434\ W$$

$$\therefore \quad \dot{Q}_t = 1683 + 434 = 2117 \text{ W}$$

$$\theta_{ri} = \frac{3}{2}\theta_{ei} - \frac{1}{2}\theta_{ai}$$

$$= 30 - 12.2$$

$$= 17.8\,°\text{C}$$

Example 7.4: Radiant system. Consider again the private office of Example 7.3, but assume that in this case the heating is to be in the form of a heated ceiling.

Again, $\Sigma A = 80 \text{ m}^2$; $V' = 48 \text{ m}^3$; $n = 1 \text{ h}^{-1}$

Note, however, that since a heated ceiling is being employed there is no need to consider a heat loss through the roof. Thus in this case

$$\Sigma\dot{Q}_f = 1683 - 221$$

$$= 1462 \text{ W}$$

The next step is the determination of the ventilation conductance, C_v

$$\frac{1}{C_v} = \frac{1}{0.33nV'} + \frac{1}{h_e\,\Sigma A}$$

$$= \frac{1}{0.33 \times 1 \times 48} + \frac{1}{4.8 \times 80}$$

$$= 0.063 + 0.0026 = 0.0656$$

$$\therefore \quad C_v = 15.2 \text{ W/°C}$$

$$\dot{Q}_v = C_v\,(\theta_{ei} - \theta_{ao})$$

$$= 15.2\,[20 - (-3)]$$

$$= 350 \text{ W}$$

$$\therefore \quad \dot{Q}_t = 1462 + 350$$

$$= 1812 \text{ W}$$

$$\theta_{ei} - \theta_{ai} = \frac{\dot{Q}_v}{4.8\,\Sigma A}$$

$$= \frac{350}{4.8 \times 80}$$

$$= 0.9\,°\text{C}$$

$$\therefore \quad \theta_{ai} = 19.1\,°\text{C},$$

and $$\theta_{ri} = \frac{3}{2} \times 20 - \frac{1}{2} \times 19.1$$

$$= 20.4\,°\text{C}$$

Note: (i) The comfort check which follows this working and the working in the previous example is discussed fully in Chapter 4.

(ii) Additional worked examples using the heat loss computation procedure are to be found in Reference (17).

7.5 MODIFICATION OF THE CALCULATED HEAT LOSS

7.5.1 Introduction

In many cases it is necessary to modify the calculated steady-state heat loss in order to establish the design heating load for a given building space. The modifications required to allow for the intermittent heating of buildings have already been discussed in Chapter 6. The following additional allowances may also be appropriate.

7.5.2 Allowance for height of space

An assumption implicit in the foregoing heat loss computation procedures is that of a uniform air temperature within the heated space. In reality, of course, significant vertical temperature gradients may exist and the higher temperatures produced near the ceiling will cause an increased heat loss through the roof and the upper parts of the walls and windows.

The extent of the vertical temperature gradient in a space (and hence the allowance necessary to take account of this effect) depends on the mode of heating employed. Convective systems generally give rise to larger gradients than radiant systems and indeed in the particular case of a heated floor there is virtually a uniform temperature from floor to ceiling. Some interesting graphs showing the vertical temperature gradients to be expected with various types of heating systems are to be found in Reference (12). Additions to the calculated heat loss to allow for the height of heated spaces are given in Reference (14).

7.5.3 Allowance for internal heat gains

An examination of the sources of heat generation within a space (lighting, occupants, machinery, etc.) may indicate that some reduction of the calculated heat loss is permissible. Clearly, only heat gains which are reasonably continuous and are of an appreciable magnitude should be taken into account in this respect. It is therefore obvious that any reduction of the calculated heat loss to allow for internal gains should be made only after careful consideration by the designer of the nature of the heat sources and the usage pattern of the space itself. The emission of heat from personnel, lighting and a wide range of appliances and equipment is well covered in Reference (15).

7.6 CONDENSATION

7.6.1 Introduction

Modern buildings (particularly dwellings) are plagued by condensation. Condensation arises out of a complexity of factors which include reduced ventilation (due to absence of open fires), design faults by architects, structural and heating engineers and in some cases improper use of the building by the occupier. The most recent edition of the Building Regulations calls for higher insulation values than before and the design of newer forms of construction to comply with these Regulations requires even more vigilance in relation to this problem. In this section the psychrometric and moisture transfer theory developed in Chapter 2 will be applied to steady-state methods of examining the possible occurrence of condensation on surfaces and within elements of the construction.

7.6.2 Estimation of dew point temperatures

Information is rather scant on the temperature/moisture levels in naturally ventilated buildings for various occupations or processes but nevertheless the designer may wish to make some estimate of the possible occurrence of condensation. The following example outlines an approach with respect to condensation on glazing.

Example 7.5 A heating engineer investigating the likelihood of condensation on the single glazed windows of a large office on a day when the outside air temperature was 1 °C obtained readings within the office of 20 °C (dry bulb) and 14 °C (wet bulb), using a sling psychrometer. Assuming the normal resistance values for the glazing apply, how does this information help him?

U value for single glazing = 5.6 W/m^2 °C

Overall thermal resistance, $\Sigma R = 1/5.6 = 0.179$ m^2 °C/ W

Inside surface resistance, $R_{si} = 0.123$ m^2 °C/ W

$$\text{Now heat transfer } \dot{Q}_f/A_s = \frac{\theta_{ai} - \theta_{ao}}{\Sigma R} = \frac{\theta_{ai} - \theta_{si}}{R_{si}}$$

$$\text{Thus } \frac{(20 - \theta_{si})}{(20 - 1)} = \frac{R_{si}}{\Sigma R} = \frac{0.123}{0.179}$$

from which $\theta_{si} = 5.6$ °C

Thus the inside surface temperature of the glazing is 5.6 °C. Next, the inside air condition of 20/14 is plotted on the psychrometric chart of Figure 2.21 and the dew point temperature for this condition is found to be 9.8 °C.

The designer now knows that the window pane is presenting a surface to the air in the office which is at a temperature less than the dew point temperature of the air, and condensation is almost certain to occur.

7.6.3 Vapour barriers

In many situations it is becoming the practice to insulate thermally the inside surfaces of wall, floor or roof elements. This has implications so far as intermittent heating is concerned since it lowers the thermal capacity of the building, but in the present context the reduction in temperatures which arises within the fabric increases the possibility of condensation occurring within the structure. This phenomenon is known as *interstitial condensation.*

We have already discussed the process of moisture transfer through the fabric in section 2.7.4. There it was seen that the vapour pressure (and therefore the dew point temperature) varies through the fabric. Thus the architect in designing the structure and the heating engineer in selecting the appropriate method of heating, must take care to ensure that no element of the structure falls below the corresponding dew point temperature.

The conventional approach to this problem is to install a vapour barrier between the room and the thermal insulation. The vapour barrier has a very high vapour resistance and a large drop in vapour pressure occurs across the barrier when moisture migration occurs through the fabric.

This in turn leads to much lower dew point temperatures within the fabric and a greatly reduced possibility of interstitial condensation. However, it also means that the moisture generated within the space has to be removed by other means, such as ventilation.

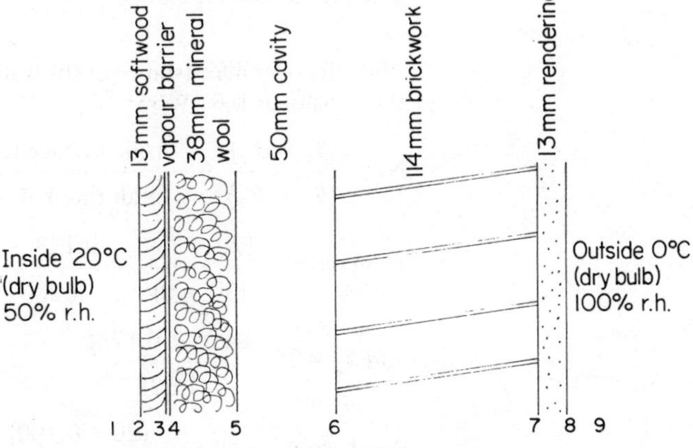

Figure 7.20
Details of construction for Example 7.6

The following example illustrates the effect which the application of a vapour barrier has on the possible incidence of interstitial condensation.

Example 7.6 In relation to the wall construction shown in Figure 7.20 and the prevailing inside and outside environmental conditions indicated, use the information given to establish the temperature and dew point temperature profiles through the wall, for the following cases:

(a) No vapour barrier employed,

(b) Barrier placed on room side of thermal insulation and consisting of a polythene sheet of permeance 0.5×10^{-11} kg/Ns.

Data:

Inside surface resistance	0.123 m^2 °C/W
Outside surface resistance	0.055 m^2 °C/W
Thermal conductivity:	
softwood	0.130 W/m °C
mineral wool	0.036 W/m °C
brickwork	1.100 W/m °C
rendering	1.200 W/m °C
Cavity resistance	0.18 m^2 °C/W
Permeability:	
softwood	0.50×10^{-11} kgm/Ns
mineral wool	7.00×10^{-11} kgm/Ns
brickwork	2.50×10^{-11} kgm/Ns
rendering	2.40×10^{-11} kgm/Ns

(a) *No vapour barrier employed*

Step 1. Determine the overall thermal resistance of the fabric.

Inside surface resistance		0.123
Softwood	13/(1000 x 0.130)	0.100
Mineral wool	38/(1000 x 0.036)	1.060
Cavity resistance		0.180
Brickwork	114/(1000 x 1.100)	0.100
Rendering	13/(1000 x 1.200)	0.010
Outside surface resistance		0.055
		$\Sigma R = 1.628$

Thus the overall resistance of the wall is 1.628 m^2 °C/W and the U value is 0.61 W/m^2 °C.

$$\text{Now } \frac{(\theta_1 - \theta_2)}{(\theta_1 - \theta_9)} = \frac{\text{inside surface resistance}}{\text{overall thermal resistance}}$$

$$\text{from which, } \frac{(20 - \theta_2)}{(20 - 0)} = \frac{0.123}{1.628}$$

$$\text{and } \theta_2 = 20 - \frac{(20 \times 0.123)}{1.628} = 18.5\ ^\circ\text{C}$$

$$\text{Similarly } \theta_3 = 18.5 - \frac{(20 \times 0.100)}{1.628} = 17.3\ ^\circ\text{C}$$

θ_4 is the same as θ_3.
In the same fashion the remaining temperatures are found to be:

$\theta_5 = 4.2\ °C$
$\theta_6 = 2.0\ °C$
$\theta_7 = 0.8\ °C$
$\theta_8 = 0.7\ °C$
and θ_9 is of course 0 °C.

The temperature profile is shown in Figure 7.21.

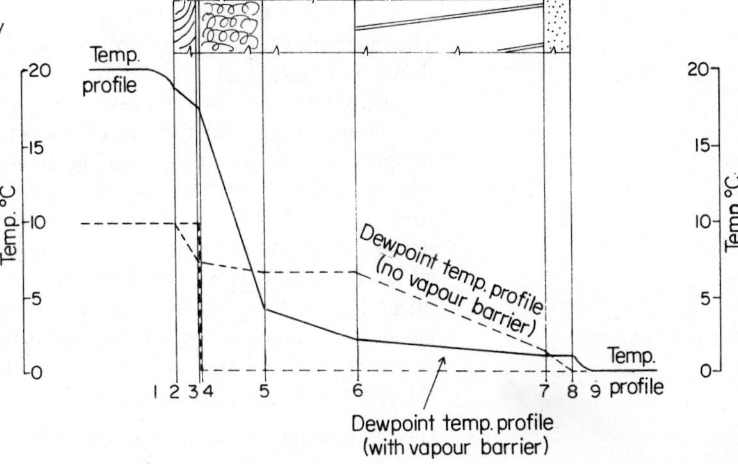

Figure 7.21 Temperature and dew point profiles for Example 7.6

Step 2. Determine the vapour resistance of the fabric.

Inside surface resistance		nil
Softwood	$13/(1000 \times 0.5 \times 10^{-11})$	26.00×10^8
Mineral wool	$38/(1000 \times 7.0 \times 10^{-11})$	5.42×10^8
Cavity resistance		nil
Brickwork	$114/(1000 \times 2.5 \times 10^{-11})$	45.6×10^8
Rendering	$13/(1000 \times 2.4 \times 10^{-11})$	5.42×10^8
Outside surface resistance		nil
		$R_{vt} = 82.44 \times 10^8$

Thus the total vapour resistance of the wall is 82.44×10^8 Ns/kg.

Recall equation 2.94 where $\dot{m}_s/A = \Delta p_s/R_v$

Thus in a similar manner to thermal resistance and temperature differences, we have:

$$\frac{(p_{s1} - p_{s2})}{(p_{s1} - p_{s2})} = \frac{\text{boundary vapour resistance}}{\text{total vapour resistance}}$$

Now at 20 °C (dry bulb) 50% r.h., $p_{s1} = 11.8$ mb
and at 0 °C (dry bulb) 100% r.h., $p_{s9} = 6.0$ mb
(from Figure 2.21)

The corresponding dew point temperatures are also obtained from the psychrometric chart, thus $\theta_{d1} = 9.8$ °C and

θ_{d9} = 0 °C. As the boundary resistance is nil, p_{s2} has the same value as p_{s1} and θ_{d2} = 9.8 °C.

Across the softwood $$\frac{(p_{s2} - p_{s3})}{(11.8 - 6.0)} = \frac{26 \times 10^8}{82.44 \times 10^8}$$

and p_{s3} = 11.8 − 1.83 = 9.97 mb
Thus θ_{d3} = 7.2 °C (from chart) and equals θ_{d4}.
Similarly, p_{s5} = 9.59 mb and θ_{d5} = 6.7 °C
p_{s6} = 9.59 mb and θ_{d6} = 6.7 °C
p_{s7} = 6.39 mb and θ_{d7} = 1.0 °C
p_{s8} = 6.00 mb and θ_{d8} = 0 °C
The dew point temperature profile may now be plotted as shown in Figure 7.21.

The temperature profiles reveal that the outermost layers of mineral wool and the brickwork and rendering are at temperatures below the prevailing dew point temperature and accordingly any moisture migrating through the structure under these steady-state conditions is liable to condense within the structure. This is of no great consequence so far as the brickwork and rendering is concerned since the outer leaf is liable to wetting due to the effects of rain. However prolonged interstitial condensation within the fibres of the insulation is liable to lead to a break-down of the thermal properties of the insulation.

(b) It is interesting now to consider the effect of a vapour barrier, in this case a polythene sheet, placed on the room side of the thermal insulation, that is, in position 3 to 4. The temperature profile remains as before since the thermal resistance of the polythene sheet is negligible.

However, the vapour resistance of the wall is increased by $1/0.5 \times 10^{-11}$ or 2000×10^8 Ns/kg.
Total vapour resistance is now $(82.44 + 2000)10^8 = 2082.44 \times 10^8$ Ns/kg. The new vapour pressures and dew point temperatures at each section are worked out in the same fashion as before and are:

$p_{s1} = p_{s2}$ = 11.8 mb; $\theta_{d1} = \theta_{d2}$ = 9.8 °C
p_{s3} = 11.73 mb; θ_{d3} = 9.8 °C
p_{s4} = 6.18 mb; θ_{d4} = 0.2 °C (etc)

It becomes quite clear on looking at Figure 7.21 that the high resistance of the vapour barrier has the effect of pulling the dew point temperature profile below that of the temperature profile. In this manner, interstitial condensation is avoided.

7.6.4 Last word on condensation

We are all familiar with condensation processes on cold window panes. However, with the complexity of modern building there are numerous other means of presenting cold surfaces to moist air within the occupied space, such as in positions where cavities are bridged (cold bridges) as for instance with lintels above windows. Link corridors or passageways may also act as tunnels for cold winds. In any given situation the designer should also give some thought to the most appropriate form of heating with respect to the possibility of condensation in the space.

This should include the consideration of such factors as the thermal patterns created by the heating system and the general movement of air within the space. Economic factors relating to the appropriate choice of fuel must not be discounted. In fact there is mounting evidence of tenants in local authority dwellings choosing to disregard the installed heating systems in certain properties and switching to the use of paraffin stoves. One current line of thought is that a base heating load should be provided independently of the tenants' wishes, the tenant being left with some degree of choice with regard to 'topping up'.

REFERENCES

(1) C.I.B.S. GUIDE, Section A3
(2) KREITH, F. *Principles of Heat Transfer*, 2nd Edn, (International Textbook Company, 1969)
(3) McADAM, W. H. *Heat Transmission*, 3rd Edn, (McGraw-Hill Book Company, New York, 1954)
(4) KIMURA, K-I. *Scientific Basis of Air Conditioning*, (Applied Science Publishers, London, 1977)
(5) FISHENDEN, M. and SAUNDERS, O. A. *An Introduction to Heat Transfer*, (Oxford University Press, 1950)
(6) BILLINGTON, N. S. *Thermal Properties of Buildings*, (Cleaver-Hume Press, London, 1952)
(7) A.S.H.R.A.E., Handbook of Fundamentals.
(8) MACEY, H. H. 'Heat Loss through Solid Ground Floors', *J. Inst. Fuel*, Vol. 22, 1949
(9) BILLINGTON, N. S. 'Heat Transmission through Solid Ground Floors', *J.I.H.V.E.* Vol. 19, 1951.
(10) Post-war Building Study No. 33, HMSO, 1955
(11) JAMIESON, H. C. 'Meteorological Data and Design Temperatures', *J.I.H.V.E.* Vol. 22, 1955
(12) FABER, O. and Kell, J. R. *Heating and Air Conditioning of Buildings*, 5th Edn. Rev. J. R. Kell and P. L. Martin, (Architectural Press, 1974)
(13) C.I.B.S. GUIDE, Section A4
(14) C.I.B.S. GUIDE, Section A9
(15) C.I.B.S. GUIDE, Section A7
(16) C.I.B.S. GUIDE, Section A5
(17) McLAUGHLIN, R. K. and McLEAN, R. C. 'Environmental Temperature Theory', *Steam and Heating Engr.*, December, 1973

SYMBOLS USED IN CHAPTER 7

A cross-sectional area
C_p specific heat at constant pressure
C_v ventilation conductance
$\dot{G}$ air infiltration rate
h_c convection heat transfer coefficient;
  h_{ci} inside surface; h_{co} outside surface

h_R radiation heat transfer coefficient;
h_{Ri} inside surface; h_{Ro} outside surface

k thermal conductivity

L length; material thickness

ℓ characteristic dimension

$\dot{m}_s$ rate of vapour tramsission

n air change rate

p_s vapour pressure

$\dot{Q}$ rate of heat loss;
$\dot{Q}_f$ fabric; $\dot{Q}_v$ ventilation; $\dot{Q}_t$ total

$\dot{q}_v$ ventilation allowance

R thermal resistance

R_a thermal resistance of air space

R_v vapour resistance

R_{si} inside surface resistance

R_{so} outside surface resistance

r.h. relative humidity

U overall heat transfer coefficient

V velocity

V' volume

ϵ emissivity

ρ density

θ temperature; θ_a air temperature; θ_r mean radiant temperature; θ_e environmental temperature; θ_s surface temperature; θ_d dew point temperature

8 Heat Emission and Emitter Selection

8.1 INTRODUCTION

In this chapter a review of the more common forms of emitter associated with hot water heating systems is undertaken. In the case of convective type systems it is broadly a case of relating the heat loss and comfort procedures of Chapters 4 and 7 to selection of the appropriate emitter type. With radiant type emission systems there is the requirement for additional checks regarding asymmetric effects. The remainder of the chapter looks at some aspects of partial load operation.

8.2 HEAT EMISSION

8.2.1 Emitter ambient temperatures

We have already seen in Chapter 7 that until around 1970 the practice in design offices was to calculate design heat losses on the basis of air-to-air temperature differences and that more recently the trend has been towards the use of environmental temperature.

Manufacturers' literature, in the main, provides information on emission rates based on air (or 'ambient' or 'surrounding') temperature. It is obvious that the emission from an appliance will vary considerably depending on the radiant temperature of the surroundings as seen by the appliance even if the air temperature were to be held at a constant value.

British Standard Tests[1] for the rating of appliances specify controlled conditions of mean radiant and air temperatures. Nevertheless the designer is urged to treat given emission rates with caution when it is intended to select emitters for situations where there is a considerable difference between the air and mean radiant temperatures. In the remainder of this chapter, unless otherwise stated, the phrase ambient or surrounding temperature will be taken to mean that the situation under discussion is one in which the air and mean radiant temperatures are so close in value as to be considered as the ambient temperature.

8.3 HEAT EMISSION FROM PIPES

8.3.1 Introduction

In most circumstances the pipework connecting the emitters to the other components of the system will be insulated to reduce the heat loss. In some cases the designer will use the bare pipe as the emitter. In all cases it is essential to calculate the emission from the pipework and this is generally computed as a heat flow per unit length of pipe.

8.3.2 Bare pipes

The emission from a bare pipe may be determined by first principles from equation 2.72 and the expression for the overall thermal transmittance given in Chapter 2. A more practical approach adopted by the Guide(2) is to give tables of emission per unit length of pipe for bare pipes (steel, bright copper and dull copper) freely exposed in ambient temperatures at 20 °C. The emission is related to the pipe size and also to the difference in temperature between the pipe surface and the surroundings. The pipe surface temperature is normally taken to be the same as the water temperature within the pipe and the air and mean radiant temperatures are taken to be 20 °C.

Variations of 5 °C in either temperature introduces errors of less than 2%. So far as convective heat transfer is concerned laminar free convection (see equation 2.49a) is taken to be the prevailing condition. A supplementary table (in the Guide) is available which provides correction factors related to air velocities across the pipes, intended for use in conditions of draught. An extract from the Guide(2) table for use with steel pipes is included here.

Example 8.1 A college gymnasium dressing room requires the emission of 1.5 kW of heat from bare pipework placed beneath slatted seats occupying two adjacent walls. The length of piping is to be 10 m and the water flow and return temperatures are to be 80 °C and 60 °C respectively. The temperature in the room is to be maintained at 18 °C. Use Table 8.1 to determine a suitable size of steel piping for this duty.

$$\text{Temp difference surface/surroundings} = \frac{(80 + 60)}{2} - 18$$

$$= 52\ °\text{C}$$

$$\text{Emission per unit length} = \frac{1.5 \times 1000}{10}$$

$$= 150\ \text{W/m}$$

Table 8.1: A 65 mm diameter pipe gives an emission of 165 W/m for a 50 °C temperature difference. This would be greater for a

Table 8.1 Heat Emission from Single Horizontal Steel Pipes ($\epsilon = 0.95$) freely Exposed in Surroundings at 20 °C

Temp. diff. between surface and surroundings (°C)	*Heat emission (W/m)*						
	Pipe nominal size (mm)						
	40	50	65	80	100	125	150
25	48	58	71	81	102	122	141
30	60	72	88	101	126	151	175
35	72	87	106	122	152	182	211
40	84	102	125	143	179	214	248
45	98	118	145	166	207	247	287
50	111	135	165	189	236	281	327

52 °C temperature difference (and greater still in draught conditions). This pipe then, although slightly oversized, is suitable.

The Guide also provides correction factors for the basic emission tables which allows their use for vertical pipes and for horizontal banks of pipes. In calculating the heat loss from pipes the designer should bear in mind the local environment of the pipework with regard to the production of draughts or the effect of, for example, an adjacent cold wall, on the radiation component of heat emission.

8.3.3 Insulated pipes

This situation is also covered by equation 2.72 and the accompanying expression for overall thermal transmittance. The most common practical case for the designer is the situation where the pipework carries one layer of insulation. If the assumption is made that the inside face of the insulation is at the water temperature then the expression reduces to

$$\dot{Q} = \frac{(\theta_{f1} - \theta_{f2})}{\dfrac{\ln \dfrac{r_o}{r_t}}{2\pi k_2} + \dfrac{1}{2\pi r_o (h_{co} + \epsilon_o h_{Ro})}} \qquad (8.1)$$

which deals with the heat transfer from the inside face of the insulation to the ambient surroundings. The Guide contains a table of emissions which takes account of three values of thermal conductivity and relates the emission to the thickness of the insulation and to the diameter of the bare pipe. An extract is produced here as Table 8.2.

Example 8.2 The sketch shows the layout of the pipework in a circuit from a boiler to some radiators. Using the data given, determine the heat emission from the pipework and express it as a percentage of the total emission of the system.

Table 8.2 Heat Emission or Absorption from Insulated Steel Pipes

Heat emission or absorption from insulated pipework/(W/m run per °C temperature difference) for given values of insulation thermal conductivity (W/m °C)

Nominal pipe size (mm)	0.040					0.055					0.070				
	Thickness of insulation (mm)														
	12.5	19	25	38	50	12.5	19	25	38	50	12.5	19	25	38	50
15	0.27	0.22	0.19	0.16	0.14	0.54	0.29	0.25	0.21	0.19	0.41	0.35	0.31	0.27	0.24
20	0.31	0.25	0.22	0.18	0.16	0.40	0.33	0.29	0.24	0.21	0.47	0.40	0.36	0.30	0.26
25	0.36	0.29	0.25	0.20	0.18	0.47	0.38	0.33	0.27	0.24	0.56	0.46	0.41	0.34	0.30
32	0.43	0.34	0.29	0.23	0.20	0.55	0.45	0.39	0.31	0.27	0.66	0.54	0.47	0.38	0.34

Note: The pipe sizes are to BS 1387 and BS 3600. It is assumed that the outside surface of the insulation has been painted, is in still air at 20 °C and the pipe outside surface resistance R_{so} = 0.1 m² °C/W.

Data:

length of pipework	40 m
diameter of pipework	25 mm
thickness of insulation	25 mm
thermal conductivity of insulation	0.055 W/m °C
total emission from radiators	5 kW
mean water temperature	70 °C
ambient temperature	20 °C

Figure 8.0

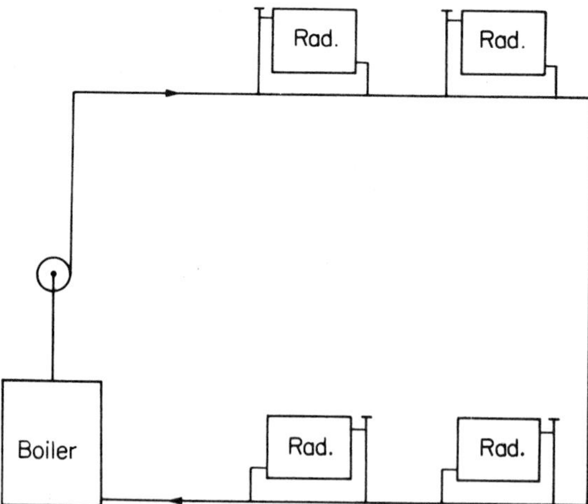

Emission = 0.33 W/m °C (from Table 8.2)

Temperature difference = (70 − 20) = 50 °C, assuming the ambient temperature to be 20 °C.

Pipework emission = 0.33 × 50 × 40 = 667 W

Pipework emission as a percentage of the total emission

$$= \frac{667}{667 + 5000} = 12\%$$

Thus the heat lost from the pipework is a considerable portion of the total emission and as shall be seen in Chapter 9, an allowance must be made for pipework emission when sizing the pipework.

Special circumstances arise when dealing with insulated pipework in underground ducts or pipework buried in the ground and this will be dealt with in the section on District Heating in Chapter 9.

8.4 HEAT EMISSION FROM PLANE SURFACES

8.4.1 Revision of basic theory

Emission by natural convection and by radiation from plane surfaces was covered in Chapter 2. Many practical emitters relate closely to the plane surface, one example being the high temperature radiant panel. This emitter consists normally of a pipe coil clipped to the back of a metal plate. Recall from Chapter 2 the expression (equation 2.59) for the radiant emission from a plane surface as:

$$\dot{Q}_R = \sigma A_1 \mathfrak{F}_{12} (T_1^4 - T_2^4)$$

where σ is the Stefan-Boltzman constant equal to 5.67×10^{-8}. We also saw earlier that when surface A_1 is enclosed by or forms an enclosure with surface A_2 then the Hottel Factor $\mathfrak{F}$ reduces to the value of ϵ_1. Thus equation 2.59 becomes for the majority of heating applications

$$\dot{Q}_R = 5.67 \times 10^{-8} A_s \epsilon (T_s^4 - T_{rm}^4) \tag{8.2}$$

where $A_s = A_1$ is the emitter surface area

It is convenient to express this equation in the linearised form:

$$\dot{Q}_R = \epsilon A_s h_R (\theta_s - \theta_{rm}) \tag{8.3}$$

after the fashion of equation 2.61 where θ_s is the emitter surface temperature and θ_{rm} is the average temperature of the room surfaces as seen by the emitter. The blackbody radiation heat transfer coefficient is obtained from equations 8.2 and 8.3 as

$$h_R = 5.67 \times 10^{-8} (T_s + T_{rm}) (T_s^2 + T_{rm}^2) \tag{8.4}$$

In relation to convection, equation 2.41 was a statement of the general form of the equation and as regards natural convection, equations 2.49 are appropriate to the solution of problems involving cylinders and flat plates. Thus convection heat transfer may be expressed as:

$$\dot{Q}_c = h_c A_s (\theta_s - \theta_a) \tag{8.5}$$

with the convection heat transfer coefficient obtained from

$$h_c = C (\theta_s - \theta_a)^{n-1} \tag{8.6}$$

Values of C and n may be obtained from the Guide[2] for specific situations or as outlined in Chapter 2 in relation to equations 2.49g and 2.49h.

Table 8.3 Heat Emission from Plane Surfaces by Radiation

Surface temp. (°C)	*Heat emission/(W/m²)*																				
	Surface emissivity 0.3							*Surface emissivity 0.6*							*Surface emissivity 0.9*						
	Enclosure mean radiant temperature (°C)							*Enclosure mean radiant temperature* (°C)							*Enclosure mean radiant temperature* (°C)						
	10	12.5	15	17.5	20	22.5	25	10	12.5	15	17.5	20	22.5	25	10	12.5	15	17.5	20	22.5	25
60	100	96	92	88	84	79	75	200	192	184	176	168	159	150	300	288	276	264	251	238	225
70	126	122	118	114	110	106	101	253	245	237	229	220	211	203	379	367	355	343	330	317	304
80	155	151	147	143	139	134	130	310	302	294	286	278	269	260	465	453	441	429	416	403	390
90	186	182	178	174	170	166	161	372	365	357	348	340	331	322	559	547	535	523	510	497	484
100	220	216	212	208	204	200	195	440	432	424	416	408	399	390	660	649	637	624	612	599	585
120	297	293	289	285	280	276	272	593	586	577	569	561	552	543	890	878	866	854	841	828	815
140	386	382	378	374	370	365	361	772	764	756	747	739	730	721	1160	1150	1130	1120	1110	1100	1080
160	489	485	481	477	473	468	464	978	970	962	954	945	936	928	1470	1450	1440	1430	1420	1400	1390
180	607	603	599	595	591	587	582	1210	1210	1200	1190	1180	1170	1160	1820	1810	1800	1790	1770	1760	1750
200	742	738	734	730	726	722	717	1480	1480	1470	1460	1450	1440	1430	2230	2220	2200	2190	2180	2170	2150

Table 8.4 Heat Emission from Plane Surfaces by Nature (Free) Convection

Surface temp. (°C)	Heat emission/(W/m^2)																				
	Horizontal looking down							*Vertical*							*Horizontal looking up*						
	Air temperature (°C)							*Air temperature* (°C)							*Air temperature* (°C)						
	10	12.5	15	17.5	20	22.5	25	10	12.5	15	17.5	20	22.5	25	10	12.5	15	17.5	20	22.5	25
60	85	80	75	69	64	59	54	255	238	221	205	189	174	158	309	289	269	249	230	211	192
70	107	101	96	90	85	80	75	324	307	289	272	255	238	221	394	372	351	330	309	289	269
80	130	124	118	112	107	101	96	398	379	361	342	324	307	289	484	461	438	416	394	372	351
90	153	147	141	135	130	124	118	476	456	436	417	398	379	361	578	554	530	507	484	461	438
100	177	171	165	159	153	147	141	556	536	516	495	476	456	436	675	651	626	602	578	554	530
120	228	222	215	209	202	196	190	726	705	683	661	640	619	598	882	856	829	803	777	751	726
140	281	274	267	261	254	248	241	907	884	861	838	816	793	771	1100	1070	1050	1020	990	963	936
160	336	329	322	315	308	301	295	1100	1070	1050	1020	1000	977	954	1330	1300	1270	1240	1220	1190	1160
180	393	386	378	371	364	357	350	1300	1270	1250	1220	1200	1170	1150	1570	1540	1510	1480	1450	1420	1390
200	451	444	437	429	422	415	407	1500	1480	1450	1420	1400	1370	1350	1820	1790	1760	1730	1700	1670	1640

As regards plane surface panel emitters, the Guide includes tables of heat emission by radiation and by natural convection. An extract of the radiant emission table is included here as Table 8.3. The radiation table is based on equation 8.2. The Guide's convective emission table (reproduced here as Table 8.4) is based on equations 8.5 and 8.6, due account being taken of variations in the values of C and n, as shown in Table 8.5. A point to be borne in mind here is that Table 8.4 relates to free convection and would therefore be unsuitable for whole floors or ceilings for instance.

Table 8.5 Values of C and n as used in equation 8.6

Situation	C	n
Warm or cold vertical planes	1.4	1.33
Warm horizontal planes facing up	1.7	1.33
Cold horizontal planes facing down	1.7	1.33
Warm horizontal planes facing down	0.64	1.25
Cold horizontal planes facing up	0.64	1.25

Many industrial heating systems employ high temperature radiant panels. In many cases these make use of vertical panels or panels angled towards the workplace or, as is becoming very common nowadays, horizontal overhead strip heaters. It is interesting to examine the percentage contribution of the radiant and convective components of the heat emission as in the next example.

Example 8.3 Determine the radiant and convective contributions of the heat emission from a radiant panel heating system which has to maintain a space at a mean radiant temperature of 17.5 °C and an air temperature of 15 °C with the panels operating at a surface temperature of 80 °C (or 100 °C or 160 °C or 180 °C) when the panels are placed:
(a) vertically (b) horizontally overhead

(a) Vertical panel at 80 °C:

radiant contribution (assume $\epsilon = 0.9$)	429 W/m^2 (Table 8.3)
convective contribution	361 W/m^2 (Table 8.4)
total emission	790 W/m^2
radiant emission	54%
convective emission	46%

Repeating the procedure for the remaining surface temperatures the results are:

Panel temp (°C)	Radiant emission (W/m^2)	Convective emission (W/m^2)	Total emission (W/m^2)	Radiant emission (%)	Convective emission (%)
80	429	361	790	54	46
100	624	516	1140	55	45
160	1430	1050	2480	58	42
180	1790	1250	3040	59	41

Intuitively it might have been expected that the percentage radiant contribution would have increased with increase in the panel surface temperature. While this is true, the increase is by no means dramatic. Panel temperatures seldom operate much above 180 °C.

(b) Horizontal panel at 80 °C:
The radiant contribution is obtained as before while the convective contribution is obtained from Table 8.4 under the heading 'horizontal looking down'. The following table can be completed for the different panel temperatures.

Panel temp (°C)	Radiant emission (W/m^2)	Convective emission (W/m^2)	Total emission (W/m^2)	Radiant emission (%)	Convective emission (%)
80	429	118	547	78	22
100	624	165	789	79	21
160	1430	322	1752	82	18
180	1790	378	2168	83	17

As before the radiant component increases with increase in surface temperature. However it is interesting also to note the much larger radiant contribution at any temperature from this type of emitter compared with the vertical panel.

This example bears out the classification of such systems by the Guide as being 'mainly radiant'.

A further point of interest here is to note that, in general, heating systems which are mainly radiant tend to produce higher mean radiant temperatures than air temperatures. In certain applications this may well lead to reductions in the heat loss from the space, especially in high spaces as frequently is the case with factories or large storage areas.

A variation on the panel emitter with radiant heating is the use of radiant strip, one form of this emitter being shown in Figure 8.1.

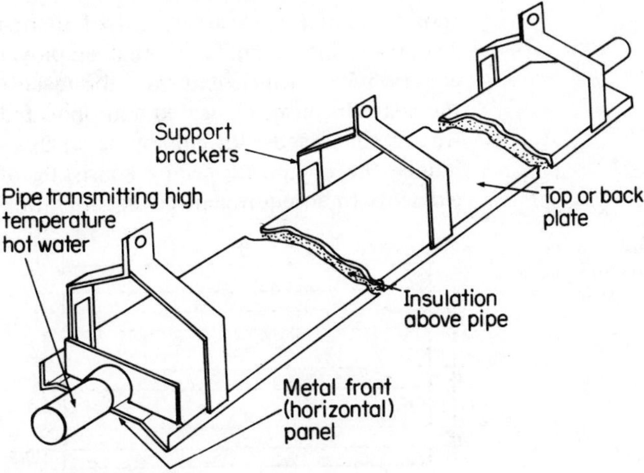

Figure 8.1
A radiant strip heater

Mounting heights range from the order of about 3 m above floor level when using low temperature hot water to about 5 m or more when using high temperature hot water. Radiant strip finds a common application in factories and warehouses and adequate provision must be made for taking up expansion in long runs of pipework. Manufacturers recommend gradients of about 1 in 600 to allow adequate venting within the pipework. It is also important to keep water velocities up around 0.5 m/s within the strip pipework.

A major consideration in relation to high temperature panel or radiant systems is the radiant flux at head level. Several options are open to the designer when dealing with this problem:

(a) Recourse may be made to the specialist manufacturers' literature which normally gives guidance on mounting heights in relation to panel surface temperatures.

(b) Use may be made of the table (rather limited) in the Guide, Book A, 1970 (given here as Table 8.6) which gives guidance on mounting heights related to panel temperatures, although in the main this is intended for radiant ceilings in offices, flats, etc.

(c) Use of the methods already explained in Chapter 4, in sections 4.7 and 4.8.

Table 8.6 Practical Limits for Ceiling Panel Systems

Room height (m)	*Maximum temperature of heated ceiling (for room temperatures of 18°C)*
2.4	33
2.6	36
2.8	39
3.0	41.5
3.2	44

If the designer chooses to adopt either of the approaches of (a) or (b) it may be useful to employ an 'equivalent ceiling' concept[3] in making a judgement of thermal comfort conditions. The reasoning is similar to that employed in the derivation of equation 7.47 which expresses the mean radiant temperature of the space in terms of heated and unheated surfaces. In this case, we are concerned with employing an area weighting technique to reduce the composite ceiling consisting of heated and unheated elements to an equivalent ceiling operating at a uniform tempera-

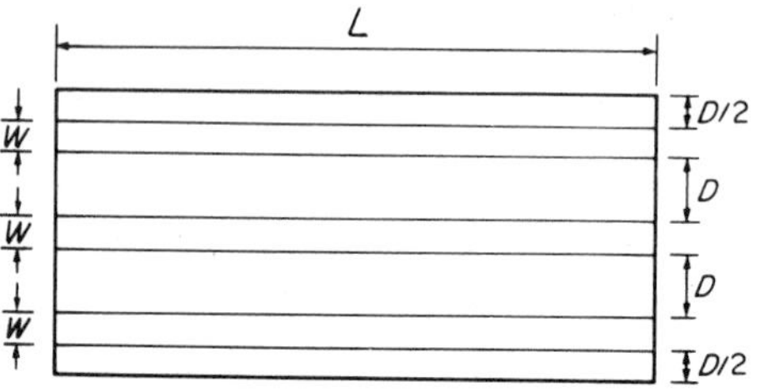

Figure 8.2
Plan of factory heated by radiant strip

ture. Consider for instance a factory L long and with radiant strip W wide and set D apart as shown in Figure 8.2.

Then, on an area weighting basis:

$$L\,(W + D)\theta_f = LW\theta_p + LD\theta_u$$

or
$$\theta_f = \frac{W\theta_p + D\theta_u}{W + D} \tag{8.7}$$

where θ_f, θ_p and θ_u represent the temperature of the equivalent ceiling, the radiant strip panel temperature and the temperature of the unheated parts of the roof, respectively.

Having estimated the temperature of the equivalent ceiling the designer may then use Table 8.6 (despite the limitations) to make some estimate of thermal comfort at head level. The following example illustrates the method.

Example 8.4 A small factory used for sedentary work is to be heated by means of radiant strip. Details of the construction are shown in Figure 8.3. Determine a suitable quantity and operating

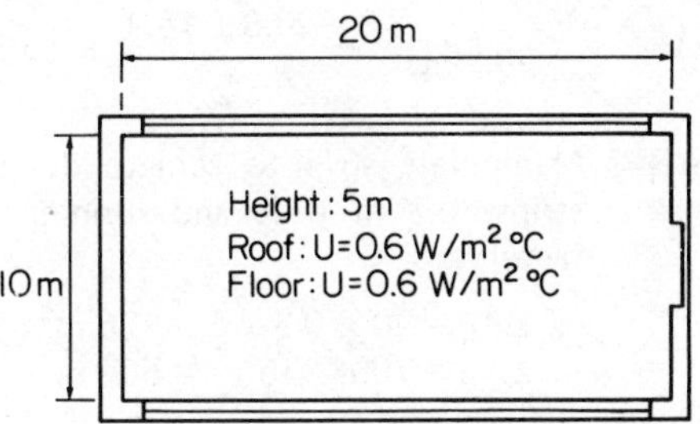

Figure 8.3
Details of factory construction

temperature of radiant strip which is 0.3 m wide. Take the inside and outside design environmental temperatures as 20 °C and −3 °C respectively.

Walls: area = 232 m^2; U value = 1.0 W/m^2 °C
Windows: area = 60 m^2; U value = 5.6 W/m^2 °C
Doors: area = 8 m^2; U value = 3.0 W/m^2 °C
Assume a ventilation allowance of 0.25 W/m^3 °C

Step 1 Determine the fabric heat loss

Item	Area (m^2)	U value (W/m^2 °C)	Temp. diff. (°C)	Heat loss (W)
Windows	60	5.6	23	7740
Walls	232	1.0	23	5340
Doors	8	3.0	23	554
Floor	200	0.6	23	2760
Roof	200	0.6	23	2760
	ΣA = 700			$\Sigma\dot{Q}_f$ = 19 154

Step 2 Determine the ventilation heat loss

$$\frac{1}{C_v} = \frac{1}{0.25 \times 1000} + \frac{1}{4.8 \times 700}$$

$$= 0.004 + 0.0003$$

$$C_v = 232 \text{ W/}^\circ\text{C}$$

Thus $\dot{Q}_v = 232\,[20 - (-3)] = 5330$ W

Step 3 Determine the total heat loss

$$\dot{Q}_t = \Sigma\dot{Q}_f + \dot{Q}_v$$
$$= 19\,154 + 5330 = 24\,484 \text{ W}$$

Step 4 Determine the air, mean radiant and resultant temperatures

$$\theta_{ei} - \theta_{ai} = \frac{5330}{4.8 \times 700} = 1.6\ ^\circ\text{C}$$

Thus $\theta_{ai} = 18.4\ ^\circ$C

Also $20 = 0.67\theta_{ri} + 0.33 \times 18.4$

Thus $\theta_{ri} = 20.8\ ^\circ$C

and $\theta_{res} = \dfrac{20.8 + 18.4}{2} = 19.6\ ^\circ$C

Step 5 Manipulate strip to achieve design requirements. Try 6 strips, 0.3 m wide and running the full length of the factory.

$$\Sigma A_h = 20 \times 0.3 \times 6 = 36 \text{ m}^2$$
$$\Sigma A_u = 700 - 36 = 664 \text{ m}^2$$

$$\theta_{ei} - (\theta_{ri})_u = \frac{19\,154}{9.6 \times 664} = 3\ ^\circ\text{C}$$

Thus $(\theta_{ri})_u = 17\ ^\circ$C

Now $\theta_{ri}\Sigma A = (\theta_{ri})_u\Sigma A_u + (\theta_{ri})_h\Sigma A_h$

$20.8 \times 700 = 17 \times 664 + (\theta_{ri})_h 36$

Thus $\theta_{ri(h)} = 92\ ^\circ$C

The design requirements are achieved by employing the radiant strip at an operating temperature of 92 °C.

Step 6 Comfort check

$$\theta_f = \frac{W\theta_p + D\theta_r}{W + D}$$

$W = 6 \times 0.3 = 1.8$ m, $D = 10 - 1.8 = 8.2$ m

Thus $\theta_f = \dfrac{(1.8 \times 92) + (8.2 \times 17)}{1.8 + 8.2} = 30.4\ ^\circ$C

The radiant strip may be considered equivalent to a heated ceiling of height 5 m operating at a surface temperature of 30.4 °C. This would be satisfactory (Table

8.6) so far as comfort restriction due to radiant flux at head level is concerned. The manufacturers' strip heater catalogue will give emission per unit length related to the flow temperature through the emitter and the ambient temperature. The designer would need to check the total available emission at 92 °C against the steady state heat loss of 24 484 W.

On the other hand, if the designer chooses to adopt method (c) two courses of action present themselves:

(i) The designer may still go through the previous stages of calculation to determine the 'equivalent ceiling' temperature of 30.4 °C. An estimate is then made of the total downward output of heat from the ceiling (W) and considering this to be spread evenly across the plan area, an irradiance (W/m^2) may be obtained in the manner explained in Chapter 4, sections 4.7 and 4.8. It is then a reasonably straight forward calculation to determine the mean radiant and vector radiant temperatures at head level in order to make a comfort judgement.

(ii) The designer establishes the panel surface temperature of 92 °C, as before. The centre position of the factory floor (maximum view factor) is taken as the worst case and view factors (see 4.7 and 4.8) are established at head level for each panel. As before, the mean radiant and vector radiant temperatures may then be obtained.

8.5 HEAT EMISSION FROM RADIATORS AND CONVECTORS

8.5.1 Radiators

Probably the most common heat emitter of all is the radiator, fed by low pressure hot water and used extensively with domestic and office space heating systems. The traditional radiator was of cast iron and this type is still available but the modern trend is towards pressed steel radiators which are available in a wide range of sizes, are very compact, and more elegant than the cast iron emitter.

The name 'radiator' is to a very large extent a misnomer as usually no more than about 30% of the emission is radiant, the remainder being convective, in relation to normal emitter surface temperatures of about 75 °C. Most pressed steel units are painted a light matt finish and the effect of painting them makes little difference to the emission at such low surface temperatures. Cast iron radiators on the other hand are frequently coated with a metallic finish. This can reduce the radiant emission by as much as 50% and therefore the total emission by about 15%.

Some thought must be given to the placing of the radiator with respect to the heated space. Most often radiators are placed below windows where there is a thermal advantage in counteracting downdraughts from the cool window surface as well as creating a more thermally uniform environment in the space. In addition, it places the radiators in a reasonably unobtrusive position. When placed against inside walls, radiators cause problems of staining of the wall surface. In the past this was often counteracted by the use of a recess or a shelf placed above the radiator but such obstructions reduced the emission from the radiator. Today, staining is more likely to be accepted together with more frequent cleaning or redecorating of the room surfaces. Radiators tend to create quite pronounced thermal gradients (of the order of about 7 °C) vertically within the space and, as was seen in Chapter 7, an allowance ought to be made for height when calculating room heat losses. Nevertheless, the lower air temperatures which are created in the region adjacent to the floor is compensated by reasonably high mean radiant temperatures within that region and the resulting effect is often a pleasantly comfortable environment.

8.5.2 Natural convectors

The major disadvantage of the radiator is the amount of wall space which it occupies. Such space becomes extremely important in public and commercial buildings and in these applications it is quite common to see the natural convector in use. Basically the natural convector is a finned pipe carrying low pressure hot water and enclosed within a metal cabinet with a low level opening allowing room air to pass into the emitter cabinet and usually a grilled outlet at a higher level to allow the warmed air to pass back into the room. It takes various forms, ranging from free-standing cabinets to continuous skirting heaters. Since the temperature of the cabinet varies little from that of other room surfaces, the emission is almost entirely due to natural convection. These emitters give even more pronounced temperature gradients than those produced by radiators.

8.5.3 Emission characteristics

The awkward geometrical shapes of radiators and natural convectors does not lend to analytical treatment as is the case with the pipe or the plane surface. Test procedures to determine emission in such cases are spelled out in BS3528[1] and manufacturers generally supply rating tables for their products on this basis. It has been found convenient to express the emission in the form:

$$\dot{Q} \propto (\Delta\theta)^n \tag{8.8}$$

Where $\dot{Q}$ is the rate of emission and $\Delta\theta$ is the difference in temperature between the radiator surface and the surrounding air. Rating tables are normally compiled on the basis of a 55 °C difference with an additional table of modifying factors to be used in situations where this difference is greater or less than 55 °C.

An obvious objection here is that radiant emission is not related to the surrounding air temperature but rather is it affected by the mean radiant temperature of those surfaces 'seen' by the emitter. However the test procedure minimises this objection by placing some control on the air and mean radiant temperatures. Also the effect of air movement across the emitter in the real rather than the test situation can have a considerable effect on the 'in situ' emission.

As a result of extensive tests[(1)] the accepted value of the index n for radiators is taken to be 1.3 with somewhat higher values being appropriate to natural convectors. The Guide suggests a figure of 1.27 for a skirting convector heater while an H.V.R.A. Report[(4)] suggests 1.43. The Guide suggests 1.35 for an 800 mm high natural convector. Different values of the index n means different emission characteristics for various types of emitter and as will be seen later this has profound implications when it comes to control of the heating system under partial load conditions.

Example 8.5 A manufacturer's catalogue indicates that a 24-section pressed steel radiator is 1.22 m long and 0.61 m high with a heating surface of 1.75 m^2. The total emission is 1009 W based on a 55 °C difference in temperature between the radiator surface and the room air. Assuming the general emission expression is valid, determine the constant for this emitter and hence estimate the emission in circumstances where the emitter surface temperature is 70 °C and the room air temperature is 20 °C. (In passing, it is interesting to note that the actual heating surface area is 1.75 m^2 while the outside area of the emitter is 1.22 × 0.61 × 2 = 1.48 m^2.)

$$\text{Now } \dot{Q} = \frac{1009}{1.75} = 580 \text{ W/m}^2$$

$$\text{Also } \dot{Q} \propto (\Delta\theta)^{1.3}$$

$$\text{Thus } \dot{Q} = k\,(\Delta\theta)^{1.3}, \text{ where } k \text{ is a constant}$$

$$\text{whence } k = \frac{580}{(55)^{1.3}} = 3.2$$

When the temperature difference is 50 °C

$$\dot{Q} = 3.2(50)^{1.3} = 518 \text{ W/m}^2$$

Now 518/580 = 0.89, thus 0.89 is a factor by which the standard emission of 580 W/m^2 can be multiplied to obtain the emission at the 'non-standard' condition. Manufacturers' catalogues, in addition to supplying tables of emission at standard conditions, normally include a table of multiplying factors for dealing with a whole range of temperature conditions.

Most often, a system is commissioned on a day when the

design load is not available. Having set the system for satisfactory operation on say an autumn day how can it be checked for its operation on a design day? Allowing certain assumptions, this can be done 'on paper' as the following example illustrates:

Example 8.6 A low pressure hot water heating system is commissioned on an autumn day when the outside temperature is 6 °C. In one office the room air temperature can be held at the design condition of 20 °C when the water flow is adjusted through the radiator to give a radiator surface temperature of 65 °C. Check this performance against the likely performance on a 'design day' when the outside temperature is –1 °C.

Assuming steady-state conditions exist both during the commissioning procedure and at the design condition and also that heat loss is proportional to the difference between inside air and outside design temperature, then during commissioning:

$$\text{rate of emission } \dot{Q}_1 = \text{rate of heat loss } \dot{Q}_{h\ell 1}$$

and at the design conditions;

$$\text{rate of emission } \dot{Q}_d = \text{rate of heat loss } \dot{Q}_{h\ell d}$$

$$\text{and } \frac{\dot{Q}_1}{\dot{Q}_d} = \frac{\dot{Q}_{h\ell 1}}{\dot{Q}_{h\ell d}}$$

$$\text{thus } \frac{(45)^{1.3}}{(75-\theta_{ai})^{1.3}} = \frac{14}{\theta_{ai}+1}$$

where θ_{ai} is the room air temperature under design conditions and it is assumed that the mean radiator temperature is to be 75 °C under design conditions. The solution of this equation is tedious and solving for θ_{ai} is perhaps best done by trial and error. Try θ_{ai} = 20 °C

$$\text{LHS} = \frac{(45)^{1.3}}{(75-20)^{1.3}} = \frac{140}{183} = 0.77$$

$$\text{RHS} = \frac{14}{20+1} = 0.67$$

Solving for θ_{ai} for other temperatures:

θ_{ai} °C	LHS	RHS
19	0.75	0.70
18	0.74	0.74

Thus it seems that in the steady state under design conditions a room temperature of 18 °C can be expected when 20 °C is obtained when commissioning under autumn conditions. This may or may not be satisfactory. Some adjustment of the flow rate through the emitter may be found necessary to increase the design temperature achieved but this of course would be accompanied by higher maintained temperatures at partial load conditions.

In some situations, repartitioning of rooms can disturb the designed heating system and affect its performance. As with the previous example, some prediction of the likely performance can be made from knowledge of the emission characteristics and of the steady state heat losses before and after the repartitioning. The following example illustrates the point.

Example 8.7 An existing office is heated by means of natural convectors with an emission index of 1.5. The office is maintained at a temperature of 20 °C when the outside (design) temperature is −1 °C. A portion of the office which does not include any emission surface is partitioned off and the remainder of the original office is left with a design heat loss of 75% per degree of the original heat loss. Assuming that the mean emitter surface temperature remains at 75 °C under design conditions, determine the air temperature which can be maintained.

Making the same assumptions as in the previous problem:

Before partitioning,

rate of emission $\dot{Q}_1$ = rate of heat loss $\dot{Q}_{h\ell 1}$

After partitioning,

rate of emission $\dot{Q}_2$ = rate of heat loss $\dot{Q}_{h\ell 2}$

$$\text{Also } \frac{\dot{Q}_{h\ell 2}}{\dot{Q}_{h\ell 1}} = \frac{0.75\,(\theta_{ai} + 1)}{(20 + 1)}$$

$$\text{Now } \frac{\dot{Q}_2}{\dot{Q}_1} = \frac{\dot{Q}_{h\ell 2}}{\dot{Q}_{h\ell 1}}$$

$$\text{thus } \frac{(75 - \theta_{ai})^{1.5}}{55^{1.5}} = \frac{0.75\,(\theta_{ai} + 1)}{(20 + 1)}$$

As before, solving for θ_{ai} by trial and error

θ_{ai} °C	LHS	RHS
18	1.06	0.68
20	1.00	0.75
22	0.93	0.78
24	0.88	0.89

Thus the adjustment to the room geometry means that a temperature of about 24 °C can be maintained under design conditions.

8.5.4 Forced convectors

Like the natural convector, the forced convector contains a heat emitter (usually a finned pipe) within a sheet steel cabinet. In addition however the unit contains a fan or fans to draw in air from the surroundings and blow it across the heat transfer surface. The fans in such units are usually capable of operating at two or three speeds giving the possibility of variable outputs. A problem which arises is noise at the higher speeds. Normally the steady state heat loss would be calculated at the middle or lower speeds but in any case advice is given on this point in manufacturers' literature.

Fan convectors are used with low pressure hot water with usually a temperature drop of about 10 °C across the emission surface. Since the space users are protected by the enclosing cabinet, advantage is sometimes taken of higher water flow temperatures obtained from medium pressure hot water systems with mean emitter surface temperatures of around 120 °C and a temperature drop of about 20 °C across the emission surface. This naturally gives more compact emitters.

The same emission characteristic equation (equation 8.8) already discussed in relation to radiators and other natural convectors also applies to forced convectors. In this latter case the value of the index n is unity.

Manufacturers' catalogues present emissions based on standard air entry temperatures and standard mean emitter surface temperatures together with tables of correction factors to be used in relation to non-standard conditions. The air entering the unit may be air recirculated from the space being heated, fresh air or a mixture of both.

Unit heaters

These are a form of forced convector used in industrial and commercial applications where, in general, it is fair to say that less priority is given to considerations of noise or aesthetics. Units may be free standing or suspended at high level. In the latter case some consideration must be given to the leaving air velocities and 'throw' which is necessary to allow the warmed air to reach the working places against its own tendency to rise due to buoyancy effects. The heat source is normally low, medium or high pressure hot water or steam or on a rare occasion electricity.

8.6 HEAT EMISSION FROM ROOM SURFACES

8.6.1 Introduction

A variety of systems are available which produce heat emission to the space from room surfaces such as ceilings, floors or walls. The emitters may be embedded water coils or electric cables or suspended ceiling systems with pipe coils clipped to the back of the tray. More recently, systems have been developed which employ electrical conducting paint films.

8.6.2 Floor systems

The basic objective in the design of a floor system (or any other of these systems) is to achieve a satisfactory balance between the floor temperature and the temperature of the other room surfaces, while creating a comfortable combination of room air and mean radiant temperatures. Apart from the requirement of satisfying ambient comfort conditions there exists the problem of achieving satisfactory temperature conditions at the floor surface. Too high temperatures create a feeling of 'tired feet' and it is normal practice to limit floor surface temperatures to about 25 °C for working areas and about 27 °C for circulating areas.

Such temperatures can be achieved normally by using water at about 55 °C in pipes embedded in a concrete floor below a screed about 50 mm thick. Some systems take advantage of higher operating temperatures of about 75 °C with the water coil within an asbestos cement sheath. However, care must be taken in selecting the correct water temperature as other problems may arise such as pattern staining and cracking of plaster. More guidance is available in this area from Reference (5).

In the case of ground floors, when determining the capacity of the system, consideration must be given to back losses in addition to the heat loss from the space above the floor. The back losses consist of horizontal edge losses from the concrete slab through the wall to outside together with downward losses to outside. The Guide (B1) gives extensive guidance on how to determine such losses taking into account such factors as the shape of the floor, insulation and thermal conductivity of the soil. In the case of upper floors, part of the heat flow will be upwards and part downwards. The proportions will depend on the relative resistances as shown in Figure 8.4.

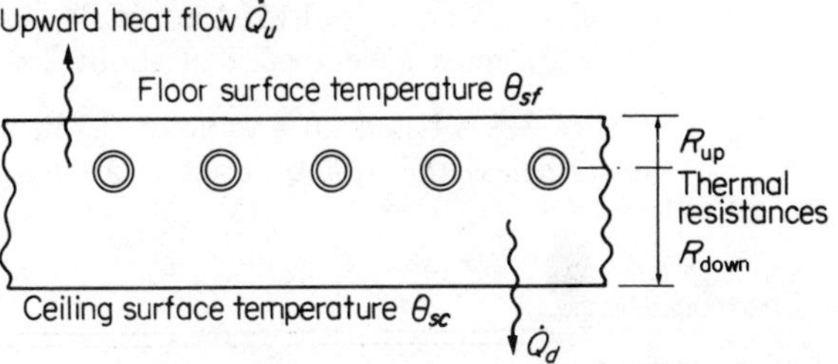

Figure 8.4
Ceiling and floor heat flows

The Guide suggests the use of an equation of the form:

$$\dot{Q}_u = k'_u\,(\theta_w - \vartheta_{ei})$$

where k'_u is a coefficient of upward heat emission related to R_{up} and R_{down} and θ_w is the mean temperature of the heating element. This has to be used in relation to graphs for selecting appropriate coil spacings in relation to the variables k'_u, R_{up} and R_{down}. Figure 8.5 is one such graph abstracted from the Guide.

Example 8.8 A block of flats of reinforced concrete construction is to be heated to a space temperature of 18 °C by means of embedded hot water coils within the floor slab. A sectional detail of the floor is shown in Figure 8.6. It is intended to run the system at a water temperature of 55 °C. A lounge in a top floor flat has a floor area of 20 m^2 and a heat loss of 2.3 kW. Determine a suitable spacing for the pipe coils, assuming they cover the

Figure 8.5

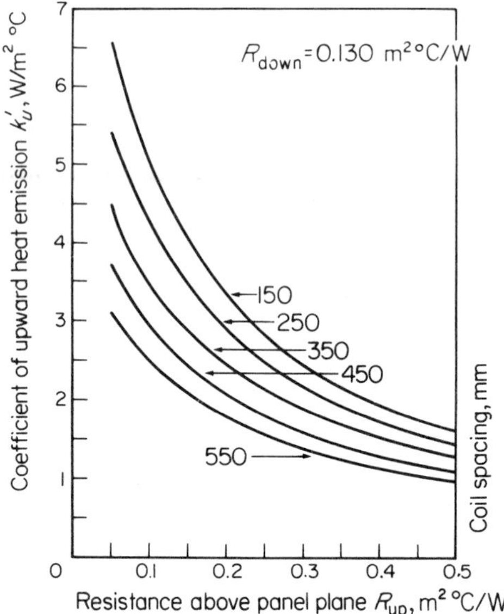

complete floor area and check the floor surface temperature for comfort. Assuming the heating pipes to be about 15 mm dia and let 8 mm of screed by associated with R_{down}.

$$\text{Then } R_{up} = (42 \times 0.0008) + 0.009 = 0.0426 \text{ m}^2\ ^\circ\text{C/W}$$
$$\text{and } R_{down} = (8 \times 0.0008) + (150 \times 0.0008) + (12 \times 0.002) = 0.1504 \text{ m}^2\ ^\circ\text{C/W}$$

Now $\dot{Q}_u$ must have a value of about $2.3 \times 1000/20 = 115$ W/m²

Figure 8.5 is based on a value of 0.130 m² °C/W for R_{down} but nevertheless this will be used as a basis for coil spacing.

Figure 8.6

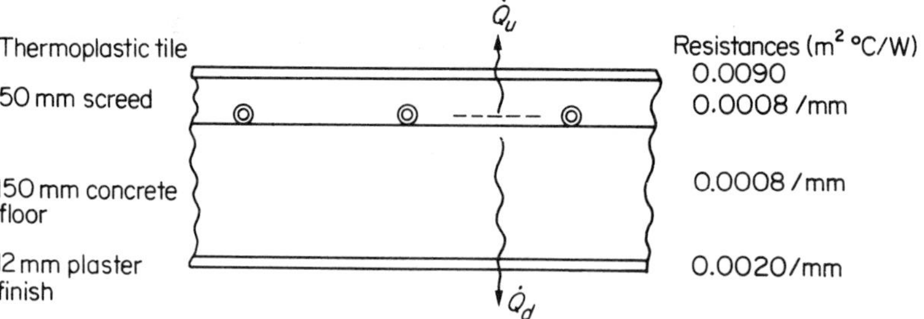

Try a spacing of 350 mm.

k'_u has a value of 5 W/m² °C (approx)
and $\dot{Q}_u = 5 \times (55 - 18) = 185$ W/m² for this spacing.

This is excessive; try a spacing of 550 mm.

k'_u has a value of 3.5 W/m² °C (approx)
and $\dot{Q}_u = 3.5 \times (55 - 18) = 129$ W/m² for this spacing.

Thus a spacing of 550 mm or perhaps a little greater would prove satisfactory.

Comfort check. If we assume that the floor surface resistance is 0.106 m^2 °C/W then:

$$\dot{Q}_u = \frac{(\theta_{sf} - \theta_{ei})}{R_s}$$

$$129 = \frac{(\theta_{sf} - 18)}{0.106}, \text{ from which } \theta_{sf} = 31.7\,°C$$

This is too high a surface temperature for comfort at foot level. The designer might well consider the effects of increasing the pipe spacing further or perhaps introducing a supplementary form of heating into the space.

8.6.3 Ceiling systems

Alternative techniques in dealing with high temperature overhead radiant panels have already been dealt with in section 8.4. Here we are considering low temperature radiant systems which usually find their application in schools, colleges, offices, flats, etc.

The design spacing of the coils may be obtained in the same manner for embedded ceiling systems as we have just considered for embedded floor systems and the reader is referred to the Guide, Book B, Section 1 for further information. Embedded ceiling systems are difficult to control because of their large thermal mass and for this (and other) reasons, suspended ceiling systems have been developed. Thus in situations where there are likely to be fluctuations in the load a suspended system would be more appropriate. The methods of determining the required operating surface temperature and of assessing thermal comfort which were explained in section 8.4 are also relevant here.

8.6.4 Wall systems

Embedded pipe systems within walls are much less common than the floor or ceiling systems already discussed. Nevertheless they do find application in situations such as the heating of swimming

Table 8.7 Heat Emission from Wall Panels

Wall surface temperature (°C)	*Surface emission* (W/m^2)
25	41
30	89
35	140
40	190

pools or hospital wards. Table 8.7 abstracted from the Guide gives useful information on the emission from wall panels in relation to the wall surface temperature for rooms assumed to be at a temperature of 20 °C.

8.7 EMITTER ARRANGEMENTS

8.7.1 Introduction

System arrangements will be examined in depth in Chapter 9 which is concerned with sizing the pipework. What is of concern here is the various piping connections which are possible to the emitter and the design water temperature drop across the emitter since both these factors influence the size of the emitter.

8.7.2 Piping connections

The three common methods of connecting the emitter to the pipework (shown in Figure 8.7) are

Figure 8.7

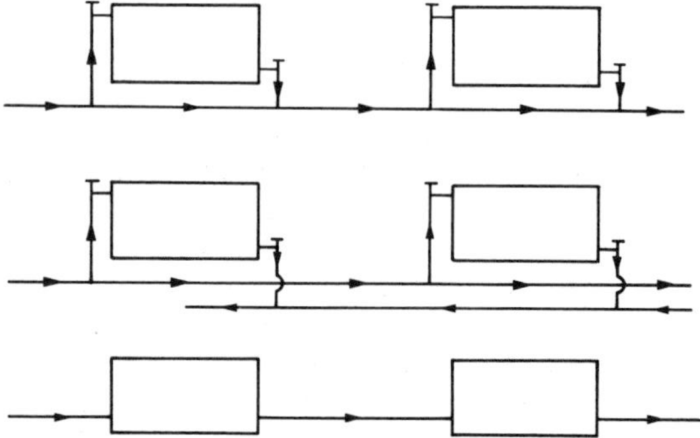

(a) The *one-pipe system,* in which the return water from the emitter mixes with the flow water, thus reducing the supply temperature to the next emitter.
(b) The *two-pipe system* in which each emitter receives flow water at about the same temperature.
(c) The *series system* in which all of the flow water passes through each emitter with each emitter thus receiving water at successively lower temperatures.

8.7.3 Design water temperature drops

In order to achieve the design temperature drop across the emitter the water flow has to be adjusted through the emitter, the larger the drop the smaller the flow. This in turn affects the mean surface temperature of the emitter. Thus the chosen temperature drop influences the emission and the water flow rate with repercussions with respect to pipe sizes, sizes of emitter connections and pump duty. Table 8.8 has been abstracted from Kell and Martin[6] in relation to design temperature drops.

Example 8.9 It is proposed to heat an office (Figure 8.8) from a low pressure hot water radiator system and the possible use of a one- or two-pipe system has to be explored. There is one external wall with three large windows on the wall and it seems reason-

Table 8.8 Design Water Temperature Drops

System	*Temperature (°C) at boiler or calorifier*	
	Flow	*Return*
Radiators (including pipe coils):		
(a) Gravity circulation	80	60
(b) Two-pipe pumped systems	80	65
(c) One-pipe pumped systems	80	70
Metal plate panels:		
(a) Two-pipe pumped systems	80	65
Convectors and unit heaters:		
(a) Pump systems	80	70
Embedded panel coils (pumped):		
(a) Two-pipe ceiling and wall systems	53	43
(b) Two-pipe floor systems	43	35
(c) Sleeved panel coils	70	60

Where radiators are served off a single pipe, successive radiators receive water at a lower temperature than those preceding. The following temperatures are recommended for design purposes:

Flow temperature to emitter (°C)	*Minimum temperature at outlet from emitter* (°C)
80	70
75	65
70	60
65	55

Figure 8.8

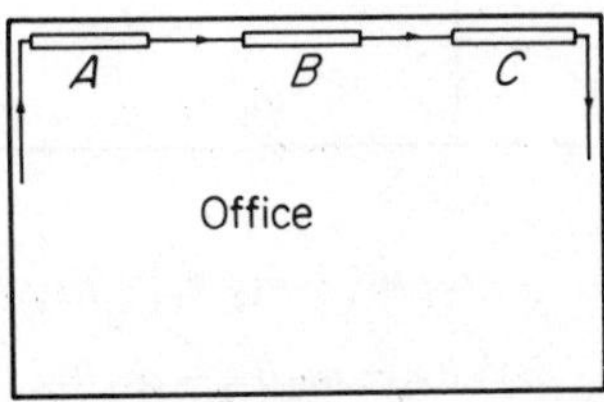

able to examine the use of one radiator beneath each window. Use Table 8.8 and the following data to determine the necessary number of sections for each emitter for each system.

Data:

Room temperature 20 °C
Steady state heat loss from the office 12 kW
Emission per radiator section per 55 °C temperature difference between emitter and room is 100 W.
Specific heat capacity of water 4200 J/kg °C.

(a) *One-pipe system* (Figure 8.9). Required emission per radiator 4kW.

$$\text{Total emission} = \dot{m}_1\, C_p\, (\theta_f - \theta_r)$$

$$\text{whence } \dot{m}_1 = \frac{12 \times 1000}{4200\,(80 - 70)} = 0.286 \text{ kg/s}$$

Figure 8.9

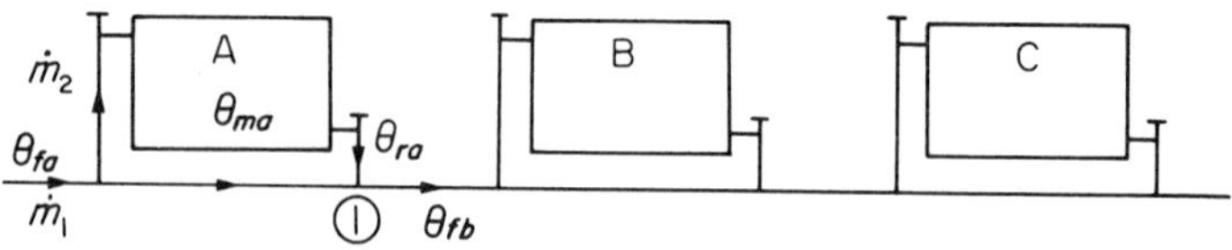

Through each radiator:

emission = $\dot{m}_2\ C_p$ (temperature drop through radiator)

Thus for θ_{fa} at 80 °C, θ_{ra} should be 70 °C and:

$$\dot{m}_2 = \frac{4 \times 1000}{4200\ (80 - 70)} = 0.095 \text{ kg/s}$$

$$\theta_{ma} = \frac{80 + 70}{2} = 75\ °\text{C}$$

and θ_{ma} – room temperature = 55 °C

Thus with the emission at 100 W per section, 40 sections are needed. Consider the mixing process at the point 1 (Figure 8.10) as an application of the steady flow energy equation (2.88).

Figure 8.10

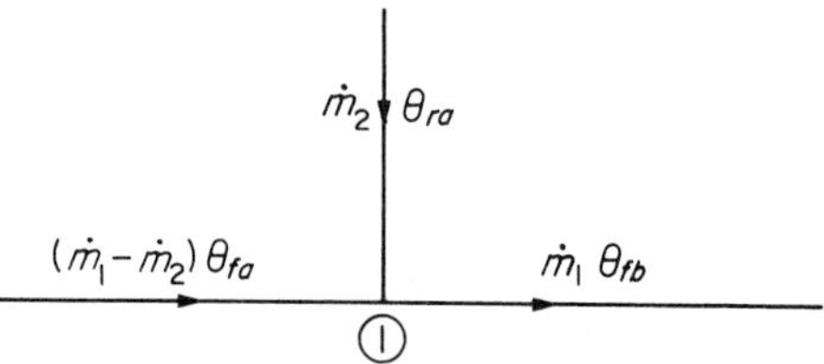

Now $(\dot{m}_1 - m_2)\ h_{fa} + \dot{m}_2\ h_{ra} = \dot{m}_1\ n_{fb}$

or $(\dot{m}_1 - \dot{m}_2)\ \theta_{fa} + \dot{m}_2 \theta_{ra} = \dot{m}_1 \theta_{fb}$ (approx), giving

$$(0.286 - 0.095)\,80 + (0.095 \times 70) = 0.286 \times \theta_{fb}$$

from which θ_{fb} = 77 °C.

Proceeding to radiator B, the flow temperature will be 77 °C and the return temperature 67 °C with a mean surface temperature of 72 °C. The emission will be less than the standard emission and can be calculated from equation 8.8 as follows:

$$\frac{\text{actual emission}}{\text{standard emission}} = \frac{(72 - 20)^{1.3}}{55^{1.3}}$$

$$\text{Thus the actual emission} = \frac{100 \times 170}{181} = 94 \text{ W per section}$$

The number of sections required for radiator B is thus

$$\frac{4 \times 1000}{94} = 42.6 \text{ or say 43 sections.}$$

In a similar manner radiator C is found to require 47 sections giving a total of 130 sections with this system coupled to increasing radiator sizes.

(b) *Two-pipe system*. Required emission per radiator 4 kW. With this arrangement each radiator will receive water at a temperature of 80 °C with a 15 degree drop through each. The mean surface temperature of each radiator will be (80 + 65)/2 or 72.5 °C. This results in an emission of 95 W per section and thus each radiator requires 42 sections giving 126 sections altogether.

A common fault with single-pipe systems is to find the last few radiators on the circuit operating at very low surface temperatures. This difficulty does not arise with the two-pipe system, for although there is more pipework employed, the preceding calculations illustrate that there is some saving made in the size of emitter surface. A further advantage is that the emitter is in series with the pump circuit whereas in the case of the single-pipe system it is in parallel with it.

(c) The series emitter arrangement referred to earlier has limited applications since there can be no occupier control allowed over individual radiators. These tend to be used in large spaces such as gymnasiums where the psychological need for control of the local environment is not so significant. Two further points are worthy of mention in relation to the system temperatures. Table 8.8 refers to flow and return temperatures at the heat source and strictly speaking an adjustment should be made to the flow and return temperatures at the emitters to allow for heat emission from the pipework. Also the temperatures quoted in Table 8.8 refer to those necessary to cope with an outside design temperature of −1 °C. It may be necessary in a prolonged spell of very cold weather to increase the supply temperature to the emitters.

8.8 PARTIAL LOAD CONDITIONS

Automatic control systems were covered in Chapter 3 but the point of interest in this section is the regulation of the emitter to cope with varying partial load conditions. Most often in the case of domestic heating systems a space thermostat controls the operation of the pump and as the load on the system diminishes the period of time that the pump is running also diminishes.

Commercial and industrial heating systems normally have the flow to the emitters regulated by means of a three-way valve which may be a mixing valve or a diverting valve. Either the flow through the emitters or the flow temperature to them is adjusted

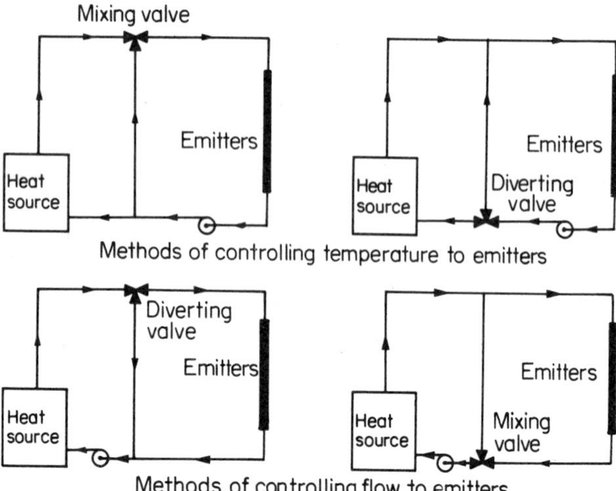

Figure 8.11 Operation of emitters at partial load

by use of the valve. Various permutations are possible and several of these arrangements are shown in Figure 8.11.

It is necessary to determine the various flows at the valve to meet the requirements at the emitter for different partial load conditions. A valve has then to be selected with a characteristic which matches reasonably with the requirements of the emitter; this was covered more fully in Chapter 3. Thereafter a schedule may be devised of valve stem movement in relation to outside temperature, or to any other suitable parameter which reflects variations in the load. The following example illustrates how the first part of this process is carried out.

Example 8.10 A college heating system employs low pressure hot water in radiators with design flow and return temperatures of 80 °C and 65 °C respectively. The rooms are to be maintained at 19 °C with an outside design temperature of −1 °C. The water temperature in the flow line from the boiler is to be maintained at 80 °C throughout the heating season but the flow temperature to the emitters has to be varied by the use of a mixing valve. Assuming the flow through the emitters' pipework is the same as that through the boiler at the design condition, determine the flow ratios at the mixing valve when the outside temperature is 5 °C.

Now heat loss rate from the building
$= \text{constant}_1\ (\theta_i - \theta_o)$
and heat emission from radiators
$= \text{constant}_2\ (\theta_m - \theta_i)^{1.3}$
Also the rate of heat flow from the circulating water
$= \dot{m} \times \text{specific heat}\ (\theta_f - \theta_r)$

Under steady state conditions at any one outside temperature these three quantities should be the same.

$$\text{At } -1\ ^\circ\text{C}, \theta_m \text{ will be } \frac{80+65}{2} = 72.5\ ^\circ\text{C}$$

Determine the value of θ_m at 5 °C outside.

$$\frac{\text{constant}_1\ (19+1)}{\text{constant}_1\ (19-5)} = \frac{\text{constant}_2\ (72.5-19)^{1.3}}{\text{constant}_2\ (\theta_m-19)^{1.3}}$$

from which $\theta_m = 59.5\ ^\circ\text{C}$

$$\text{Also } \frac{\text{constant}_1\ (19+1)}{\text{constant}_1\ (19-5)} = \frac{\dot{m} \times \text{specific heat } (80-65)}{\dot{m} \times \text{specific heat } (\theta_f - \theta_r)}$$

from which $(\theta_f - \theta_r) = 10.5\ ^\circ\text{C}$

$$\text{Thus } \theta_f = 10.5 + \theta_r \tag{8.9}$$

$$\text{But } \theta_m = 59.5 = \frac{\theta_f + \theta_r}{2}$$

and substituting from equation 8.9,

$$59.5 = \frac{(10.5 + \theta_r) + \theta_r}{2}$$

from which $\theta_r = 54.25\ ^\circ\text{C}$ and thus $\theta_f = 64.75\ ^\circ\text{C}$.

Now examining the mixing process at the valve, shown in Figure 8.12:

Figure 8.12

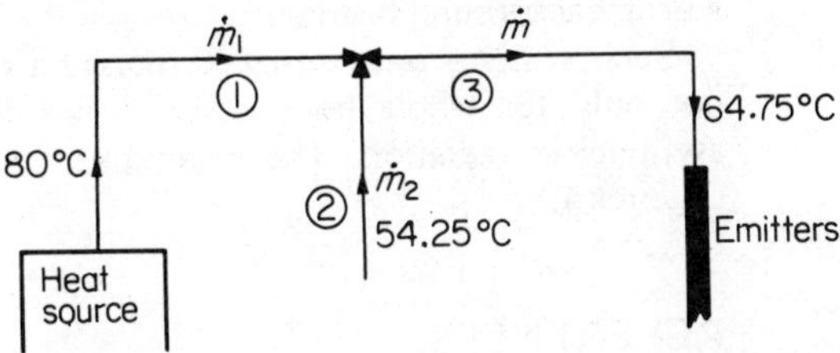

$$\dot{m}_1 h_1 + \dot{m}_2 h_2 = \dot{m} h_3$$

giving $80\dot{m}_1 + 54.25\dot{m}_2 = 64.75\dot{m}$ (approx)
and of course $\dot{m} = \dot{m}_1 + \dot{m}_2$, thus

$$80(\dot{m} - \dot{m}_2) + 54.25\dot{m}_2 = 64.75\dot{m}$$

$$\text{from which } \frac{\dot{m}_2}{\dot{m}} = 0.114 \text{ and thus } \frac{\dot{m}_1}{\dot{m}} = 0.886$$

In other words when the outside temperature is 5 °C the flow through the boiler drops to 88.6% of the design flow and 11.4% of the flow goes through the bypass.

The exercise can be repeated for other outside temperatures and a similar approach can be used when dealing with other valve arrangements.

It is worth noting at this point that if the exercise were to be repeated for emitters with different emission characteristics (e.g. natural convectors with n = 1.45) then the flow ratio to maintain an inside temperature of 19 °C against an outside temperature of 5 °C would be significantly different from the earlier calculations. Thus it is bad practice to mix the emitter types on a circuit in which partial load control is achieved by regulating the mass flow rate or the flow temperature to the emitters.

8.9 OTHER FORMS OF EMISSION

8.9.1 Conventional forms

There are many other forms of heat release within the space which have not been covered in this book. These include direct heaters (solid fuel, oil and gas appliances), electric heaters (embedded floor systems, convectors, radiators, storage heaters) and radiant strip heaters heated by combustion gases. Many of the emitters we have dealt with (e.g. radiant panels and unit heaters) could equally well be used with steam rather than with hot water as the energy source.

8.9.2 High intensity radiant systems

These have not been dealt with in this chapter. Basically they consist of small high intensity sources of infra-red radiation which are beamed on to localised areas. An example could be a small work area within a large warehouse which otherwise only requires a little background heating.

Such systems once designed require a thermal comfort check not only for whole body comfort but for the effects of the asymmetric radiation. The techniques for this are explained in Chapter 4.

REFERENCES

(1) BS 3528, Convection-type Space Heaters Operating on Steam or Hot Water, British Standards Institution.
(2) C.I.B.S. Guide, Section C3.
(3) McLAUGHLIN, R. K. and McLEAN, R. C. 'Environmental Temperature Theory', *Steam and Heating Engr.*, December, 1973.
(4) H.V.R.A. Report.
(5) C.I.B.S. Guide, Section B1.
(6) FABER, O. and KELL, J. R. *Heating and Air Conditioning of Buildings*, Rev. J. R. Kell and P. L. Martin (Architectural Press, London, 1974).

SYMBOLS USED IN CHAPTER 8

A area
C_p specific heat at constant pressure
C_v the ventilation coefficient
D distance between elements

h specific enthalpy
h_c convection heat transfer coefficient
h_R radiation heat transfer coefficient
k thermal conductivity; a constant
k'_u coefficient of upward heat transfer
L length
$\dot{m}$ rate of mass flow
n emission characteristic index
$\dot{Q}$ rate of heat transfer; $\dot{Q}_c$ by convection, $\dot{Q}_R$ by radiation
R thermal resistance
r radius
T absolute temperature
U overall thermal transmittance
W width of element
ϵ emissivity
θ temperature
θ_f water flow temperature
θ_m mean water (emitter surface) temperature
θ_r water return temperature
$\mathfrak{F}$ Hottel Factor
σ Stefan-Boltzman constant

9 Water Heating Systems

9.1 INTRODUCTION

In this chapter the theoretical fluid mechanics and heat transfer content developed in Chapters 1 and 2 are applied to the design of low pressure and pressurised hot water heating systems. This chapter also includes a section on pumps and pump selection and also a section on district heating.

9.2 SYSTEM LAYOUTS

9.2.1 Low pressure natural circulation systems

Many early water-type heating systems relied on natural convection to transfer heat from the source to the emitters. The pipework was of large diameter and often was unsightly. The circulation was produced by a rising column of hot water in the flow main and a descending column of cooler water in the return main, in motion as a result of their different densities. Figure 9.1 shows

Figure 9.1
The basic gravity natural circulation system

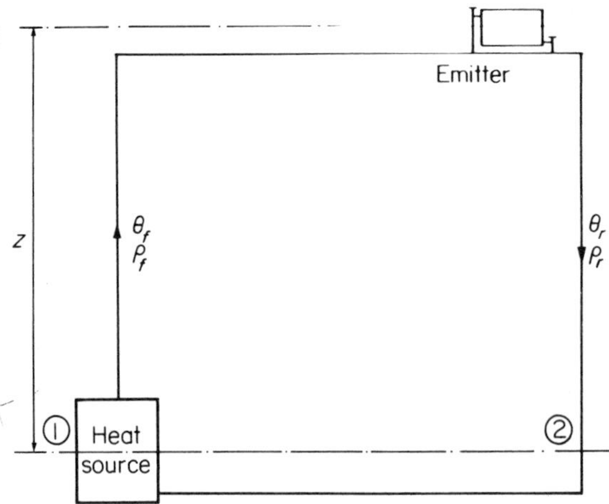

the basic circulation circuit with the mean difference in vertical height of the water columns being z and with the flow and return temperatures and densities being θ_f, ρ_f and θ_r, ρ_r respectively. The pressure difference available between the sections 1 and 2 is given by the expression:

Table 9.1 Density of Water at Different Temperatures

Temperature (°C)	0	10	20	40	50	55	60	65	70	75	80	85
Density (kg/m³)	1000	1000	998	992	988	986	983	980	978	975	972	969

$$p_1 - p_2 = (\rho_r - \rho_f)gz \qquad (9.1)$$

Table 9.1 gives values for the density of water at a variety of temperatures with the water at atmospheric pressure.

Example 9.1 In relation to the system shown in Figure 9.1, given that the flow and return temperatures are 80 °C and 60 °C respectively and that the vertical distance between the heat source and the emitter is 6 m, determine the available pressure between the sections 1 and 2. From Table 9.1, ρ_r = 983 kg/m³ and ρ_f = 972 kg/m³

$$\text{Thus } p_1 - p_2 = (983 - 972) \times 9.81 \times 6$$
$$= 648 \text{ N/m}^2$$

Identification of Index Circuit

As we shall see later in this chapter, pipes are sized on the basis of the pressure drop per metre length of pipe and the flow rate through the pipe. A problem arises when the system has more than one circuit and some of the pipework is common to two or more circuits. The common pipework has then to be sized on the basis of the circuit with the smallest available pressure per metre length, known as the *index circuit.* The ability to identify the index circuit is a skill which the heating designer must develop.

Figure 9.2
Locating the index circuit

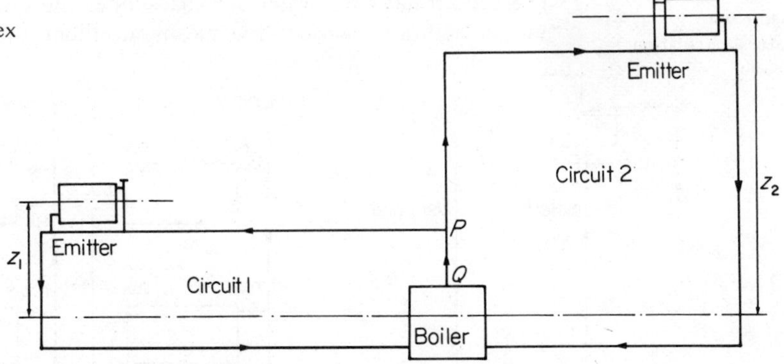

Consider the situation in Figure 9.2. Circuits 1 and 2 have about the same lengths of pipework but the available pressure head in circuit 1 is much smaller than that of circuit 2. The index circuit is quite easily identified as circuit 1 in this case and the common pipe section PQ is sized on the basis of the available pressure per metre in circuit 1. Identification of the index circuit is normally more complex than is shown in this case, especially when (as we shall discuss later) the equivalent lengths of bends, tees and fittings are taken into account.

Although natural (or gravity) circulation systems are out of fashion, one place at least exists where such effects must be taken into account. This is in the sizing of connections from the main to emitters on a one-pipe circuit (Figure 9.3).

Figure 9.3

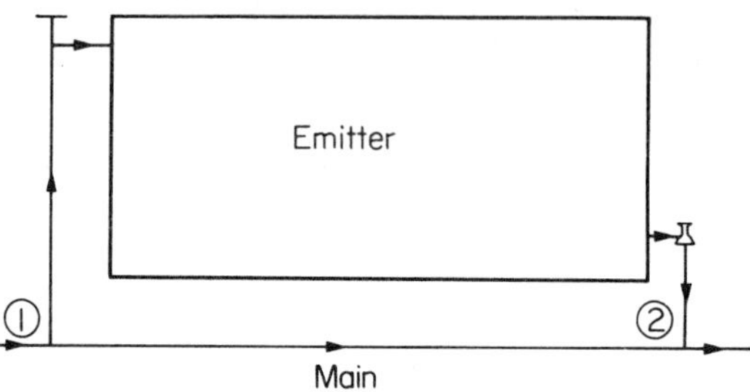

Here there are two effects to be considered.

1 Forced circulation produced by the pump which pushes the water through the emitter in the direction 1 to 2.
2 Natural circulation due to density differences causing the flow water to rise into the emitter at connection 1 and return to the main at connection 2.

In this case the resulting pressure available for circulation through the emitter is the sum of the separate effects. If the emitter had been placed below the main the resultant pressure would have been the difference of these effects.

Figure 9.4 shows a two-pipe up-feed gravity system suitable for a house or perhaps a small school building.

The cistern takes the water of expansion as the system heats up and in addition provides the means for filling the system and

Figure 9.4
Natural circulation system

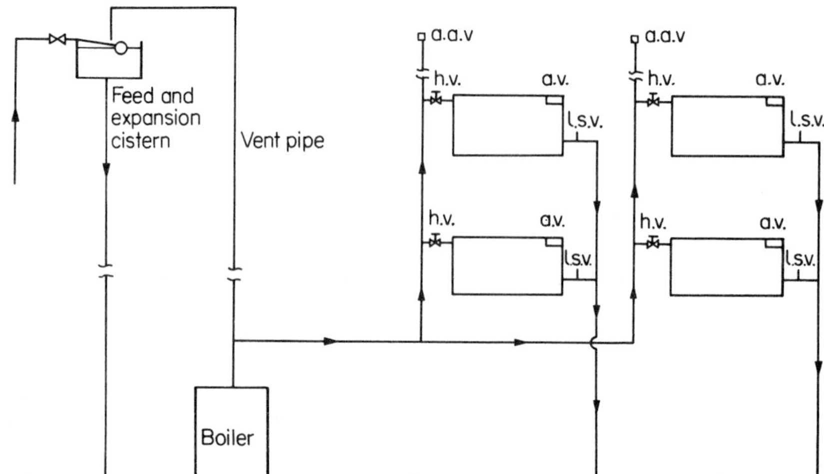

topping up. Air or gases within the system can escape via the open vent or air bleed valves at the top of each radiator. There is normally a hand controlled valve on the flow side at entry to each radiator and a system balancing valve on the return side leaving each radiator.

9.2.2 Low pressure forced circulation systems

In such systems, known also as accelerated systems, the circulation of the water is achieved usually by means of an electrically driven pump. Various system layouts are possible and Figure 9.5 represents a typical domestic situation using a single-pipe arrangement.

Figure 9.5
Domestic heating system

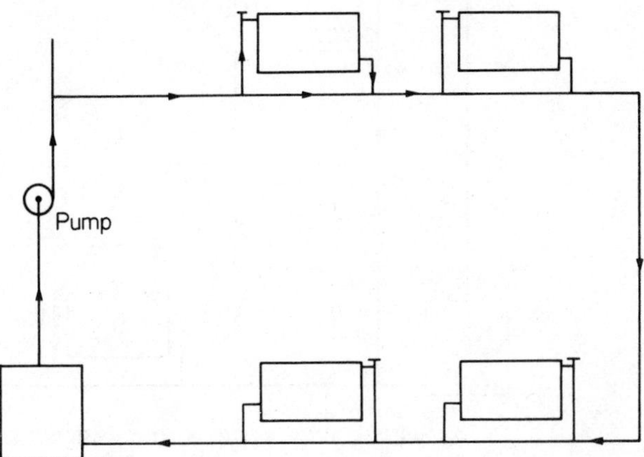

The larger types of system in factories, offices and public buildings are more likely to employ a two-pipe arrangement. Figure 9.6 shows such a system which also includes a mixing valve to control the temperature of the water supplied to the emitters.

Figure 9.6
Heating system for large building

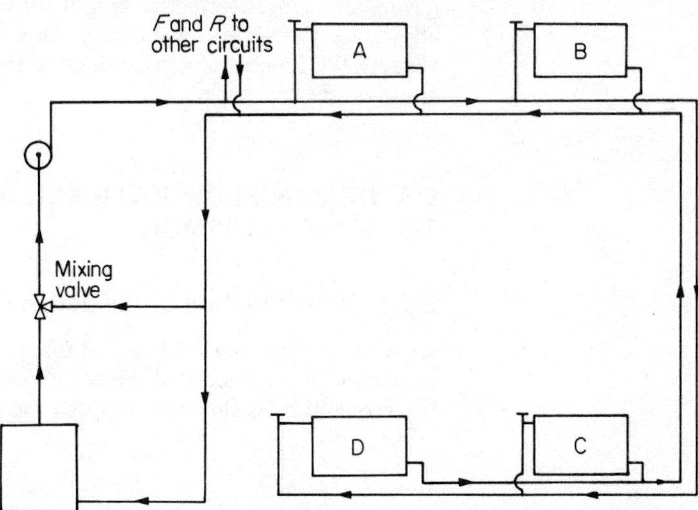

With the two-pipe layout there is in effect a separate circuit to each emitter and a disadvantage of the arrangement is that emitter circuits close to the boiler (for example, A) have much lower resistances than those farther away (for example, D). One solution to this problem is to connect the pipework 'reverse return' as shown in Figure 9.7 which keeps the individual emitter circuits of about equal resistance 'length'.

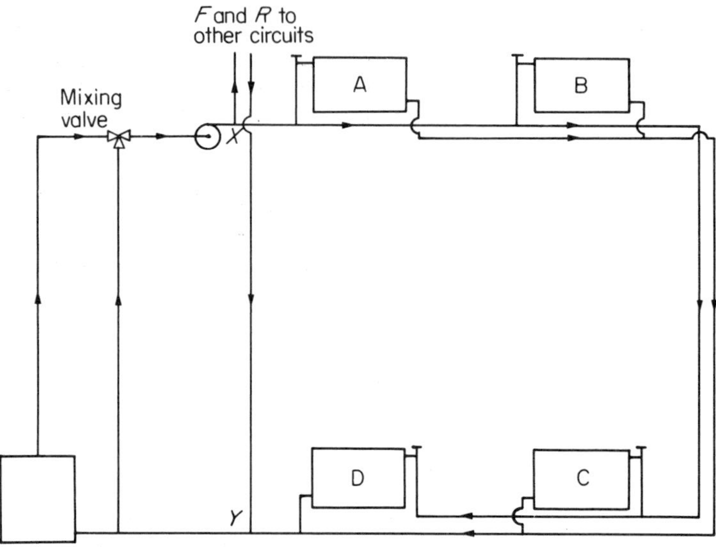

Figure 9.7
The reversed return system

As with gravity systems, the designer has to be able to identify the index circuit of a forced circulation system. In the system shown in Figure 9.7 there are four separate circuits (A,B,C and D) on one floor level and presumably similar layouts on higher floors. The pressure difference between X and Y will be available to each group at each floor level and the index circuit will be the one having the longest effective length (physical length plus length allowance for other resistances). This is very often (but not always) the longest pipe run which in this case would be to the group on the top storey.

9.3 DESIGN FLOW RATE AND APPORTIONING OF THE MAINS EMISSION

9.3.1 Determination of the design flow rate

We have already seen in Chapter 8 that for an emission $\dot{Q}$ from an emitter with an associated water temperature drop of $(\theta_f - \theta_r)$ the required mass flow rate through the emitter, $\dot{m}$, is given by:

$$\dot{m} = \frac{\dot{Q}}{C_p(\theta_f - \theta_r)} \tag{9.2}$$

If there were no heat emission from the pipework then it would be a simple matter to determine the design mass flow rate carried by each section of the pipework based on the flow rates required at each emitter. The complication arises due to emission of heat from pipework. This is a function of the pipe diameter and is therefore not precisely known until the pipework is sized. The flow rate must be increased to allow for this emission but this is an area where the designer is presented with a chicken and egg situation.

Essentially what is required is an assessment of the heat loss from the pipework, which necessitates an increased water flow rate, then the flow rates require to be adjusted in each section according to the additional heat emission they require to carry. This is called apportioning of the mains emission. The process is very tedious and time consuming but it is absolutely essential that the designer grasps the concept of this adjustment. It is hoped that the worked example presented in the next section will make this clear.

9.3.2 Apportioning the mains emission

In order to facilitate the exercise of apportionment of mains emission Kell & Martin[1] deploy the graph presented here as Figure 9.8.

The emission is calculated from the expression:

$$\frac{0.18\,(\theta_f + \theta_r - 2\theta_a)^{1.3}\,\dot{Q}^{0.28}}{(\theta_f - \theta_r)^{0.28}},$$

with the curves drawn for θ_f and θ_a taken as 80 °C and 20 °C respectively and the temperature ranges noted on the curves being $(\theta_f - \theta_r)$.

The graph is based on normal design conditions of 80 °C flow temperature and 20 °C ambient temperature with a sizing on a pressure drop of somewhere in the range of 100–150 N/m² per

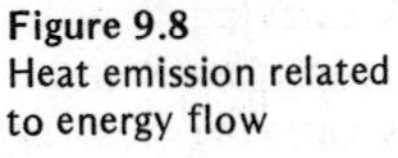

Figure 9.8
Heat emission related to energy flow

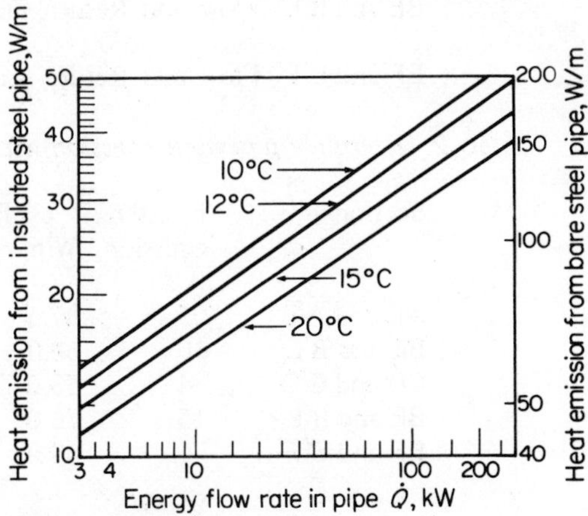

metre. The likely pipe size then depends on the heat load supplied by the water flow and the temperature drop between the flow and return pipework (since each of these factors influences the flow rate in the emitter). The graph rather than yielding pipe size, presents emission in watts per metre for bare or insulated pipe in relation to the heat load and the design temperature drop.

Example 9.2 Figure 9.9 shows the layout of a low pressure hot water heating system which employs radiators with a temperature

Figure 9.9

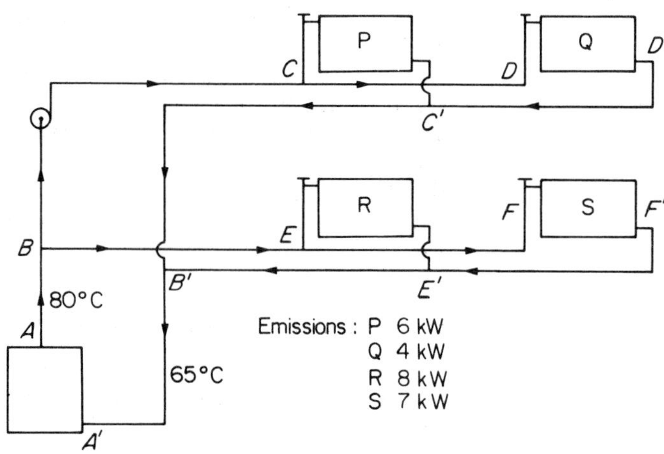

difference between flow and return of 15 °C used to maintain a space temperature of 20 °C. Assuming the pipework to be uninsulated apportion the pipework heat loss to the relevant sections and assess the necessary flow rate in each section.

Step 1 Identify and label pipe sections

AB and A′B′ Flow and Return sections carrying total flow
BC and B′C′ Flow and Return sections carrying flow to and from P and Q
CD and C′D′ Flow and Return sections carrying flow to and from Q
BE and B′E′ Flow and Return sections carrying flow to and from R and S
EF and E′F′ Flow and Return sections carrying flow to and from S

Step 2 Determine pipework emission with the aid of Figure 9.8

Section	Emitter emission (kW)	Pipework emission W/m	Pipe length (m)	Total (kW)
AB and A′B′	25	87.5	7	0.610
BC and B′C′	10	68.0	14	0.950
CD and C′D′	4	53.0	10	0.530
BE and B′E′	15	76.0	10	0.760
EF and E′F′	7	62.0	8	0.496
	25 kW			Total 3.346

In passing it is interesting to observe that the mains emission is 3.346/25 or approximately 13.5% of the emitter emission.

Step 3 Apportioning the mains emission

Consider the upper circuit. The 0.53 kW load from section CD/C′D′ is credited to emitter Q.

However the 0.95 kW load from section BC/B′C′ must be shared between emitters P and Q and in the ratio of their emissions. Thus P is credited with 6/10 of 0.95 or 0.57 kW and Q with 4/10 of 0.95 or 0.38 kW. This apportioning process is carried out for each section all the way back to the boiler. The results may usefully be set out as follows:

Source of Load	*Emitter Load* (kW) P	Q	R	S
Emitter Rated Emission	6.000	4.000	8.000	7.000
CD/C′D′	—	0.530	—	—
BC/B′C′	0.570	0.380	—	—
EF/E′F′	—	—	—	0.496
BE/B′E′	—	—	0.405	0.355
AB/A′B′	0.146	0.098	0.196	0.170
Totals	6.716	5.008	8.601	8.021

The increase is 0.716 + 1.008 + 0.601 + 1.021 = 3.346 kW and checks.

Step 4 Determination of design flow rates

Since there is a temperature drop in the pipework due to pipework emission, this step must be carried forward from the boiler or energy source. In this case consider section AB/A′B′.

$$\text{Water flow rate} = \frac{\text{emission from all emitters}}{C_p\,(\theta_f - \theta_r)}$$

$$= \frac{6.716 + 5.008 + 8.601 + 8.021}{4.200\,(80 - 65)}$$

$$= 0.45 \text{ kg/s}$$

Now estimate the temperature drop in these pipes.

$$\text{Emission from AB/A'B'} = (\text{water flow rate in AB/A'B'})\,C_p\,(\text{temperature drop})$$

$$0.610 = 0.45 \times 4.2\,(\text{temperature drop})$$

$$\text{Thus temperature drop} = \frac{0.610}{0.45 \times 4.2} = 0.32\ ^\circ\text{C}$$

The temperature drop available from B to B′ is thus 15 – 0.32 = 14.68 °C.

Now consider section BC/B′C′

$$\text{Water flow rate} = \frac{6.716 + 5.008}{4.2\,(14.68)} = 0.19 \text{ kg/s}$$

$$\text{Again temperature drop} = \frac{0.950}{0.19 \times 4.2} = 1.19\ ^{\circ}\text{C}$$

and the temperature drop available from C to C′ is thus 14.68 − 1.19 = 13.49 °C. The remaining sections can be dealt with in a similar manner and the results are as presented:

Section	*Design Mass Flow Rate* (kg/s)	*Temperature Drop*	*Available to Next Section* (°C)
AB/A′B′	0.450		14.68
BC/B′C′	0.190		13.49
CD/C′D′	0.088	12.05	across emitter Q
BE/B′E′	0.260		13.98
EF/E′F′	0.136	13.11	across emitter S

With the design flow rates established for each section of pipework, the next step in the design procedure is then the sizing of the pipework.

Mean surface temperature of emitters

The use of Figure 9.8 does not allow the temperature drop in each pipe to be determined, however if we assume that the section temperature drop is shared between the flow and return pipes equally it is possible then to examine the mean surface temperatures of the emitters, as the next example illustrates.

Example 9.3 Apply the assumption just stated to the previous example in order to establish the temperatures at all important points in the system and to estimate the mean surface temperature of each emitter. Consider once again section AB/A′B′.
Overall temperature drop is 0.32 °C. Assuming this drop is shared equally between flow and return pipes then the temperature at B is (80 − 0.16) or 79.84 °C. Similarly the temperature at B′ is (65 + 0.16) or 65.16 °C. This checks with the available drop of 14.68 °C from B to B′.
Continuing in this manner, the remaining temperatures may be found as shown in Figure 9.10.
Consider emitter P.
Flow and return temperatures to and from the emitter are 79.26 °C and 65.74 °C respectively, with a temperature drop across the emitter of (79.26 − 65.74) or 13.52 °C (13.49 °C approx)

$$\text{Emitter mean surface temperature} = \frac{79.26 + 65.74}{2} = 72.5\ ^{\circ}\text{C}$$

If a similar calculation is carried out for each emitter then the mean surface temperature in each case is found to be 72.5 °C,

Figure 9.10

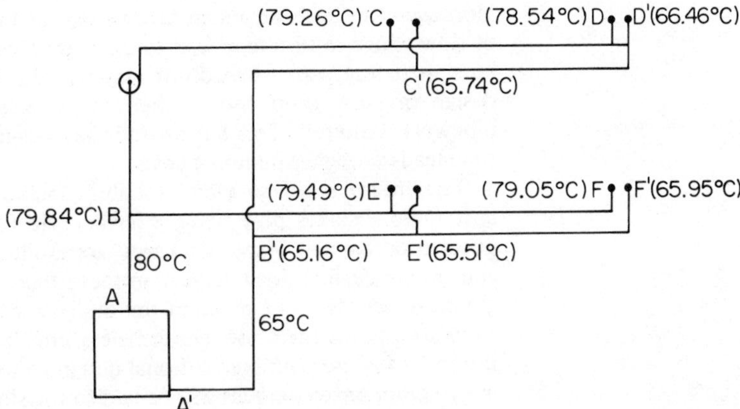

which is the value that would be expected, based on flow and return temperatures of 80 °C and 65 °C if there had been no heat loss from the pipework. Thus the extra water flowing in the system to account for mains losses has the effect of maintaining the proper mean surface temperatures in the emitters. This further underlines the necessity of proper apportioning of the mains emission.

Other methods exist of apportioning mains emission such as the direct ratio method and a graphical representation method but the reader is referred to Reference (2) for further information in this area.

9.4 SIZING THE PIPEWORK

9.4.1 Introduction

Pipework, bends, tees, components of the system, etc., all offer resistance to the flow of water as already discussed in Chapter 1. We shall see, in what follows, that the resistances of fittings are reduced to an equivalent length of straight pipe. This allows the designer to reduce the real (complex) circuit to an abstract equivalent circuit of straight pipework (Figure 9.11) in which a certain mass rate $\dot{m}$ can flow if a pressure difference ΔP is applied across the pipe of diameter D over a length L.

Figure 9.11

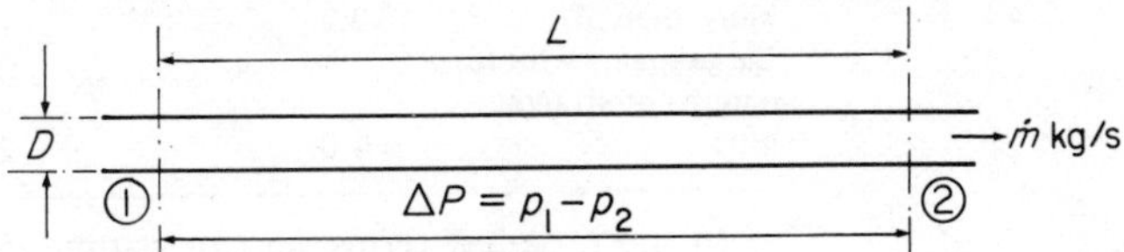

In order to achieve the desired flow rate some degree of choice is open to the designer in relation to the pressure difference (or the pressure difference per metre) and the pipe diameter. Constraints which have to be borne in mind here are the range of pipe diameters available allied to fittings and also the range of pumps.

Moreover the decision taken on this matter influences velocities of flow which in turn may lead to other problems such as noise or erosion or corrosion. Broadly it may be said that the higher the design pressure drop chosen then the smaller the diameter of pipework required. This has obvious advantages but at the same time leads to higher pumping costs.

The interplay of so many variables including economic and cost factors makes pipe sizing a fertile field for the use of computer programmes. Many 'packages' are available to the designer and when dealing with large contracts there is no doubt that great savings are to be made in the designer's time and the company's costs, in their use. Nevertheless, this book is intended in the main for use with conventional design techniques and accordingly computer techniques will be laid to the side.

9.4.2 Allowable pressure drop

Care must be taken in the selection of a suitable pressure drop when sizing pipework, not only because of its influence on the cost of pumping but also because of secondary effects arising from the flow velocities thus imposed on the system. In some situations the designer must ensure that he has provided sufficient velocity to keep any air, which tends to collect in certain parts of the system, moving. Thus, for example, on the upper floors of high rise buildings, minimum velocities of about 1.25 m/s have been recommended(3) for pipes greater than 50 mm diameter and of about 0.75 m/s for pipes less than 50 mm diameter.

On the other hand, excessive velocities will result in unacceptable levels of noise being generated, especially at valves or other fittings. Moreover high velocities may give rise to erosion or corrosion processes within pipework and fittings due to the breakdown of fluid boundary films. Table 9.2, which has been abstracted from the Guide (B1) provides information on maximum allowable velocities.

Table 9.2 Limiting Maximum Water Velocities

Pipe diameter (mm)	*Non-corrosive water* (m/s)	*Corrosive water* (m/s)	*Copper pipework* (m/s)
50 and less	1.5	1.0	1.0
More than 50	3.0	1.5	1.5
Large mains with long lengths of straight pipe	4.0	2.0	–

In the broadest terms we can assume design pressure drops to be somewhere in the range of about 150 to 300 N/m^2 per metre in conjunction with water velocities of about 1.0 to 1.5 m/s.

9.4.3 Pipe sizing tables

Recall the following equations developed in Chapter 1:

$$h_f = \frac{4fLV^2}{2gD} \quad \text{(the Darcy equation)}$$

$$f = \frac{16}{Re} \quad \text{(laminar flow)}$$

$$\frac{1}{\sqrt{f}} = -4\log\left(\frac{k_s}{3.7D} + \frac{1.255}{Re\sqrt{f}}\right) \quad \text{(turbulent flow)}$$

The first of these equations specifies the pressure head loss of a fluid flowing at a velocity V in a circular pipe of diameter D and of length L. The remaining equations allow determination of the appropriate friction factor f, depending on the nature of the flow. The Moody Chart is a graphical presentation of these equations (shown in Chapter 1) and is of quite general application. The design engineer tends to be making a pipe selection within reasonably specific temperature regimes, for example around 10 °C for cold water pipework and around 75 °C for low temperature hot water and heating pipework. It then becomes a matter of convenience to distil from these equations a pipe sizing table relating the pressure drop per metre length, the velocity of flow, the mass flow rate related to a specific diameter of pipe, this information being restricted to a particular pipe material with water flowing at a particular mean temperature.

Section C4 of the Guide provides pipe sizing tables for the following materials and at the following temperatures:

Heavy grade steel	75 °C
Medium grade steel	75 °C
Large size steel	75 °C
Galvanised steel	75 °C
Light gauge copper	75 °C
Thin wall copper	75 °C
Galvanised steel	10 °C
Light gauge copper	10 °C
Cast iron	10 °C

Table 9.3 is an abstract of the Guide table for the flow of water at 75 °C in black (medium grade) steel pipes, which conform to BS 1387:1967. V is the water velocity (m/s); Δp is the pressure drop (N/m^2 per metre, or N/m^3); $\dot{m}$ is the water flow rate (kg/s) and EL (as we shall see later) is the equivalent length allowance for fittings related to straight pipe.

Example 9.4 Consider the pipework section AB/A′B′ of Example 9.2 and assuming a reasonable water velocity in the pipework is about 1.5 m/s, size the section on the basis of the design flow rate and determine the pressure drop across the section.

The design flow rate = 0.45 kg/s

From Table 9.3 several options are available, e.g.

Table 9.3 Flow of Water at 75 °C in Black Steel Pipes

Δp (N/m²/m)	V (m/s)	10 mm		15 mm		20 mm		25 mm		32 mm		40 mm		50 mm		65 mm		V (m/s)	Δp (N/m²/m)
		$\dot{m}$	EL	$\dot{m}$	EL	$\dot{m}$	EL	$\dot{m}$	EL	$\dot{m}$	EL	$\dot{m}$	EL	$\dot{m}$	EL	$\dot{m}$	EL		
72.5		0.026	0.3	0.051	0.5	0.116	0.7	0.217	1.0	0.455	1.4	0.683	1.7	1.280	2.4	2.550	3.4		72.5
120.0	0.3	0.034	0.3	0.069	0.5	0.152	0.7	0.284	1.0	0.595	1.5	0.893	1.8	1.670	2.4	3.320	3.4		120.0
140.0		0.037	0.3	0.075	0.5	0.165	0.8	0.308	1.0	0.646	1.5	0.968	1.8	1.810	2.5	3.600	3.5	1.0	140.0
160.0		0.040	0.4	0.081	0.5	0.178	0.8	0.331	1.0	0.693	1.5	1.040	1.8	1.940	2.5	3.860	3.5		160.0
180.0		0.042	0.4	0.086	0.5	0.189	0.8	0.353	1.0	0.738	1.5	1.110	1.8	2.060	2.5	4.100	3.5		180.0
200.0		0.045	0.4	0.091	0.5	0.200	0.8	0.373	1.1	0.780	1.5	1.170	1.9	2.180	2.5	4.330	3.5		200.0
220.0		0.047	0.4	0.096	0.5	0.211	0.8	0.392	1.1	0.820	1.5	1.280	1.9	2.290	2.5	4.550	3.5		220.0
240.0		0.050	0.4	0.100	0.5	0.221	0.8	0.411	1.1	0.858	1.5	1.290	1.9	2.400	2.5	4.760	3.5		240.0
260.0		0.052	0.4	0.105	0.5	0.230	0.8	0.428	1.1	0.895	1.5	1.340	1.9	2.500	2.5	4.960	3.6		260.0
280.0		0.054	0.4	0.109	0.5	0.239	0.8	0.445	1.1	0.931	1.5	1.390	1.9	2.600	2.6	5.160	3.6		280.0
300.0		0.056	0.4	0.113	0.5	0.248	0.8	0.462	1.1	0.965	1.5	1.440	1.9	2.690	2.6	5.340	3.6	1.5	300.0
320.0	0.5	0.058	0.4	0.117	0.5	0.257	0.8	0.478	1.1	0.998	1.6	1.490	1.9	2.780	2.6	5.520	3.6		320.0
340.0		0.060	0.4	0.121	0.5	0.265	0.8	0.493	1.1	1.030	1.6	1.540	1.9	2.870	2.6	5.700	3.6		340.0
360.0		0.062	0.4	0.125	0.5	0.273	0.8	0.508	1.1	1.060	1.6	1.590	1.9	2.960	2.6	5.870	3.6		360.0
380.0		0.064	0.4	0.128	0.5	0.281	0.8	0.523	1.1	1.090	1.6	1.630	1.9	3.040	2.6	6.060	3.6		380.0
400.0		0.065	0.4	0.132	0.5	0.289	0.8	0.537	1.1	1.120	1.6	1.680	1.9	3.120	2.6	6.200	3.6		400.0
420.0		0.067	0.4	0.135	0.5	0.297	0.8	0.551	1.1	1.150	1.6	1.720	1.9	3.200	2.6	6.360	3.6		420.0
440.0		0.069	0.4	0.139	0.5	0.304	0.8	0.564	1.1	1.180	1.6	1.760	1.9	3.280	2.6	6.510	3.6		440.0
460.0		0.070	0.4	0.142	0.5	0.311	0.8	0.578	1.1	1.210	1.6	1.800	1.9	3.360	2.6	6.660	3.6		460.0
480.0		0.072	0.4	0.145	0.5	0.318	0.8	0.591	1.1	1.230	1.6	1.840	1.9	3.430	2.6	6.810	3.6		480.0
580.0		0.080	0.4	0.160	0.6	0.351	0.8	0.652	1.1	1.360	1.6	2.030	1.9	3.780	2.6	7.500	3.6		580.0
940.0		0.103	0.4	0.207	0.6	0.452	0.8	0.838	1.1	1.750	1.6	2.610	2.0	4.850	2.6	9.600	3.7		940.0
960.0		0.104	0.4	0.209	0.6	0.457	0.8	0.847	1.1	1.760	1.6	2.640	2.0	4.900	2.6	9.710	3.7		960.0

20 mm dia. at close to 1.5 m/s with a pressure drop of 940 N/m^3;
25 mm dia. at about 0.75 m/s with a pressure drop of 280 N/m^3;
32 mm dia. at about 0.50 m/s with a pressure drop of 72.5 N/m^3.

Here we enter an area of judgement. The disadvantage of selecting a 20 mm dia. pipe is the resulting very high pressure drop. The designer must bear in mind however that this pressure drop is related to a relatively short run of pipework. On the other hand the disadvantage of selecting either of the remaining options is that low water velocities arise from such a choice. On balance the 20 mm diameter pipe may in this case be the best choice. Thus the pressure drop across the section which is 7 m long will be $940 \times 7 = 6580$ N/m^2.

9.4.4 Equivalent length and velocity pressure (K) factors

The usual method of dealing with fittings is to express the pressure drop across the fittings as the product of a 'K' factor (already dealt with in Chapter 1) and the velocity pressure. Thus across the fitting:

$$p_l = K\rho \frac{V^2}{2} \tag{9.3}$$

Example 9.5 Given that the K value for a gate valve is 0.2, determine the pressure drop across the valve for a water velocity in the pipework of 1.5 m/s when the flow is (a) cold water at 10 °C, (b) hot water at 80 °C.

At 10 °C the water density is 998 kg/m^3 (Table 9.1)

Thus $p_l = 0.2 \times 998 \times (1.5)^2 = 224$ N/m^2

A similar calculation for a density of 972 kg/m^3 yields a pressure drop for hot water of 218 N/m^2. The following Table 9.4, based on the table in the Guide (C4) gives K factors for some common forms of resistance.

These of course are only typical values of K and the designer should refer to manufacturers' data regarding K values or other means of ascertaining pressure drop through their products, be they small items such as bends, tees or valves or larger items such as emitters, calorifiers or boilers.

It has been mentioned earlier that the resistances of other items of the system can be reduced to the resistance of equivalent lengths of straight pipe. In order to understand how this is done, consider a section of straight pipe of length L between the cross-sections 1 and 2 in Figure 9.12.

Given that the total pressure drop across the section is ΔP_{12} and that the pressure drop per metre length is Δp, then:

$$p_1 - p_2 = \Delta P_{12} = \Delta p \times L$$

Suppose now that a fitting such as a valve is installed in the

Table 9.4 Velocity Pressure (K) Factors

Tees or junctions (based on velocity pressure of combined flow)
(a) To or from a 90° branch: 0.5 + bend factor + reduction or enlargement factor
(b) To or from a run, that is, a straight-through tee: 0.2 + reduction or enlargement factor

Reductions and Enlargements (based on velocity pressure in the smaller pipe):

Reductions			Enlargements		
Diameter ratio	3/2	K 0.3	Diameter ratio	3/2	K 0.4
	2/1	0.4		2/1	0.7
	3/1	0.4		3/1	0.9
	4/1	0.5		4/1	1.0

Bends			Valves	
Malleable cast iron or screwed mild steel:				
Diameter	10–25 mm	K 0.7	Gate valve	K 0.2
	32–50 mm	0.5	Angle valve	5.0
	65–90 mm	0.4	Tap or stopcock	10.0
Elbows			**Components**	
Copper pipe				
Diameter	10–25 mm	K 1.0	Radiator	K 5.0
	32–50 mm	0.8	Entry to vessel	1.0
	65–90 mm	0.5	Exit from vessel	0.4

Figure 9.12

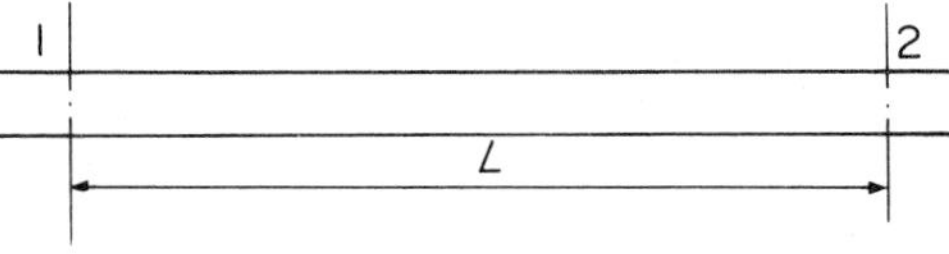

Figure 9.13

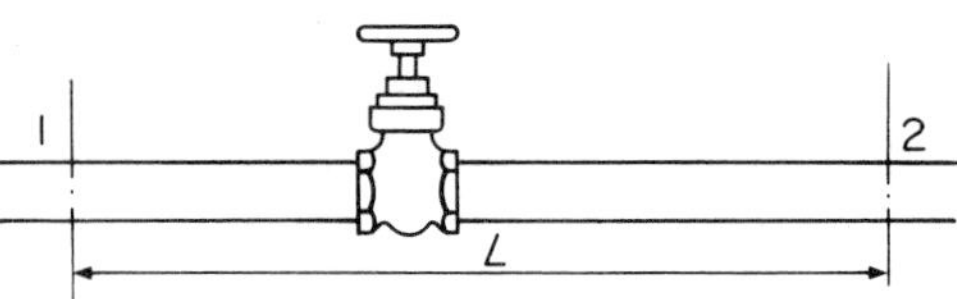

section of pipework (Figure 9.13) the valve having a loss factor of K.

The pressure drop now occurs partly across the pipework and partly across the valve and may be expressed as:

$$\Delta P_{12} = \Delta p \times L + K\rho \frac{V^2}{2}$$

The expression may now be manipulated as follows:

$$\Delta P_{12} = \Delta p \left[L + \frac{K\rho \frac{V^2}{2}}{\Delta p} \right]$$

Now let EL replace $\dfrac{\rho \frac{V^2}{2}}{\Delta p}$, then

$$\Delta P_{12} = \Delta p \ [L + \text{EL} \times K]$$

Thus the product of EL and K may be regarded as the *equivalent length* of straight pipe offering the same resistance as the valve. In the case of a run of pipework containing several fittings, the K factors are summed and then:

$$\Delta P_{12} = \Delta p \ [L + \text{EL}\Sigma K] \tag{9.4}$$

We have already noted that values of EL are contained within the pipe sizing tables.

Example 9.6 Check the value of EL in Table 9.3 for water flowing at a rate of 3.2 kg/s with a velocity of 1.5 m/s and a pressure drop of 420 N/m³ in a 50 mm diameter pipe. The density of water at 75 °C is 974.9 kg/m³.

$$\text{EL} = \frac{\rho \dfrac{V^2}{2}}{\Delta p} = \frac{974.9 \times 1.5 \times 1.5}{2 \times 420} = 2.6 \text{ m}$$

A useful skill which the designer develops is the ability to add a percentage to the physical length of a circuit to allow for the resistance of fittings and components (referred to in Chapter 1 as minor losses). This 'feel' is developed by doing calculations such as the one which follows.

Example 9.7 Determine with the aid of Table 9.3 the percentage addition to the length of the circuit to radiator X in Figure 9.14 to allow for the resistance of fittings and components. Take the K factor for the boiler as 4.0. Circuit length is 56 m.

Figure 9.14

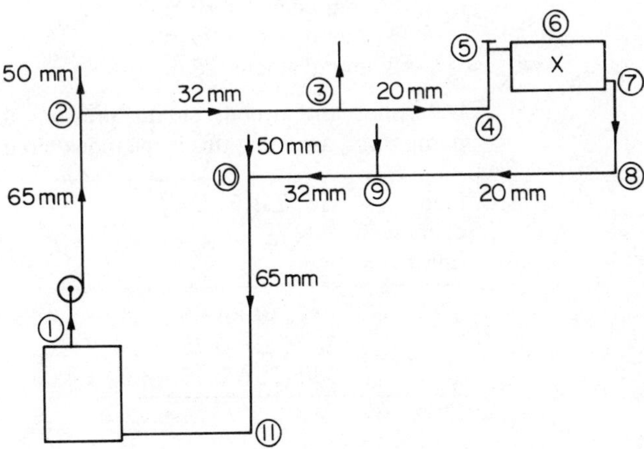

Note: The flows in the various pipes are as follows:

65 mm diameter pipe, 2.59 kg/s
50 mm diameter pipe, 1.81 kg/s
32 mm diameter pipe, 0.78 kg/s
20 mm diameter pipe, 0.35 kg/s

From Table 9.3 the relevant EL values are 3.4 m for 65 mm diameter; 2.5 m for 50 mm diameter; 1.5 m for 32 mm diameter and 0.8 m for 20 mm diameter pipe. Dealing with each position in turn and using Table 9.4:

Position	Calculation of K	EL x K
1	K 4.0 for boiler/EL 3.4 m	13.60
2	0.5 + 0.4 + 0.4/EL 3.4	4.42
3	0.2 + 0.3/EL 1.5	0.75
4	0.7 for bend/EL 0.8	0.56
5	5.0 for angle valve/EL 0.8	4.00
6	5.0 for radiator/EL 0.8	4.00
7	5.0 for lockshield valve/EL 0.8	4.00
8	0.7 for bend/EL 0.8	0.56
9	0.2 + 0.4/EL 1.5	0.90
10	0.5 + 0.4 + 0.7/EL 3.4	5.44
11	0.4 for bend/EL 3.4	1.36
	Total	39.59 m

Thus the addition to the actual length is $\frac{39.59 \times 100}{56} = 71\%$

For a layout such as this the term 'minor loss' is hardly appropriate. With this information the designer is now in a position to determine the necessary pump duty. The following extension of the previous example will illustrate the point.

Example 9.8 The actual lengths of pipework for the circuit in the previous example are as follows:

65 mm diameter 10 m
32 mm diameter 20 m
20 mm diameter 26 m

Determine the overall circuit pressure drop and the required pump duty, assuming this is the index circuit.

Pipe diameter (mm)	$\Delta p\ [L + \text{EL}\Sigma K]$	ΔP
65	75 (10 + 13.6 + 4.42 + 5.44 + 1.36)	2610
32	200 (20 + 0.75 + 0.9)	4340
20	580 (26 + 0.56 + 4.00 + 4.00 + 4.00 + 0.56)	22600
	Total	29550 N/m^2

A pump able to develop a pressure difference of 29550 N/m^2 while delivering water at a flow rate of 2.59 kg/s is needed for this situation. Some comments on selection of pumps will be made later in this chapter.

9.4.5 Allowances for fittings and mains emission

We have discussed already the difficulty which arises in relation to heat losses from mains at a stage where the diameter of the main is unknown. A similar problem could arise where a heating designer is adding an extension to a system. In this case he will know the total pressure which is available but will have to make a percentage addition to the length of new pipework in his calculations in order to arrive at a pressure drop ($N/m^2/m$) for the performance of preliminary pipe sizing. On selection of pipe sizes he then has a firmer base for determining the effective length of the pipe, from which he can then proceed to a second round of pipe sizing. The procedure could be repeated a third or a fourth time, refining the solution, but the designer develops a 'feel' for such matters and usually only a second sizing is necessary. The percentage added is usually in the region of 10% to 70% depending on the complexities of the piping system.

A similar technique is applied to account for heat losses in the main when making a preliminary estimate of the design flow rate in the pipework. In this case the percentage allowance is usually in the range 10% to 30% of the loss from the emitters, depending on how compact or extensive the runs of pipework are and whether they are bare or insulated. One further area where the designer is 'whistling in the dark' is in a choice of pressure drop. This is normally taken to be somewhere in the region of 150 to 300 $N/m^2/m$ which hopefully keeps the velocities in the range around 1 m/s or a little higher.

9.4.6 Preliminary pipe sizing

Several points have been made in relation to preliminary pipe sizing and the steps of the design process are perhaps best understood by applying them to a typical situation.

Example 9.9 Consider once again the system of Example 9.2. Suppose we start from the situation where the design flow and return temperatures have been chosen and the emitters have been

Figure 9.15

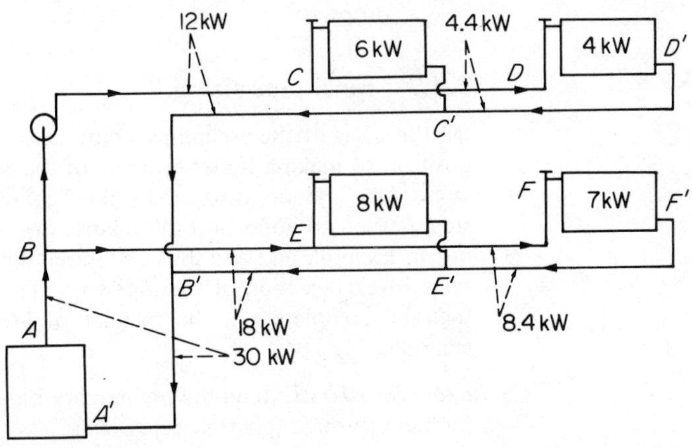

sized. The next stage in the design process is to perform a preliminary pipe sizing.

Step 1 Redraw the sketch of Example 9.2 (Figure 9.15) and note the required emission rates on the emitters.

Step 2 Decide an allowance for pipework emission. Since pipework is bare but runs are not particularly long, try 20%.

Step 3 Note on each pair (flow and return) of pipes for each section the total emission (pipes plus emitters) carried by the pair.

Step 4 Determine preliminary design flow rates on the basis of the emissions carried by the pipes and the design temperature drop.

For pipe CD/C′D′ $$\dot{m} = \frac{\dot{Q}}{C_p\,(\theta_f - \theta_r)} = \frac{4.4}{4.2\,(80 - 65)}$$

$$= 0.07 \text{ kg/s}$$

Similarly for AB/A′B′ flow rate is 0.475 kg/s
BC/B′C′ 0.190 kg/s
BE/B′E′ 0.285 kg/s
EF/E′F′ 0.134 kg/s

Step 5 Complete the preliminary pipe sizing on the basis of a pressure drop of say around 200 N/m^3 using Table 9.3.

Pipe AB/A′B′ 25 mm diameter pressure drop 320 N/m^3 with velocity of about 0.8 m/s
BC/B′C′ 20 mm diameter pressure drop 180 N/m^3 with velocity of about 0.5 m/s
CD/C′D′ 10 mm diameter pressure drop 460 N/m^3 with velocity of about 0.6 m/s
BE/B′E′ 20 mm diameter pressure drop 380 N/m^3 with velocity of about 0.75 m/s
EF/E′F′ 15 mm diameter pressure drop 420 N/m^3 with velocity of about 0.7 m/s

Notice how difficult it is in this situation to get sufficiently high velocities without resorting to excessive pressure drops.

9.4.7 Final pipe sizing

On the basis of the preliminary pipe sizes the designer is now in a position to make a better estimate of the emission from the pipework. This is done using tables like Tables 8.1 and 8.2. The next step is then to apportion the mains emission in the manner laid out in Example 9.2 and thus determine the necessary design flow rates in each section of the pipework. The final pipe sizing may then be completed in the manner of Step 5 of the preceding example.

Example 9.10 Extend the preliminary pipe sizing of the previous problem through to a final pipe sizing.

Step 6 Determine the emission from the pipework sections.
At this stage we must assume that the flow temperature to all the emitters must be 80 °C and the return temperature from them must be 65 °C (although Example 9.3 shows this to be not quite true). Thus using Table 8.1:

AB/A′B′ has an emission of 102 W/m for a 60 °C drop.

From the lengths given in Example 9.2 we find:

Emission from AB/A′B′ = 102 x 7 = 714 W

The emissions for the remaining pipes are found in a similar manner, the one difficulty being that an emission for a 10 mm diameter pipe can only be obtained by extrapolation. Use 57 W/m for a 10 mm diameter pipe. Then:

Emission from BC/B′C′ = 84 x 14 = 1170 W
Emission from CD/C′D′ = 57 x 10 = 570 W
Emission from BE/B′E′ = 84 x 10 = 840 W
Emission from EF/E′F′ = 69 x 8 = 552 W

Step 7 Apportioning the mains emission
Carry out the procedure detailed in Example 9.2 (Step 3)

Source of load	Emitter Load (kW)			
	P	Q	R	S
Emitter Rated Emissions	6.000	4.000	8.000	7.000
CD/C′D′	–	0.570	–	–
BC/B′C′	0.700	0.470	–	–
EF/E′F′	–	–	–	0.552
BE/B′E′	–	–	0.450	0.390
AB/A′B′	0.172	0.114	0.228	0.200
	6.872	5.154	8.678	8.142

Step 8 Recalculation of design flow rates
This is carried out in a similar manner to Example 9.2 (Step 4) and gives the following results:

Section	Design Mass Flow Rate (kg/s)	Temperature Drop Available to Next Section (°C)	
AB/A′B′	0.460		14.62
BC/B′C′	0.196		13.20
CD/C′D′	0.093	11.74	across emitter Q
BE/B′E′	0.275		13.90
EF/E′F′	0.140	12.96	across emitter S

Step 9 Final pipe sizing
In exactly the same manner as with the preliminary sizing:

Section	Diameter (mm)	Pressure Drop (N/m^3)	Velocity (m/s)
AB/A′B′	25	300	about 0.75
BC/B′C′	20	190	just over 0.5

CD/C′D′	15	210	just under 0.5
BE/B′E′	20	360	about 0.75
EF/E′F′	15	440	about 0.7

As a result of the final pipe sizing there has been a minor adjustment to the size of pipework but arising from this stage of the calculations is a better estimate of the pressure drop across the system.

Earlier a special case was mentioned where an extension is to be made to an existing system. In this situation the designer usually has a fixed pressure difference available (from an existing pump) and is constrained so far as the pressure drop parameter N/m^3 is concerned. A problem will illustrate a possible approach to this situation.

Example 9.11 An existing LTHW system is to be extended to include an additional three radiators, the extension being as indicated on Figure 9.16. Carry out a preliminary pipe sizing of the extension pipework given that the emissions of A, B and C are 11 kW, 13 kW and 17 kW respectively. The pressure difference available from the existing pump at PP′ is 15 kN/m^2. All pipework is uninsulated. The length of the pipe runs are 1, 8 m; 2, 9 m; 3, 14 m; 4, 7 m; 5, 9 m.

Figure 9.16

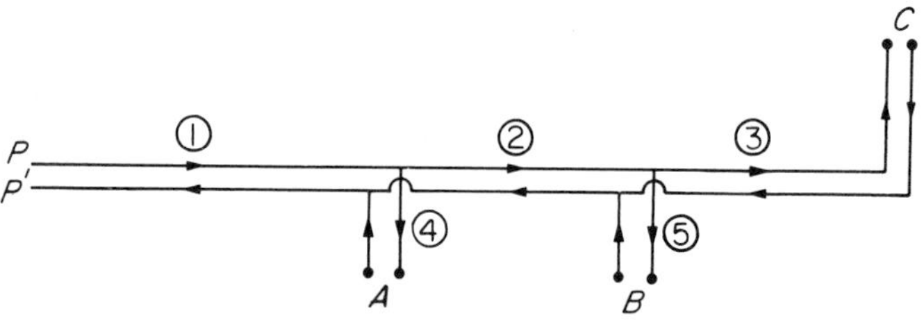

Making an allowance of say 20% for mains emission, the following flow rates are obtained in the same manner as in Example 9.9:

Pipe section	Flow rate (kg/s)
1	0.742
2	0.545
3	0.310
4	0.198
5	0.236

From PP′ there are three circuits and of these the run to emitter C and back is almost certainly the index circuit. The pipework runs look reasonably long and straight, so assume a 20% allowance to the actual length for the resistance of fittings, etc.

$$\text{Thus available pressure drop} = \frac{15 \times 1000}{31 \times 1.2} = 404 \text{ N/m}^3$$

Now use Table 9.3 to size the system, giving:

Pipe section	Diameter (mm)	Pressure drop (N/m^3)	Velocity—approx. (m/s)
1	32	180	0.75
2	25	410	1.00
3	20	460	0.80

Check the pressure drop.

$$\text{Pressure drop} = (8 \times 1.2 \times 180) + (9 \times 1.2 \times 410) + (14 \times 1.2 \times 460) = 13\,860\ N/m^3$$

Accept these values bearing in mind that the system is slightly oversized.

Now examine the circuit to emitter A. If a 20 mm diameter pipe were chosen for section 4 this would give a pressure drop of 200 N/m^3 at a velocity of about 0.5 m/s. The total pressure drop for this circuit would then be $(8 \times 1.2 \times 180) + (7 \times 1.2 \times 200) = 3400\ N/m^3$. This is much too low. A possibility is to try a smaller pipe size. For a 15 mm diameter pipe in section 4 the total pressure drop for the circuit would be 8920 N/m^3. This is still short of the value of 15 000 N/m^3 available and the usual way out of this difficulty is to fit a balancing valve in section 4 to take up the additional pressure. It is left to the reader to examine the possibilities of using a 20 mm or 15 mm diameter pipe for section 5.

9.4.8 Last word on *K* factors

Whilst Table 9.4 and the corresponding (more detailed) table of the Guide are very helpful when estimating the effective length of a system, nevertheless, there are certain difficulties which face the designer. One such problem is where on large systems the flow water from the boilers enters a manifold before passing on to various sub-circuits (Figure 9.17).

How does one deal with the resistance of the manifold? One possibility is to use the *K* factors (Table 9.4) for entry into and exit from a large vessel applied to the entry and exit pipes from

Figure 9.17

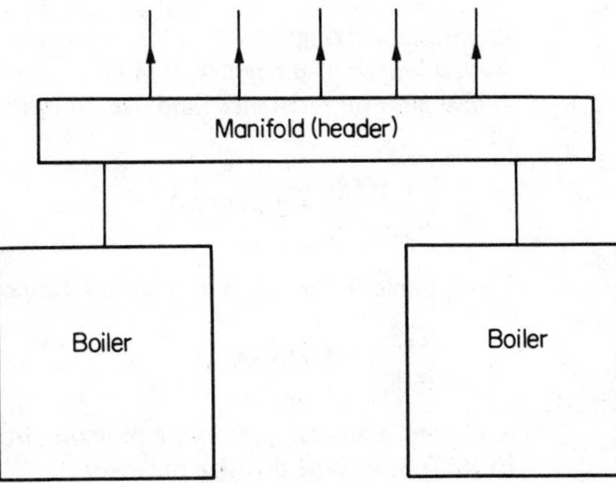

Table 9.5 Pressure Drop Across a Radiant Heating Panel

Panel size	*Panel type*		*Water flow rate and pressure drop*				
		l/h	91	136	182	227	273
723/4	B, B_1, A	mm W.G. Pressure drop	76	152	250	366	488
	AX	mm W.G.	31	55	85	119	153

the manifold. The alternative strategy is to seek guidance from the boiler manufacturer's literature. While on the subject of manufacturers' literature it is best when sizing the system to seek guidance from the manufacturer's catalogue with respect to pressure drops across components such as boilers, emitters, valves, etc. Normally this information is given as a pressure drop related to a possible range of flows through the component. Table 9.5 is reasonably typical of the way in which such information is presented and the following example illustrates how it may be used.

Example 9.12 A factory is heated by a radiant panel system and Figure 9.18 shows the proposed run of pipework from a point where a pressure difference of 9 kN/m^2 is available (across flow and return connections) to an emitter of type B as shown in Table 9.5. Assuming a flow rate of 273 l/hour is required, size the pipework for this run, for a flow and return length of 24 m.

Figure 9.18

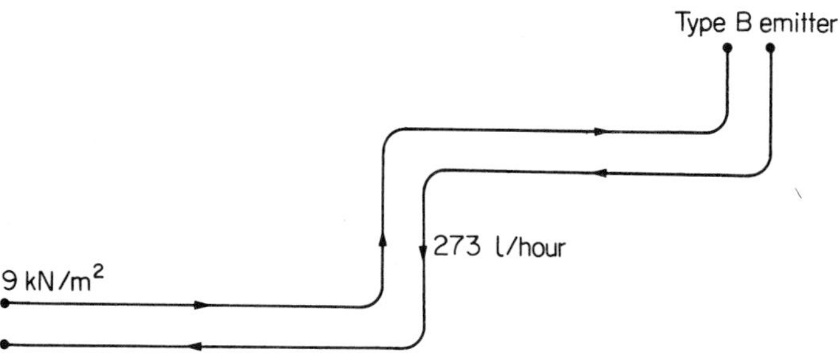

Preliminary sizing:
Actual length of pipework = 24 m
Allow 50% for resistances and size on basis of an available pressure of

$$\frac{9 \times 1000}{24 \times 1.5} = 250 \text{ N/m}^3$$

From Table 9.3 in relation to a flow rate of

$$\frac{273}{3600} = 0.076 \text{ kg/s},$$

a 15 mm diameter pipe has a pressure drop of 140 N/m^3 with an EL of 0.5. Accept this size meanwhile.

Final sizing:
Now determine the pressure drop based on a 15 mm diameter pipe.
Pressure drop across pipework, $\Delta P = 140 \times 24 = 3350$ N/m^2.
There are six bends each with a K factor of 0.7
Thus pressure drop across local resistances = $\Delta p \text{EL}\Sigma K$

$$= 140 \times 6 \times 0.7 \times 0.5$$
$$= 295 \text{ N/m}^2$$

The pressure drop across the panel for a flow rate of 273 l/hour is 488 mm W.G. Now 1 mm W.G. is approximately equal to 10 N/m^2 and so the pressure drop across the panel is 4880 N/m^2.

Hence the total pressure drop is 3350 + 295 + 4880 = 8525 N/m^2.

This is slightly less than the available pressure drop and is thus a satisfactory situation. Notice however that our incorrect allowance for resistance (which is nearer 150%) is balanced by the lower pressure drop of 140 N/m^3 for the 15 mm diameter pipe in relation to the estimated drop of 250 N/m^3.
Thus a 15 mm diameter flow and return pipe to the emitter is acceptable.

9.5 OTHER COMPONENTS OF THE SYSTEM

9.5.1 The feed and expansion tank

This tank is used for filling the system via a ballvalve and also for supplying any makeup water which may be necessary (e.g. due to leakage at a pipe joint). In addition it has to be sized to accommodate the expansion of the water in the system as the system warms up on firing the boiler. In addition it places the system under some pressure and often acts as a disposal area for the contents of the vent pipe. A further important consideration is that the point of connection of the feed and expansion tank pipework into the closed heating system is a point of known pressure (due to the gravity head imposed).

To reinforce this point consider a typical domestic circuit as shown in Figure 9.19, with the feed and expansion tank connected adjacent to the pump on the return.

Typically the dimensions of the feed tank at the free water surface would be about 0.6 m by 0.4 m and the diameter of the vent from the boiler about 19 mm. With the system cold the water level in the tank would be the same as that in the vent at level AA. If the reader refers back to the calculation in Example 9.8, relating to Figure 9.14 in Example 9.7, it will be seen that a pump pressure of about 30 kN/m^2 will be fairly typical.

If the pump is now energised to circulate cold water round the system then the difference in 'heads' in the left and right hand limbs will be equal to the pump pressure less the pressure drop through the boiler. Assume a drop through the boiler of about 1 kN/m^2. Hence the residual pressure of 29 kN/m^2 causes the water movement of $h_1 + h_2$. Assuming a nominal water density of 1000 kg/m^3 then:

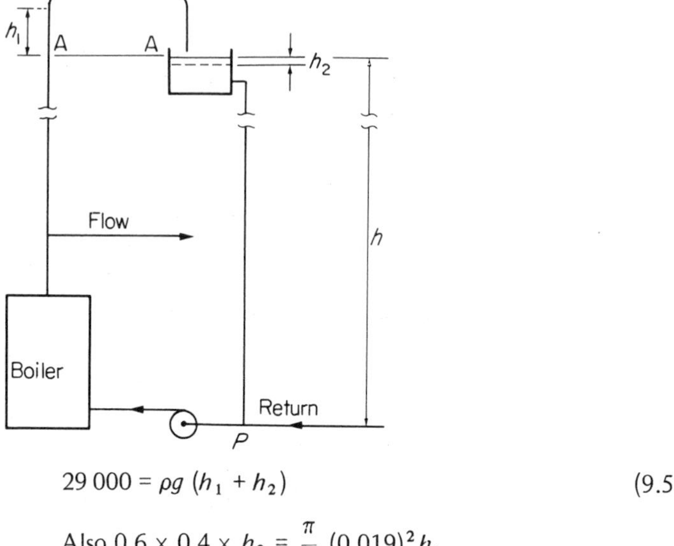

Figure 9.19
Connection of feed tank to system

$$29\,000 = \rho g\,(h_1 + h_2) \tag{9.5}$$

$$\text{Also } 0.6 \times 0.4 \times h_2 = \frac{\pi}{4}\,(0.019)^2 h_1$$

$$\text{From which } h_1 = 845 h_2$$

and substituting for h_1 in equation 9.5 then

$$29\,000 = 1000 \times 9.81\,(845 h_2 + h_2)$$

from which h_2 = 3.5 mm and thus h_1 = 2.96 m

Thus it is reasonable to assume that the pressure at the point P is constant and equal to the gravity head h whether the pump is running or not.

When the system is warmed up the water column in the vent pipe will increase since the density of this column is less than that of the column from the feed tank to the connection P. This means that with the pump connected into the circuit in the position shown, the water level h_1 will then be well in excess of 2.96 m. A basic objection to the arrangement in Figure 9.19 is that there is a great danger of simply pumping water from the vent pipe back into the feed and expansion tank. On the other hand there are certain advantages in relation to the distribution of pressure throughout the system. Consider the system layout of Figure 9.20 and plot the variations of hydrostatic, circulating and total pressure through the system as shown in Figure 9.21.

It can be seen from the plot of resultant pressure that the system is under a pressure well above atmospheric pressure at all points. This aids anti-cavitation by ensuring that in all parts of the system the actual pressure is well in excess of the prevailing saturation pressure related to the temperature at that part of the system. A further advantage of this arrangement is that such pressures in the system facilitate venting (such as occurs at emitter bleed valves).

Consider the changes which arise in pressure distribution due

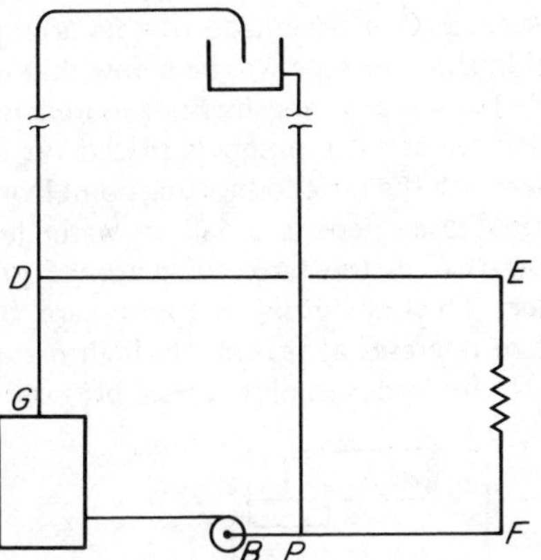

Figure 9.20 Layout of a LTHW system

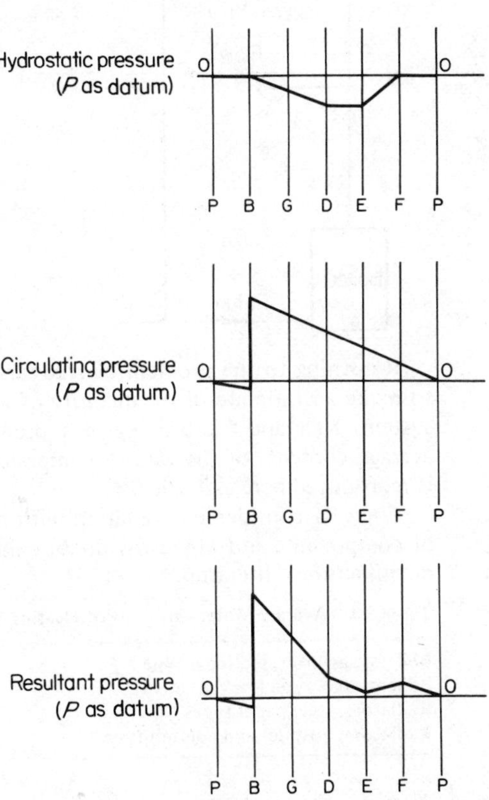

Figure 9.21

to an adjustment to the system which places the feed connection point P between the pump and the boiler. In this case the pump discharge pressure equals the head due to the feed tank and the pump suction pressure will be considerably less than this. The pressures prevailing in the high level mains may even be sub-atmospheric (with the possibility of drawing air into the system). Actual working pressures will be closer to saturation pressures especially in high level mains with an increased possibility of cavitation. One advantage of this arrangement is that the water level in the vent pipe will be below that in the feed tank.

Perhaps the arrangement favoured most by heating designers is that where the pump is placed on the flow adjacent to the boiler with the tank connecting point P on the return (Figure 9.22). In this case there is a fall in water level (h) in the vent pipe below that in the tank equal to the pressure drop through the boiler. This obviously requires care in siting the pump. The system operates at reasonably high pressures and once again it is left to the reader to plot typical pressure variations.

Figure 9.22

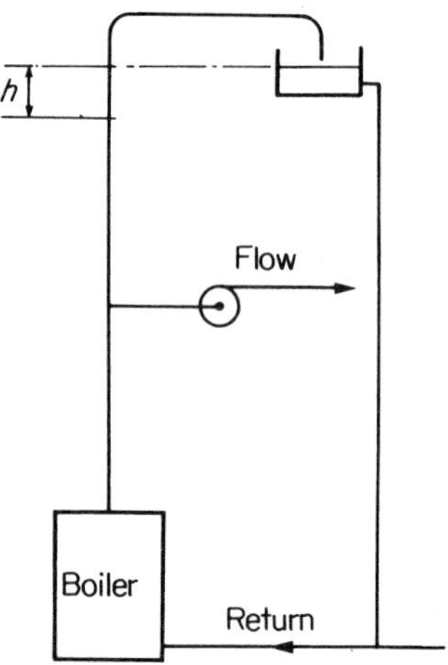

Returning to the problem of feed tank sizing, the designer has to make an estimate of the quantity of water stored in the closed system. Kell and Martin[(1)] give a useful little table of typical average contents of the various components of the system and it is reproduced here as Table 9.6.

There is considerable variation with respect to water content of components and where any doubt exists it is as well to refer to manufacturers' literature.

Table 9.6 Average Water Contents of Heating System Components

Boilers, cast iron, sectional type	1.5– 2.0 l/kW
Boilers, steel type (vary greatly)	0.5– 3.0 l/kW
Radiators, steel panel types	5.0 l/kW
Radiators, hospital and column type	10.0–15.0 l/kW

Table 9.7 Water Contents of Heating System Pipework

Steel	*Diameter* (mm)											
	10	15	20	25	32	40	50	65	80	90	100	125
Medium	0.121	0.205	0.367	0.586	1.016	1.376	2.205	3.700	5.115	–	8.680	13.250
Heavy	0.098	0.175	0.326	0.518	0.926	1.271	2.070	3.530	4.905	–	8.380	13.050

	Diameter (mm)									
Copper	13	19	25	32	38	51	63	76	102	127
(Light)	0.095	0.321	0.540	0.837	1.234	2.095	3.245	4.211	8.68	13.35

A considerable volume of water is stored in the pipework and Table 9.7 will be useful in this context.

These two tables used in conjunction with Table 9.1, which gives water densities at various temperatures, will allow the designer to make an educated guess of the likely expansion of the water within the system.

Example 9.13 A domestic central heating system comprises the parts indicated. Determine the expansion water to be accommodated over a range 10 to 75 °C.

Cast iron sectional boiler, rated output 15 kW
Pressed steel radiators, total output 13 kW
36 metres of 19 mm diameter light gauge copper pipe
34 metres of 13 mm diameter light gauge copper pipe

$$\begin{aligned}\text{Total water content} &= (1.75 \times 15) + (1.75 \times 13) \\ &\quad + (36 \times 0.321) + (34 \times 0.095) \\ &= 63 \text{ litres when cold } (10\,^{\circ}\text{C})\end{aligned}$$

Thus on filling, the system holds 63 kg of water.

An interesting aside here is that 40% of the water is in the boiler, 36% in the radiators and 24% in the pipework. At 75 °C, the volume of 63 kg of water = 63/975 = 0.0645 m^3 or 64.5 litres.

Thus the expansion tank must accommodate 1.5 litres of water from the system on heating up.

The reader is referred to the Guide (Book B) for appropriate data on feed tank sizing. For example, in relation to a system up to 20 kW load a nominal system capacity of 40 litres would be suitable with a 19 mm diameter cold feed pipe to the boiler. The cold supply to the tank is likely to be 13 mm diameter (depending on mains pressure) and a 25 mm diameter overflow is required in addition to a 25 mm diameter open vent from the boiler carried back up to the feed tank.

A modern alternative to the use of the open feed and expansion tank is the introduction of a closed pressurising cylinder fitted with a flexible diaphragm as shown in Figure 9.23.

This system overcomes two basic disadvantages of the open feed and expansion tank which are the gradual absorption of oxygen into the system and neglect of remotely sited tanks. The

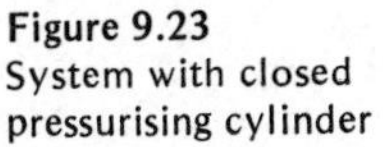

Figure 9.23
System with closed pressurising cylinder

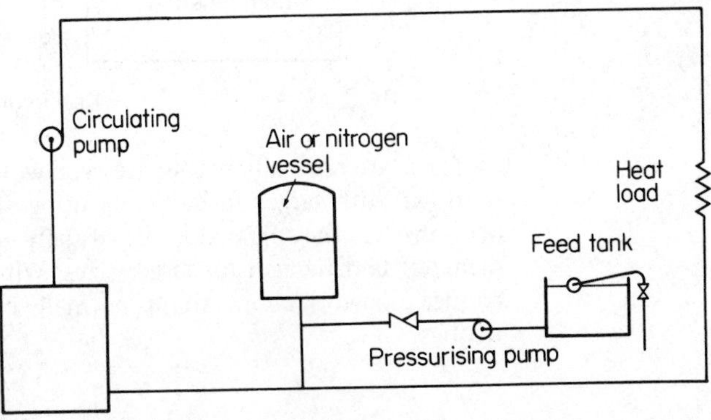

sizing of the pressurising cylinder is a function of the ability of the flexible diaphragm to accommodate the water of expansion and the designer is advised to consult the relevant manufacturers' literature. More will be said later about pressurising cylinders in relation to pressurised heating systems.

9.5.2 Valves

Valves fitted in heating systems have a variety of functions associated with the filling and emptying of the system, control of the flow while in operation, removal of air from the system and safety of operation and isolation of sections for maintenance. Sizing and characteristics of control valves has already been discussed in Chapter 3. A few words about the other forms of valve will be appropriate here.

Ball valve

This is a float valve used to maintain a constant level of water in the feed tank. The nominal seat bore of the valve will be 13 mm up to a boiler rating of 150 kW, 19 mm up to 450 kW and 25 mm up to 1500 kW. Initial filling of the system is through the ballvalve.

Draincock

This is a tapered plug which rotates in the valve body and is placed at appropriate positions (e.g. at the bottom of the boiler) for draining down the whole (or part) of a system.

Isolating valves

Components of the system such as pumps and boilers will be fitted with isolating valves to facilitate removal, maintenance or repair (Figure 9.24).

Figure 9.24

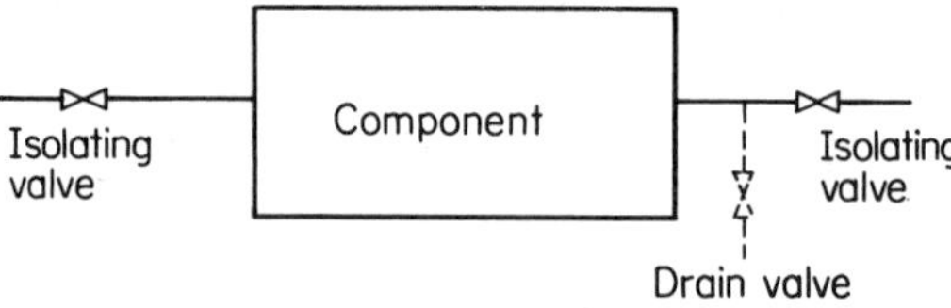

These are normally of the gate type which have a low resistance to flow. With larger installations using steel pipework, the connections to the pipework are usually screwed up to 80 mm diameter and flanged for larger sizes. With domestic systems and copper pipework connection is normally by means of compression fittings.

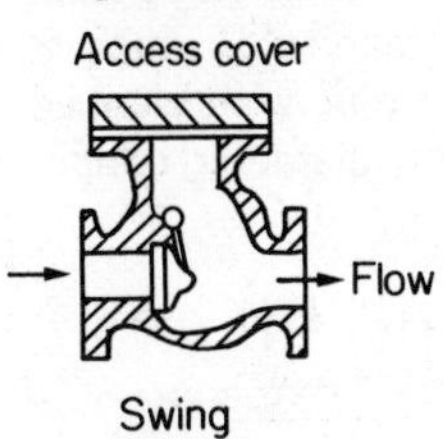

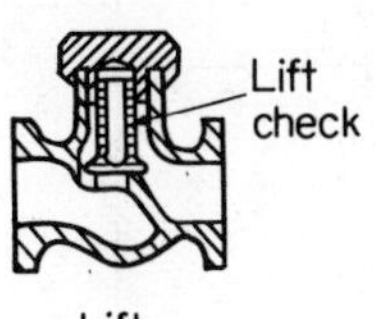

Figure 9.25
Swing and lift valves

Check or non-return valves

These are designed to allow flow in one direction only and are usually of either swing or lift type as shown in Figure 9.25.

Safety valves

These must be incorporated on all pressure vessels and boilers. Normally the valve is set to open at a certain allowable pressure within the vessel and that pressure will then open the valve against the compression effect of the valve spring. The Guide, (Book B) provides information on the sizing of safety valves. For instance, for boilers up to 260 kW a 19 mm diameter valve is required where solid fuel firing is used and 25 mm diameter where oil firing is used.

9.6 DOMESTIC FORMS OF LOW PRESSURE (TEMPERATURE) HOT WATER HEATING SYSTEMS

9.6.1 Small bore system

This consists generally of a back boiler or an independent boiler providing the energy input to a hot water cylinder for domestic hot water and also supplying sufficient energy to feed a single pipe radiator system as shown in Figure 9.26. Control of emission

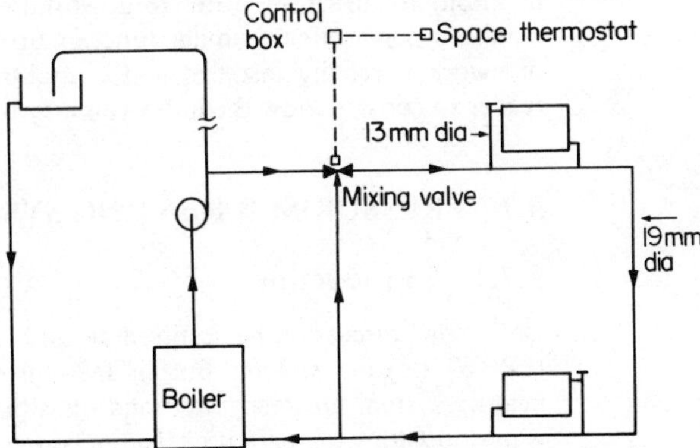

Figure 9.26
The small bore heating system

is often by on-off control of the pump through a time clock and space thermostat or less commonly, as shown here, thermostatic regulation of a mixing valve.

Typical pipe sizes for such a system are indicated.

9.6.2 Microbore system

Although there is no great difficulty involved in the installation of small bore systems in existing properties, nevertheless the trend has been towards even smaller pipework and the most recent development has been the elegant manifold system known as microbore. This consists of conventional flow and return pipework between the boiler and a header or manifold which is normally a 19 mm or 25 mm diameter copper pipe with blanked ends and connections for 6.2, 9.5 or 13 mm diameter copper pipe. Figure 9.27 shows a typical layout.

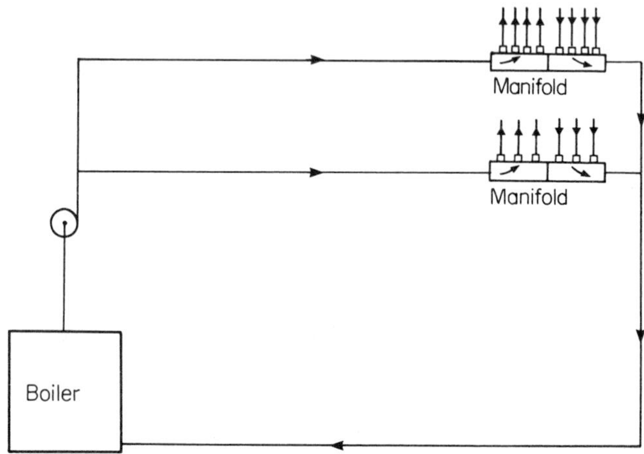

Figure 9.27
The microbore heating system

The manifold has two compartments, one accommodating flow and the other, return water. Each flow outlet from a manifold runs to an emitter and returns to the same manifold on the return side. In a conventional dwelling there would be a single manifold to distribute water to downstairs emitters and another manifold performing a similar function upstairs. With this system pipework is readily installed and is unobtrusive. It is left to the reader to consider how the index run may be identified.

9.7 PRESSURISED HEATING SYSTEMS

9.7.1 Introduction

Until now, discussion has centred around low pressure hot water (LPHW) heating systems but in fact three broad categories of heating system are recognised and classified with respect to the water operating temperatures (and pressures) as follows:

Low temperature hot water	LTHW (LPHW)	up to 100 °C
Medium temperature hot water	MTHW (MPHW)	100 to 120 °C
High temperature hot water	HTHW (HPHW)	over 120 °C

Various methods are available for pressurising the water systems but a whole new range of problems confronts the designer, including design of pipework and components to withstand higher pressures, means of accommodating expansion and even more stringent anti-cavitation measures than those mentioned earlier in Section 9.5.1. These matters are considered in the following sections.

9.7.2 Working range

It was shown previously that with LTHW the mean system temperature is around 75 °C with a temperature drop across the emitters of about 10 to 15 °C. With MTHW systems the corresponding figures are around 110 °C and a 20 to 25 °C drop and with HTHW systems around 150 °C with a 30 to 35 °C drop. The designer must resort to the use of property tables for water to discover the pressure implications of operating the system at such temperatures. A condensed form of such tables is presented here as Table 9.8.

The working pressure for a MTHW system is seen from Table 9.8 (due allowance being made for anti-cavitation in relation to the flow temperature) to be somewhere around 3 bar while the corresponding figure for a HTHW system is around 8 bar. This is summarised in Table 9.9.

Table 9.8 Properties of Water at Saturation

Temperature (°)	100	110	120	130	140	150	160	170	180
Saturation pressure (bar)	1.00	1.43	1.98	2.70	3.61	4.76	6.18	7.92	10.0
Density (kg/m³)	958.3	951.0	943.1	934.8	926.1	916.9	907.4	897.3	886.9

Table 9.9 Working Ranges of Hot Water Heating Systems

System	*Flow temperature* (°C)	*Return temperature* (°C)	*Temperature drop* (°C)	*Emitter mean temp* (°C)	*Pressure* (bar)
LTHW	about 80	60–70	10–15	about 75	1–2
MTHW	about 125	about 100	20–25	about 110	about 3
HTHW	about 165	about 135	30–35	about 150	5–8

The other means of achieving such high emitter temperatures is to use a steam system but this will be discussed later in the chapter.

9.7.3 Pipe sizing adjustments

Operation at higher water temperatures must take account of certain changes in properties which may influence pipe sizes. It was seen earlier that the equation used to establish the flow rate of water in the pipework to the emitters was

$$\dot{Q} = \dot{m}\, C_p\, (\theta_f - \theta_r)$$

but mass flow rate = density x volume flow rate ($\dot{m} = \rho\dot{G}$)

thus $$\dot{Q} = \rho\dot{G}C_p\, (\theta_f - \theta_r)$$

Now the specific heat capacity C_p increases with increase of temperature while the density ρ decreases and to a considerable extent the two effects are self correcting. Of greater significance is the possible increase in friction factor arising mainly from changes in viscosity with temperature. This was discussed in Chapter 1.

The normal method of dealing with such difficulties is to make corrections to the standard pipe sizing tables to allow their use for higher operating temperatures. We have already used a truncated version of the Guide pipe sizing tables for the flow of water at a mean temperature of 75 °C in steel pipes. The same source provides an additional table of correction factors to be applied to these tables for use when water is flowing at 150 °C. Table 9.10 is an abstract of the correction factor table from the Guide.

Table 9.10 Correction Factors for Pressure Loss and Equivalent Length as read from Table 9.3 and applied to Water Flowing at 150 °C

Pressure Loss as read from table (N/m^3)	*OP Correction Factor for pressure loss as read* *OL Correction Factor, Equivalent Length as read*								*V* (m/s)
	15 mm		25 mm		50 mm		100 mm		
	OP	OL	OP	OL	OP	OL	OP	OL	
2	0.81	1.3	0.84	1.3	0.87	1.2	0.90	1.2	
10	0.85	1.2	0.88	1.2	0.91	1.2	0.94	1.2	0.3
20	0.87	1.2	0.90	1.2	0.93	1.1	0.96	1.1	
50	0.90	1.2	0.92	1.2	0.95	1.1	0.98	1.1	
100	0.92	1.2	0.94	1.2	0.97	1.1	0.99	1.1	1.0
200	0.95	1.2	0.96	1.1	0.98	1.1	1.00	1.1	

Correction factors can be interpolated or extrapolated without serious errors for temperatures less than or greater than 150 °C. Adjustments should be carried out strictly in relation to *K* factors for fittings but such an exercise would be awkward and is not normally undertaken.

9.7.4 Methods of pressurisation

Various methods are available and these include head tank, vapour pressurisation (steam) and gas pressurisation (air or nitrogen). These methods shall be examined in turn.

Head tank

This was a method used in earlier systems in which the feed and expansion tank was used to pressurise the system by means of the imposed head of water. When one considers that 1 bar pressure is about 11 m head of water then the limitation of such methods can be seen.

Vapour pressurisation

A conventional steam boiler is used to generate steam which places the water in the drum under the appropriate pressure. Instead of circulating steam (as would normally be the case) the pressurised water is circulated round the heating circuit. Figure 9.28 shows a typical arrangement. A major consideration here is the possibility of the hot water within the pipework flashing into steam, since the water within the boiler is at the saturated (boil-

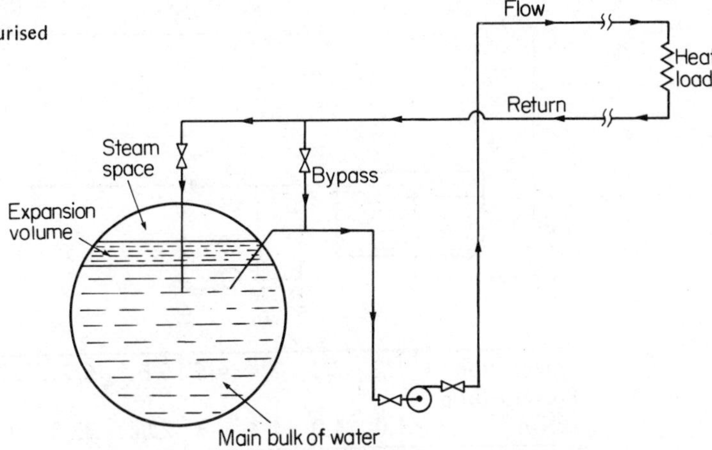

Figure 9.28
System pressurised by steam

ing) state and circulation within the pipework involves a drop in pressure. Injection of cooler return water from the bypass section of the pipework into the flow line from the boiler drops the water temperature sufficiently to remove this possibility. The necessary amount of mixing is determined on the basis of an anti-cavitation temperature margin of somewhere in the region of 10 to 15 °C below the boiling condition at the most vulnerable part of the system.

In making the connection from the pipework to the boiler the designer has to consider the possibility of draining water from the boiler through the system in the case of major leakage at some point within the system of piping. The normal solution here is to make the flow and return connections through dip pipes within the boiler which draw and return water below the water level but at a position in the upper portion of the drum as indicated in the sketch. With multiple boiler installations it is important to equalise the water levels and the pressures within the steam spaces. Balance pipes are used for this purpose and the boiler manufacturers should be consulted in relation to this problem.

The following example illustrates a method of locating the vulnerable point in the system and the determination of an appropriate flow temperature.

Example 9.14 The pressurised hot water system in Figure 9.29 has had the emitters sized on the basis of flow and return temperatures of 143 and 103 °C. The pipework has also been sized and the pressure drops across each section of pipework are as shown in the table with relation to the index circuit. It is proposed to generate pressure by means of steam and circulate the water from the boiler. Plot sketches of variations in hydrostatic and circulating pressures round the circuit and identify the point of lowest static pressure. On the basis of an anti-cavitation margin of 15 °C, establish a suitable working steam pressure and proportion of bypass water.

Figure 9.29

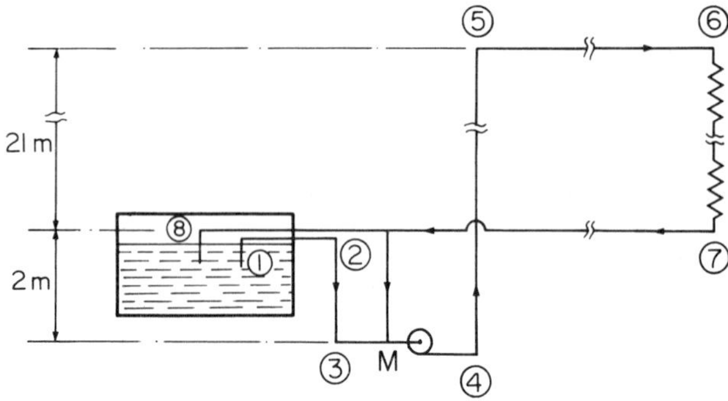

Pipe section	1/2	2/3	3/4	4/5	5/6	6/7	7/8	Total
Pressure drop (kN/m^2)	1.0	1.0	1.2	7.3	22.4	42.3	19.5	94.7

Taking the steam pressure in the drum as datum (and as the pressure at the points 1 and 8) the other circuit pressures may be related to this pressure.

Step 1 Establish densities:

Density in section 4/5 is 923 kg/m^3 (for 143 °C)
Density in section 6/7 is 941 kg/m^3 (for 123 °C)
Density in section 2/3 will have to be assumed. Suppose meanwhile a temperature of 160 °C in the boiler. Then density in section 2/3 will be 907.4 kg/m^3.

Step 2 Determine hydrostatic pressure variations:

$p_2 = p_1 = 0$ (datum)
$p_3 = p_2 + \rho gh = 0 + 907.4 \times 9.81 \times 2 = 17\,800\ N/m^2$
(assuming length of section 2/3 is 2 m)
$p_3 = p_M = p_4 = 17\,800\ kN/m^2$
$p_5 = p_4 - \rho gh = 17\,800 - 923 \times 9.81 \times 23 = -190\,200\ kN/m^2$
$p_6 = p_5 = -190\,200\ kN/m^2$
$p_7 = p_6 + \rho gh = -190\,200 + 941 \times 9.81 \times 21 = 3800\ N/m^2$
$p_8 = p_7 = 3800\ N/m^2$. Now plot the variations (Figure 9.30).

Figure 9.30

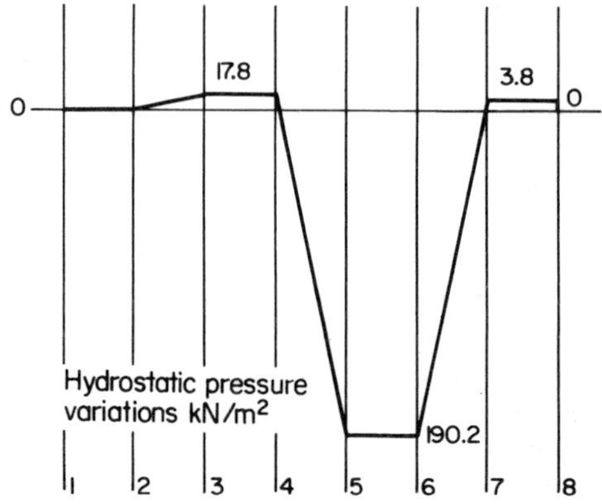

Step 3 Plot variations in circulating pressure (Figure 9.31):
The preceding table gives the loss in pressure in each section due to friction in pipes and fittings. The pump provides the total pressure loss as a lift of 94.7 kN/m^2 across the pump.

Figure 9.31

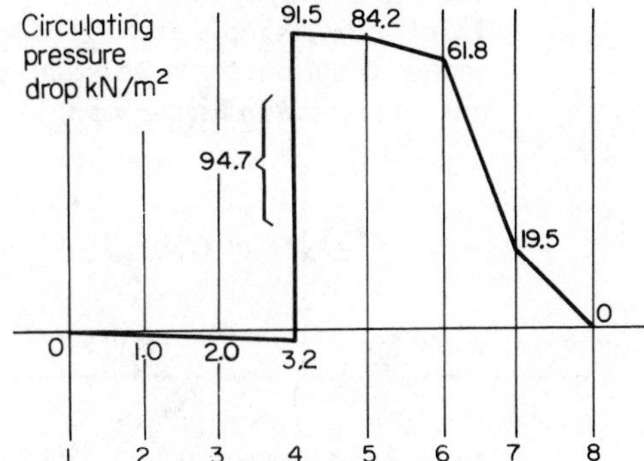

Step 4 Plot composite pressure diagram (Figure 9.32):
This diagram combines the effects of hydrostatic and circulating pressures.

Figure 9.32

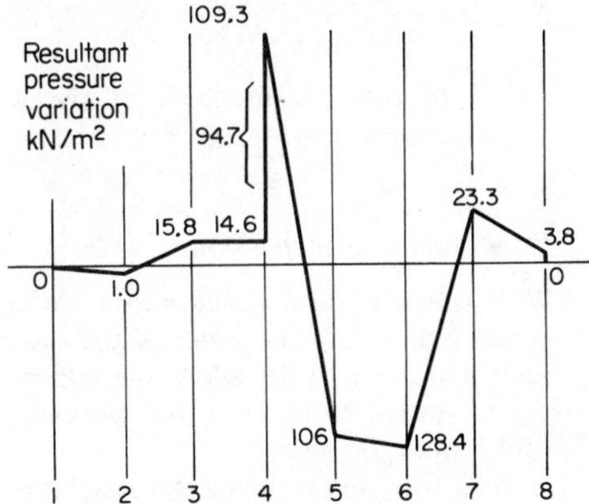

Step 5 Location of vulnerable point(s):
The vulnerable point occurs at the farthest end of the high level main where the pressure is 128.4 kN/m^2 below that of the steam in the boiler drum, when the pump is running. The designer must also consider the possibility of pump failure in which circumstances this is still the weakest point in the system but with a pressure now 190.2 kN/m^2 below that of the steam in the drum. Thus the pump fail condition takes priority.

Step 6 Establishing the system pressures:
The flow temperature is to be 143 °C and the anti-cavitation margin of 15 °C requires a pressure at point 6 related to a saturation temperature of 158 °C. From tables[4] this is found to be 600 kN/m^2 (approx). Thus the steam pressure in the boiler drum is 600 + 190.2 = 790.2 kN/m^2.

This means that the water temperature in the boiler is about 169 °C. Recall that this temperature was previously assumed to be 160 °C. A fresh round of calculations could be made but this is hardly necessary since the anti-cavitation margin is quite generous.

Step 7 The bypass proportion:

This is now a simple exercise of applying the steady flow energy equation for a constant pressure process to the mixing point M in Figure 9.33.

Figure 9.33

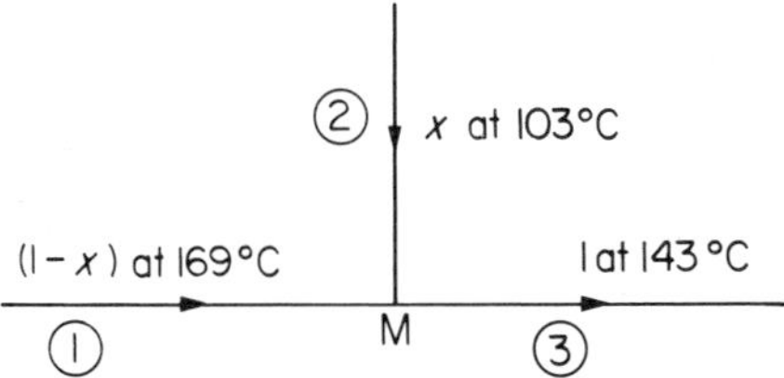

Thus for 1 in the flow and x in the bypass, the flow through the boiler is $(1 - x)$.

Then $$(1 - x)h_1 + xh_2 = 1h_3$$

$$(1 - x)716.4 + x429 = 604.7$$

$$\therefore x = 0.39$$

In other words 39% of the water flows through the bypass.

Figure 9.34

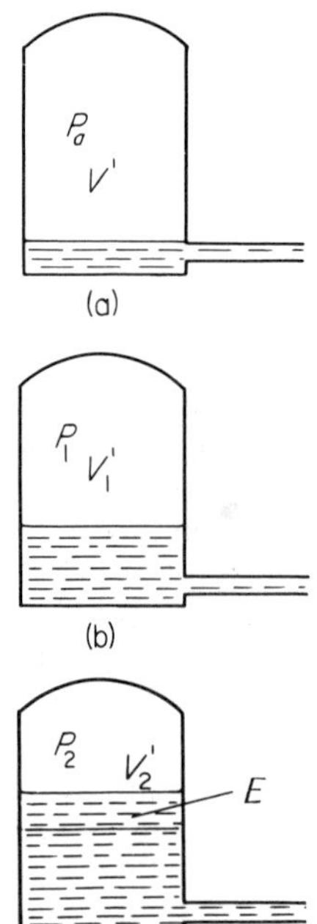

Pressurisation by expansion of water

With this method a sealed expansion vessel is employed which contains air above the water of the system. The vessel is sized in such a manner as to allow the expansion of the water in the closed system to generate the working pressure by compression of the air in the vessel.

As a first step in sizing the vessel it is assumed that isothermal compression occurs during both the initial filling and the warming up processes. Let the volume of the air in the vessel be V' when during filling the level of water in the vessel is the same as that in the pipe. Under these conditions the air is at atmospheric pressure P_a (Figure 9.34a).

As filling continues, the air is eventually compressed isothermally to a pressure P_1 which is more or less the static pressure imposed by the feed cistern. Let the volume of the air at this stage be V_1' (Figure 9.34b).

The boilers are now fired and the system warms up. The expansion water is accommodated in the expansion vessel. Let the final air pressure be P_2 and the final air volume V_2'. Also let the expansion volume of water be E (Figure 9.34c).

Now for the assumed isothermal expansions.

$$P_a V' = P_1 V_1' \tag{9.6}$$

and $$P_1 V_1' = P_2 V_2' \tag{9.7}$$

also $$V_2' = V_1' - E \tag{9.8}$$

substituting from equation 9.8 in equation 9.7,

$$P_1 V_1' = P_2 (V_1' - E)$$

thus $$P_1 V_1' = P_2 V_1' - P_2 E$$

and on rearranging, $$P_2 V_1' - P_1 V_1' = P_2 E$$

from which $$V_1' = \frac{P_2 E}{(P_2 - P_1)} \tag{9.9}$$

Substituting for V_1' from equation 9.9 in equation 9.6, then

$$P_a V' = P_1 \frac{P_2 E}{(P_2 - P_1)} \text{ from which}$$

$$V' = \frac{P_1 P_2 E}{P_a (P_2 - P_1)} \tag{9.10}$$

and for the situation where the atmospheric pressure is taken as 1.013 bar, then:

$$V' = \frac{P_1 P_2 E}{1.013\,(P_2 - P_1)} \tag{9.11}$$

An adjustment may be made however for the effect of temperature changes of the air during the warming up process. If equations 9.6 and 9.7 are replaced by

$$\frac{P_a V'}{T_1} = \frac{P_1 V_1'}{T_1}$$

and $$\frac{P_1 V_1'}{T_1} = \frac{P_2 V_2'}{T_2}$$

and the same analysis performed as before, then it can be shown that

$$V' = \frac{T_1 P_1 P_2 E}{1.013\,(T_1 P_2 - T_2 P_1)} \tag{9.12}$$

where T_1 and T_2 are the absolute filling and working temperatures respectively.

Another calculation of interest to the designer is the change in water level during expansion. Assuming the expansion vessel in Figure 9.35a to be cylindrical and of diameter d, then for isothermal compression of an initial column of air of height h metres by an amount of height x metres, then

$$P_1 V_1' = P_2 V_2'$$

thus $$P_1 \frac{\pi d^2}{4} h = P_2 \frac{\pi d^2}{4} (h - x)$$

and $$P_1 h = P_2 h - P_2 x$$

thus $$h(P_2 - P_1) = P_2 x$$

or $$x = h \frac{(P_2 - P_1)}{P_2} \tag{9.13}$$

or again, if temperature effects are allowed for then

$$x = h \frac{(P_2 T_1 - P_1 T_2)}{P_2 T_1} \tag{9.14}$$

Figure 9.35

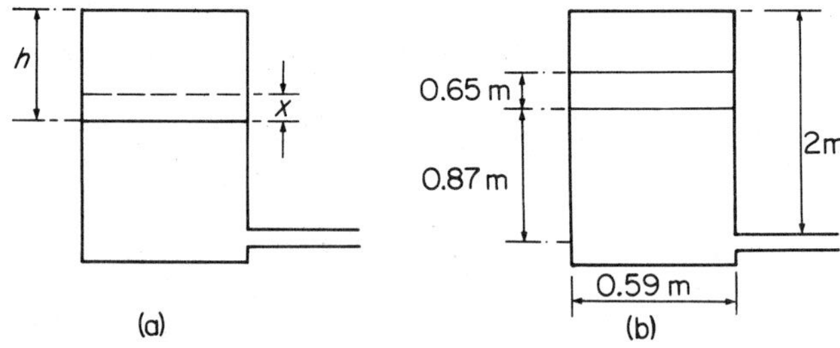

Now E has already been defined as a contraction from V_1' to V_2' on the air side but it could equally well be defined in terms of an expansion on the water side. Suppose the initial system water fill is C at T_1 with a density ρ_1. On heating to T_2 with a density ρ_2 the water expands by an amount $\rho_1 C/\rho_2$.

$$\text{Thus } E = \frac{\rho_1 C}{\rho_2} - C = \frac{(\rho_1 - \rho_2)}{\rho_2} C \tag{9.15}$$

Example 9.15 A college pressurised heating system has a fill capacity of 3.5 m^3. It is filled with water at a temperature of 10 °C from a feed tank situated 8 m vertically above the expansion vessel. The system has been designed to run with flow and return temperatures of 130 °C and 90 °C respectively. Size the vessel on the basis of a height of about 2 m and the use of an anti-flash margin of about 15 °C.

At 10 °C water density = 999.7 kg/m^3
At 110 °C water density = 951.0 kg/m^3

$$\text{Now } E = \frac{(999.7 - 951.0)}{951.0} \times 3.5 = 0.178 \text{ m}^3$$

Take P_a = 1.013 bar

Now $P_1 = 1.013 + 999.7 \times 9.81 \times 8 \times 10^{-5} = 1.798$ bar

Strictly speaking the anti-cavitation margin should be applied at the most vulnerable point in the system identified in the manner shown in the previous example. In fact the 15 °C margin employed in that problem was rather generous and it is more usual practice to use the 15 °C margin in this vessel sizing procedure and relate it to the flow temperature (taking no account at this stage of pressure drop due to friction or to rising mains).

Thus P_2 is selected from tables[4] as the saturation pressure corresponding to a temperature of 130 + 15 = 145 °C. Thus P_2 = 4.2 bar and substituting in equation 9.11,

$$V' = \frac{1.798 \times 4.2 \times 0.178}{1.013\,(4.2 - 1.798)} = 0.55\ \text{m}^3$$

Assuming a height above the fill pipe of 2 m, then

$$\frac{\pi d^2}{4} \times 2 = 0.55, \text{ from which } d = 0.59\ \text{m}$$

During the filling process,

$$x = \frac{2\,(1.798 - 1.013)}{1.798} = 0.87\ \text{m}$$

and during the heating process,

$$x = \frac{1.13\,(4.2 - 1.798)}{4.2} = 0.65\ \text{m}$$

These results are indicated in Figure 9.35b.

It is left to the reader to rework the problem taking account of temperature changes during the heating process.

Diaphragm expansion tanks

A disadvantage of the sealed expansion tank is the gradual absorption of the air from the tank into the water in the system. A recent development which overcomes this problem is the diaphragm tank. The tank contains a rubber or rubber composition diaphragm which acts as the interface between air or nitrogen in the upper portion and the system water in the lower portion of the tank. The same sizing principles already discussed may be applied to the sizing of such vessels.

Gas pressurisation

A method which is receiving increasing use is pressurisation by a gas and normally nitrogen is the gas employed, supplied from a pressure bottle. The two main variants of this system depend on either retaining or spilling the water of expansion in or from the system.

(a) Retention system

The components of the pressurising system are as shown in Figure 9.36. The concept is to operate at all times as a constant pressure and the starting point is thus a fully pressurised system when cold. As the system heats up the nitrogen pressure in the expansion vessel starts to increase and the pressure relief valve is adjusted to allow spillage of nitrogen to the low pressure receiver. A pressure sensor on the low pressure receiver brings the compressor in from time to time to return the nitrogen to the high pressure receiver.

Figure 9.36
Gas pressurisation with the retention system

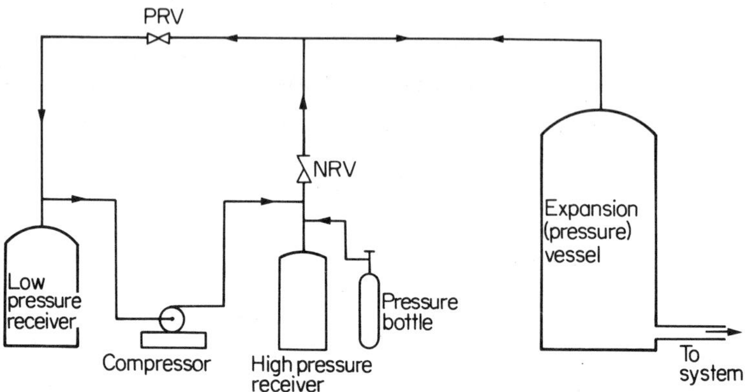

As the system cools down again the nitrogen is returned from the high pressure receiver to the expansion vessel via the non-return valve. There is also provision for recharging the high pressure receiver from the pressure bottle.

(b) Spill tank system

Figure 9.37
Gas pressurisation with the spill system

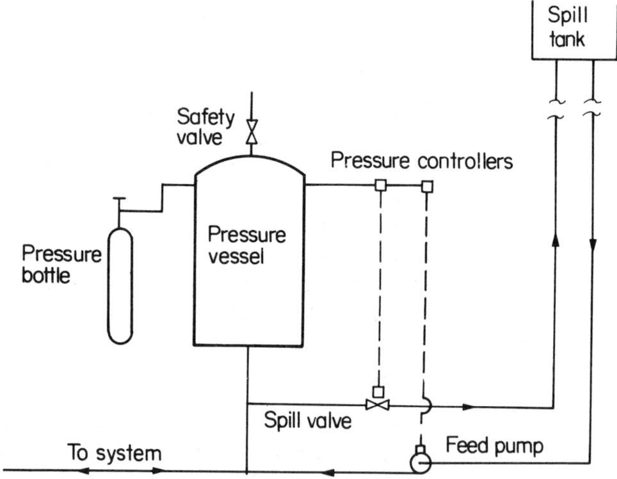

The components of this system are as shown in Figure 9.37. As with method (a) the system is pressurised when cold by the nitrogen from the bottle. As the system heats up and the nitrogen pressure begins to rise, water is allowed to spill from the system to the overhead spill tank via the spill valve. As the minimum system operating temperature is reached the spill valve pressure controller begins to close the spill valve and the remaining water of expansion, as the system comes up to full temperature, is retained within the system. As the system cools down the feed pump pressure controller, set at a lower pressure setting, brings in the feed pump to return the spill water to the system. Considerable care is required in setting the two controllers to avoid the possibility of the spill valve being open at the same time as the feed pump is running.

9.7.5 Emitters for use with high temperature systems

As with low temperature systems the water may be fed to natural convectors or to forced convectors such as unit heaters or to pipe coils. Wider application is made of radiant systems with the emitters consisting usually of plain panels or overhead strip heaters. This latter method is especially suitable for the heating of large spaces such as factories or storage areas. There may be considerable energy saving due to the lack of temperature gradient and lower acceptable air temperatures which are a feature of such systems.

High temperature radiant panels consist usually of a pipe coil of about 15 mm diameter welded to a steel plate. The panels may be suspended vertically or at an appropriate angle overhead within the workshop. They may be designed to release heat from both sides or may have one side insulated to reduce back losses. Strip heaters in essence are similar to radiant panels. Several proprietary systems are available and consist normally of a pipe coil clipped to strips of aluminium plate with the top side of the plate insulated. They are provided in standard lengths which are joined to give a continuous emitter running say the length of a factory bay.

A method for checking the system for cavitation has already been demonstrated in example 9.14 but additional checks may be necessary in the case of partial load operation of the emitters. This is perhaps best explained by worked examples. Consider first the case of a unit heater system operating by on/off control from space thermostats as outlined in the following example:

Example 9.16 Figure 9.38 shows the layout of a steam pressurised HTHW system employing unit heaters coupled up in a two pipe arrangement and operating on/off from space thermostats. The system has already been sized and the following information is available:

Figure 9.38

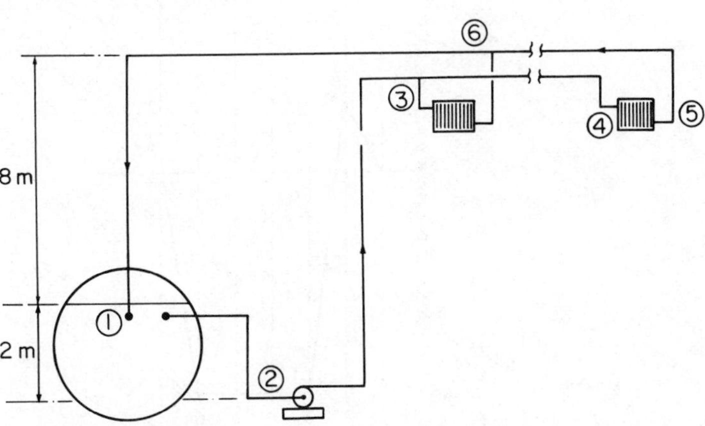

Pipe section	1/2	2/3	3/4	4/5	5/6	6/1	Total
Pressure drop (kN/m^2)	9	8	42	15	39	8	121
Pipe emission (kW)	1	19	63	3	52	11	149

The system includes 36 unit heaters each with an output of 24 kW with the fan running and 3 kW with the fan off. The system design flow and return temperatures are 150 °C and 110 °C. On the basis of an anti-flash margin of 10 °C, establish a suitable steam pressure in the boiler drum. Also check for cavitation in circumstances where:

(a) the pump stops running,
(b) under design conditions the space thermostats cut off power to the fans.

Hydrostatic pressure
Pressure increase from 1 to 2

$$= \rho g h = \frac{917 \times 9.81 \times 2}{1000} = 18 \text{ kN/m}^2$$

for water of density 917.9 kg/m^3 at 150 °C.

Pressure decrease from 2 to 3

$$= \frac{917.9 \times 9.81 \times 10}{1000} = 90 \text{ kN/m}^2$$

Pressure increase from 6 to 1

$$= \frac{951.0 \times 9.81 \times 8}{1000} = 74.5 \text{ kN/m}^2$$

for water of density 951.0 kg/m^3 at 110 °C.

A plot can now be made of pressure variations throughout the circuit (Figures 9.39, 9.40 and 9.41).

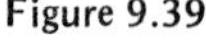
Figure 9.39

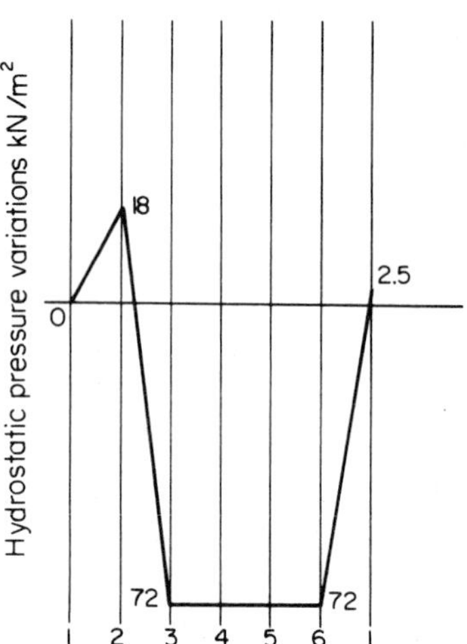

Figure 9.40

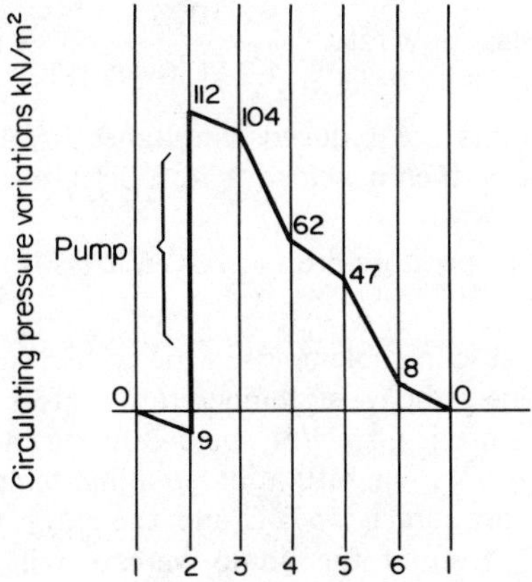

Figure 9.41

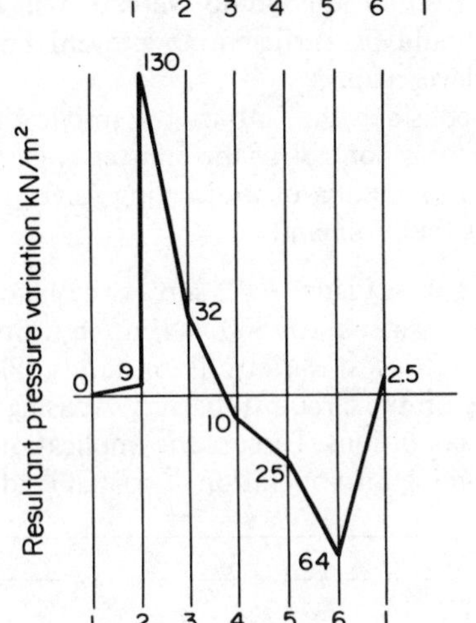

Required flow temperature = 150 °C
Anti-flash margin = 10 °C
Pressure corresponding to 160 °C = 620 kN/m^2
(from tables of properties of steam).

For the normal situation of the system operating on design load the weakest point in the system for water at 150 °C is at 4 (the last emitter on the line) where the pressure is 10 kN/m^2 below the drum pressure. Applying the anti-cavity margin to this point, the required pressure in the boiler drum is 620 + 10 = 630 kN/m^2.

Case (a)

When the pump ceases to run, the pressure in the high level mains falls 72 kN/m^2 below the pressure in the steam drum to 558 kN/m^2. The saturation (boiling) temperature corresponding to this pressure is 156 °C which cuts the anti-cavity margin to 6 °C.

Case (b)

When the emitters are operating under full load conditions the temperature in the return main is 110 °C. The design water mass flow rate is now calculated.

System heat load = 36 × 24 + 149 = 1013 kW

$$\text{Mass flow rate} = \frac{1013}{4.2\,(150 - 110)} = 6 \text{ kg/s}$$

If all units are reduced simultaneously to emission by natural circulation then new load = 36 × 3 + (149 − 11) = 246 kW

$$\text{Temperature drop across emitters is } \frac{246}{4.2 \times 6} = 9.8\,^\circ\text{C}$$

The most vulnerable point in the system is now at 6 on the return main where the water temperature is (150 − 9.8) or 140.2 °C and the pressure is 64 kN/m^2 below the steam drum pressure, that is, 586 kN/m^2. The saturation (boiling) temperature corresponding to this pressure is 157 °C and the anti-cavitation margin is thus 16.8 °C. Even if the pump were to fail with the fans off, the margin should be sufficient to prevent boiling occurring within the high level mains.

Now consider the anti-cavity implications of other arrangements such as control of the emitter at partial load conditions by the use of a mixing or a diverting valve. The following example illustrates the problem.

Example 9.17 Figure 9.42 shows the layout of a college MTHW system pressurised to 500 kN/m^2 by nitrogen and feeding air heater batteries at various floor levels. Plot the total pressure variations on the circuit to battery A, using the data provided, for full load conditions. Discuss the implications of the diverter valve shown coming into operation at partial load conditions.

Figure 9.42

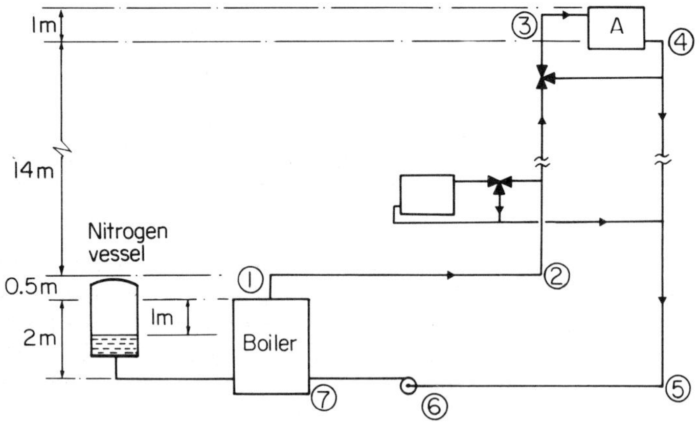

Pipe section	1/2	2/3	3/4	4/5	5/6	6/7	Total
Pressure drop kN/m^2		10	25	12	11	7	65

The 25 kN/m^2 drop across 3/4 includes the resistance of the air heater battery. Allow a drop of 5 kN/m^2 through the boiler. Assume a mean water density of 950 kg/m^3.

From the total pressure variation diagram (Figure 9.43c), it can be seen that under design conditions the most vulnerable point

Figure 9.43a
Hydrostatic pressure variation

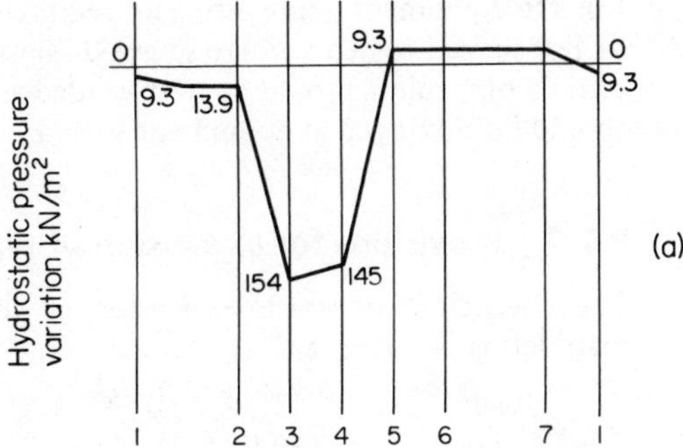

Figure 9.43b
Circulating pressure variation

Figure 9.43c
Resultant pressure variation

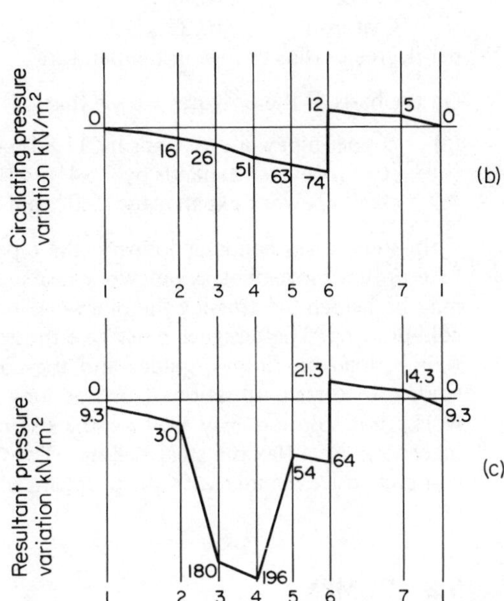

in the system is at 3 where water at design flow temperature enters the topmost heater battery (at a pressure of 320 kN/m^2). The saturation temperature at this pressure is about 135 °C and if an anti-cavity margin of 10 °C is to be applied then the supply temperature should be limited to 125 °C.

Under partial load operation as the valve begins to divert water from the air battery through the bypass, the temperature of the water rises in the return main and the effect is to reduce the anti-cavity margin in the return main in a similar fashion to the case dealt with in the previous problem.

9.7.6 Pipework for heating systems

The range of pipe materials used in heating systems has already been mentioned in section 9.4.3. With LTHW systems it is normal practice to use compression fittings for making joints in relation to copper pipework and to use screwed joints in relation to steel pipework. This applies to pipe-to-pipe joints and also at connections to equipment. More stringent requirements apply to jointing of pressurised systems where screwed joints are not suitable. Here pipe-to-pipe joints should be butt-welded and flanged connections should be employed at equipment such as valves or pumps.

9.7.7 Provisions for expansion of pipework

The coefficient of linear expansion of the common pipework materials is:

Steel	11.34×10^{-6}
Copper	16.92×10^{-6}
Cast iron	10.22×10^{-6}

per degree Celsius change in temperature.

On the basis of these figures we see that a 10 m run of:

(a) copper pipework expands by 11.84 mm (from 10 °C to 80 °C),
(b) steel pipework expands by 7.94 mm (from 10 °C to 80 °C),
(c) steel pipework expands by 17.01 mm (from 10 °C to 160 °C).

In some cases bends or loops in the pipework may be used to take up the movement as pipework heats up or cools down. This may be helped by stressing the pipework in the cold condition by cold-drawing. The designer must give thought to such matters as fixings, load on fixings, guides and the constraints imposed by guides. In certain situations (such as long straight runs of pipework) the solution may well be the use of an expansion joint incorporating a flexible steel bellows. The Guide (B16) deals in a comprehensive manner with the problems of expansion.

9.8 PUMPS

9.8.1 Pump and system characteristics

The pump has a pressure/volume flow rate characteristic and so also has the system of pipework and heating apparatus. The designer has to exert his skill in choosing a pump which is appropriate to the task in hand, will operate efficiently and which will interact with the characteristic of the system to provide the required pressure drop through the system while creating the necessary water flow rate within the system.

9.8.2 Pump laws

From tests carried out with pumps, the following relationships have been developed between the pump rotational speed (N rev/s) and the characteristics of volume flow rate ($\dot{G}$ m^3/s), of pressure developed (p N/m^2) and of power ($\overline{P}$ watts).

(a) Volume flow rate is directly proportional to speed

$$\dot{G} \propto N$$

(b) Pressure developed is directly proportional to the square of speed

$$p \propto N^2$$

(c) Power developed is directly proportional to the cube of speed

$$\overline{P} \propto N^3$$

The power $\overline{P}$ is the power imparted to the water and this is related to the pressure P developed and the volume flow rate $\dot{G}$ as follows:

$$\overline{P} = p\dot{G}$$

There are of course hydraulic losses at the pump and mechanical and electrical losses at bearing and motor. If the overall efficiency is η_o then

$$\text{power supplied} = \frac{p\dot{G}}{\eta_o}$$

The power characteristic is usually a rising one while the efficiency characteristic usually peaks then drops at large volume flow rates. Figure 9.44 shows typical characteristics for a centrifugal pump.

Figure 9.44
Pump characteristics

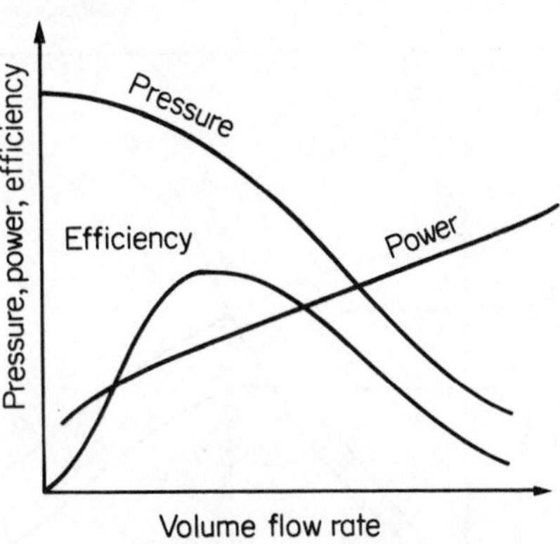

It may be inferred from the discussion on system resistance in Chapter 1 that the pressure drop in overcoming resistance is proportional to the square of the velocity or indeed of the volume flow rate. Thus the characteristic of the system is of the form

$$p \propto \dot{G}^2$$

The shape of the characteristic is parabolic and it interacts with the p–$\dot{G}$ characteristic of the pump to indicate the pressure which is developed by the pump in providing the volume flow rate, as indicated in Figure 9.45.

It is interesting to observe the effect of fitting the same pump to a different system, in fact one with a lower resistance. In such a situation the developed pressure would be less and the volume flow rate would be greater.

Figure 9.45
Interaction of system and pump characteristics

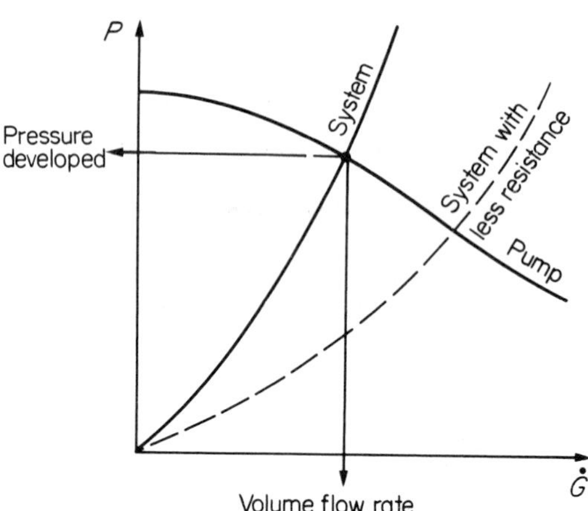

Figure 9.46

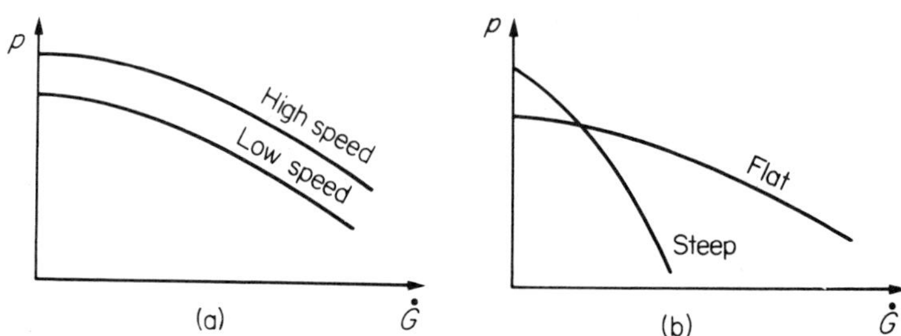

Figure 9.47
Good and bad pump selections

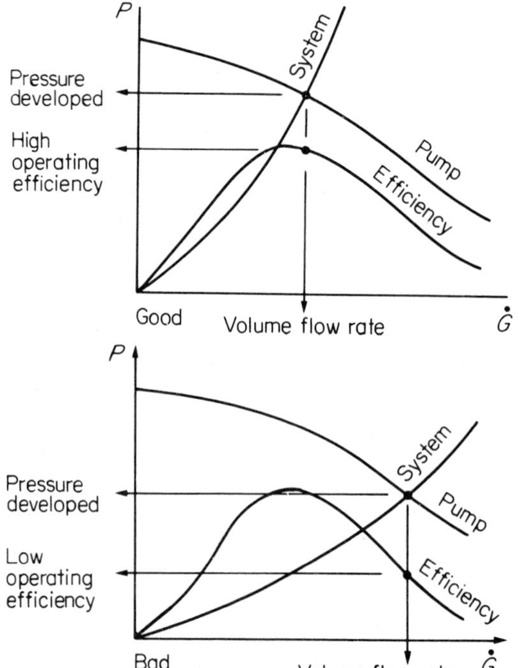

In making his selection the designer has some flexibility at his disposal. He can for instance add resistance to the system to vary the system characteristic. The pump characteristic can also be manipulated by varying the pump speed (Figure 9.46a).

The designer may also be free to choose a pump with a steep or with a flat characteristic (Figure 9.46b).

Thought must also be given to the operating efficiency. Figure 9.47 illustrates the difference between a good and a bad selection.

Example 9.18 Recall Example 9.7 where the total pressure drop in the index circuit was 29.55 kN/m^2. It is proposed to locate the pump on the return near the boiler and the duty will be a circulation of 2.59 l/s against a head of 29.55 kN/m^2. A manufacturer's catalogue yields the following information for a centrifugal pump operating at 12 rev/s:

Pressure (N/m^2)	56	54	48	37	0
Volume flow (i/s)	0	1	2	3	4
Power (W)	94	180	192	227	105
Overall efficiency	0	30	50	49	0

Plot the pump and system characteristics on the same graph and determine the resulting operation point. Hence determine the adjustment which would be required to the speed of rotation in order that this pump would give the required duty.

Step 1 Determine the system characteristic:

Now $p \propto \dot{G}^2$ or $p = k'\dot{G}^2$ for the system where k' is a constant. Obtain k' from the duty.

$$\text{Thus } k' = \frac{p}{\dot{G}^2} = \frac{29.55}{2.59^2} = 4.4$$

The law of the system is thus $p = 4.4\,\dot{G}^2$.

The following figures are now obtained for the system:

p (N/m^2)	0	4.4	17.6	29.55	39.6	70.5
$\dot{G}$ (l/s)	0	1.0	2.0	2.59	3.0	4.0

Step 2 Plot the pump and system characteristics:

From Figure 9.48 the operating condition is seen to be

Figure 9.48

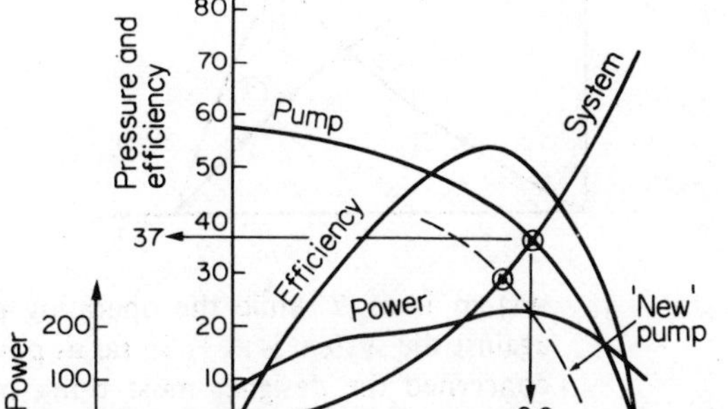

2.9 l/s at 37 N/m² (point 1). It is interesting also to note that the efficiency is good at this condition. The operating point could be adjusted to the design requirement (point 2) by lowering the operating speed.

Step 3 Determine the new speed:

$$\text{Since } \dot{G} \propto N \text{ then } \frac{\dot{G}_{des}}{\dot{G}} = \frac{N_{des}}{N}$$

$$\text{Thus } N_{des} = \frac{2.59}{2.9} \times 12 = 10.7 \text{ rev/s}$$

If the pump speed is adjusted to 10.7 rev/s then this pump is capable of giving the exact design duty with a reasonably high efficiency. The same solution would have been obtained using the relationship $p \propto N^2$. The new pump characteristic is shown 'ghosted' on the graph.

9.8.3 Pump combinations

Pumps may be run in series or in parallel as pairs. The majority of heating systems will require only one pump but cases do occur (for example cold water supply in tall buildings or where the developed pressure exceeds about 3 bar) where combinations are used.

(a) *Two pumps in series.* The pressure developed is twice that for the single pump, corresponding to each volume flow rate. However when the pumps operate on a system the situation is a little more complex. Consider the situation in Figure 9.49. The operating point for one pump against the

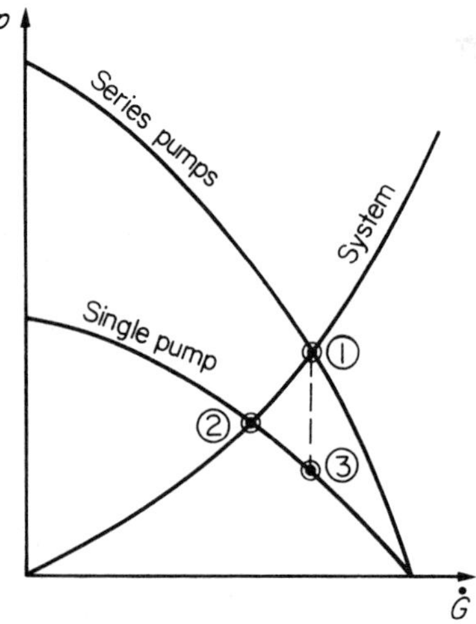

Figure 9.49
Pumps in series

system is at 2 while the operating point for two pumps against the system is at 1. So far as power and efficiency are concerned the designer must think of the performance of the individual pumps in relation to point 3.

(b) *Two pumps in parallel.* In this case the pressure developed remains the same and the volume flow rate is doubled. Again the effect of one or two pump operation against a system is best understood by examination of the p–$\dot{G}$ diagram of Figure 9.50.

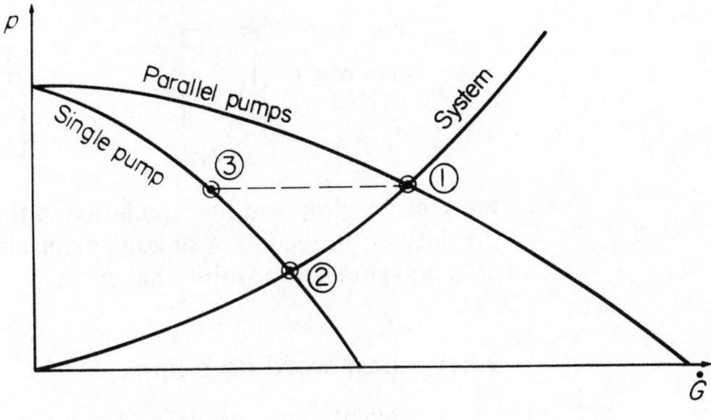

Figure 9.50
Pumps in parallel

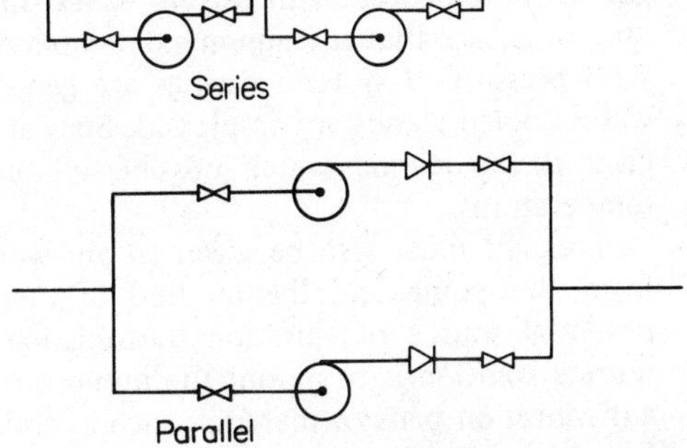

Figure 9.51
Combination arrangements

Point 1 represents both pumps operating against the system while point 2 represents one pump operating against the system. The power and efficiency of each pump (when both are running) is evaluated at point 3. Isolating valves require to be fitted for maintenance of the pumps and non-return valves are required to prevent the possibility of short-circuiting. Suitable combination arrangements are shown in Figure 9.51.

9.8.4 More complex arrangements

With larger buildings the heating system may be divided into zones, possibly with temperature control to each zone by the use of mixing valves. In such circumstances there is usually a pump dealing with the primary mains circulation and in addition each secondary or zone circuit has its own circulating pump. Such an arrangement is shown in Figure 9.52.

The duty of the primary circulation pump is related only to

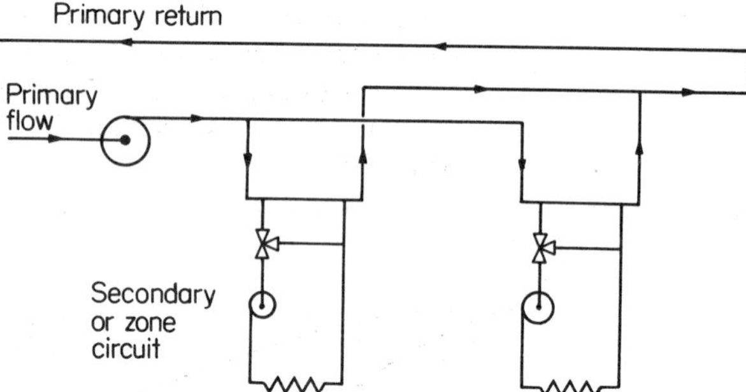

Figure 9.52
Primary and secondary pumped circuits

the pressure drop and the circulation in the primary circuit, while the duty of the secondary or zone pumps is related to the pressure drop and circulation within that zone.

9.8.5 Last word on pumps

Some thought must be given to the quality, temperature and pressure of the water passing through the circulating pumps. The normal materials for pump construction are cast iron casings and gun-metal impellers. With certain water supplies it may be necessary to ensure that any gun-metal components are free from zinc. With pressurised systems casings are generally of cast steel and water cooled glands are employed. Special consideration must be given to the bearings which must be suitable for operation at high temperatures.

Thought must also be given to pipework connections to and from the pump and the method of mounting since this is a potential source of vibration transmission to the building. The normal solution is to mount the pump on a concrete plinth with anti-vibration pads of materials such as cork or stiff rubber, interfacing between the pump and the concrete.

9.9 GROUP AND DISTRICT HEATING

9.9.1 Introduction

District heating is the provision of space heating and hot water from a centralised boiler house feeding all the buildings in the 'district' which may be an area of a city or perhaps a whole town. Smaller versions, providing this facility for say a hospital or university complex or a shopping centre may be termed Group or Block Heating schemes. District Heating was developed on the Continent and in the United States and it is only in recent years with the growth of New Towns and the creation of many housing developments that District Heating has emerged to any extent in the United Kingdom.

The advantages usually claimed for DH are a more attractive and healthy environment coupled with the provision of a convenient and clean supply of heating and hot water. A wider advantage is the more efficient use of fuel associated with a central boiler house with, in many cases, the ability to burn poorer grades of fuel. Some schemes recover heat from waste products and refuse incineration is employed in several large schemes.

At the present time, a wider debate is taking place with regard to the possibilities of Combined Heat and Power (CHP). This is the extraction of useful heat for space heating and hot water in addition to the creation of electrical energy in a power station (similar in concept to the Battersea Power Station/Pimlico Heating Scheme). An efficient modern power station plant converts about 35% of the fuel energy to electrical energy while about 55% of the fuel energy is thrown away as heat in the cooling water at temperatures in the range 25 to 40 °C. Such plant could be converted to CHP by releasing the steam from the turbine at a higher pressure and temperature giving about 25% conversion to electrical energy and about 65% conversion to heat energy in water at a temperature of about 90 °C. This energy would then be available for District Heating.

There is probably no economic case for CHP/DH at the present time but such schemes will gradually become economically more viable as resources of oil and natural gas become depleted. This is one of the long-term energy options being considered by the Department of Energy at the present time, and may be the next major development in the field of District Heating.

Conventional District Heating Schemes may be thought of as consisting of a heat source, a means of distribution and a means of utilisation and each of these aspects will now be considered in turn.

9.9.2 Heat source

Waste heat from power stations (CHP/DH) or from the burning of refuse poses an exciting challenge to designers and planners. A major difficulty in either of these cases is load balancing. On the one hand there is the balance between the electrical load and the heating load and on the other between the refuse for incineration (which tends to be reasonably uniform) and the heat load. The system has to be designed to cope with a winter design heating load which is prevalent for a surprisingly short period of the year.

Systems of these types do exist, the Pimlico Scheme already mentioned and an incineration scheme at Nottingham. However by far the more usual situation is in the use of conventional coal, oil or gas fired boilers by arrangement between on the one hand the energy supplier such as the National Coal Board, Gas Board or oil companies and the customer who is likely to be a private developer or a local authority. With such schemes the customer is normally able to take advantage of favourable tariffs and also obtains the benefit of efficient, well run and well retained boiler plant with suitable provision for the removal of atmospheric pollutants from the flue gases. Such plant has been discussed more fully in Chapter 5.

9.9.3 Heat distribution

Many of the problems which arise in relation to the operation of District Heating systems are associated with the distribution system. These in the majority of cases centre around the ingress of water into the trenches which carry the distribution mains and this water may give rise to spoiling of the insulation by wetting and even worse, corrosion of pipework and components. Traditional methods involved excavation of the trench and the construction of *in situ* brick or concrete passages within which the insulated mains were run. Figure 9.53 is a typical example of this type of arrangement.

Figure 9.53
Section of an In-situ construction

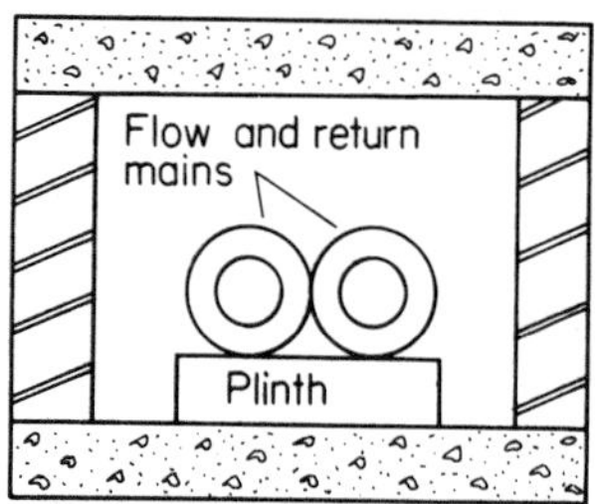

Increasingly the trend has been towards the introduction of prefabricated systems involving for example pipe-in-pipe units, which have the twofold advantage of cutting back on time on site while with good workmanship and supervision eliminating (more or less!) the problems of spoiling of the insulation and corrosion of the mains. Figure 9.54 is an illustration of such a system in which an air gap exists between the insulated mains and the outer casing.

Figure 9.54
Section of prefabricated construction

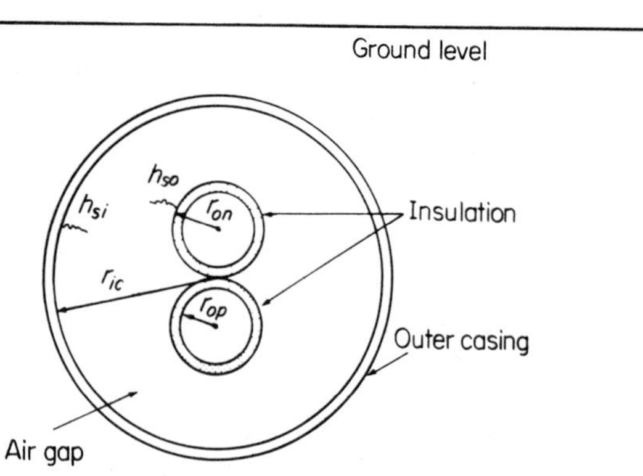

Other forms of pipe-in-pipe unit have very much smaller diameter outer casings with all the space between the outer casing and the two mains completely filled with insulation. Special techniques have to be developed to make the joints in the lengths of trunking and at the tees which allow the connections from the mains to and from the buildings.

Estimation of heat loss from the mains

It was seen in Chapter 2 that the heat loss per unit length from an insulated pipe could be expressed by equation 2.73, as

$$\dot{Q} = UA_2\ (\theta_{f1} - \theta_{f2})\ \text{W/m}$$

where in this case A_2 represents a perimeter multipled by unit length of pipe. The Guide suggests the use of this equation in the form

$$\dot{Q} = \pi d_{op} U\ (\theta_s - \theta_e)\ \text{W/m}$$

where d_{op} represents the outside diameter of the bare pipe and θ_s the outside surface temperature of the bare pipe (assumed to be the same value as the water temperature within the pipe). The temperature at ground level is taken to be θ_e. The units of U in this case would then be W/m^2 °C related to the surface area of the bare pipe. Further,

$$U = \frac{1}{R_n + R_a + R_e}$$

where the three thermal resistance values relate to the insulation, the air gap (if any) and the earth respectively.

The insulation resistance (as seen in Section 2.5.3) takes the form

$$R_n = \frac{r_{op}}{k_n}\ \ell n\ (r_{on}/r_{op})$$

where r_{op} and r_{on} represent the radii of the bare and insulated pipe respectively and k_n the thermal conductivity of the insulation.

The thermal resistance of the air gap is basically the sum of two surface resistances and these must be related to the bare pipe surface area yielding

$$R_a = \frac{r_{op}}{h_{so} r_{on}} + \frac{r_{op}}{h_{si} r_{ic}} \quad \text{(see Figure 9.54)}$$

Estimation of the thermal resistance of the earth is more difficult and the most recent edition of Section C3 of the Guide suggests the use of the empirical expression

$$R_e = \frac{d_{op}}{2k_e}\ \ell n\ (2b/d_{ic}) \left\{1 + [1 - (d_{ic}/2b)^2]\ 0.5\right\}$$

where b is the depth of the centre of the main below finished ground level and k_e is the thermal conductivity of the earth. If the burial depth is more than twice the insulated pipe diameter (as is usually the case) this latter equation reduces to

$$R_e = \frac{d_{op}}{2k_e}\ \ell n\ (4b/d_{ic})$$

Section C.3 of the Guide provides sufficient information to allow evaluation of the component resistances. The design steps in estimating heat loss from a single pipe arrangement would be:

Step 1 Determine component resistances R_n, R_a and R_e.

Step 2 Determine the overall thermal transmittance (U value).

Step 3 Estimate the heat loss in watts per metre run in terms of the previous items and the design temperature difference.

The procedure is similar with twin-pipe systems, the calculation being done once for each pipe assuming the other is absent and the total heat being taken as the sum of the two.

Apportioning the mains emission

In Example 9.2 we saw the importance of increasing the flow rate and apportioning properly in the various sections of the pipework in order that the design mean temperatures can be maintained at the appliances. This was in relation to low pressure hot water heating systems but similar techniques are required in relation to district heating mains.

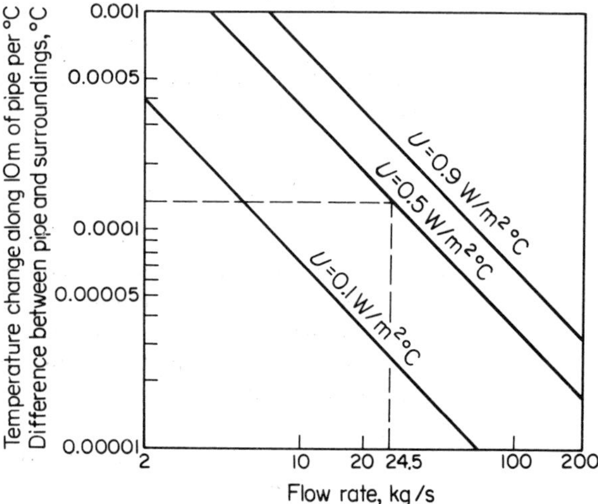

Figure 9.55
Temperature changes along buried pipes

To aid the designer in assessing variations in temperature in the mains the Guide[5] includes a graph which is presented here (Figure 9.55) in truncated form.

Example 9.19 A district heating system has a connected load of 3000 kW and employs a twin pipe circulating high temperature hot water, with flow and return temperatures at the boiler of 120 °C and 90 °C respectively. The thermal transmittance value from bare pipe surface to earth has been assessed as 0.5 W/m^2 °C through insulation, air and earth resistances.

Estimate:

(i) The design mass flow rate based on the load.

(ii) The mass flow rate adjusted for mains losses.

(iii) The temperature in the mains at connection to the first and last customers with the aid of Figure 9.55 and with relation to Figure 9.56.

Figure 9.56

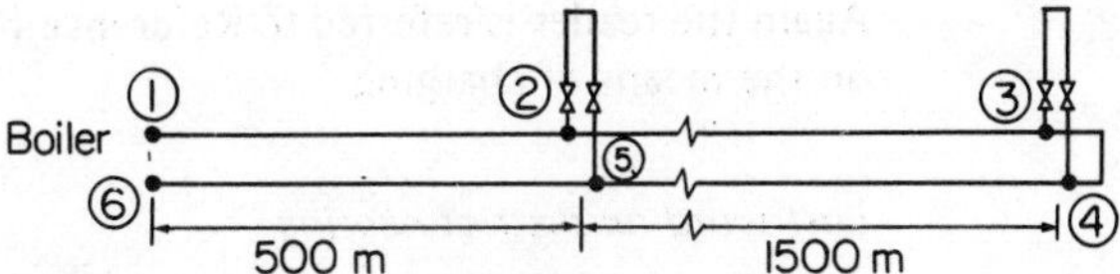

The pipes are 150 mm diameter and the earth temperature may be taken as 5 °C.

From information given, $\theta_1 = 120\,°C$; $\theta_6 = 90\,°C$ and the mean temperature difference at the boiler is 105 °C.

(i) $\dot{Q} = \dot{m}C_p\,(\theta_1 - \theta_6)$, from which

$$\dot{m} = \frac{3000 \times 1000}{4200 \times 30} = 23.8 \text{ kg/s, i.e. design mass flow rate.}$$

(ii) Heat loss from main $= U\pi d_{op}\,(\theta_s - \theta_e)$

$$= 0.5 \times \pi \times 0.15(120 - 5) + 0.5 \times \pi \times 0.15\,(90 - 5)$$

$$= 27 + 20 = 47 \text{ W/m}$$

Mains heat loss = 47 x 2000 = 94 kW

$$\text{Thus adjusted mass flow rate} = \frac{3094 \times 1000}{4200 \times 30} = 24.5 \text{ kg/s}$$

(iii) From the plot shown on Figure 9.55 the temperature change reading shows 0.00014 °C for 10 m/°C. Thus $\theta_1 - \theta_2 = 0.00014 \times 50(120 - 5) = 0.8\,°C$ and $\theta_2 = 119.2\,°C$. In a similar fashion the other temperatures are obtained: $\theta_5 = 90.6\,°C$; $\theta_3 = 116.8\,°C$; $\theta_4 = 92.4\,°C$

Heat utilisation

At the consumer end of the system the energy transmitted from the generating station is consumed for heating and hot water services in much the same way as in conventional systems. One problem of course is to transform the energy from the mains temperatures down to around 80 °C for heating and hot water services within the users' premises. A variety of methods are possible such as the use of calorifiers or mixing arrangements. This is well covered by Kell and Martin(1).

Consumption charging methods

We have looked briefly at means of generation, distribution and utilisation of heat energy. At the end of the day the consumer has to pay for the energy used. Again a variety of means of charging is employed, the commonest of these being:

(a) Flat rate charge as part of total rent of flat.
(b) Flat rate charge based on total heating bill.
(c) Unit charge arising from metering.
(d) Fixed charge plus charge from metering.

Again the reader is referred to Reference (1) for further discussion on the means of charging.

Last word on district heating

As further pressure is placed on our fast depleting oil resources there will be a greater drive towards district heating. One of the major difficulties with installations in this country has been corrosion within the below-ground runs of heating main. The designer must pay good attention to the constructional details of the buried pipework, including methods of jointing and of connection from the branches to the main. This has also to be followed up with close supervision of the installation as the constructional work proceeds.

REFERENCES

(1) FABER, O. and KELL, J. R. (rev. Kell, J. R. and Martin, P. L.) *Heating and Air Conditioning of Buildings* (Architectural Press, London, 1979)
(2) SHAW, E. W., *Heating and Hot Water Services* (Crosby Lockwood)
(3) C.I.B.S. Guide, Book B
(4) Thermodynamic and Transport Properties of Fluids by Mayhew and Rogers
(5) C.I.B.S. Guide, Book C, Section C3

SYMBOLS USED IN CHAPTER 9

A area
C_p specific heat at constant pressure
D diameter
E expansion volume of water
EL equivalent length
f friction factor
g acceleration due to gravity
h enthalpy of water
h_s surface heat transfer coefficient
h_f head loss in pipe due to flow
K pressure loss factor
k thermal conductivity
k' system characteristic constant
L actual length of pipework
$\dot{m}$ rate of mass flow
N speed of rotation
p pressure
p_l pressure loss across fitting
P gas pressure
ΔP available pressure across length of pipework
$\underline{\Delta p}$ pressure drop per metre length of pipework
$\overline{P}$ pump power

$\dot{Q}$	rate of heat transfer
R	thermal resistance
r	pipe radius
Re	Reÿnold's Number
T	absolute temperature
U	overall thermal transmittance
V	velocity
V'	volume
$\dot{G}$	volume flow rate
z	potential head
θ	temperature
θ_a	ambient temperature
ρ	density
η_o	overall efficiency

10 Water Supply and the Design of Cold and Hot Water Services

10.1 INTRODUCTION

This chapter examines those aspects of water supply which are of most concern to the designer, namely, hardness, formation of scale and sludge and corrosion tendencies in relation to the normal range of materials used. Layouts of cold water services to buildings are followed by some examples of system sizing. There follows a descriptive treatment of water services for fire protection and the final section deals with the design of hot water services.

10.2 WATER SUPPLY

10.2.1 Its effect on the formation of scale and sludge

All water supplies originate from rainfall. As rain, hail or snow falls through the air, gases and solid impurities are absorbed and the purity on reaching the ground will depend on the quality of the atmosphere. Thus in addition to oxygen and carbon dioxide which are general impurities of rainfall, there may also be present sulphur dioxide or soot or other impurities absorbed within urban areas. As the rainwater flows on the surface, or percolates through the ground, it dissolves any soluble mineral salts present. For example, rainwater containing carbon dioxide, which flows over chalky ground, will react with the chalk (insoluble calcium carbonate, $CaCO_3$), the carbon dioxide (CO_2) present combining with the chalk to form the soluble calcium bicarbonate. This is indicated by the chemical reaction:

$$\underset{\text{insoluble}}{CaCO_3} + CO_2 + H_2O \rightarrow \underset{\text{soluble}}{Ca(HCO_3)_2} \qquad (10.1)$$

The extent of the process depends on the amount of carbon dioxide present in the water.

Salts of carbon, magnesium and sodium are common in rainwater and the concept of 'hardness' or 'softness' of water deals with such impurities in the supply. Thus rainwater (collected, for example, directly from roofs) or water from upland surfaces, is soft, while river, spring or well water will vary in degree of hardness depending on the nature of the ground over which it flows.

10.2.2 Hardness of water

It has been practice to quantify only those salts which have soap destroying properties (those of calcium and magnesium) and the Degree Clark is the traditional unit of hardness. One grain of calcium carbonate per imperial gallon of water represents one degree of hardness on the Clark Scale. More recent practice is to replace the 'grain per gallon' by 'parts per million' which is more or less equivalent to mg/litre. Table 10.1 is a generally accepted classification of water according to hardness.

Table 10.1 Water Classification According to Hardness

Designation	*Parts per Million* (ppm or mg/l)	*Degrees Clark* (grains/gall)
Soft	0– 50	0– 3.5
Moderately soft	50–100	3.5– 7.0
Slightly hard	100–150	7.0–10.5
Moderately hard	150–200	10.5–14.0
Hard	200–300	14.0–21.0
Very hard	over 300	over 21.0

10.2.3 Temporary and permanent hardness

Temporary hardness is that portion of the dissolved solids which may be precipitated by heating the water. Such hardness is common in areas having a chalky subsoil. In such areas, heating of the water causes the breakdown of the soluble bicarbonate into the insoluble carbonate at the same time releasing carbon dioxide gas. This is indicated by the chemical reaction

$$Ca(HCO_3)_2 + \text{heat} \rightarrow CaCO_3 + H_2O + CO_2 \qquad (10.2)$$

Temporary hardness can present difficulties in hot water and heating systems. If precipitation processes like the one just mentioned occur in such systems then the insoluble salts will deposit as a scale on the inside of pipework and units within the system or collect as a sludge at places in the system (boilers, calorifiers, radiators, unit heaters) where the speed of the water reduces sufficiently to allow settlement to take place. In the sense that a certain very limited amount of scaling can protect pipes and fittings from corrosion, it may be to some extent beneficial. However, scaling reduces heat transfer which in turn leads to waste of energy. Excessive scaling may lead to blockage within narrow passages, which could produce dangerous operating conditions and may result in plant requiring to be shut down for descaling and maintenance.

Permanent hardness is due normally to the presence of mineral salts such as lime sulphates (gypsum or spar) which remain soluble when the water is heated. Such dissolved solids are normally removed by chemical treatment.

10.2.4 Information with respect to water supply

General information relating to the properties of water supplies in the United Kingdom is obtainable from the Water Engineers' Handbook[1] and from the Guide, Book B[2]. This gives the relevant Water Authorities for each town and the water characteristics in terms of temporary, permanent and total dissolved solids.

Specific information is best obtained by consultation with the relevant Water Authority who will generally be only too pleased to give information and instructions in relation to the contents of the supply and the appropriate materials for use in the design of systems.

10.2.5 The pH value

In addition to the hardness characteristics, the engineer is concerned about the nature of the water with respect to acidity or alkalinity. The pH value is a means of measuring this characteristic. Pure (neutral) water contains hydrogen (H^+) and hydroxyl (OH^-) ions produced by the dissociation of some of the water molecules. The concentration of each ion in pure water is 10^{-7} gram-ions per litre at a temperature of 21 °C. The concept of pH is simple in the sense that the power (p) of the index of the hydrogen ion concentration (H) in gram-ions per litre is used to convey the nature of the solution. Pure water at 21 °C would then have a pH of 7.

An increase in the hydrogen ion concentration from 10^{-7} to 10^{-2} gram-ions per litre would mean that the pH of the (acidic) solution was 2. Thus acid solutions are characterised by low pH values and alkaline solutions by high pH values. Strong acid solutions would be about pH 0 while strong alkalis (having around 10^{-14} gram-ions of hydrogen ion per litre) would be pH 14.

10.3 CORROSION

10.3.1 Introduction

Corrosion is often the life determining factor of plant or equipment in practical use and although this chapter is concerned principally with the effects of corrosion within water supply systems consideration must also be given to the effects of corrosion on the outside of these systems. To cite one example of this, the external corrosion of buried pipes in some district heating schemes is so severe that the pipes are perforated within a year, whereas internal corrosion is much less common and much less severe.

10.3.2 Mechanism of corrosion

The rate of corrosion of a particular metal depends on several factors including temperature, but it is established by the presence of a reactant which supports the basic corrosion reaction (really half of a reaction)

metal → metal ion + electrons

Any reaction which consumes electrons can act as the supporting reaction to this basic corrosion reaction. Some common supporting reactions are:

(a) *Oxygen consumption reaction.* Oxygen dissolved in the water reacts with the water and the electrons (from the metal) to produce hydroxyl ions. This encourages the production of more metal ions and electrons and a balance is struck between the two halves of the reaction with the establishment of a corrosion rate. The supporting half reaction may be stated as

oxygen + water + electrons → hydroxyl ions

Of course the two half reactions could be stated within one equation as

metal + oxygen + water ± electrons → metal ions +
+ hydroxyl ions ± electrons

(b) *Hydrogen evolution reaction.* Hydrogen ions present in solution absorb the electrons from the metal producing hydrogen gas, thus

hydrogen ions + electrons → hydrogen gas

Once again a balance is struck between the production of electrons and their consumption by the hydrogen ions. It is worth remembering that hydrogen ions are present in greater numbers in acidic solutions so that this is the most usual supporting reaction in acids.

(c) *Reduction of sulphate ions by anaerobic bacteria.* The problem of buried pipelines was raised in the introduction. External corrosion of pipelines is quite common, particularly when the pipes are buried in clay soils which are so impervious as to exclude oxygen but which are colonised by anaerobic bacteria (which thrive in certain environments which exclude oxygen) capable of reducing sulphate ions (common in soils) in an environment where hydrogen ions and electrons are available, thus

sulphate ions + hydrogen ions + electrons → sulphide
ions + water

Such conditions can produce very rapid corrosion of the pipes and can usually be recognised by the smell of hydrogen sulphide and the appearance of the black slimy sulphide covering the metal surface when the suspect portion of the pipeline is excavated.

10.3.3 Bimetallic or galvanic couples

As already indicated, the *rate* of corrosion is governed often by the rate of arrival of the supporting reactant (e.g. oxygen), but the *intensity* of attack and thus the rate of perforation of say a pipe is often accelerated by the presence of another metal.

It is important to realise that any metal immersed in or in contact with an electrolyte (such as impure water) will undergo the basic corrosion process of

metal → metal ion + electrons

and if no supporting reaction is available the corrosion process ceases at an equilibrium condition at which the metal (covered with electrons) has developed a certain potential, usually measured in mV (Figure 10.1a).

Figure 10.1

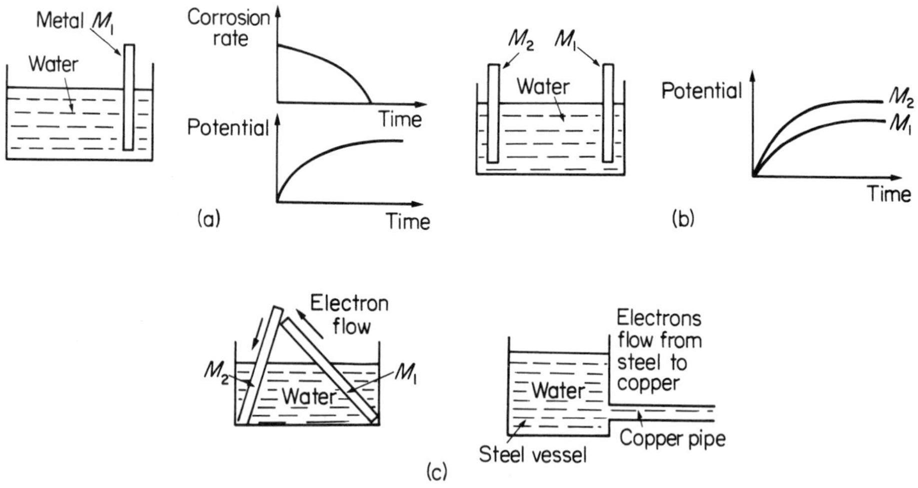

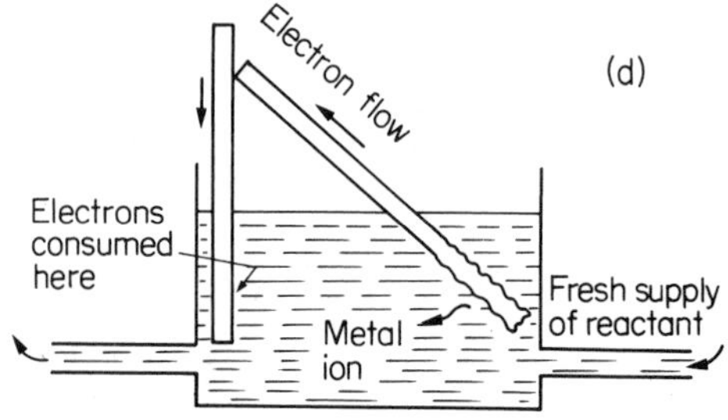

This is true of any metal but each metal reaches the equilibrium condition at its own potential. The same phenomenon would occur with two (or more) metals immersed in the electrolyte (Figure 10.1b).

In a situation where the two metals were connected outside the electrolyte the difference in potentials would create a flow of current (electrons) in the external circuit from the metal at the lower potential to the metal at the higher potential (Figure 10.1c).

As before, if the supporting reactant (e.g. oxygen) is not present then the corrosion process would soon cease. This is precisely the sort of situation which occurs in a water heating system which is closed. Since no fresh water is entering the system carrying oxygen to provide the supporting reaction then no supporting reactant is available and little or no corrosion occurs.

However, the picture is considerably different in the case of open systems (e.g. direct hot water heating) where the supporting reactant is constantly being provided as freshly aerated water is drawn into the system. In such a situation the metal having the lower potential (anode) corrodes to metal ion plus electrons and the electrons pass round the external circuit to the metal having the higher potential (cathode) to be consumed on re-entering the solution, by the supporting reactant (Figure 10.1d).

Thus with bimetallic or galvanic couples or cells, the corrosion process is typified by the gradual dissolution of the metal having the lower potential (the anode), always provided that a supporting reaction is available. Various potential scales exist[3] and Figure 10.2 is a truncated version of the Galvanic Series presented in the Guide, Book B.

Figure 10.2
Galvanic series of metals and alloys

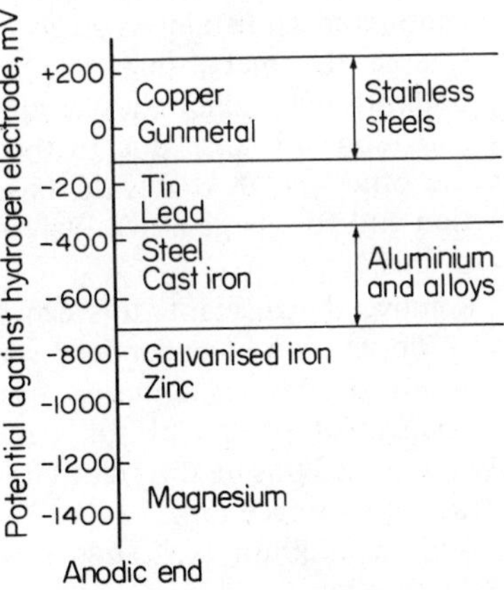

The designer must be very wary of such a series since wrong inferences may be drawn from it. For instance, the stainless steels and aluminium and its alloys take up the positions shown only because of the passivity effects of coherent oxide layers on the metal or alloy surface. Place aluminium in a water solution containing chlorides and it then behaves in a quite different manner, rather more like the active magnesium at the anodic end of the series. This arises from the break-down of the oxide film. A useful introductory series of papers on this subject is given in Reference (3).

Galvanic series such as indicated in Figure 10.2 must thus be treated with caution but nevertheless we see that where, for instance, a piece of steel is coupled to a piece of copper, the steel is more susceptible to corrosion, provided that a supporting reactant such as oxygen or hydrogen ion is available. Generally the corrosion occurs at the junction of the metals and a common misconception is to say that 'the copper causes the steel or iron to corrode'. Of course, the main culprit is the supporting reactant and in its absence iron-copper couples will live quite happily with each other. This is the case (already mentioned) with closed water heating and indirect hot water supply systems and of course, with boilers supplied with de-aerated water.

10.3.4 Prevention of corrosion

Corrosion can often be avoided or prevented at the design stage by the use of suitable materials or by the incorporation of a suitable protection system. The practical details are best found in specialist textbooks but the methods involved are usually based on one or other of the following principles:

(a) Use a material which is less willing to corrode in a specific environment—copper and its alloys (and plastics) are obvious examples.

(b) Use a metal which reacts with its environment to produce a layer of oxide which is so coherent and unreactive that no further reaction will take place—aluminium and stainless steels depend on this effect.

(c) Couple the metal (e.g. iron) to another of lower potential (e.g. zinc) thus ensuring that any resulting corrosion will arise due to the anodic behaviour of the other metal—this is the basis of cathodic protection and to a large extent galvanising is a special case of this.

(d) Remove the reactant—this can be done for example by de-aeration of boiler feed water using mechanical means or by dosing using hydrazine or sulphate. Adjustment of the pH of water would also be an example of this method since it lowers the concentration of hydrogen ions.

(e) Add an inhibitor (e.g. amines) which interferes with the reactions involved.

(f) Add a chemical (e.g. phosphate solution) which encourages the formation of a protective scale.

(g) Cover the metal with a layer of material which separates the reactants and presents a resistance to the electron path. This may well be an inert organic material such as a coating of PVC or it may be a layer of paint which, particularly in the case of specialist primers, contains an inhibitive pigment.

10.3.5 Materials selection in system design

(a) *Pipework*

Lead was the traditional material for cold water systems. Unfortunately its salts are poisonous so that corrosion not only produces leaks in pipework but gives rise to serious contamination. Whether it corrodes in relation to a particular water supply depends very much on the nature of the protective film which usually forms on the surface of the lead, so that it often performs better with hard rather than soft water supplies.

It must also be remembered that lead compounds which are normally insoluble at room temperature, can be very much more soluble at higher temperatures. This makes lead completely unsuitable for hot water pipework.

Although the use of lead has declined very sharply in recent years it is still available particularly in the form of tin-lined lead pipe. Lead pipework is available in the range up to 50 mm diameter but it has poor mechanical strength even at room temperature and the designer must give careful thought to the problems of support and fixing.

Copper is widely used for the range of piped services and has a high resistance to corrosion. However in certain situations (e.g. combination of acidity and chloride content) it is liable to pitting which can lead to failure by pinhole perforation.

Cold water pipework may be liable to pitting where the supply is hard (e.g. boreholes) and contains no organic impurities which tend to form a protective scale on the inside of the pipework. Hot water pipework may be liable to pitting where the supply is soft (e.g. moorland water) and contains manganese ions.

Certain waters are cupro-solvent and while this may present little problem for the copper pipework, deposition of the copper on galvanised components where mixed fittings are used may give rise to corrosion of these components. It seems worth mentioning in passing that this is also the cause of the perforation of aluminium components such as kettles with such waters even though there is no direct connection to pipework.

Steel is widely used because it is readily available in a large range and is relatively inexpensive. This is often offset in practice due to early failure by corrosion and the resulting high maintenance costs inherited by the building owner. It can give good service when used in the correct situations.

Stainless steels refers to a family of steels, each member of which contains sufficient chromium to form a layer of an extremely protective oxide over its surface. This film does not always resist solutions containing chlorides, particularly when hot, and in such circumstances failure may occur by pitting or insidious cracking referred to as 'stress-corrosion' cracking. The corrosion properties are also influenced by the amount of carbon present in the alloy. This can be affected drastically by the process of oxy-acetylene welding and this method of jointing is unsuitable for stainless steel pipework.

The presence of other alloying agents, particularly nickel, alters the mechanical properties of the stainless steels. This is exemplified by the well-known 18Cr/8Ni alloy which is quite malleable and ductile. Other alloys such as those without nickel have good corrosion resistance but have poor workability characteristics.

There is an increasing trend towards the use of stainless steel pipework as a substitute for copper.

Plastics are now widely used for cold water services and have a high resistance to corrosion. Unplasticised polyvinyl chloride (UPVC) is most common but its use is restricted to cold water service pipework since it is a thermoplastic and is not recommended for use at temperatures above 70 °C. Polypropylene can withstand temperatures up to 120 °C and is now being introduced for hot water service pipework.

(b) *Components*

Cast iron is the traditional material for boilers and is of course susceptible to corrosion. One possible form of protection is to cover the material with vitreous enamel.

Steel. It is customary to galvanise steel components such as cisterns and cylinders with a coating of zinc which is anodic to steel. Hard water tends to build up a protective layer on such components while soft water does not and thus great care should be taken when choosing such components when the supply water is soft.

Another difficulty may arise in situations where the water is cupro-solvent, thereby causing deposition of copper from pipework on zinc galvanised steel components. Since both zinc and steel are anodic to copper this increases the possibility of corrosion failure of the component. In the case of cold water components such as cisterns a possible solution is to paint the inside surface with a bituminous paint. With hot water components such as cylinders the likely solution is the use of a sacrificial anode such as magnesium.

One further comment may be of interest in relation to the effect of temperature. In temperature regimes above about 60 to 70 °C the galvanic cell may well reverse and the steel then corrodes, protecting the zinc. This tends to promote very intense local attack, first of the zinc, then of the steel, which may produce a perforation of the tank. A similar effect occurs if particles of metallic copper are allowed to fall into the cistern during pipefitting (even if the water is not cupro-solvent). Thus the use of galvanised steel tanks for hot water storage is unwise but may not result in failure, depending on the total combination of circumstances which includes the temperature effect.

Copper. As with copper pipework, copper components have a high resistance to corrosion. Even in areas where pitting of the copper pipework is observed, little difficulty arises with copper components. Nevertheless, in cases where corrosion tendencies are suspected, a possible solution is the use of a sacrificial anode such as aluminium.

Aluminium, as stated earlier, is a very reactive metal and relies for its protection on an easily formed, readily repaired, strong and coherent film of aluminium oxide. When damaged, the metal will normally react to produce a rapid perforation and this tendency increases with the chloride content of the water. Its potential on the galvanic series reflects the characteristics of the oxide film rather than the metal. As is the case with zinc and steel, aluminium is anodic to copper. Thus any deposition of copper ions on the metal surface will produce conditions (providing, as always, that the supporting reactant is available) which may allow local attack. Nevertheless aluminium is becoming increasingly popular in several applications, notably for the construction of domestic radiators.

Plastics have a high resistance to corrosion and are becoming increasingly popular in certain applications such as the fabrication of cisterns and ballvalves as well as being used for a wide range of sanitary ware.

(c) *Fittings*

Duplex Brasses. Brass is a general name for alloys of copper and zinc. Certain soft water supplies having a high chloride content make certain brasses prone to corrosion attack by a process known as dezincification, in which there is preferential removal of the zinc from the alloy leaving weakened porous copper. Increased temperatures accelerate this form of corrosion which is most likely with 'once through' hot water systems. In some cases the corrosion product is quite voluminous and the phenomenon is known as meringue dezincification, which can cause blockage of the system at fittings such as valves, tees or elbows. The Guide, 1970, Book B, gives useful guidance as to the potentiality of this hazard for different parts of the United Kingdom. The addition of arsenic to brass alloys reduce the possibility of dezincification but gunmetal fittings may be used as an alternative.

In the selection of materials for pipework, fittings and components, there are of course other considerations such as cost, availability, strength, weight and handling, available lengths, thermal expansion, methods of fixing, supporting, jointing, local byelaws and regulations and so on, to be taken into account. Thus considerable skill and knowledge are required to produce a good selection in any particular circumstance.

10.4 WATER TREATMENT OF THE PUBLIC WATER SUPPLY

10.4.1 Traditional method

The traditional method of softening the public water supply to remove temporary hardness is known as Clark's process and consists of adding lime water (calcium hydroxide, $Ca(OH)_2$) to the supply. The soluble calcium bicarbonate is converted to the insoluble carbonate which is then allowed to settle out. The equation is

$$Ca(HCO_3)_2 + Ca(OH)_2 \rightarrow 2CaCO_3 + 2H_2O \qquad (10.3)$$

If the temporary hardness is due to the bicarbonate of magnesium rather than calcium, a similar reaction produces the partially soluble magnesium carbonate and additional lime water converts this to insoluble magnesium hydroxide which is then allowed to settle out.

10.4.2 Lime-soda process

Where permanent hardness is present in the supply (usually due to sulphates such as calcium, $CaSO_4$, or magnesium, $MgSO_4$) and requires to be reduced in quantity, soda ash (sodium carbonate, Na_2CO_3) and lime may be added. The soluble calcium sulphate reacts with the soluble soda ash to produce a precipitate of calcium carbonate

$$CaSO_4 + Na_2CO_3 \rightarrow CaCO_3 + Na_2SO_4 \qquad (10.4)$$

With magnesium sulphate, the soluble sulphate reacts with both the soluble lime and soda ash, to produce insoluble calcium carbonate and magnesium hydroxide which is again precipitated. The equation is

$$MgSO_4 + Ca(OH)_2 + Na_2CO_3 \rightarrow Mg(OH)_2 + CaCO_3 + Na_2SO_4 \tag{10.5}$$

In either case the remaining product of the reaction is sodium sulphate which is soluble and remains in the supply water.

10.4.3 Base exchange process

One problem with the precipitation processes already discussed is the need to remove the precipitated sludge. The base exchange process gets round this difficulty and can be used on the public supply as well as by individual customers. The traditional method used certain silicates of sodium, calcium and alumina, known as Zeolites, and in the reaction there was a change of base between the zeolite and the salts present in the water supply. Thus in the case of a supply having temporary hardness due to the presence of calcium bicarbonate the sodium zeolite in the base exchange bed is converted to calcium zeolite in the bed and the bicarbonate which passes through in the water supply is the harmless one of sodium, as shown by the equation

$$Na_2Al_2Si_2O_8 + Ca(HCO_3)_2 \rightarrow CaAl_2Si_2O_8 + 2NaHCO_3 \tag{10.6}$$

A similar exchange process occurs where permanent hardness exists. Thus in the case of permanent hardness due to the presence of magnesium sulphate the exchange process is

$$Na_2Al_2Si_2O_8 + MgSO_4 \rightarrow MgAl_2Si_2O_8 + Na_2SO_4 \tag{10.7}$$

Ultimately, the sodium of the zeolite is absorbed in the water supply and the zeolite becomes 'exhausted'. Regeneration occurs by closing off the water supply through the bed and allowing a brine solution (sodium chloride) to pass slowly through the bed. The zeolite returns to its original form (sodium silicate) and the chlorides of calcium and magnesium thus formed are removed with the waste water. Modern base exchange plants employ the same principles as outlined here but in place of the naturally occurring zeolites, synthetic resins are employed.

10.4.4 Soft waters

The problems of plumbo-solvency and corrosion associated with certain soft waters have already been raised. The usual remedy with such water supplies is to treat with alkalis to neutralise the acidic nature. Normally, controlled amounts of lime or soda are used for this purpose.

10.5 WATER TREATMENT FOR HEATING AND HOT WATER PLANT

10.5.1 Hot water plant

With direct hot water systems (explained in more detail later) the water for consumption is heated directly in the boiler, the system being open, or once through. Accordingly there is great potential for the build of scale in the boiler. If such systems are employed in areas where the water supply has scale forming properties then external treatment such as base exchange may be required. An alternative would be to dose the supply with sodium metaphosphate to minimise scaling. The metaphosphate also acts as a corrosion inhibitor.

With indirect hot water systems, the water recirculates between the boiler and the calorifier (or hot water cylinder) and the likelihood of scale formation or corrosion is greatly reduced. In such circumstances it is normally sufficient to dose the system with a corrosion inhibitor at the initial and any ensuing fillings of the recirculation system.

10.5.2 Heating plant

While heating plant does not properly come within the scope of this chapter it seems appropriate to cover it at this point with respect to water treatment. Treatment depends broadly on the type of system.

With steam systems, although these are recirculatory in nature, the evaporation process within the boiler leads to a concentration of dissolved solids in the boiler which must be kept to a permissible level (as laid down by the boiler manufacturer) which depends on the level within the supply water and governs the blowdown procedure for the boiler. Generally the water must be softened and deoxygenated and maintained in an alkaline condition. The treatment plant may range from simple dosing equipment where the supply water is soft, to base-exchange or perhaps demineralisation plant in situations where the water supply is hard. With demineralisation plant, the raw make-up water is run through resin beds to remove the solid content and leaves with high purity. Normally a small hydrazine injection plant then doses the water in circulation to the boiler to deoxygenate it. There is always some risk of corrosion in the condensate lines due to the presence of carbon dioxide. Dosing with volatile amines tends to produce a non-wettable film which alleviates this type of problem.

With pressurised hot water systems, make-up water losses may amount to as much as one per cent per week of system capacity. It is common practice in such cases to use demineralisation plant and hydrazine injection as previously mentioned. For more specific information on the range of treatments available the Guide, 1970, Book B should be consulted.

10.6 COLD WATER SUPPLY

10.6.1 Mains distribution

Normally the outlets from service reservoirs are divided into a variety of trunk mains to service different areas, with at least two trunk mains serving each area to reduce the disruptive effect of a fracture in the main. Within each area being served, it is customary to have secondary mains arranged as ring mains connected between the trunk mains. From these secondary mains, the branch, street or service mains are taken and it is from these latter mains that the service pipes are run to individual premises.

Most mains are laid at varying gradients and problems may arise due to the collection of air at high points or of impurities at low points in the pipework system. The provision of air release and washout valves overcomes these difficulties. The service mains are divided into sections and sluice valves (of the gate type) are used to isolate each section for repair, maintenance or any other purpose. In laying water mains, consideration must be given to frost protection and structural considerations such as load due to traffic, backfill, etc. Normally the minimum depth is 0.9 m under roads and 0.75 m where frost is the primary consideration, although local variations may have to be taken into account.

10.6.2 Pressure and flow in mains

Service mains must provide sufficient pressure and quantity of water to the various premises connected to them. For normal low-rise housing and other low properties, a pressure of about 30 m head of water is thought to be sufficient for fire-fighting, while pressure above 70 m will lead to waste and excessive noise in the pipework within the premises. Taller buildings may require to use additional pumps to boost the mains supply pressure. Thus pressure considerations affect the relative siting of the service reservoir and the areas to be served by it.

The mains must also meet the maximum demand of the outlets which are connected to it. Most often the deciding factor in sizing the main are the requirements for fire-fighting services. Service mains are normally in the range of 75 to 300 mm diameter with 100 and 150 mm diameter pipes probably being the most common sizes.

Cast iron, spun iron, steel, asbestos cement, reinforced concrete and most recently plastic, are all materials used for the construction of cold water mains. These materials, and their assembly, must conform to the relevant British Standards. The pipework requires to be tested to twice its working pressure.

10.6.3 The service pipe

Copper, lead, galvanized mild steel and plastic are the materials most often used for the service pipe which is run from the water main to the building. The water authority normally makes the connection to the main and terminates their connecting pipe at a stop valve which is most often placed below the footpath just outside the premises, as indicated in Figure 10.3. The remainder of the supply pipe or 'rising main' runs from the stop valve usually to a cold water storage cistern within the building.

Figure 10.3
The service pipe

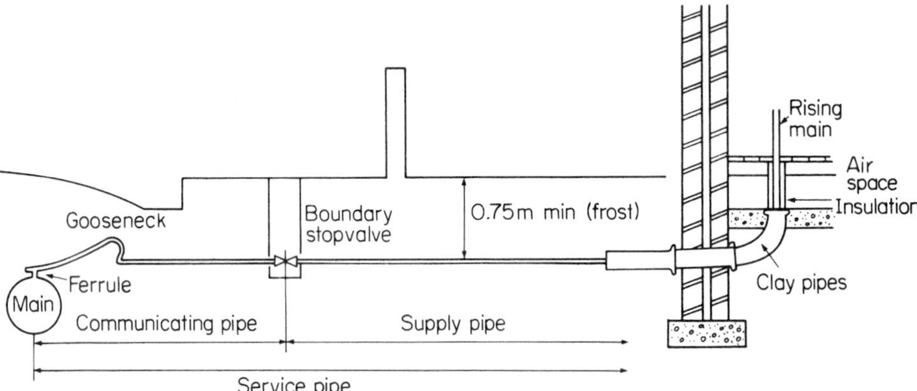

In the case of domestic properties or where demand is small, most often connection is made to the water main without shutting off the mains supply. This is done by clamping a special watertight box to the main, then drilling and tapping the main and introducing a ferrule into the hole, the operation being achieved by a special revolving tool box which sits atop the watertight box. The connecting or communicating pipe is then connected to the ferrule with a 'goose neck' if lead is used or a large radius or slow bend in the case of other materials. This allows some play in the event of settlement. An internal screw plug within the ferrule can now be raised and this allows water to flow from the main into the connecting pipe.

With larger buildings or in cases where a 100 mm diameter connection is needed for fire-fighting purposes, it is necessary to shut off the relevant section of the main and introduce a tee into the run of the main.

10.6.4 Cold water storage

Although there is no specific requirement for the storage of cold water within the Model Water Byelaws, storage is general practice. In any case there are situations (a feed storage cistern is required for a hot water system when the hot water storage capacity exceeds 70 litres) where it is necessary, to comply with regulations. Again (as in the case of nature of water supply, type of treatment and type of materials used) the designer, in relation to storage, will benefit greatly by discussing the problem with the water authority.

So far as domestic premises are concerned, CP310 Water Supply[4] gives guidance on the storage (litres per person) based

Table 10.2 Domestic Storage to Cover 24 hours Interruption

Type of Building	*Storage* (litres per person)
Dwelling houses and flats	90
Hostels	90
Hotels	135
Nurses' homes and medical quarters	115
Offices with canteens	45
Offices without canteens	35
Restaurants, per meal	7
Schools: Boarding	90
Day	30

on 24 hour interruption of the supply. The table is reproduced here as Table 10.2.

Water usage is so much more diverse with industrial processes that it is difficult to give guidance. The main difficulty is likely to be in relation to processes with intermittent requirements for large quantities of water. On calculation of the demand in such cases, consultation with the water authority will be helpful in achieving a suitable storage quantity.

Benefits which accrue from the storage of water include:

(i) reserve against failure or necessary interruption of supply,

(ii) economy in the size of water mains and service pipes since demand is partially met from storage,

(iii) reduced pressures on the distribution pipework within the building resulting in less noise and less waste of water and use of lighter gauges of pipework and components.

On the other hand, storage consumes space in the building and introduces structural implications. Also, since a reduced pressure is available, the distribution pipework is generally of larger diameter.

10.6.5 Storage cisterns

Domestic cisterns traditionally have been fabricated in galvanised sheet steel but modern materials include asbestos cement and plastic. The Model Water Byelaws require a minimum cistern capacity of 115 litres in situations where the storage is for cold water supplies only. Where the cistern has in addition to act as a feed for hot water supply the requirement is for a minimum capacity of 230 litres. This capacity can be achieved by the use of more than one cistern.

In non-domestic situations, the size and capacity of the cistern must be determined in agreement with the water authorities in relation to the nature, number and type of outlets to be serviced. Large storage cisterns are often constructed from iron or steel panels or in some cases formed in concrete and lined with bitumen.

Precautions must be taken in the siting of the cistern. It must be placed in a reasonably thermally uniform environment and it is desirable that it should have a loose cover. Insulation of the walls and cover of the cistern is necessary for frost protection. Figure 10.4 shows a typical detail of a domestic cold water storage cistern.

Figure 10.4
Domestic cold water storage cistern

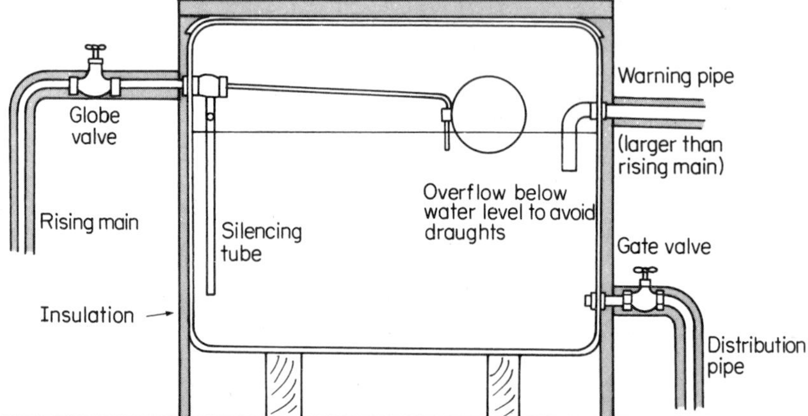

In larger installations, two or more cisterns may be employed, each having its own inlet supply via a ball valve. The outflows pass to a manifold from which the distribution pipes carry the cold water to the various outlets. By judicious siting of valves it is then possible to isolate any one cistern for routine maintenance or cleaning without causing too great a disturbance to the supply of water within the building.

10.6.6 Domestic cold water supply system

Normal practice is to serve directly from the main, one sanitary fitting and one outlet for drinking purposes—normally the kitchen sink. The remainder of the supply is taken to the storage tank. There is usually a consumer's stop valve adjacent to the entry of the pipe within the building with a draincock close to it. All pipework should be firmly fixed in straight runs preferably with even gradients. This reduces the possibilities of water hammer and/or airlocks and also of dips which could retain water in the event of the system being drained for frost precautions. Figure 10.5 shows a typical layout of a cold water service system.

The cold water distribution system consists of pipework which is run from the cistern to the various sanitary appliances within the building and to the hot water storage container. It is useful to be able to isolate branches for maintenance purposes and since the head available is limited, it is customary to use gate valves rather than globe valves, as they have a smaller resistance to flow, in such circumstances. Figure 10.6 shows a typical layout of a cold water distribution system.

Similar principles relate to the layout of distribution pipework

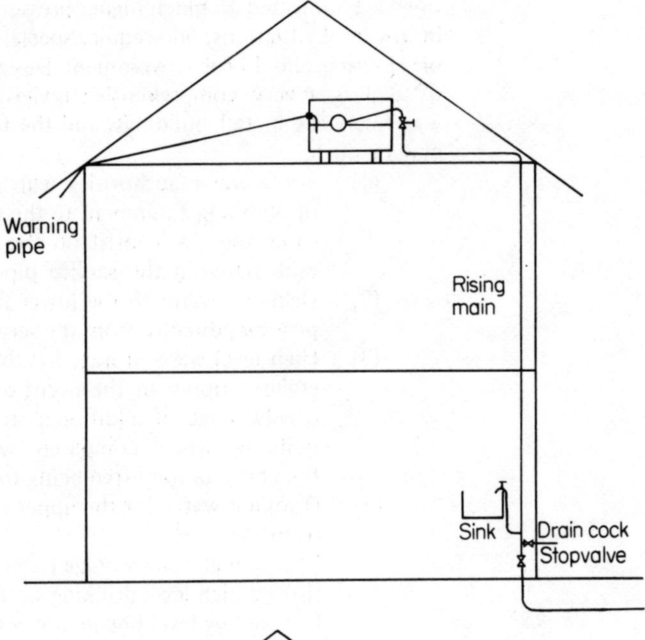

Figure 10.5
Cold water service system

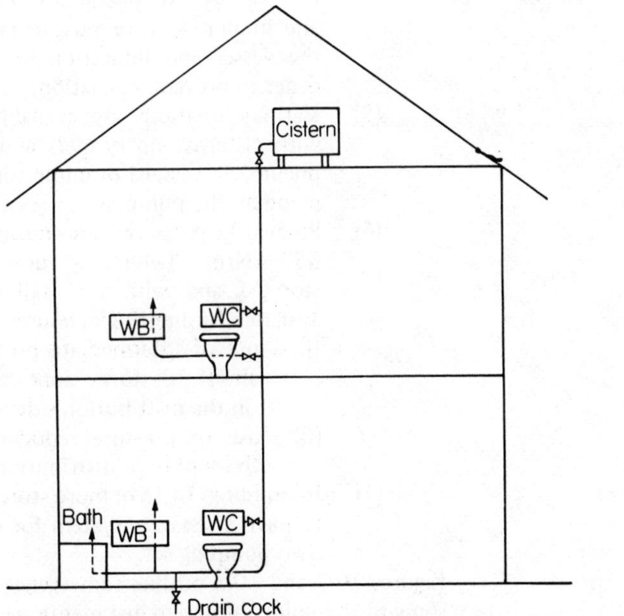

Figure 10.6
Cold water distribution system

in larger buildings with more extensive arrangements. In the case of tall buildings special problems arise and these will now be considered.

10.6.7 Cold water supply to tall buildings

In many cases pumps will be required within the building to raise the water to the upper levels. In these buildings, pipework and fittings are subjected to much higher pressures than are experienced in low-level situations, and require special solutions. The Ministry of Housing and Local Government Design Bulletin No. 3, Part 6[(5)], gives a very comprehensive review of the design of cold water services in tall buildings, and the following general points may be made:

(1) Some water authorities will allow direct connection of pumping equipment to the service pipe but in most cases they will insist on the installation of a break tank between the service pipe and the pumping set.

(2) Drinking water to the lower floors is supplied, where possible, directly from the service pipe.

(3) High level water storage has the advantage of allowing gravity supply in the event of power failure but the disadvantage of additional structural loading on the building when compared with low-level storage, the compromise often being to have both.

(4) Drinking water for the upper storeys may be supplied from:
   (i) a high level storage cistern,
   (ii) a high level drinking water header,
   (iii) a low level pneumatic vessel,

   and in all cases care must be taken as to cleanliness of the vessel and limitation of the quantity stored in order to prevent stagnation,

(5) Various methods are available (water level control with cisterns and headers and pressure control with pneumatic vessels) of minimising the number of starts made by the pump(s).

(6) Pipework pressures are normally limited to about 350 kN/m^2 (which is equivalent to just over 10 storeys) and with very tall buildings the methods available for limiting pressures are:
   (i) use of intermediate break tanks or cisterns in about 10 storey zones either on the supply or on the distribution side,
   (ii) use of pressure reducing valves within either the supply or distribution pipework.

(7) In buildings of 15 or more storeys it is normal practice to have a separate system for drinking water with its own pumping set.

Figures 10.7 and 10.8 outline some general systems which cover some of the points of design just mentioned.

Pump starting is minimised in the system shown in Figure 10.7 in that a predetermined quantity of water must be consumed in either the storage cistern or the drinking water header to cause operation of the float switch to energise the pump. With the system shown in Figure 10.8, the pump runs for a period and pressurises a quantity of air contained in the pneumatic air vessel and at the same time raises the water level within the cylinder between predetermined limits (of the order of about 4.5 to 9

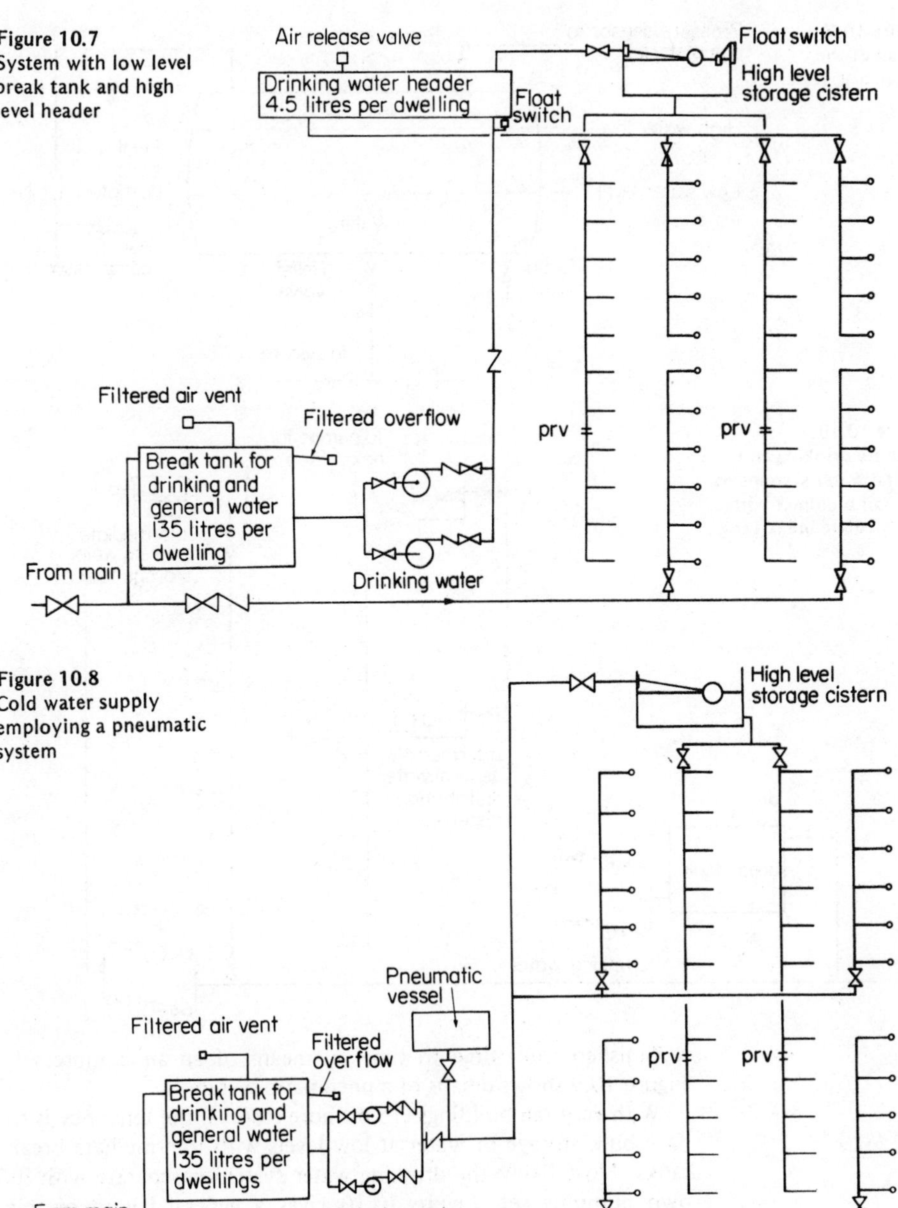

Figure 10.7
System with low level break tank and high level header

Figure 10.8
Cold water supply employing a pneumatic system

litres per dwelling). The pump may then be switched off either by means of a float switch operating between these water levels or by means of a pressure switch operating from the air pressure within the vessel. As cold water is consumed for drinking or general purposes, it is replenished by virtue of the potential energy stored in the air, which gradually becomes exhausted and requires the pump to re-start after a period of time. Air is gradually consumed within the water in the vessel and requires to be

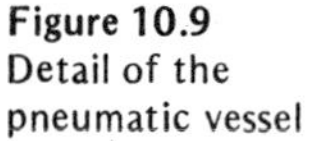

Figure 10.9
Detail of the pneumatic vessel

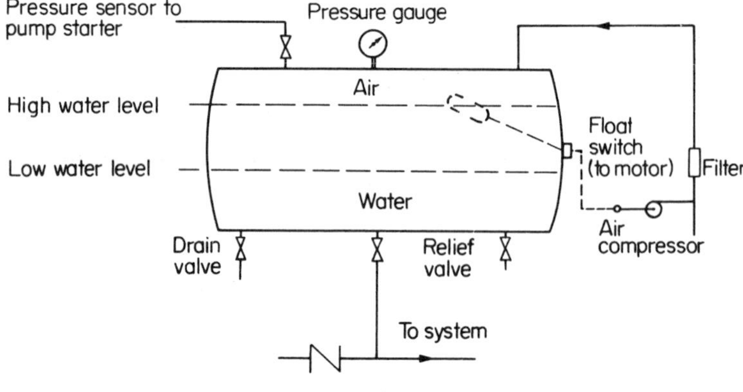

Figure 10.10
Separate drinking and general water systems for very tall buildings with intermediate break tank

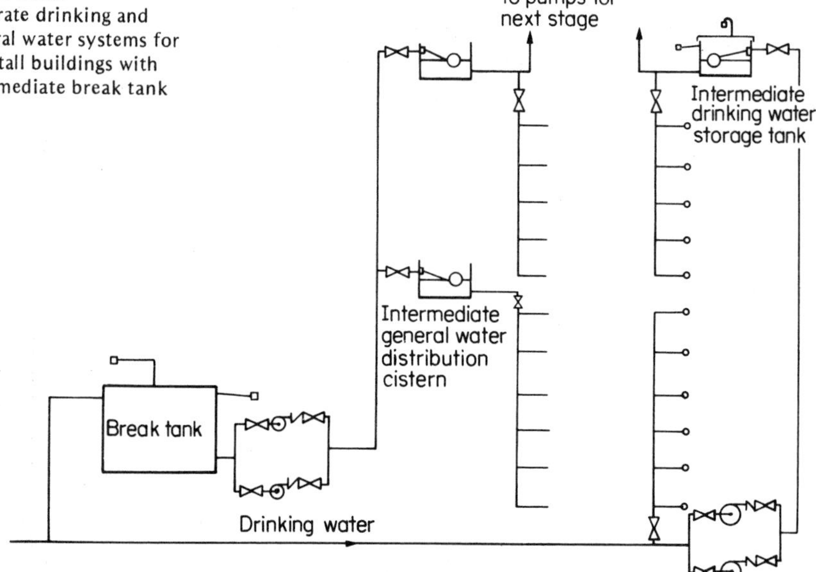

replenished from time to time by means of an air compressor. Figure 10.9 shows details of a pneumatic system.

With very tall buildings (15 or more storeys) the tendency is to have bulk storage of water at low levels with intermediate break tanks. Most likely the drinking water system is separate with its own pumping set. Figure 10.10 gives a general layout of this approach.

10.7 SIZING THE COLD WATER SYSTEM

10.7.1 Cisterns

The system consists, as we have seen, of storage tanks or cisterns, pumps, pipework and various fittings. Information on quantities to be stored has already been given and BS 417[(6)] gives a table of cisterns which indicates for each cistern the nominal capacity derived from the stated length, width and height, together with a capacity taken at the water line which takes into account the true stored quantity. Thus in the case of a domestic system for a single dwelling, a cistern having a BS number C10, being 0.81 m long by 0.66 m wide by 0.61 m high, has a nominal capacity of 318 litres and a working capacity of 232 litres. This may be selected in any one of a variety of materials, and other factors such as the nature of the water supply may determine which is the appropriate material in a specific case.

The sizing of pumps has already been dealt with in Chapter 9 and we shall now turn our attention to the problem of diversity.

10.7.2 Flow rates at sanitary fittings

As a prelude to pipe sizing the first thing to be considered is the flow rate in the pipework supplying the sanitary fittings. The major difficulty in assessing the flow rate in the system is the variety of flow possible as a result of the combinations of sanitary fittings which may be in use at any one time. In the case of individual dwellings it is customary to make no allowance for diversity and to design on the basis that all fittings will be in operation at the same time. On larger systems the design flow is assessed making an appropriate allowance for simultaneous demand. In any case, the starting point is the flow rate which is required at the sanitary fittings and both CP 310, and the Guide give similar tables of such information. The table in the Guide is included here (Table 10.3).

Table 10.3 Flow Rates at Sanitary Fittings

Fitting	*Flow Rate* (l/s)
Basin (spray)	0.05
Basin (tap)	0.15
Bath (private) (18 mm tap)	0.30
Bath (public) (25 mm tap)	0.60
Flushing cistern	0.10
Shower (nozzle)	0.15
*Shower (100 mm rose)	0.40
Sink (15 mm tap)	0.20
Sink (20 mm tap)	0.30
Wash fountain	0.40

*The use of shower roses results in wasteful consumption of water.

10.7.3 Diversity in estimation of design flow rate

Over the years a considerable volume of information has been assembled in relation to usage patterns of sanitary fittings and, for the purposes of design, three broad areas emerge. These are domestic, public and congested, the last of these areas relating to such situations where there is 'organised' use of sanitary fittings such as occurs at washing-up time in a factory or works. The Guide (B4) outlines two methods of dealing with this problem.

Method of binomial and poisson distributions

This approach is based on the use of the theory of probability and relates to a system consisting of a range of the one type of fitting. The number of fittings simultaneously discharging is obtained from the expression:

$$m \simeq np' + x\,[2np'(1-p')]^{0.5} \tag{10.8}$$

where m = number of fittings simultaneously discharging,
n = total number of connected fittings,
p' = usage ratio t/T,
t = time of inflow in seconds,
T = time between use in seconds,
x = coefficient equivalent to an Error Function.

A value of 1.8 is normally taken for x and this represents an acceptable level of error of around 1.1%.

The method is really suitable only for systems containing one type of fitting. It can be adapted to cover systems with two types of fitting but the procedure becomes very cumbersome and recourse is normally taken to the use of the method of demand units.

Method of demand units

This approach is a method of ascertaining the load which each fitting places on the system. It takes account of the inflow rate and the capacity of the fitting in addition to the frequency of use. This information is presented initially as Table 10.4. The information is then reformed into a scale of theoretical demand units taking the cold water supply to a lavatory basin at a frequency of once per five minutes as being representative of one demand unit. This table is presented as Table 10.5. A practical scale of demand units is obtained broadly by multiplying the values in Table 10.5 by ten, and rounding up to the nearest whole number. This table is presented as Table 10.6. It is used with a table linking total demand units to design flow rate. The design flow rate table is presented as Table 10.7. Tables 10.4 to 10.7 are reproduced from the Guide, Book B.

In order to assess the design flow rate for any portion of the pipework, the designer adds the demand units for each fitting served by that pipe. The appropriate design flow rate is then obtained from Table 10.7. CP 310 Water Supply proposes a similar approach to this problem. It lists loading units for different types of sanitary fittings and when the loading units are added for a particular portion of pipe, the design flow rate is obtained from a graph which relates loading units to flow rate. The following examples illustrate the techniques discussed.

Table 10.4 Sanitary Fitting Demand Data

Fitting	Detail				Times		Usage ratio
	Capacity (litre)	Temp. (°C)	Flow (litre/s)	Quantity (litre)	t	T	p'
Basin	5	45	0.15	2.5	17	1200	0.014
						600	0.028
						300	0.056
Basin	5	10	0.15	5.0	33	1200	0.028
						600	0.056
						300	0.112
Bath	80	45	0.3	40	133	4800	0.028
						2400	0.056
						1200	0.112
Bath	80	10	0.3	80	266	4800	0.056
						2400	0.112
						1200	0.224
Sink	12	10	0.2	12	60	1200	0.05
		or				600	0.1
		70				300	0.2
Sink	18	10	0.2	18	60	1200	0.05
		or				600	0.1
		70				300	0.2
Urinal	4	10	*	4.5	1200	1200	1.0
WC	13.5	10	0.1	13.5	135	1200	0.112
						600	0.224
						300	0.448
WC	9	10	0.1	9	90	1200	0.075
						600	0.15
						300	0.3

*Negligible (0.003 litre/s) but running continuously.

Table 10.5 Theoretical Demand Units

Fitting		Interval of use (minutes)				
Type	*Detail*	5	10	20	40	80
Basin	Hot	0.50	0.26	0.11	–	–
Basin	Cold	1.00	0.50	0.25	–	–
Bath	Hot	–	–	2.45	1.23	0.60
Bath	Cold	–	–	4.70	2.45	1.23
Sink	Small	4.28	2.18	1.09	–	–
Sink	Large	4.28	2.18	1.09	–	–
Urinal	Per stall	–	–	–	–	–
WC	13.5 litre	3.45	1.50	0.75	–	–
WC	9 litre	2.18	0.99	0.48	–	–

Table 10.6 Practical Demand Units

Fitting	Type of application		
	5 min Congested	*10 min Public*	*20 min Private*
Basin*	10	5	3
Bath†	47	25	12
Sink	43	22	11
Urinal cistern‡	—	—	—
WC (13.5 litre)	35	15	8
WC (9 litre)	22	10	5

*These data apply to tap supply only. If spray taps be used, demand may be continuous at 0.05 litre/s per tap.
†If a shower spray nozzle be used over the bath, demand may be continuous at 0.1 litre/s per nozzle.
‡Demand will be continuous at 0.003 litre/s per stall.

Example 10.1 A cold water system in a factory has a distribution to 35 washbasins and a separate distribution to 35 WCs. In each case determine:

(i) the number of fittings in simultaneous use with relation to the design flow rate,
(ii) the design flow rate.

Washbasins

$p' = 0.112$ (from Table 10.4)

$m \simeq np' + x\,[2np'(1 - p')]^{0.5}$ and $x = 1.8$

$\simeq (35 \times 0.112) + 1.8\,[2 \times 35 \times 0.112(1 - 0.112)]^{0.5}$

$\simeq 3.92 + 4.75$

$\simeq 8.67$, say 9 fittings

inflow rate per fitting = 0.15 l/s (from Table 10.4)

Thus design flow rate = 1.35 l/s

WCs

As above, and for 9 litre cisterns, 17.4 say 18 fittings, giving a design flow rate of 1.8 l/s.

Example 10.2 A cold water supply cistern feeds one bathroom on each floor of a four-storey block of flats. The fittings served in each bathroom are a WC, washbasin, shower and bath, as shown in Figure 10.11. Determine the design flow rates for each of the pipe sections numbered on the sketch by the demand unit method.

From Table 10.6, the practical demand units per fitting are: WC 5; washbasin 3; bath 12. For the showers assume a continuous use of 0.1 l/s.

Next add the demand units carried by the numbered pipe sections:

Section 1 carries 80 demand units plus 0.4 l/s
Section 2 carries 20 demand units plus 0.1 l/s
Section 3 carries 15 demand units plus 0.1 l/s
Section 4 carries 12 demand units

The design flow rate can now be obtained with reference to Table 10.7, as follows:

Table 10.7 Design Flow Rates based on Demand Units

Demand units	*Design Demand* (litre/s) 0	50	100	150	200	250	300	350	400	450	500	550	600	650	700	750	800	850	900	950
0	0.0	0.3	0.5	0.6	0.8	0.9	1.0	1.2	1.3	1.4	1.5	1.6	1.7	1.9	2.0	2.1	2.2	2.3	2.4	2.5
1 000	2.6	2.7	2.8	2.9	3.0	3.1	3.2	3.3	3.4	3.5	3.6	3.7	3.8	3.9	4.0	4.1	4.2	4.3	4.4	4.5
2 000	4.6	4.7	4.8	4.9	5.0	5.1	5.1	5.2	5.3	5.4	5.5	5.6	5.7	5.8	5.9	6.0	6.1	6.2	6.3	6.4
3 000	6.4	6.5	6.6	6.7	6.8	6.9	7.0	7.1	7.2	7.3	7.4	7.4	7.5	7.6	7.7	7.8	7.9	8.0	8.1	8.2
4 000	8.3	8.3	8.4	8.5	8.6	8.7	8.8	8.9	9.0	9.1	9.1	9.2	9.3	9.4	9.5	9.6	9.7	9.8	9.8	9.9
5 000	10.0	10.1	10.2	10.3	10.4	10.5	10.5	10.6	10.7	10.8	10.9	11.0	11.1	11.2	11.2	11.3	11.4	11.5	11.6	11.7
6 000	11.8	11.9	11.9	12.0	12.1	12.2	12.3	12.4	12.5	12.5	12.6	12.7	12.8	12.9	13.0	13.1	13.1	13.2	13.3	13.4
7 000	13.5	13.6	13.7	13.7	13.8	13.9	14.0	14.1	14.2	14.3	14.3	14.4	14.5	14.6	14.7	14.8	14.9	14.9	15.0	15.1
8 000	15.2	15.3	15.4	15.5	15.5	15.6	15.7	15.8	15.9	16.0	16.0	16.1	16.2	16.3	16.4	16.5	16.6	16.6	16.7	16.8
9 000	16.9	17.0	17.1	17.2	17.2	17.3	17.4	17.5	17.6	17.7	17.7	17.8	17.9	18.0	18.1	18.2	18.2	18.3	18.4	18.5
10 000	18.6	18.7	18.8	18.8	18.9	19.0	19.1	19.2	19.3	19.3	19.4	19.5	19.6	19.7	19.8	19.8	19.9	20.0	20.1	20.2
11 000	20.3	20.3	20.4	20.5	20.6	20.7	20.8	20.8	20.9	21.0	21.1	21.2	21.3	21.3	21.4	21.5	21.6	21.7	21.8	21.8
12 000	21.9	22.0	22.1	22.2	22.3	22.3	22.4	22.5	22.6	22.7	22.8	22.8	22.9	23.0	23.1	23.2	23.3	23.3	23.4	23.5
13 000	23.6	23.7	23.8	23.8	23.9	24.0	24.1	24.2	24.3	24.3	24.4	24.5	24.6	24.7	24.7	24.8	24.9	25.0	25.1	25.2
14 000	25.2	25.3	25.4	25.5	25.6	25.7	25.7	25.8	25.9	26.0	26.1	26.2	26.2	26.3	26.4	26.5	26.6	26.6	26.7	26.8
15 000	26.9	27.0	27.1	27.1	27.2	27.3	27.4	27.5	27.6	27.6	27.7	27.8	27.9	28.0	28.0	28.1	28.2	28.3	28.4	28.5
16 000	28.5	28.6	28.7	28.8	28.9	29.0	29.0	29.1	29.2	29.3	29.4	29.4	29.5	29.6	29.7	29.8	29.9	29.9	30.0	30.1
17 000	30.2	30.3	30.3	30.4	30.5	30.6	30.7	30.8	30.8	30.9	31.0	31.1	31.2	31.2	31.3	31.4	31.5	31.6	31.7	31.7
18 000	31.8	31.9	32.0	32.1	32.1	32.2	32.3	32.4	32.5	32.6	32.6	32.7	32.8	32.9	33.0	33.0	33.1	33.2	33.3	33.4
19 000	33.5	33.5	33.6	33.7	33.8	33.9	33.9	34.0	34.1	34.2	34.3	34.3	34.4	34.5	34.6	34.7	34.8	34.8	34.9	35.0

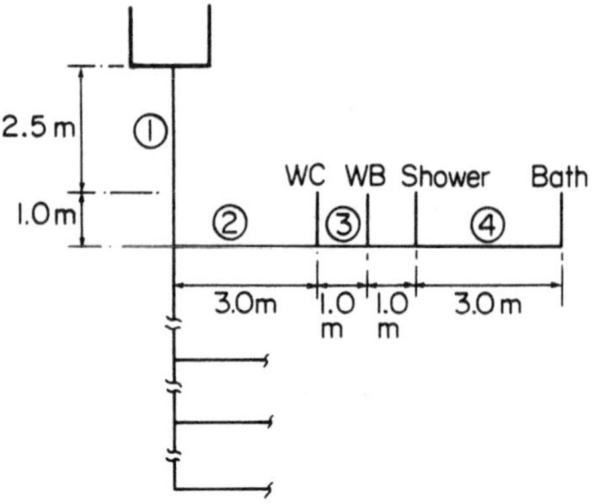

Figure 10.11 Layout of the system for Example 10.2

Section 1: 0.3 + 3/5(0.5 – 0.3) + 0.4 = 0.82 l/s
Section 2: 2/5(0.3 – 0) + 0.1 = 0.22 l/s.

Now notice that the discharge rate for a bath is 0.3 l/s which is in excess of the suggested design flow rate of 0.22 l/s calculated for pipe 2. In fact there is little point in applying diversity when the demand units fall below about 50. Thus:

Section 2: 0.1 + 0.15 + 0.1 + 0.3 = 0.65 l/s
Section 3: (0.65 – 0.1) = 0.55 l/s
Section 4: supplies only bath and gives 0.3 l/s.

10.7.4 Estimation of the available pressure

Cold water supply

With smaller buildings the supply pipework consists normally of a service pipe and rising main providing the supply from the water main to a storage cistern (see Figures 10.3, 10.4 and 10.5). With tall buildings the supply may well be drinking water to a group of fittings directly from the main together with water for other purposes supplied by pump from a low-level storage cistern to a high-level storage cistern (see Figures 10.7 and 10.8). Consider the system in Figure 10.12 as being broadly representative of a cold water supply system with the pressure in the water main being p_m above atmospheric pressure.

Now apply the energy equation between 1 and 2,

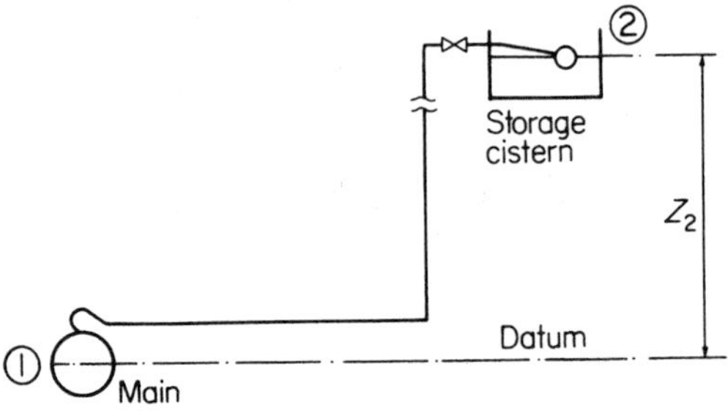

Figure 10.12

$$\frac{p_1}{\rho} + \frac{V_1^2}{2} + gz_1 = \frac{p_2}{\rho} + \frac{V_2^2}{2} + gz_2 + \Sigma \text{ losses}_{1\rightarrow 2}$$

Since p_2 is atmospheric pressure then $p_1 - p_2 = p_m$
Also V_1 and V_2 may be considered to be negligibly small.
Since z_1 is also zero, then

$$\frac{p_m}{\rho} = gz_2 + \Sigma \text{ losses}_{1\rightarrow 2}$$

or the available pressure for pipesizing

$$= p_m - \rho g z_2 = \rho \, \Sigma \text{ losses}_{1\rightarrow 2} \tag{10.9}$$

Cold water distribution

Consider the system in Figure 10.13 with an available head from the free water surface in the tank to the fitting outlet of H and from the base of the tank to the fitting outlet of h.

Figure 10.13

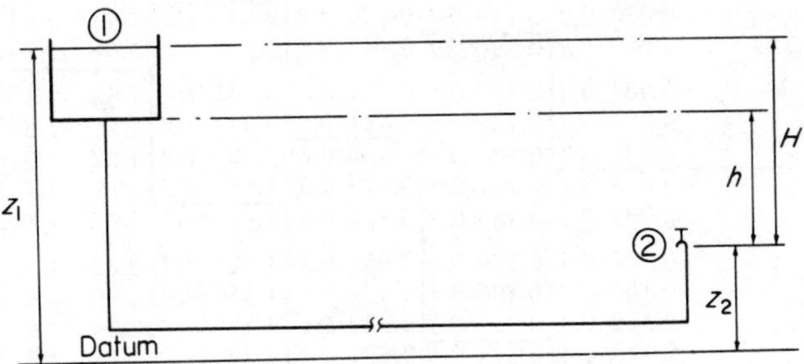

Now apply the energy equation between 1 and 2,

$$\frac{p_1}{\rho} + \frac{V_1^2}{2} + gz_1 = \frac{p_2}{\rho} + \frac{V_2^2}{2} + gz_2 + \Sigma \text{ losses}_{1\rightarrow 2}$$

and $g(z_1 - z_2) = gH$, thus

$$gH = \frac{V_2^2}{2} + \Sigma \text{ losses}_{1\rightarrow 2}$$

It is normal practice to apportion the available head from the base of the tank to cater for the losses, so that

$$gh = \Sigma \text{ losses}_{1\rightarrow 2}$$

or the available pressure for pipesizing

$$= \rho g h = \rho \, \Sigma \text{ losses}_{1\rightarrow 2} \tag{10.10}$$

leaving the head within the tank to provide the leaving velocity at the fitting, thus:

$$g(H - h) = \frac{V_2^2}{2} \tag{10.11}$$

The designer has to select the appropriate sizes of pipework for the system in order that with the pressure differences available

Table 10.8 Flow of Water at 10 °C in Light Gauge Copper Pipes

		12 mm		*15 mm*		*22 mm*		*28 mm*		*35 mm*		*42 mm*			
		3/8 in		*1/2 in*		*3/4 in*		*1 in*		*1 1/4 in*		*1 1/2 in*			
Δp (N/m²/m)	V (m/s)	ṁ	EL	ṁ	EL	ṁ	EL	ṁ	EL	ṁ	EL	ṁ	EL	V (m/s)	Δp (N/m²/m)
475		0.047	*0.3*	0.088	*0.4*	0.260	*0.7*	0.527	*1.0*	0.952	*1.4*	1.61	*1.8*		475
500		0.048	*0.3*	0.091	*0.4*	0.268	*0.7*	0.542	*1.0*	0.980	*1.4*	1.66	*1.8*		500
550		0.051	*0.3*	0.096	*0.4*	0.283	*0.7*	0.572	*1.0*	1.04	*1.4*	1.75	*1.8*		550
600		0.054	*0.3*	0.101	*0.4*	0.297	*0.7*	0.601	*1.0*	1.09	*1.4*	1.83	*1.8*	1.5	600
650		0.056	*0.3*	0.106	*0.4*	0.311	*0.7*	0.629	*1.0*	1.14	*1.4*	1.92	*1.9*		650
700		0.059	*0.3*	0.110	*0.4*	0.324	*0.7*	0.656	*1.1*	1.19	*1.4*	2.00	*1.9*		700
750		0.061	*0.3*	0.115	*0.4*	0.337	*0.7*	0.682	*1.1*	1.23	*1.4*	2.08	*1.9*		750
800		0.063	*0.3*	0.119	*0.4*	0.350	*0.7*	0.708	*1.1*	1.28	*1.5*	2.16	*1.9*		800
850		0.066	*0.3*	0.123	*0.4*	0.362	*0.8*	0.732	*1.1*	1.32	*1.5*	2.23	*1.9*		850
900		0.068	*0.3*	0.127	*0.4*	0.374	*0.8*	0.757	*1.1*	1.37	*1.5*	2.30	*1.9*		900
950		0.070	*0.3*	0.131	*0.4*	0.386	*0.8*	0.780	*1.1*	1.41	*1.5*	2.37	*1.9*		950
1000		0.072	*0.3*	0.135	*0.4*	0.398	*0.8*	0.803	*1.1*	1.45	*1.5*	2.44	*2.0*	2.0	1000
1100		0.076	*0.3*	0.143	*0.4*	0.420	*0.8*	0.847	*1.1*	1.53	*1.5*	2.58	*2.0*		1100
1200		0.080	*0.3*	0.150	*0.4*	0.441	*0.8*	0.890	*1.1*	1.61	*1.5*	2.71	*2.0*		1200
1300		0.084	*0.3*	0.157	*0.5*	0.461	*0.8*	0.931	*1.1*	1.68	*1.6*	2.83	*2.0*		1300
1400		0.088	*0.3*	0.164	*0.5*	0.481	*0.8*	0.971	*1.2*	1.75	*1.6*	2.95	*2.0*		1400
1500	1.0	0.091	*0.3*	0.171	*0.5*	0.500	*0.8*	1.00	*1.2*	1.82	*1.6*	3.06	*2.1*		1500
1600		0.095	*0.3*	0.177	*0.5*	0.519	*0.8*	1.05	*1.2*	1.89	*1.6*	3.18	*2.1*		1600
1700		0.098	*0.3*	0.184	*0.5*	0.537	*0.8*	1.08	*1.2*	1.95	*1.6*	3.29	*2.1*		1700
1800		0.101	*0.3*	0.190	*0.5*	0.555	*0.8*	1.12	*1.2*	2.02	*1.6*	3.39	*2.1*		1800
1900		0.104	*0.3*	0.196	*0.5*	0.572	*0.8*	1.15	*1.2*	2.08	*1.6*	3.50	*2.1*		1900
2000		0.108	*0.3*	0.201	*0.5*	0.589	*0.8*	1.19	*1.2*	2.14	*1.6*	3.60	*2.1*	3.0	2000
2250		0.115	*0.4*	0.215	*0.5*	0.629	*0.9*	1.27	*1.2*	2.28	*1.7*	3.84	*2.2*		2250
2500		0.122	*0.4*	0.229	*0.5*	0.668	*0.9*	1.35	*1.2*	2.42	*1.7*	4.07	*2.2*		2500
2750		0.129	*0.4*	0.242	*0.5*	0.705	*0.9*	1.42	*1.3*	2.55	*1.7*	4.29	*2.2*		2750
3000	1.5	0.136	*0.4*	0.254	*0.5*	0.740	*0.9*	1.49	*1.3*	2.68	*1.7*	4.51	*2.2*		3000
3250		0.142	*0.4*	0.266	*0.5*	0.774	*0.9*	1.56	*1.3*	2.80	*1.7*	4.71	*2.2*		3250

across the pipework, the design flow rates may be achieved at the fittings.

The losses indicated in the preceding analysis are in part due to the resistance of pipework and in part due to the resistance of fittings. In Chapter 9 we saw how the effective length of a pipe could be obtained from the sum of the physical length and the equivalent length due to fittings (equation 9.4). Thereafter it is a simple matter to determine the available pressure drop per metre length and then go on to size the pipe in conjunction with the design flow rate. Table 10.8 is an abstract of a pipesizing table from the Guide, Book C[7].

10.7.5 Pipe sizing for cold water supply systems

The techniques explained in Section 10.7.3 are used to assess the design flow rates and the available pressure per metre length of pipe is obtained by the method explained in Section 9.4.4. The designer is then in a position to perform a preliminary pipesizing.

The approach is perhaps best understood by recourse to worked examples.

Example 10.3 The drinking water supply to the six lower storeys of a tall building is taken straight from the main to kitchen sinks (one per storey). In relation to Figure 10.14, size pipework (preliminary sizing) assuming that the mains pressure is 300 kN/m^2.

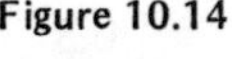

Figure 10.14

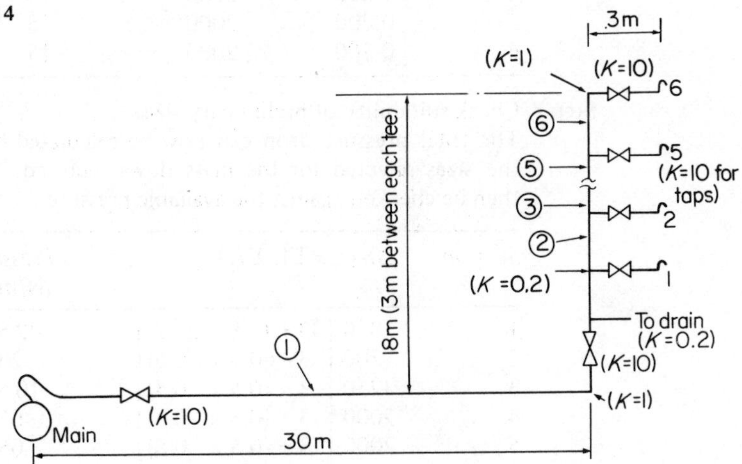

Step 1 Establish design flow rates, l/s

Demand units for a sink = 11 (Table 10.6)

Demand units, pipe 1 = 66

$$\text{Design flow rate, pipe 1} = 0.3 + \frac{16}{50}(0.5 - 0.3)$$

$$= 0.364 \text{ l/s}$$

The remaining design flow rates are obtained in the same manner thus:

Pipe 2: 0.32 l/s; pipe 3: 0.265 l/s; pipe 4: 0.2 l/s;
Pipe 5: 0.2 l/s; pipe 6: 0.2 l/s

Step 2 Establish the available pressure drop

Basically there are six 'circuits' in the layout and the index circuit is from the main to the topmost sink.

Use equation 10.9:

$$\text{available pressure} = p_m - \rho gz$$
$$= (300 \times 1000) - (1000 \times 9.81 \times 18)$$
$$= 124\,000 \text{ N/m}^2$$

Step 3 Obtain the pressure drop per metre length. With regard to equation 9.4, Δp and EL are still unknown but some restrictions exist in relation to pipework velocities. A trial and error approach is normally now used and for cold water flowing in copper pipework the velocity of flow should be in the region of 1.0 to 1.5 m/s. Thus from the chart (Table 10.8) the following information may be assembled:

Section	*Mass flow rate* (kg/s)	*Pressure drop* ($N/m^2/m$)	*Size* (mm)	EL (m)
1	0.364	850	22	0.8
2	0.321	700	22	0.7
3	0.265	3250	15	0.5
4	0.200	2000	15	0.5
5	0.200	2000	15	0.5
6	0.200	2000	15	0.5

Step 4 Check suitability of preliminary size.
The total pressure drop can now be calculated based on the sizes selected for the mass flows required. This can then be checked against the available pressure.

Section	$\Delta p\,[L + \text{EL}\,\Sigma K]$	*Pressure drop* (N/m^2)
1	850 [33 + (0.8 × 21.2)]	42 500
2	700 [3 + (0.7 × 0.5)]	2340
3	3250 [3 + (0.5 × 0.5)]	10 580
4	2000 [3 + (0.5 × 0.5)]	6500
5	2000 [3 + (0.5 × 0.5)]	6500
6	2000 [6 + (0.5 × 21.5)]	33 500
	Total pressure drop	101 920

This is well within the available pressure of 124 000 N/m^2 and these sizes are suitable.

There are many different ways in which the information in relation to the resistance of fittings is presented. We have already seen that the equivalent length of a particular fitting in relation to a particular pipe diameter is obtained as the product of EL and *K*. In the case of ball valves for storage cisterns it is more likely that the information supplied will be the working pressure drop across the valve. In such cases an adjustment is made to the available pressure rather than to the effective length of pipework so that:

$$\text{available pressure} = p_m - \rho g z_2 - \text{pressure drop (valve)}$$

as a modification of equation 10.9.

Pipe sizing techniques are very similar on the distribution side to those just illustrated on the supply side but of course in this case the available pressure is limited by the position of the storage cistern. Again with more complex arrangements the identification of the index circuit becomes important. Consider the cold water distribution system for a block of flats shown in Figure 10.15.

If it is assumed that identical fittings are employed on the four horizontal runs in this system then circuit 1 (from tank to bath 1) is shortest but has the smallest pressure head available. On the

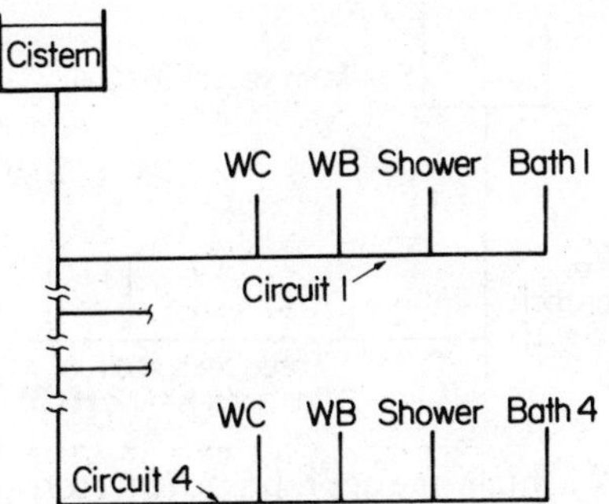

Figure 10.15
Cold water system for a block of flats

other hand circuit 4 (from tank to bath 4) is longest but has the greatest pressure head available. It is likely in this case that circuit 1 is the index circuit and the design of this circuit would dictate the size of the vertical pipe from the tank, which is common to all circuits. The designer very quickly develops the skill of identifying the index circuit which in many cases is obvious but in some cases can only be identified after some calculation.

Example 10.4 Size the pipework for the index circuit of the cold water distribution system for the four storey block of flats already considered in Example 10.2.

Step 1 Identify the index circuit.
The top floor pipework will be most critical and even here there exist four separate circuits, one for each fitting. The index circuit will run from the tank to the bath.

Step 2 Obtain the required mass flow rates.
There is little point in applying diversity within a single dwelling although it does apply to the main distribution pipe. From the information in Example 10.2 and the table of design flow rates mark the information on a sketch as Figure 10.16a.

Figure 10.16a

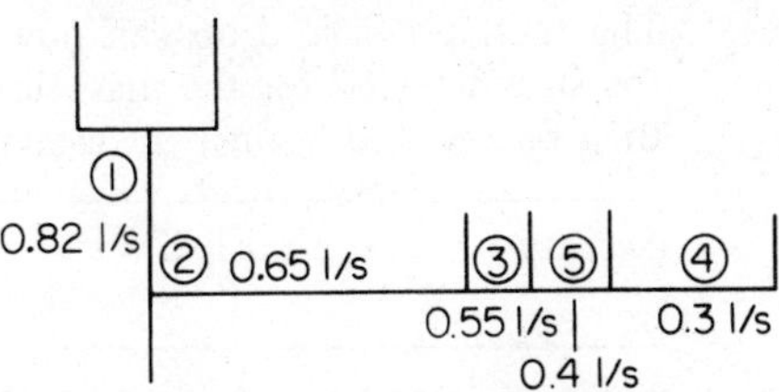

Step 3 Obtain the available pressure drop.
For the reasons outlined earlier this is taken in relation to the head available from the base of the cistern to the outlet of the fitting. In this case the pressure head is 2.5 m and

$$\begin{aligned}\Delta P &= \rho g h \\ &= 1000 \times 9.81 \times 2.5 \\ &= 24\,525 \text{ N/m}^2\end{aligned}$$

Figure 10.16b

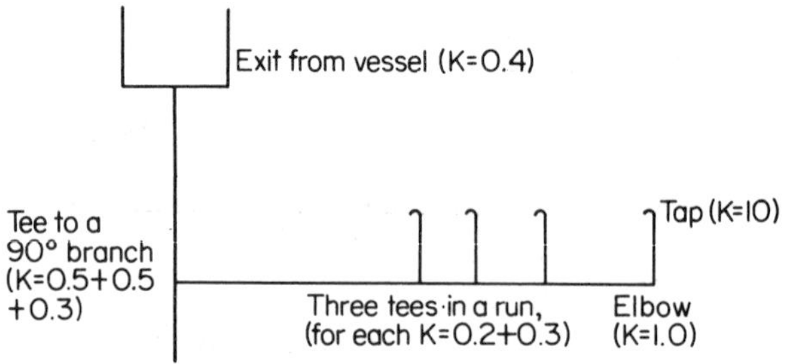

Step 4 Obtain the appropriate K factors (see Table 9.4) and mark on a sketch as Figure 10.16b.

$$K = 0.4 + (0.5 + {*0.5} + {*0.3}) + 3\,(0.2 + {*0.3}) + {*1.0} + 10.0 = 14.2$$

*signifies a guess at this stage since diameters are as yet unknown.

Step 5 Obtain the pressure drop per metre length.

As in the previous example, choose preliminary pipe sizes which satisfy a velocity restriction of about 1 to 1.5 m/s from which pressure drops per metre may be obtained. Then complete the following table.

Section	*Mass flow rate* (kg/s)	*Pressure drop* ($N/m^2/m$)	*Size* (mm)	EL (m)
1	0.82	1050	28	1.1
2	0.65	700	28	1.1
3	0.55	500	28	1.0
5	0.40	1000	22	0.8
4	0.30	600	22	0.7

Step 6 Check suitability of preliminary sizes.

The total pressure drop can now be calculated based on the sizes selected for the mass flows required and this can then be checked against the available pressure, as follows:

Section	$\Delta p\ [L + \text{EL}\ \Sigma K]$	*Pressure drop* (N/m^2)
1	1050 [3.5 + (1.7 × 1.1)]	5640
2	700 [3.0 + (0.5 × 1.1)]	2485
3	500 [1.0 + (0.5 × 1.0)]	750
5	1000 [1.0 + (0.5 × 0.8)]	1400
4	600 [4.0 + (11.0 × 0.7)]	7000
	Total pressure drop	17 275 (N/m^2)

This is well within the available pressure drop estimated at 24 525 N/m^2. Thus these pipe sizes are suitable.

Thus we see that the sizing of cold water pipework is in most respects similar to the sizing of heating pipework. The designer must consider many other matters in relation to the design of cold water systems. The method of jointing pipework must be considered in relation to the pressures to which the joints will be subjected. This is especially so in the case of pipework installations in tall buildings. Thought must also be given to the resistance of fittings in relation to available pressures. For instance, gate valves (with their low resistance) are often used on the distribution side of cold water systems where available pressure may be limited.

10.8 COLD WATER FIRE PROTECTION SERVICES

10.8.1 Legislation, systems and connection

In many modern buildings, for example tall buildings or those used for industry or commerce, special consideration must be given to the possibility of outbreak of fire. There is a great deal of information available in this area and two of the best sources are British Standard CP 3, Chapter 4, Precautions Against Fire[8] and publications available from the Fire Offices Committee[9]. The designer must take cognizance of general legislation such as the Factories Act and the Offices, Shops and Railway Premises Act as well as various bye-laws of individual local authorities. It is thus essential that the designer consults with the local fire authority at a very early stage of the design.

Systems available for firefighting may be classified as:

(i) First aid systems.
(ii) Systems for fire brigade use.
(iii) Automatic systems.

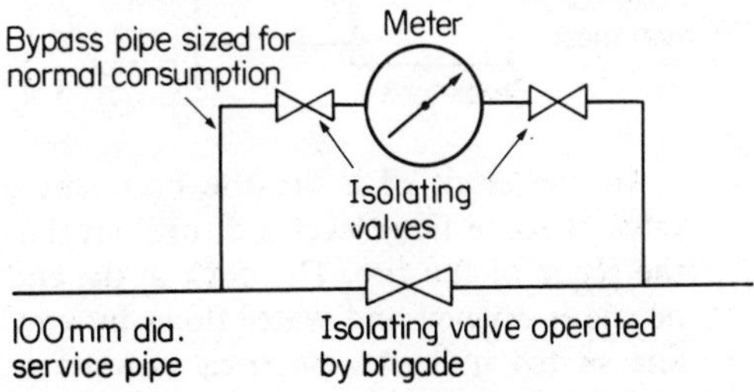

Figure 10.17 Water metering for large buildings

These shall be examined in turn, with the emphasis on the design of the cold water systems, but an initial point of consideration is that of the water supply to the building. In the case of water supply to individual dwellings, the size of the supply pipe is chosen in relation to the normal consumption pattern in such dwellings. However in the types of premises considered here the size of the supply pipe is chosen in relation to the needs of water supply in the event of fire. This usually means the use of a 100 mm diameter supply or service pipe. This can raise a difficulty in situations where the water is metered, but the usual solution to this problem is to fit a by-pass containing the meter to and from the main pipe as shown in Figure 10.17.

10.8.2 First-Aid systems

These are systems designed to allow the occupants to attempt containment of the fire while the brigade is on its way. Such systems may be subdivided into portable and fixed systems. Portable systems includes the range of water and sand buckets and a wide variety of hand extinguishers for dealing with special risks. The FOC lays down rules for the size of extinguisher related to floor area covered. Glass fibre and asbestos blankets are also available where people are at risk or there is the possibility of a fire within a container.

Hose reel systems are fixed first-aid systems intended for use by the building occupants. Some local authorities allow direct connection of the system to the water main while others insist on the use of a break tank. Figure 10.18 shows a system employing a low-level break tank.

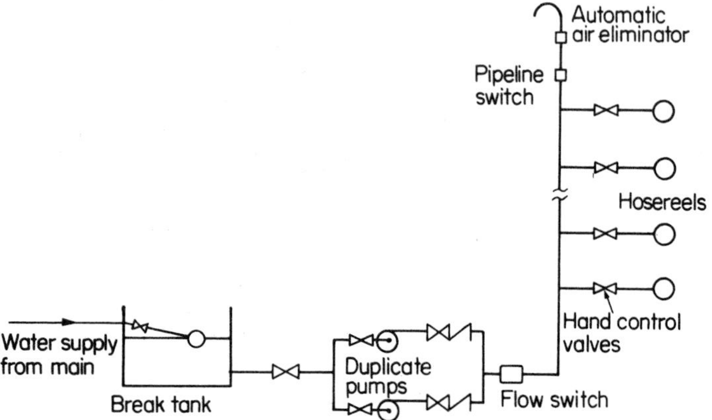

Figure 10.18
Cold water supply system for hosereel installation

In the event of a fire the occupant opens the hand control valve at some floor level and runs out the hose through a guide to the scene of the fire. The cock at the end of the host next to the nozzle is opened and water flows by gravity, emptying the pipeline switch unit. An electrical contact is made within this unit which brings on the pump to supply water at the hosereel from the break tank. The flow switch detects flow in the line and when the hand control valve at the reel is closed then the flow switch cuts out the pump. Variations on this arrangement are possible such as the use of a high level break tank, with a high level pump set if the gravity head available is insufficient.

Hosereel systems must be capable of producing a flow rate of 0.4 l/s at the nozzle with a throw of 6 m with three of the hosereels in simultaneous operation. They must also of course be capable of covering the complete floor area on each floor. The hose is normally rubber and of either 19 mm or 25 mm diameter. Unreinforced hose is available up to 23 m and reinforced hose up to 37 m, wound on the drum. The terminal nozzle is normally 5 mm or 6 mm diameter. To achieve the desired flow and throw, the designer must provide a minimum pressure at the nozzle of about 200 kN/m^2. The Guide (B5) provides a table of flow rates at the nozzle for a range of pressures related to nozzles of different diameters. Some authorities allow a hydraulic sizing of the system while others insist on the sizes as laid down in the FOC Rules.

10.8.3 Systems for Fire Brigade use

These are usually external hydrants or internal hydrants fed by wet or dry systems or foam installations. The hydrants are 65 mm diameter with a valve and a special coupling for the brigade's hoses. Ordinary street water mains have external hydrants sited at regular spacings along the length of the mains but in situations where the building is extensive or sited at a considerable distance (90 m) from the mains supply the designer will have to consider the problems of a reliable supply of water and the spacing of internal hydrants.

Dry riser systems

Buildings in the height range of about 18 m to 60 m normally have a dry riser installation; below 18 m, the hosereels are fed from external hydrants. With such systems the tender pump is connected to an external hydrant on the street main and a connection run from the tender usually to a double inlet breeching piece at the foot of the dry riser. The breeching piece and its two 65 mm connections are set within a box openable only by the brigade and sited on the outside wall of the building in a conspicuous position. The box is positioned 760 mm above ground level and within 12 m of the vertical riser. The necessary flow and

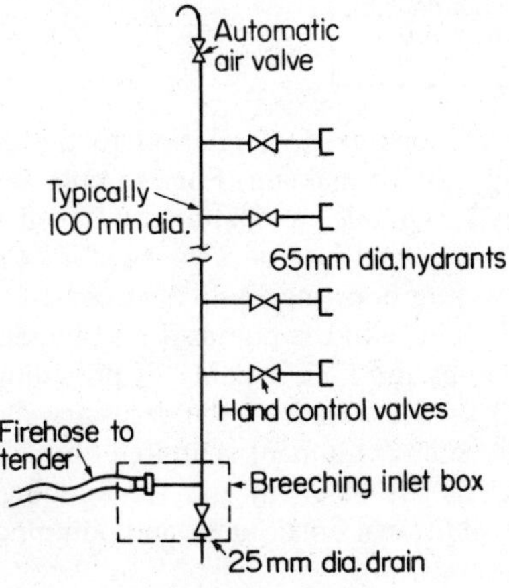

Figure 10.19
Typical layout of a dry riser system

pressure of water drawn from the main and supplied to the system is provided by the tender pump.

On each landing, and usually contained within a box, there is a 65 mm diameter valved outlet or landing valve to which the brigade's hosereels can be connected. In addition there is a valved outlet on the roof used for testing the system and this latter outlet incorporates a 25 mm diameter air release valve.

Incorporated in the breeching inlet is a 25 mm diameter drain valve for removing any residual water within the system after the fire. A typical layout is shown in Figure 10.19.

Wet riser system

Buildings above about 60 m in height are outside the range of the tender pumps and recourse has to be made to a system of wet risers which are fed from a separate water supply usually through pressure boosting equipment. Such systems require to make available a supply of water at a pressure of not less than 410 kN/m^2 at each outlet. Also the outlet valves contain orifice plates to ensure that the outlet pressure to the hosereels does not exceed 520 kN/m^2. A further precaution is to limit the static pressure within the pipework to 690 kN/m^2 by having a return pipe from each landing valve connected back to the source of water supply which comes into operation if this pressure is exceeded. A typical layout is shown in Figure 10.20.

Figure 10.20
Typical layout of a wet riser system

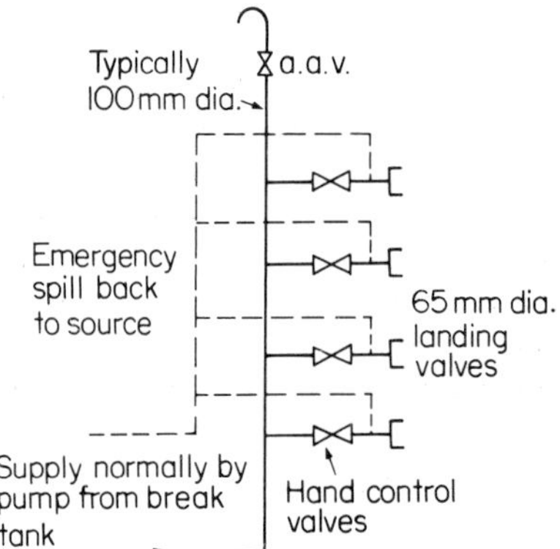

Restrictions exist in relation to the quantity of stored water and the rate of makeup. For example an 11.4 m^3 storage break tank is acceptable in relation to a feed rate from the mains of 27 l/s. At a feed rate of 8 l/s the storage requirement is 45.5 m^3. The pressure boosting equipment consists normally of a duplicate pumping set with the pumps driven by separate sources of power. The pumps must be capable of providing a flow of between 15 and 23 l/s while creating the necessary static pressure at the valve outlets. Such equipment is limited normally to operation within a vertical height of 60 m and for buildings in excess of 60 m in height additional break tanks and pumping sets will be required at higher levels.

Foam installations

These are provided for specialised situations such as boiler or plant rooms, oil storage tank rooms or areas which have difficult access. The brigade's foam tender is connected by hosereel to a foam inlet which supplies a 65 mm or 75 mm diameter pipeline leading to the protected space. Within the space a series of fixed foam inlets allows distribution of the foam. The concept is similar to that of the dry riser.

10.8.4 Automatic sprinkler systems

These are automatic systems which are activated by the fire itself. Outbreak of fire generates heat which bursts sprinkler heads local to the fire and generates a stream of water which may put out or contain the fire until the arrival of the brigade. The flow of water in the system then gives rise to an audible alarm and may also set off an alarm at the brigade headquarters.

In certain situations there is a statutory obligation to provide such systems and in these cases the installation must be to some prescribed standard such as those laid down by the Fire Offices Committee or other relevant bodies. In other cases such systems are desirable and in any case the designer must weight the reduction in insurance premium obtained, against the cost of the system. Figure 10.21 shows the layout of a typical sprinkler system. It consists of a supply of water via installation control

Figure 10.21
Typical layout of an automatic sprinkler system

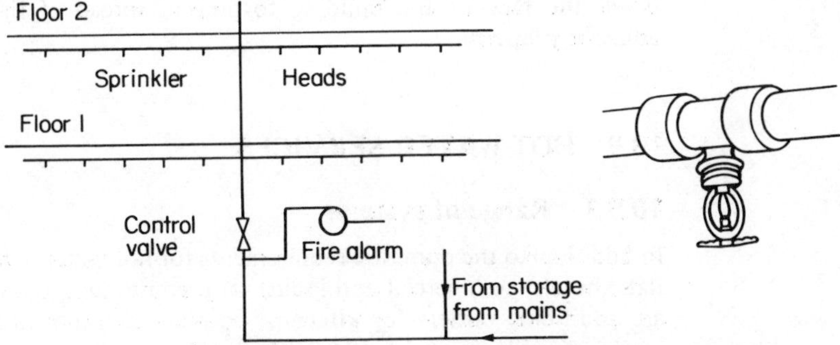

valves to a network of piping containing automatic valves, otherwise known as sprinkler heads. These are held closed by a strut or a quartz bulb either of which collapses or bursts in the event of fire and discharges water on the fire. A variety of heads are available for protection at different temperatures although the most common temperature rating is 68 °C.

In small premises it would normally be sufficient to have one installation but for large complexes there are arguments for providing several systems with individual alarms. Installations may be classified as dry systems, wet systems, alternate wet and dry systems or preaction systems. Water must never be allowed to freeze within a system and where this risk is always present a dry system should be used. The system is kept dry by having it charged with air at a gauge pressure of about 2.5 atmospheres. In such systems the sprinkler heads are upright to facilitate drainage. Where there is no likelihood of freezing, a wet system, fully charged with water may be employed. The heads in this case may be upright or pendant. The alternate wet and dry system is used where there is a possibility of freezing during the winter months. With the preaction system, a dry system is used in conjunction with either heat or smoke detectors which on sensing the fire admit water to the pipework prior to any response at the sprinkler head.

The FOC Rules lays down standards in relation to water supplies, spacing and location of sprinkler heads and discharge water densities related to different categories of risk. Useful information in this area is given in the Guide, Book B[(10)].

Other systems

Highly specialised automatic systems exist such as Deluge or Drencher systems. Deluge systems are systems which cover internal areas with their sprinkler heads unsealed. The system is filled via a quick-opening valve opened usually by the action of heat or smoke detectors. Such systems are employed in situations where there is a possibility of very rapid spread of fire such as may occur within an aircraft hanger. With Drencher systems a vertical discharge of water is provided over an external opening to prevent spread of fire to adjacent property or in certain cases down the face of the building to prevent spread of fire from adjacent property.

10.9 HOT WATER SERVICES

10.9.1 Range of systems

In addition to the domestic requirements for hot water in houses, flats, hotels, commercial and industrial premises and so on, there are additional outlets for kitchens, hospitals and industrial processes. A wide range of systems (as well as fuel) is in use and perhaps this range can be most usefully subdivided into two broad categories of local and central systems. Before discussing types of system, however, it will be as well as consider the questions of supply temperature, flow rate, consumption, storage capacity, and design flow rate.

10.9.2 Temperature

In relation to storage systems, the stored water temperature affects the heat lost from the storage vessel and supply pipework and also the recovery time (the time to heat up a tank of cold supply water to the storage temperature) for the storage vessel. In addition to these considerations is the need for a suitable temperature at the sanitary fitment or outlet, where for example, the possibility of scalding must be taken into account. Accordingly, normal storage temperatures are around 65 °C although in some cases (where showers only are being supplied, temperatures as low as 44 °C may be suitable) lower temperatures are acceptable.

10.9.3 Flow rate at sanitary fittings

In relation to the normal range of sanitary fitments the required hot water flow rate is the same as that for cold water, as presented earlier in Table 10.4.

10.9.4 Consumption

As a result of extensive examinations of usage patterns, the Guide, Book B, contains a table of likely maximum daily demand

Table 10.9 Hot Water Consumption

Building	*Maximum daily demand per person* (litre)
Colleges and schools:	
Boarding	115
Day	15
Dwellings and flats:	
Low rental	70
Medium rental	115
High rental	140
Factories	15
Hospitals:	
General	135
Infectious	225
Infirmaries	70
Infirmaries (with laundry)	90
Maternity	230
Mental	90
Nurses' homes	135
Hostels	115
Hotels:	
First class	135
Average	115
Offices	15
Sports pavilions	35

(litres per person) which may be used in a wide variety of situations where the storage capacity of the domestic hot water tank is required. This information is reproduced here as Table 10.9.

10.9.5 Storage capacity

In relation to reasonably standard situations such as the needs of an individual house it is practice to use a standard hot water cylinder capacity. This is a 114 litre cylinder. In other situations it is better to make an estimate of the demand throughout the whole day, hour by hour, and compare this with the energy input to the cylinder to ensure that there is adequate capacity and that the recovery rate is high enough.

10.9.6 Design flow rate

Advantage may be taken of diversity with hot water systems in the same way as with cold water systems. Thus the tables previously given for estimating the demand units and the design flow rates in cold water systems may also be used for sizing pipework for hot water supply.

10.9.7 Central hot water systems

With such systems water is normally heated within a boiler fired by oil or gas and circulated to a hot water storage cylinder or calorifier. This is often carried out in conjunction with space heating. With systems within individual dwellings, where the water supply is soft, it may be possible to install a direct system, as shown in Figure 10.22, in which the water is heated directly in the boiler.

Figure 10.22 Direct hot water system

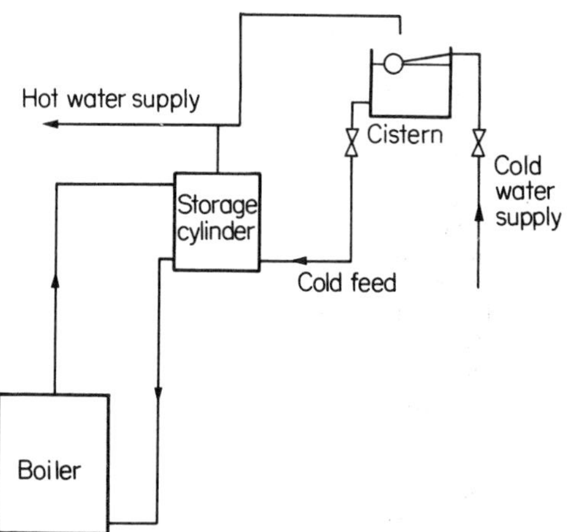

The difficulty with such systems is that oxygen and scale-forming products are constantly being introduced to the system and corrosion and scale formation may arise. Thus in certain areas the local authority may forbid the use of such systems. An alternative to this system is the indirect system in which water is circulated from the boiler to the storage tank or cylinder and there gives up heat to cold water fed from a storage cistern. The

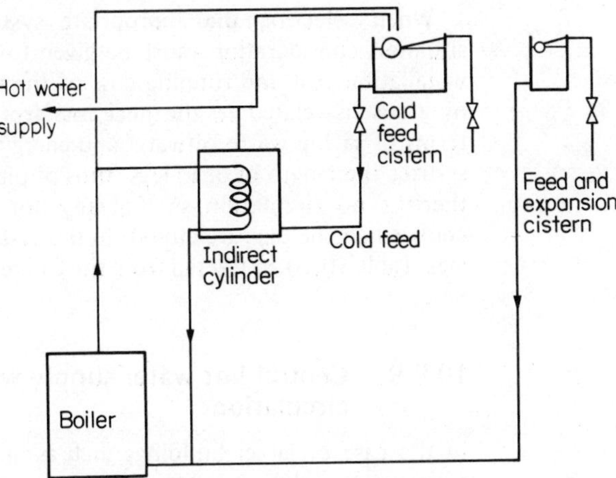

Figure 10.23
Indirect hot water system

heat transfer is not so effective but in the main, problems of scale and corrosion are overcome. The indirect system is illustrated in Figure 10.23.

10.9.8 Local hot water systems

With local systems the water is heated adjacent to the outlet and usually by either gas or electricity. This may involve some storage but often an instantaneous heater is employed. An illustration of an instantaneous system to a set of showers is shown in Figure 10.24.

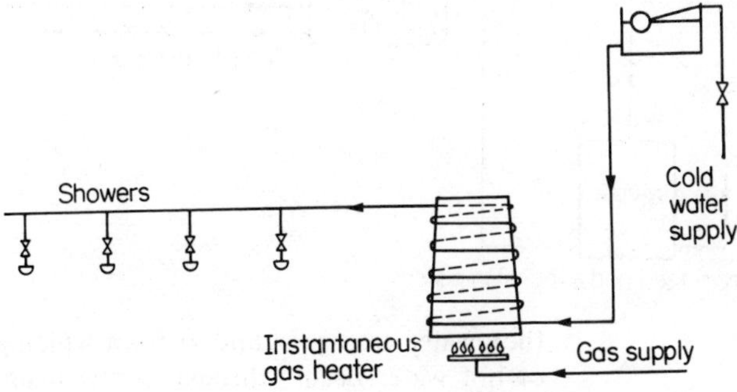

Figure 10.24
Local hot water supply to a group of showers

Table 10.10 Maximum Permitted Length of Dead-leg

Largest internal diameter of pipe	*Nominal pipe size (mm)*		*Maximum length (m)*
	Steel	*Copper*	
Not exceeding 19 mm	15	15	12
Exceeding 19 mm but not exceeding 25 mm	20	22	8
Exceeding 25 mm	25	28	3

When selecting the appropriate system for any particular situation consideration must be given to the cost of equipment, installation cost and running cost of the system. Part of the running cost is related to the heat loss from the pipework and in terms of saving waste of water and energy, Model Water Bye-laws restrict the length of dead-legs (runs of pipework to taps in which there is no circulation so that any hot water in the pipework cools when the taps are closed) in the system. In relation to dead-legs, Table 10.10, extracted from the Guide, Book B, is reproduced here.

10.9.9. Central hot water supply with pumped circulation

In the case of larger buildings such as blocks of flats, offices or colleges, the solution is often to have a central boiler house providing a primary circulation of hot water (or steam) to a calorifier

Figure 10.25
Central hot water supply with pumped secondary circulation

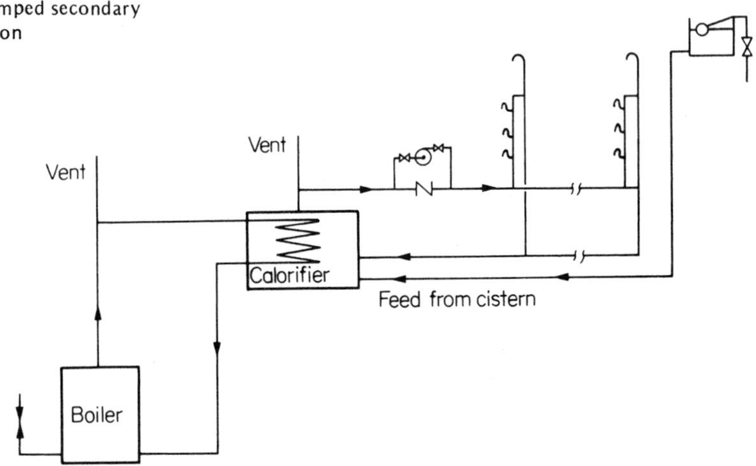

(hot water storage cylinder) from which a secondary circulation of hot water occurs through a ring main by means of a pump. Local mains are then drawn from the ring as illustrated in Figure 10.25.

10.9.10 Frost precautions

The designer must give some thought to the protection of water services from frost. Positioning of pipework and equipment such as cisterns in roof spaces must take such protection into account. Insulation of such items must be considered. It is surprising how a chilling draught up an overflow pipe can produce freezing within a cistern and it is thus a good idea to place the cistern end of an overflow pipe below the waterline within the cistern. If pipework is to run on outside walls it should be placed on wooden battens. A useful practice with pumped hot water (and space heating) systems is to arrange for a frost thermostat to energise the pump when the outside temperature drops to a predetermined level. The circulation of the unheated water helps to reduce the possibility of freezing within the system.

10.10 PIPE SIZING FOR HOT WATER SYSTEMS

10.10.1 The demand unit method

The techniques available for sizing the hot water pipework are the same as those for cold water pipework. Thus the method of Binomial and Poisson Distribution may be employed with the same limitations as discussed previously. The more common method, as with cold water design, is the system of demand units. The technique is best illustrated again by example.

Example 10.5 Figure 10.26 shows the layout of the cold feed and secondary pipework for the supply of hot water to the sanitary fittings of a block of flats. Determine for the index circuit:

Figure 10.26

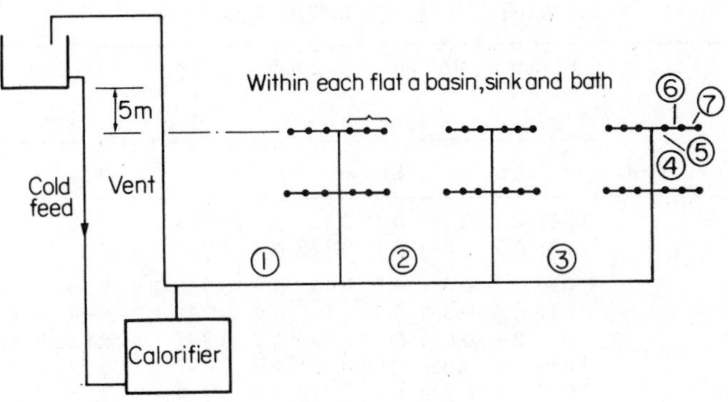

(i) The design flow rate in each section of the pipework, based on the method of demand units.
(ii) A preliminary size for the pipework.
(iii) A final size for the pipework.

Use the table (extracted from the Guide, Book C) for thin wall copper tube presented here as Table 10.11.

Step 1 Identify the index circuit:
Recall that the index circuit is that circuit having the smallest pressure available per total length of pipework. In this case the circuit identified as 1–7 is the index circuit.

Step 2 Obtain the required design mass flow rates in each section:

Section	*Demand units*	*Design flow rate* (kg/s)
1	312	1.05
2	208	0.82
3	104	0.51
4	52	0.30
5	not applicable	0.30
6	not applicable	0.30
7	not applicable	0.30

Step 3 Obtain the available pressure drop.
As explained previously this is measured from the base of the tank to the tap outlet. In this case it is 5 m.

$$\Delta P = \rho g h = 1000 \times 9.81 \times 5 = 49\,050 \text{ N/m}^2$$

Step 4 Obtain the appropriate K factors (see Table 9.4 and Figure 10.27).
The asterisks (*) in Figure 10.27 signify a guess at this stage, since the pipework sizes are as yet unknown.

Step 5 Obtain the pressure drop per metre length.

Table 10.11 Flow of Water at 75 °C in Thin Wall Copper Pipes

		12 mm		15 mm		22 mm		28 mm		35 mm		42 mm			
		$\frac{3}{8}$ *in*		$\frac{1}{2}$ *in*		$\frac{3}{4}$ *in*		1 in		$1\frac{1}{4}$ *in*		$1\frac{1}{2}$ *in*			
Δp (N/m²/m)	V (m/s)	$\dot{m}$	EL	$\dot{m}$	EL	$\dot{m}$	EL	$\dot{m}$	EL	$\dot{m}$	EL	$\dot{m}$	EL	V (m/s)	Δp (N/m²/m)
80.0		0.021	*0.3*	0.041	*0.4*	0.120	*0.8*	0.237	*1.1*	0.437	*1.6*	0.717	*2.0*		80.0
82.5		0.021	*0.3*	0.041	*0.5*	0.122	*0.8*	0.241	*1.1*	0.445	*1.6*	0.729	*2.0*		82.5
140.0		0.029	*0.3*	0.056	*0.5*	0.164	*0.9*	0.325	*1.2*	0.598	*1.7*	0.979	*2.1*		140.0
160.0		0.031	*0.4*	0.061	*0.5*	0.177	*0.9*	0.351	*1.2*	0.644	*1.7*	1.06	*2.2*		160.0
180.0		0.034	*0.4*	0.065	*0.5*	0.189	*0.9*	0.375	*1.3*	0.688	*1.7*	1.13	*2.2*		180.0
200.0		0.036	*0.4*	0.069	*0.5*	0.201	*0.9*	0.397	*1.3*	0.730	*1.7*	1.19	*2.2*	1.0	200.0
220.0		0.038	*0.4*	0.073	*0.5*	0.212	*0.9*	0.419	*1.3*	0.770	*1.7*	1.26	*2.2*		220.0
240.0		0.040	*0.4*	0.076	*0.5*	0.223	*0.9*	0.440	*1.3*	0.808	*1.8*	1.32	*2.3*		240.0
260.0		0.041	*0.4*	0.080	*0.5*	0.233	*0.9*	0.460	*1.3*	0.845	*1.8*	1.38	*2.3*		260.0
280.0		0.043	*0.4*	0.083	*0.5*	0.243	*0.9*	0.480	*1.3*	0.880	*1.8*	1.44	*2.3*		280.0
300.0	0.50	0.045	*0.4*	0.087	*0.5*	0.252	*0.9*	0.498	*1.3*	0.915	*1.8*	1.50	*2.3*		300.0
320.0		0.047	*0.4*	0.090	*0.5*	0.262	*0.9*	0.517	*1.3*	0.948	*1.8*	1.55	*2.3*		320.0
340.0		0.048	*0.4*	0.093	*0.6*	0.271	*1.0*	0.535	*1.4*	0.980	*1.8*	1.60	*2.4*		340.0
360.0		0.050	*0.4*	0.096	*0.6*	0.280	*1.0*	0.552	*1.4*	1.01	*1.8*	1.66	*2.4*		360.0
380.0		0.052	*0.4*	0.099	*0.6*	0.288	*1.0*	0.569	*1.4*	1.04	*1.9*	1.71	*2.4*		380.0
400.0		0.053	*0.4*	0.102	*0.6*	0.297	*1.0*	0.585	*1.4*	1.07	*1.9*	1.75	*2.4*		400.0
420.0		0.055	*0.4*	0.105	*0.6*	0.305	*1.0*	0.601	*1.4*	1.10	*1.9*	1.80	*2.4*		420.0
440.0		0.056	*0.4*	0.108	*0.6*	0.313	*1.0*	0.617	*1.4*	1.13	*1.9*	1.85	*2.4*	1.5	440.0
460.0		0.057	*0.4*	0.110	*0.6*	0.321	*1.0*	0.633	*1.4*	1.16	*1.9*	1.89	*2.4*		460.0
480.0		0.059	*0.4*	0.113	*0.6*	0.328	*1.0*	0.648	*1.4*	1.19	*1.9*	1.94	*2.4*		480.0
500.0		0.060	*0.4*	0.116	*0.6*	0.336	*1.0*	0.663	*1.4*	1.21	*1.9*	1.98	*2.4*		500.0
520.0		0.062	*0.4*	0.118	*0.6*	0.344	*1.0*	0.677	*1.4*	1.24	*1.9*	2.03	*2.5*		520.0

Figure 10.27

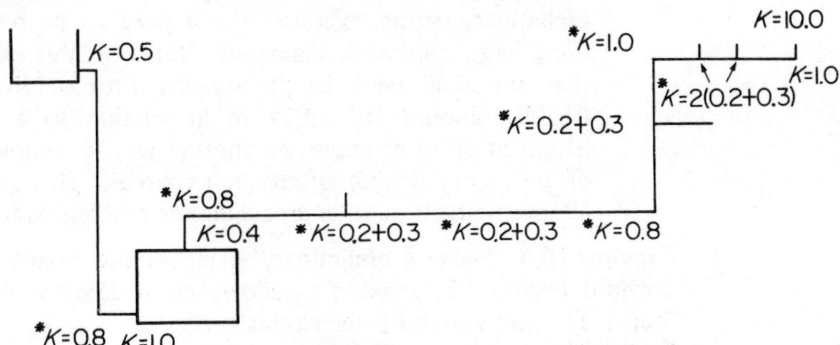

As with cold water design, pipework velocities should be in the region of 1.0 to 1.5 m/s. Thus with the constraints of mass flow rate and velocity in mind, a preliminary pipe sizing can be done in relation to Table 10.11.

Section	*Mass flow rate* (kg/s)	*Pressure drop* ($N/m^2/m$)	*Size* (mm)	EL (m)
1	1.05	380	35	1.9
2	0.82	250	35	1.8
3	0.51	310	28	1.3
4	0.30	400	22	1.0
5	0.30	400	22	1.0
6	0.30	400	22	1.0
7	0.30	400	22	1.0

Step 6 Check the suitability of the preliminary sizes.
The total pressure drop can now be calculated on the basis of the preliminary sizes and checked against the available pressure, as follows:

Section	$\Delta p\ [L + \text{EL}\ \Sigma K]$	*Pressure drop* (N/m^2)
1	380 [30 + (1.9 × 3.5)]	13 950
2	250 [20 + (1.8 × 0.5)]	5240
3	310 [20 + (1.3 × 1.3)]	6700
4	400 [4 + (1.0 × 0.5)]	1800
5	400 [2 + (1.0 × 1.0)]	1200
6	400 [2 + (1.0 × 0.5)]	1000
7	400 [2 + (1.0 × 11.5)]	5400
Note: EL ΣK totals 22.74 m		Total 35 290 N/m^2

The pressure drop in the system as a result of using the preliminary pipe sizes is 35290 N/m^2 and the available head of 49050 N/m^2 exceeds this figure. Thus the preliminary sizes could be used as the final sizes for the system. In a situation where the preliminary sizing left no available head or worse, exceeded the available head, then the preliminary sizing exercise would have to be repeated using larger pipework diameters. Notice in this example that the equivalent length for the fittings (given by $EL\Sigma K$) amounts to 22.74 m in relation to a circuit length of 80 m or expressed another way it is about 28% of the actual length of the index circuit. This gives an alternative method of approaching the problem as follows.

Example 10.6 Make a preliminary sizing of the circuit in the previous example by making an allowance of 25% for fittings.

Step 1 The index circuit is the circuit 1-7.

Step 2 The design mass flow rates are determined as previously.

Step 3 The available pressure drop is determined as previously.

Step 4 The available pressure drop per metre run is obtained by:

Available pressure drop = 49 050 N/m^2
Actual length of pipework = 80 m
Effective length of pipework = 80 + 0.25 × 80
= 100 m
Available pressure drop = 49 050/100
= 490.5 N/m^2

Step 5 Select preliminary pipe sizes on the broad basis of this pressure drop, keeping in mind the required mass flow rates and acceptable pipe velocities.

Section	*Mass flow rate* (kg/s)	*Pressure drop* ($N/m^2/m$)	*Size* (mm)	EL (m)
1	1.05	380	35	1.9
2	0.82	250	35	1.8
3	0.51	310	28	1.3
4	0.30	400	22	1.0
5	0.30	400	22	1.0
6	0.30	400	22	1.0
7	0.30	400	22	1.0

The remaining steps are exactly as given in Example 10.5 so far as final pipe sizing is concerned.

10.10.2 Sizing the secondary circuit of a recirculation system

As we have already seen, the reason for having a pumped secondary circuit is to reduce waste of water by having to run taps until 'the water runs hot'. An interesting problem which arises is the basis on which the ring main is to be sized. There are two distinct types of flow which may occur.

(a) An open tap somewhere in the system (Figure 10.28a): In these circumstances the head of water in the feed tank provides the energy to feed the tap and the bulk of this water by-passes the pump. The normal design method is to size the flow pipe to provide the required flow at the taps, in exactly the manner outlined in the previous examples.

Figure 10.28

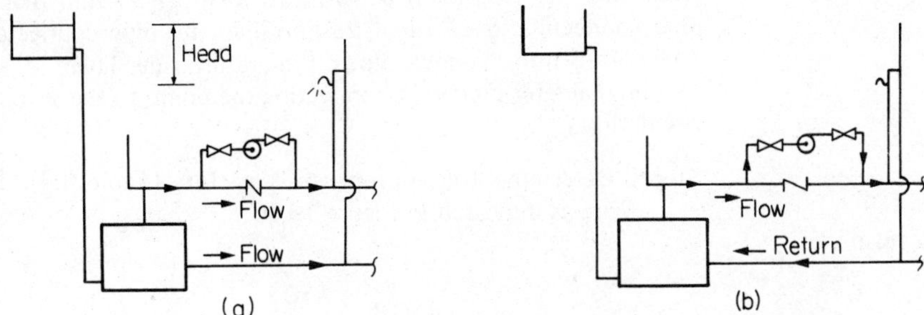

(b) All taps in the system closed (Figure 10.28b):
In these circumstances the pump provides the energy to circulate the water through the ring main and back to the calorifier. Normally flow and return temperatures of 65 °C and 55 °C are adopted and the return pipe is sized on the basis of the flow rate required to offset the heat emission for a ten degree drop in temperature.

Example 10.7 Determine a suitable size for the return pipework of the secondary circuit of Example 10.6, assuming a circulation heat emission of 700 W with a ten degree drop in temperature (Figure 10.29).

Figure 10.29

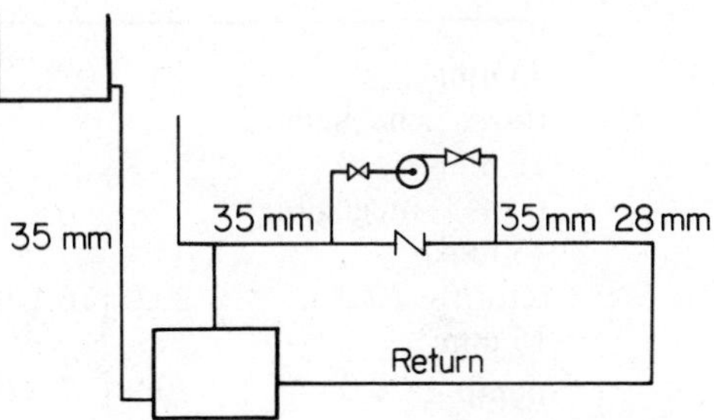

Step 1 Determine the required mass flow rate in the ring main when all taps are closed:

$$\dot{Q} = \dot{m} C_p (\theta_{flow} - \theta_{return})$$
$$700 = \dot{m} \times 4200 \times 10$$
$$\dot{m} = 700/(4200 \times 10)$$
$$= 0.017 \text{ kg/s}$$

Step 2 Select a suitable return pipe diameter:
From Table 10.11, a return pipe of 15 mm diameter would be satisfactory and at a guess would produce a pressure drop of 20 $N/m^2/m$.
Note also that the pipework connecting the pump to the

secondary circulation has also to deal with a flow rate of 0.017 kg/s and thus this pipework would also be 15 mm in diameter.

Example 10.8 Determine the duty of the pump for this system, given that the ring main consists of 40 m of 35 mm diameter pipe connected to 20 m of 28 mm diameter pipe connected to 70 m of 15 mm diameter pipe on the return side. There is 7 m of 15 mm diameter pipework connecting the pump to the secondary circulation.

Step 1 Determine the appropriate K factors (Table 9.4). These are as indicated in Figure 10.30.

Figure 10.30

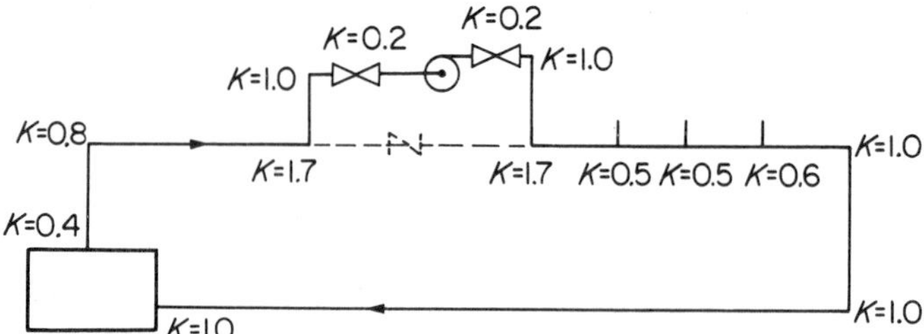

Step 2 Determine the pressure drop in each section of pipework related to a flow of 0.017 kg/s and tabulate as shown.

Section	*Pressure drop* (N/m²/m)	$\Delta p\ [L + \text{EL}\ \Sigma K]$	*Pressure drop* (N/m²)
35 mm flow	negligible	—	negligible
28 mm flow	negligible	—	negligible
15 mm return	20.0	20.0 [70 + (0.4 x 3.6)]	1430
15 mm pump	20.0	20.9 [7 + (0.4 x 4.1)]	174
		Total pressure drop	1604 N/m²

Thus a pump must be selected which is capable of delivering 0.017 kg/s against a static pressure loss of 1604 N/m².

From time to time the designer is required to size pipework based on a circulation pressure which arises due to differences in density in the flow and return legs of a primary or a secondary (gravity) circulation. Apart from the means of estimating the circulation pressure (explained in Section 9.2.1), the problem is little different from those already covered. The designer would have to consult manufacturers' literature to obtain guidance on resistance to the flow of water through components such as boilers or calorifiers.

This chapter together with Chapter 9 has established the means by which the designer can undertake the sizing of various water services. There are of course variations on the theme and the individual designer will quickly develop an approach to sizing which suits him (or her) best.

REFERENCES

(1) *The Water Engineers' Handbook* (Fuel and Metallurgical Journals Ltd)
(2) C.I.B.S. Guide, Section B4
(3) STEWART, D. 'The electrochemistry of corrosion and its prevention', *Anti-Corrosion Methods and Materials J.*, March 1979
(4) BS CP 310, Water Supply
(5) Design Bulletin No. 3, Part 6. Ministry of Housing and Local Govt.
(6) BS 417
(7) C.I.B.S. Guide, Section C4
(8) BS CP3, Chapter 4, Precautions Against Fire
(9) Fire Offices Committee Publications
(10) C.I.B.S. Guide, Section B5

SYMBOLS USED IN CHAPTER 10

C_p specific heat at constant pressure
EL equivalent length
g acceleration due to gravity
h available head from base of cistern
H available head from water surface in cistern
K velocity pressure loss factor
L actual length of pipework
m number of fittings simultaneously discharging
$\dot{m}$ rate of mass flow
n total number of connected sanitary fittings
p pressure
p' usage ratio
p_m pressure above atmospheric
ΔP available pressure across length of pipework
Δp pressure drop per metre length of pipework
$\dot{Q}$ rate of heat transfer
t time of inflow to sanitary fittings
T time between use of sanitary fittings
V average velocity of flow
x coefficient related to an error function
z potential height above a datum
θ temperature
ρ density

Index

Absorptivity, 59
Admittance, 234
Admittance procedure, 235
Air change rate, 279
Air space resistance, 66
Air temperature, 143, 265, 301, 314
Air-radiant temperature, 177, 187
Air within system, 333
Allowable pressure drop, 340
Allowance for fittings, 347
Aluminium components, 399
Ambient temperature, 301
Anti-cavitation, 354, 363
Anti-cavitation margin, 365, 368, 374
Anti-vibration pads, 382
Apportioning mains emission, 335, 386
Authority, control valve, 92
Automatic control, 83
Automatic sprinkler systems, 427
Available pressure,
  cold water systems, 416
  forced circulation, 353
  hot water systems, 434
  natural circulation, 331

Back losses, 319
Balance pipes, 363
Base exchange process, 401
Bernoulli equation, 15
Bimetallic couples, 394
Binomial and Poisson distributions, 412
Black body, 60, 184
Blasius, 25
Block heating, 382
Boiler,
  control, 121
  types, 208
Borda-Carnot loss, 33
Bulk modulus, 2
Burner,
  control, 123
  types, 208

Calorifiers,
  heat transfer surface, 213
  non-storage, 127, 211, 387
  storage, 130, 211, 432
Cast-iron components, 398
Cavitation, 6, 356, 371
Charging for heat consumption, 387
Clark scale, 391
Clark's process, 400
Climate, 227
Coefficient of contraction, 34
Coefficient of cubical expansion, 57
Coefficient of upward heat emission, 319
Coefficient of viscosity, 3
Coil spacing, 320
Colebrook-White equation, 28, 341
Cold water supply,
  automatic sprinkler systems, 427
  cisterns, 405
  demand units, 412
  distribution, 406
  dry riser systems, 425
  fire protection services, 423
  hosereel systems, 424
  mains distribution, 403
  mains pressure, 403
  pipe materials, 403
  pneumatic vessel, 410
  service pipe, 403
  sizing cisterns, 410
  storage requirements, 404
  tall buildings, 407
  wet riser systems, 426
Combined heat and power, 383
Comfort, thermal, 138
Commissioning, 315
Compression fittings, 375
Condensation, 294
Conduction,
  description, 39
  differential equation, 44, 46
  one-dimensional,
    in cylindrical layers, 49
    in plane slabs, 47
Conductivity, thermal, 40
Control,
  areas, 111
  cascade, 107
  ceiling systems, 321
  central plant, 121
  commissioning, 109
  dead time, 87
  definitions, 84
  demand side capacity, 86
  detector lag, 88
  domestic heating, 133, 359
  domestic hot water service, 131
  feedback, 84
  feedforward, 84
  modes, 94
  non-storage calorifiers, 128

Control *cont.*
  on–off, 371
  partial load, 325
  storage calorifiers, 130
  time clock, 359
  time lag, 87
  quality, 110
  system lag, 87
Control valves,
  characteristics, 91
  types, 91
Convection,
  coefficient, 54, 268, 305, 384
  description, 52
  dimensional analysis, 55
  empirical relations, 57
  forced, 53
  natural, 53
Copper pipework, 397
Copper components, 399
Corrosion, 392
  inhibitors, 402
  of pipework, 384, 388, 391
  prevention, 396
  rate, 393
Critical velocity, 7

Darcy equation, 19, 341
Dead leg, 431
Deluge and drencher systems, 428
Demand units, 412, 434
Demineralisation, 402
Density of water, 331, 361, 364
Design flow rate, 334, 337, 345, 386, 434
Design process, the, 200
Design team, the, 202
Design temperatures, 278, 283
Design water temperature drop, 323, 361
Dew-point temperature, 76, 78
Diaphragm expansion tanks, 369
Diffusion,
  coefficient, 78
  Fick's law, 78
Dimensional analysis, 20, 55
Dip-pipe, 363
Direct hot water systems, 402, 430
Discomfort, localised, 173
District heating, 382
Diversity, 411
Draining, 363
Draughts, 314
Dry riser system, 425
Duplex brasses, 399

Eddy viscosity, 12
Emissivity, 61
Energy consumption, 222, 259
Environmental temperature, 283
Equivalent ceiling, 310
Equivalent dead time, 91
Equivalent grain size, 27, 29
Equivalent length, 37, 339, 343, 345
Equipotential characteristic, 92
ESP, 244
Euler equation, 14
Expansion of pipework, 376

Fanger,
  comfort criteria, 149
  comfort equation, 149
Feed and expansion tank, 353
Fick's law, 78
Filling the system, 332
First law of thermodynamics, 72
Flow rate, design, 334, 337, 345, 386, 434
Flow rate, sanitary fittings, 411, 429
Flow velocities, 340
Foam installations, 427
Fourier equation, 40
Friction factor, 21
Frost protection,
  cisterns, 406, 433
  cold water mains, 404
  heating systems, 128
Fully rough zone, 26

Galvanic couples, 394
Galvanic series, 395
Gas pressurisation, 370
Grashof number, 57
Grey body, 185
Group heating, 382

Hardness of water, 391
Harmonic method, 229
Head loss to friction, 18, 21, 341
Heat balance method, 242
Heat distribution, 217
Heat emission,
  bare pipes, 68, 302
  ceilings, 321
  convective component, 308, 313
  embedded pipe, 318
  emission characteristics, 314
  floors, 319
  forced convectors, 317
  high intensity, 328
  high temperature panels, 308, 371
  insulated pipes, 68, 303
  mains, 385
  natural convectors, 314
  plane surface panels, 306
  radiant component, 308, 313
  radiant strip, 309
  radiators, 313
  unit heaters, 318
  walls, 321

Heat flow paths, 285
Heat generation, 207
Heat losses,
  air temperature method, 265
  computation procedure, 290
  design temperatures, 278, 283
  environmental temperature method, 280
  height allowance, 293, 314
  steady-state, 244, 265, 316
  ventilation, 279
Heat transfer coefficient,
  convection, 54, 63, 305
  inside surface, 270
  outside surface, 268
  overall, 66, 69, 267, 278
  radiation, 63, 305
Heating zones, 381
High pressure hot water, 360
Hosereel systems, 424
Hot water services,
  central systems, 432
  dead leg, 431
  demand units, 434
  direct systems, 430
  flow rates, 429
  indirect systems, 431
  local systems, 431
  operating temperatures, 428
  pipe sizing, 433
  storage capacity, 430
  water consumption, 429
Hottel, 63, 305
Hydrazine injection, 402
Hydrogen evolution reaction, 393

Ideal fluid, 1, 14
Index circuit, 331, 334, 421, 433
Indirect hot water systems, 402, 431
Inherent regulation, 87
Inside surface convection coefficient, 270
Integral control, 102
Intermittent heating, 246
Irradiance, 313
Isothermal compression, 366

K-factor 31, 343
Kata thermometer, 144
Kinematic viscosity, 5
Kirchhoff's law, 62

Lag, control, 87
Laminar flow, 7, 10, 21, 341
Laminar sub-layer, 26, 52
Lead pipework, 397
Lime-soda process, 400
Linear characteristic, 92
Load balancing, 383
Load profiles, 205, 258
Local control methods, 116
Localised discomfort, 173
Localised discomfort criteria, 177
Logarithmic mean temperature difference, 70, 213
Longwave radiant heating, 178
Loss factor, 31, 343
Low pressure hot water,
  forced circulation, 333
  microbore, 359
  natural circulation, 330
  reverse return, 334
  small bore, 359
  two-pipe upfeed, 332
Lowest possible percentage dissatisfied, 150

Mains emission, 385
Mean radiant temperature, 145, 174, 186, 282, 301, 314
Methods of connection,
  one-pipe, 322
  two-pipe, 322
  series, 322
Mineral salts, 392
Minor losses, 31, 343
Mixing valves,
  non-thermostatic, 132
  thermostatic, 132
  three-way, 93, 325, 359
Model Water Byelaws, 405, 432
Modes, control, 94
Moody diagram, 29
Multi-stage systems, 90

Neutral pressure point, 353
Nikuradse, 24
Notional system time-constant, 91
Numerical methods, 240
Nusselt number, 56

Outside surface convection coefficient, 268
Overall heat transfer coefficient, 66, 69, 267, 278
Oxygen absorption, 357
Oxygen consumption reaction, 393

Panel, mounting height, 310
Parabolic characteristic, 92
Permanent hardness, 391
Permeability, 79
Permeance, 80
PH value, 392
Pipe-in-pipe systems, 384
Pipe materials, 341
Pipe sizing,
  cold water, 418
  final, 349

Pipe sizing *cont.*
- hot water, 342, 433
- preliminary, 347, 350, 418, 435
- tables, 341, 342, 362, 418, 434

Pipework, 375
Planck's law, 61
Plane radiant temperature, 175
Plastic components, 399
Plastic pipework, 398
Prandtl number, 56
Predicted mean vote, 150
Predicted percentage dissatisfied, 150
Pressure,
- circulating, 355, 373, 375
- hydrostatic, 355, 364, 372, 375
- loss in fittings, 31, 344
- loss to friction, 19, 343
- resultant, 355, 365, 373, 375
- steam, 364

Pressure distribution, 354
Pressure drop, 340, 347
Pressurising cylinder, 357
Pressurising methods,
- air, 366
- head tank, 362
- nitrogen, 369
- vapour, 362

Pressurised systems, working range, 361
Process reaction rate, 85
Proportional control, 98
Proportional/integral control, 103
Proportional/integral/derivative control, 106
Psychrometry, 75
Psychrometric chart, 78
Pumps,
- characteristics, 377, 378
- duty, 346, 381, 411, 438
- head, 19
- in parallel, 381
- in series, 380
- laws, 376
- materials, 382
- position, 354
- power, 377
- pressure, 19, 378
- speed, 378

Radiation,
- coefficient, 63
- laws, 61
- thermal, 59

Rating tables, 314
Rangeability, control valve, 93
Reflectivity, 59
Relative humidity, 76
Resistance,
- airspace, 66
- fittings, 420
- pipework, 418
- thermal, 48, 51, 66
- system, 377

Resistance *cont.*
- vapour, 80

Response factor, methods, 235
Resultant temperature, 148
Reynolds, 6
Reynolds number, 8, 21, 24, 56
Roughness, surface, 20, 24

Second law of thermodynamics, 73
Scale, 391
Screwed joints, 375
Shear stress, 3, 11
Shortwave radiant heating, 184
Sizing,
- emitter, 301
- feed tank, 356
- pipework, 339
- pressure cylinder, 358
- pressure vessel, 368

Smooth zone of flow, 25
Soft water, 401
Solid angle factor, 174, 189
Spill system, 370
Staining, 314
Stainless steel pipework, 398
Stanton diagram, 25
Steady-flow energy equation, 73
Steady-state heat losses, 244, 265
Steam boilers, 362
Steel components, 398
Steel pipework, 397
Stefan-Boltzmann,
- constant, 61, 305
- law, 61

Step control, 94
Storage calorifiers, control, 130
Sulphate ions, 393

Temperature,
- air, 143, 265, 301, 314, 315
- air-radiant, 177, 187
- ambient, 301
- ceilings, 310
- dew point, 76, 78
- drop, 318, 323
- environmental, 283
- floor, 319, 321
- mean radiant, 145, 174, 186, 282, 301, 314
- mean surface, 322, 338
- panel, 310
- pressurised systems, 361
- resultant, 148
- surrounding, 301, 315
- vector radiant, 177, 186
- walls, 321

Temporary hardness, 391
Thermal comfort, 138
Thermal conductivity, 40
Thermal diffusivity, 44
Thermal gradients, 314

Thermal indices, 147
Thermal resistance, 48, 51, 66
Transition zone, 26
Transmissivity, 59
Tuning methods, 112
Turbulent flow, 7, 10, 24

U-value, 66, 70, 267

Valves,
  air, 333
  air, automatic, 333
  authority, 92
  balancing, 333, 351
  ball, 358
  characteristics, 94, 326
  check, 359
  control, 91
  diverting, 325
  drain, 350
  gate, 358, 406, 423
  globe, 406
  hand-controlled, 333, 426
  isolating, 358, 381
  landing, 426
  mixing, 124, 325, 359
  non-return, 359, 381
  safety, 359
  spill, 370
  three-port, 91
  two-port, 91
Vapour barrier, 295
Vapour pressure, 5, 75, 77, 144
Vapour resistance, 80
Vapour transmission, 79
Vector radiant temperature, 177, 186
Velocity head, 16, 31
Velocity pressure factor, 31, 343
Velocity, water, 340
Vena contracta, 34
Ventilation allowance, 279
Ventilation conductance, 286
Ventilation heat loss, 279
Venting, 354
View factor, 63, 175, 189, 191, 313
Viscosity,
  absolute, 2
  kinematic, 5

Water content, 356
Water Engineers' Handbook, 392
Water hardness, 391
Water of expansion, 332
Water supply, 390
Water treatment, 402
Water velocities, 340
Weather, 227
Weather compensated control, 124
Wet riser systems, 426
Wein's displacement law, 61

Zones, heating, 381